VOYAGES
TO THE PLANETS

VOYAGES

TO THE PLANETS

Volume 1 of *Voyages Through the Universe*

Second Edition

ANDREW FRAKNOI

Chair, Astronomy Department
Foothill College

DAVID MORRISON

Director, Astrobiology and Space Research
NASA Ames Research Center

SIDNEY WOLFF

Director
National Optical Astronomy Observatories

SAUNDERS COLLEGE PUBLISHING

A DIVISION OF HARCOURT COLLEGE PUBLISHERS

FORT WORTH   PHILADELPHIA   SAN DIEGO   NEW YORK   ORLANDO   AUSTIN

SAN ANTONIO   TORONTO   MONTREAL   LONDON   SYDNEY   TOKYO

Publisher: Emily Barrosse
Acquisitions Editors: Kelley Tyner, Jennifer Bortel
Marketing Strategist: Erik Fahlgren
Developmental Editor: David Robson
Project Editor: Sarah Fitz-Hugh
Managing Editor: Sally Kusch
Production Manager: Alicia Jackson
Text Designer, Cover Designer, and Art Director: Ruth A. Hoover

Frontispiece: This beautiful image of the giant planet Saturn was taken with the Hubble Space Telescope in September 1996. One of the planet's moons, Enceladus, is casting a shadow on the clouds of Saturn as it passes just above the ring system. Saturn's rings are so large that if they were placed next to the Earth, they would stretch almost to the Moon. Like the other giant planets, Saturn is made of mostly gas and liquid; its overall density is less than that of water. (Hubble Heritage Project, STScI/NASA)

Voyages Through the Universe, 2/e
ISBN: 0-03-026793-5
Library of Congress Catalog Card Number: 99-64251

Address for domestic orders:
Saunders College Publishing, 6277 Sea Harbor Drive, Orlando, FL 32887-6777
1-800-782-4479

Address for international orders:
International Customer Service, Harcourt Brace & Company
6277 Sea Harbor Drive, Orlando FL 32887-6777
(407) 345-3800
Fax (407) 345-4060
e-mail hbintl@harcourtbrace.com

Address for editorial correspondence:
Saunders College Publishing, Public Ledger Building, Suite 1250, 150 S. Independence Mall West, Philadelphia, PA 19106-3412

Web Site Address
http://www.harcourtcollege.com

Printed in the United States of America

9012345678 032 10987654321

ABOUT THE AUTHORS

Andrew Fraknoi is the Chair of the Astronomy Department at Foothill College near San Francisco and an Educational Consultant for the Astronomical Society of the Pacific (where he directs Project ASTRO, a national program to bring astronomers into elementary and junior high school classrooms). From 1978 to 1992 he was Executive Director of the Society, as well as Editor of *Mercury* Magazine and the *Universe in the Classroom* Newsletter. He has taught astronomy and physics at San Francisco State University, Cañada College, and the University of California Extension Division. He is author of *The Universe in the Classroom*, co-author of *Effective Astronomy Teaching and Student Reasoning Ability*, and scientific editor of *The Planets* and *The Universe*, two collections of science and science fiction. For five years he was the lead author of a nationally syndicated newspaper column on astronomy, and he appears regularly on radio and television explaining astronomical developments (most recently as astronomy correspondent for *Weekend All Things Considered.*) In addition, he has organized three national symposia on teaching introductory astronomy at the college level, and over 20 national workshops on improving the way astronomy is taught in earlier grades. He has received the Annenberg Foundation Prize of the American Astronomical Society and the Klumpke-Roberts Prize of the Astronomical Society of the Pacific for his contributions to the public understanding of astronomy. Asteroid 4859 was named Asteroid Fraknoi in 1992 in recognition of his work in astronomy education.

David Morrison is the Director of Astrobiology and Space Research at NASA's Ames Research Center, where he manages basic and applied research programs in the space, life and Earth sciences, with emphasis on astrobiology — the study of the living universe. He received his Ph.D. from Harvard University. Prior to joining NASA, he was Professor of Astronomy at the University of Hawaii, where he also directed the Infrared Telescope Facility of Mauna Kea Observatory and served as the university's Vice-Chancellor for Research. Internationally known for his research on small bodies in the solar system, he is the author of more than 120 technical papers and has published a dozen books, including five textbooks and a number of popular trade books on space science topics. He chaired the official NASA study of impact hazards that recommended that a Spaceguard Survey be carried out to search for potentially threatening asteroids and comets and in 1995 received the NASA Outstanding Leadership Medal for this work. He has served as President of the Astronomical Society of the Pacific, Chair of the Astronomy Section of the American Association for the Advancement of Science, and President of the Planetary Commission of the International Astronomical Union. Among other awards, he has received the Klumpke-Roberts Prize of the Astronomical Society of the Pacific for contributions to public understanding of science. Asteroid 2410 Morrison is named in his honor.

Sidney C. Wolff received her Ph.D. from the University of California at Berkeley and then joined the Institute for Astronomy at the University of Hawaii. During the 17 years she spent in Hawaii, the Institute for Astronomy developed Mauna Kea into the world's premier international observatory. She became Associate Director of the Institute for Astronomy in 1976 and Acting Director in 1983. During that period, she earned international recognition for her research, particularly on stellar atmospheres and how they can help us understand the evolution, formation, and composition of stars. In 1984, she was named Director of the Kitt Peak National Observatory and in 1987 became Director of the National Optical Astronomy Observatories. She is the first woman to head a major observatory in the United States. As Director of NOAO, she and her staff oversee facilities used annually by nearly 1000 visiting scientists. During its early phases, she was Director of the Gemini Project, which is an international program to build two state-of-the-art 8-m telescopes. She has served as President of the Astronomical Society of the Pacific and is the second woman to be elected President of the American Astronomical Society. She is also a member of the Board of Trustees of Carleton College, a liberal arts school that excels in science education. The author of more than 70 professional articles, she has written a monograph, *The A-Type Stars: Problems and Perspectives,* as well as several astronomy textbooks for Saunders College Publishing.

PREFACE FOR THE STUDENT

In college textbooks, there is a long tradition that the preface of the book is read by the instructor and the rest of the book by the student. Still, many students start reading the preface (it does come first) and then wonder why it doesn't say much to them.

So we begin our book with a preface specifically for student readers. It's not a preface about the subject matter of astronomy, which is introduced in the Prologue. Rather, we want to tell you a little about the book and give you some hints for the effective study of astronomy. (Your professor will probably have other, more specific suggestions for doing well in your class).

Astronomy, the study of the universe beyond the confines of our planet, is one of the most exciting and rapidly changing branches of science. Even scientists from other fields often confess to having had a lifelong interest in astronomy, though they may now be doing something like biology, chemistry, engineering, or writing software. There are fewer than 10,000 professional astronomers in the entire world. However, astronomy has a large group of amateur devotees who spend many an evening with a telescope under the stars observing the sky, and who occasionally make a discovery, such as a new comet or a developing storm on Saturn.

Many other people are "armchair astronomers" — fascinated just to read about bizarre worlds and processes that astronomers are uncovering. Some are intrigued by the scientific search for planets or life in other star systems. Others like to follow the challenges of space exploration, such as the refurbishing of the Hubble Space Telescope by the Shuttle astronauts, or the results from the Galileo mission to probe the giant planet Jupiter and its intriguing moons. Hearing about an astronomical event in the news media may be what first sparked your interest in taking an astronomy course.

But some of the things that make astronomy so interesting also make it a challenge for the beginning student. The universe is a big place, full of objects and processes that do not necessarily have familiar counterparts here on Earth. Like a visitor to a new country, it will take you a while to feel familiar with the territory or the local customs. Astronomy, like other sciences, has its own special vocabulary, some of which you will have to learn if you want to have a good voyage with us. (Since the title of this text includes the word Voyages, we will use voyaging metaphors from time to time. Humor your authors; we mean well.)

To assist students taking their first college-level course in astronomy, we have built a number of special features into this book, and we invite you to make use of them:

- All technical terms are printed in boldface type the first time they are used and clearly defined in the text; their definitions are listed alphabetically in Appendix 3 (the glossary), so you can refer to them at any time. The summaries at the end of each chapter also include these boldface terms as a review.

- The book begins with a historical summary of astronomy and then surveys the universe, starting at home and finishing with the properties of the entire cosmos. But don't worry if your instructor doesn't assign all the chapters or doesn't assign the chapters in order. Throughout the book "directional signs" lead you to earlier material you need to know before tackling the current section.

- We use tables to bring together numerical data for your convenience. For example, some tables summarize the important properties of each planet in the solar system (including the Earth). Students who want to see more of the data that astronomers use can investigate the appendices at the back of the book, which give the latest information on many aspects of astronomy.

- Figure captions clearly describe what phenomena or objects students are looking at. In many textbooks, captions are afterthoughts, with only a few words of description, but in this book, we have scrutinized each figure and asked what would help clarify the diagram or image for a reader just being introduced to astronomy. Plus, we have to tell you that we are very proud of the many beautiful images we have been able to assemble (with the help of colleagues around the world) to show you the cosmic phenomena we are describing in full color.

- Each chapter ends with a summary of the essential points in the chapter, plus small-group activities (more about these in a minute), review questions, thought questions, and problems to help you "process" what you have learned. Some simple multiple-choice quiz questions to help you review are also available on our web site: **www.saunderscollege.com/astro/fraknoi**

- Suggestions for further reading are included at the end of each chapter, in case you want to learn more about a particular topic. These books and articles are written at the same introductory level as this text. Appendix 2 contains a listing of general sources of astronomical information.

- Each chapter also contains a guide to some of the more interesting astronomy sites on the World Wide Web. Because Web listings change even faster than college tuition does, we try to keep an updated list of good sites on the Voyages Web site. Appendix 1 provides some general Web sites that contain information on a wide range of astronomical topics and some of the best new astronomical images.

- In this edition, we have added a series of small group activities to each chapter. These are mostly questions that take off in new directions from the material in the chapter and are meant to foster debate and discussion among students. Your instructor or discussion section leader may use these in class, or you may want to discuss questions that interest you with the other members of your study group. ("What study group?," you may be asking. See below.)

How can you increase your chances of doing well in your astronomy class? Here are a few suggestions for studying astronomy that come from good teachers and good students from around the country:

- First, the best advice we can give you is to be sure to leave enough time in your schedule to study the material in this class regularly. It sounds obvious, but it is not very easy to catch up with a subject like astronomy by trying to do everything just before an exam. Try to put aside some part of each day, or every other day, when you can have uninterrupted time for reading and studying astronomy.

- Try to read each assignment in the book twice, once before it is discussed in class, and once afterwards. Take notes or use a highlighter to outline ideas that you may want to review later. Also, take some time to coordinate the notes from your reading with the notes you take in class. Many students start college without good note-taking habits. If you are not a good note-taker, try to get some help. Many colleges and universities have student learning centers that offer short courses, workbooks, or videos on developing good study habits. Take a little time and find out what your school has to offer. Good note-taking skills will be useful for many jobs or hobbies you may get involved with after college.

- Form a small astronomy study group with people in your class; get together as often as you can and discuss the topics that may be giving group members trouble. Make up sample exam questions and make sure everyone in the group can answer them confidently. If you have always studied alone, you may at first resist this idea, but don't be too hasty to say no. Study groups are a very effective way of digesting new information, learning a foreign language (which astronomy words sometimes resemble), or studying a broad field like law or astronomy.

- Before each exam, do a concise outline of the main ideas discussed in class and presented in your text. Compare your outline with those of other students as a check on your own study habits.

- If you find a topic in the text or in class especially difficult or interesting, don't hesitate to make use of the resources in your library for additional study.

- Don't be too hard on yourself! If astronomy is new to you, many of the ideas and terms in this book may be unfamiliar. Astronomy is like any new language; it may take a while to become a good conversationalist. Practice as much as you can, but also realize that it is natural to be overwhelmed by the vastness of the universe and the variety of things that are going on in it.

We hope you enjoy reading this text as much as we enjoyed writing it. We are always glad to hear from students who have used the text and invite you to send us your reactions to the book and suggestions for how we can improve future editions. We promise you we will read and consider every serious letter we receive. You can send your comments to Andrew Fraknoi, Astronomy Department, Foothill College, 12345 El Monte Rd., Los Altos Hills, CA 94022, USA. (Please note that we will not send you the answers to the chapter problems or do your homework for you, but all other thoughts are welcome.)

Andrew Fraknoi, David Morrison, and Sidney Wolff
(*June 1999*)

PREFACE FOR THE INSTRUCTOR

Voyages is an astronomy text written with today's students in mind — it is designed for non-science majors who may even be a little intimidated by science, and who approach astronomy with more interest than experience. With features that make it appropriate for everyone from university business majors to first-time community college students, *Voyages* is written to draw in and engage all readers, while preserving the accuracy and timeliness that our colleagues expect of us.

In thinking through how best to present astronomy to students whose science and math background may be limited, we have tried to make the language friendly and inviting and to use examples and analogies drawn from everyday experience. We very much hope this approach, together with our vignettes from the lives of astronomers, references to other fields that are influenced by astronomy, and occasional touches of humor, can help make this a book that students will actually enjoy reading. There is, we believe, nothing less useful than a textbook the instructor opens more often than the student!

Organization and Special Features

The book you are holding is the first volume in a two-volume sequence that separates the solar system from stars, galaxies, and cosmology. The two volumes each include introductory material on astronomical history, sky phenomena, radiation and spectra, and telescopes. As befits its special role in astronomy, the Sun's chapter is included in both books. As the field of astronomy has grown, more and more instructors find that they cannot cover every branch of our field in one semester or quarter. We hope the two volumes can assist you in providing students with a lower-cost text for the topics your courses cover.

While your authors are proud of the books we have produced, we realize that some instructors will want to teach topics in a different order or leave some out altogether. It's fair to say that in astronomy — much more than any other introductory science class — there are as many approaches to the basic course as there are instructors teaching it. Accordingly, we have tried to write the text so that the chapters and even main sections within chapters often stand on their own. We will confess that even the lead author does not cover all the chapters in the book in his own courses, nor does he always stick with the order of topics as they are given.

Since we do not expect students to remember every concept introduced in previous chapters, we have inserted unobtrusive verbal "sign posts" throughout to help students find where a concept was defined or explained in detail. We sometimes review a key idea that may have been introduced many chapters ago. Our aim is that your student should be able to use the book as an easy navigational tool, no matter how you teach your course. A complete glossary is supplied in Appendix 3.

Our book is not an encyclopedia of astronomy (which is why students do not need to work out for six weeks in a gym before they can carry it around.) We tried not to overwhelm the reader with detail, but instead to focus on the major threads and overarching ideas that illuminate the relationships among the various branches of astronomy. We probably have a bit less jargon than most textbooks, but we have worked hard not to sacrifice any of the key concepts that we believe students should learn in a basic course.

We include many of the latest ideas and discoveries in astronomy, not merely for their novelty, but for their value in advancing the quest for a coherent understanding of the solar system. In each case, we have tried to fit the latest research results into a wider context and to explain clearly why they mean. Among the recent topics included in this edition are: the discovery of a good number of new planets around others stars, results from the Galileo probe investigating the satellites of Jupiter, and images and ideas from the Mars Pathfinder and Global Surveyor missions.

We portray astronomy as a human endeavor and have tried to include descriptions and images of some of the key men and women who have created our science over the years. Illustrations also include many of the latest images from the Hubble Space Telescope and other space instruments, as well as an up-to-date collection of color images from ground-based observatories around the world. Many of the planetary images are second- or third-generation corrected views, not merely the first releases rushed out for the news media. Full-color diagrams are used as teaching tools, not as cosmetic devices. Figure captions contain full explanations of what the student should be seeing and understanding, not just the cryptic one-liners featured in so many other books.

In this second edition, we have also introduced two new features to help instructors and students:

Collaborative Group Activities as a Teaching Tool

Recent research has shown that listening to an hour- or hour-and-a-half lecture on some topic in astronomy is not always conducive to effective student learning. Students learn far better when they are active participants rather than passive listeners. Unfortunately, many of us teach huge classes, where it might seem that student participation could not possibly be practical. Yet, science instructors in many fields are beginning to come up with ways that even large lectures can involve small group activities. To keep up with this work as it pertains to astronomy, we recommend the Web sites from:
°The Astronomical Society of the Pacific (www.aspsky.org/subpages/education.html) and especially the annotated list of Web sites for college astronomy instructors found there
°The American Astronomical Society Education Office (www.aas.org/education/).

Here we will focus on one approach, which involves small groups and collaborative group activities either during lecture or in discussion section. The instructor divides the class into groups of three or four students. The smaller you make the groups, the less likely it is that some students will be able to remain uninvolved, but groups of two are typically not sufficient for good discussions to evolve. (Opinion is divided about whether it is better to assign students to groups randomly, or to allow students to form their own group.)

Each group is given a task and a time limit. The section is each chapter of *Voyages* called *Inter-activity* presents some examples of tasks that do not involve any outside materials or data. In addition, some instructors like to present each group with some simple data to analyze (on a handout or an overhead.) After group members discuss or do the activity, they elect a spokesperson, who shares their results with the whole class. If your class is really large, perhaps only a representative share of groups will report on any given day.

Based on what they hear in the reports, groups can then return to their discussion and change or refine their answers. Some instructors require each person or the whole group to write a short summary of their conclusions; others feel that the experience of interacting with their peers is sufficient.

This approach works well even with showing slides. Put up a high-resolution Galileo image of the surface of Europa, for example, and just ask the groups to discuss what they are seeing, how large an area they think it is, and how such features might originate. The important thing is not for students to get a "right answer," but to get them thinking, talking, interacting, and participating in their own education.

We should warn you that many of the suggested group activities in *Voyages* are open-ended, and may engender a great deal of discussion. You should probably have a strategy for coming to closure in a way that still leaves students feeling that their opinions are valued. A short paper (perhaps for extra credit), an interactive Web site, or a chance for further debate in a discussion section are possible ways of capturing student enthusiasm but still being able to move on in the main lecture.

- each chapter contains a series of annotated Web links for further exploration. These are kept up-to-date on our Web site: www.saunderscollege.com/astro/fraknoi. Appendix 1 gives more general Web links and also points to sources for the latest astronomical images.

- each chapter has a new section called *Inter-activity*, consisting of a series of interactive group exercises (see box) that can be a change of pace during a lecture, form the basis of a discussion section, or serve as homework to teams of students.

The chapters in *Voyages* feature a number of highlighted boxes designed to help non-science students appreciate the breath of astronomy without distracting them from the main narrative.

- *Making Connections.* These special boxes show how astronomy connects to students' experiences with other fields of human endeavor and thought, from poetry to engineering, from popular culture to natural disasters.

- *Voyagers in Astronomy.* These profiles of noted astronomers focus not only on their work, but on their lives and times.

- *Astronomy Basics.* In these short sections we try to explain fundamental science ideas and terms that other texts often assume (incorrectly) that students know.

- *Seeing for Yourself.* Here students get familiar with the sky and everyday astronomical phenomena through observations using simple equipment.

Ancillary Materials

We are happy to offer instructors who adopt *Voyages* a series of learning and teaching aids to make their lives a little bit easier. These include:

- The Redshift™ College Edition CD-ROM. Saunders offers a special educational version of the award-winning Redshift™ software, published by Maris Multimedia, packaged with *Voyages* for a very low price. Students can enjoy this dual platform CD-ROM as a learning aid, a lab tool (see below), or for recreation during or after the course. Among its many features, it allows students to:

Discover astronomy's origins through Bill O. Walker's interactive history book
view realistic models of the planets and main satellites in the solar system
identify over 300,000 stars, nebulae, and galaxies
simulate astronomical events over the course of 15,000 years
view more than 700 full-screen photographs
navigate through surface maps of Earth, Moon, and Mars.

- Workbook to Accompany Redshift™ College Edition. Bill O. Walker of Tyler College has written a lab manual that contains many practical tips for making good use of Redshift™ College Edition. It can be packaged with *Voyages* or ordered as a separate text for your lab.

- The *Voyages* Instructor's Manual/Test Bank.
The instructor's manual test bank contains answers to thought questions and problems in the textbook, as well as a host of multiple-choice test questions for use in a variety of classroom settings. We're proud to say, that unlike most texts, our multiple-choice questions have five answers, and are far more than just regurgitation of factoids. Many of the questions require applying the principles students learned in new situations, or finding similarities or differences among disparate phenomena. There is also a liberal sprinkling of humor among the answers, to cheer up the suffering students during the exam.

- ExaMaster Computerized Test Bank for Windows and Macintosh.
ExaMaster features all the questions from the printed test bank in a format that allows instructors to edit them, add questions, and print assorted versions of the same test.

- The *Voyages* Transparency (or Slide) Collection.
This set contains many of the images in the book for your use in the classroom. Instructors have told us that they have found the full-color diagrams throughout the book especially useful, and all of these are included (together with a good number of photographs). The set is available either as transparencies or as slides.

- The *Voyages* Instructor's Resource CD-ROM.
This CD-ROM includes virtually all the diagrams and paintings, and many of the photographs from *Voyages*, together with several other images that we wanted to use in the book but couldn't find room for (especially recent images from planetary exploration and the Hubble Space Telescope.) Images on the CD-ROM are PICT images and can be used with our Presentation Manager or can easily be exported and used with another commercial software presentation package such as Microsoft PowerPoint.

Saunders College Publishing may provide complementary instructional aids and supplements or supplement packages to those adopters qualified under our adoption policy. Please contact your sales representative for more information. If as an adopter or potential user you receive supplements you do not need, please return them to your sales representative or send them to

Attn: Returns Department
Troy Warehouse
465 South Lincoln Drive
Troy, MO 63379

The Registered Adopters Program

With many textbooks, the time you place your order with the publisher is the last time you hear from the publisher or the authors. In fact, many authors and publishers have no idea who adopted their book and never communicate with the instructors except through sales reps. We hope to be a little bit different; we'd like to know who our adopters are and to stay in touch with them, providing new materials, updated information, etc.

Thus we have created a "Registered Adopters" program for *Voyages;* to join, you merely fill out a brief questionnaire. We will be in touch with you from time to time, mostly to let you know when we are publishing something that may help your teaching. There is no cost involved, and we promise not to release the names for any other purpose.

If you are not yet a registered adopter, we urge you to contact Saunders College Publishing and get on our list. You can do so via:

telephone: 1-800-939-7377
e-mail: voyages@saunderscollege.com,
particle mail: David Robson, Saunders College Publishing, Public Ledger Building, Suite 1250, 150 S. Independence Mall West, Philadelphia, PA 19106.

Let Us Hear from You

Unlike stars and planets, textbooks in astronomy do not exist in a vacuum. All three authors have benefited tremendously over the years from the advice of colleagues and students who teach astronomy. We want to be sure that we continue to make changes and updates in *Voyages* that will be the most useful to you. We also welcome your comments for improving the *Voyages* Web site, which is at: www.saunderscollege.com/astro/fraknoi.

We welcome your comments and suggestions about the text and the ancillaries, and are always grateful for your ideas for how future materials can be made more effective. Address your cards and letters to: Andrew Fraknoi, Astronomy Dept., Foothill College, 12345 El Monte Rd., Los Altos Hills, CA 94022 or e-mail: fraknoi@admin.fhda.edu.

Acknowledgments

We would like to thank the many colleagues and friends who have provided information, images, and encouragement, including: Charles Avis, Charles Bailyn, Bruce Balick, Roger Bell, Roy Bishop, Chip Clark, Grace Deming, Richard Dreiser, Doug Duncan, George Djorgovski, Alex Filippenko, Debra Fischer, Dale Frail, Lola Fraknoi, Kathy Fransham, Michael Friedlander, Alan Friedman, Ian Gatley, Paul Geissler, Margaret Geller, Cheryl Gundy, Heidi Hammel, Todd Henry, George Jacoby, William Keel, Geoff Marcy, Jeff McClintock, Michael Merrill, Jacqueline Mitton, Janet Morrison, Donald Osterbrock, Michael Perryman, Carle Pieters, Edward Purcell, Axel Quetz, Dennis Schatz, Rudy Schild, Maarten Schmidt, Joe Tenn, Ray Villard, Althea Washington, Adrienne Wasserman, and Richard Wolff.

We are grateful to Lynette Cook, Bill Hartmann, John Spencer, Don Davis, and Don Dixon for permission to reproduce their wonderfully illustrative astronomical paintings, and to David Malin for his assistance and his superb astronomical photographs. Michael Briley has been our *Voyages* Webmaster and we want to thank him for his ideas and the skillful way he puts them into execution.

We appreciate the guidance of colleagues who reviewed all or part of the manuscript.

First Edition:

Grady Blount
Texas A&M University

Michael Briley
University of Wisconsin, Oshkosh

David Buckley
East Strohdsburg University

John Burns
Mt. San Antonio College

Paul Campbell
Western Kentucky University

Eugene R. Capriotti
Michigan State University

George L. Cassiday
University of Utah

John Cunningham
Miami-Dade Community College

Grace Deming
University of Maryland

Miriam Dittman
DeKalb College

Gary J. Ferland
University of Kentucky

George Hamilton
Community College of Philadelphia

Ronald Kaitchuck
Ball State University

William C. Keel
University of Alabama

Steven L. Kipp
Mankato State University

Jim Lattimer
SUNY, Stonybrook

Robert Leacock
University of Florida

Terry Lemley
Heidelberg College

Bennett Link
Montana State University

Charles H. McGruder III
Western Kentucky University

Stephen A. Naftilan
Claremont Colleges

Anthony Pabon
DeAnza College

Cynthia W. Peterson
University of Connecticut

Andrew Pica
Salisbury State University

Terry Richardson
College of Charleston

Margaret Riedinger
University of Tennessee, Knoxville

Jim Rostirolla
Bellevue Community College

Michael L. Sitko
University of Cincinnati

John Stolar
West Chester University

Charles R. Tolbert
University of Virginia

Steve Velasquez
Heidelberg College

David Weinrich
Moorhead State University

David Weintraub
Vanderbilt University

Mary Lou West
Montclair State University

Dan Wilkins
University of Nebraska, Omaha

J. Wayne Wooten
Pensacola Junior College

Second Edition:

Mitchell C. Begelman
University of Colorado

Stephen Danford
University of North Carolina, Greensboro

Richard French
Wellesley College

Catharine Garmany
University of Colorado, Boulder

Owen Gingerich
Harvard University

Edward Harrison
University of Massachusetts

Scott Johnson
Idaho State University

Mark Lane
Palomar College

John Patrick Lestrade
Mississippi State University

Michael C. LoPresto
Henry Ford Community College

Anthony Marston
Drake University

Ronald A. Schorn
Texas A&M University

Vernon Smith
University of Texas, El Paso

Ronald Stoner
Bowling Green State University

Jack W. Sulentic
University of Alabama

Stephen Walton
California State University, Northridge

Warren Young
Youngstown State University

No book project of this complexity could succeed without the diligent efforts of many people at Saunders College Publishing. We very much appreciate the assistance of:

Emily Barrosse, Vice President/Publisher
Kelley Tyner and Jennifer Bortel, Acquisitions Editors
Erik Fahlgren, Marketing Strategist
David Robson, Developmental Editor
Sarah Fitz-Hugh, Project Editor
Sally Kusch, Managing Editor
Alicia Jackson, Production Manager
Carol Bleistine, Manager of Art and Design
Ruth Hoover, Art Director and Text and Cover
    Designer
Kim Menning, Illustration Supervisor
Jane Sanders, Photo Researcher
Mary Beth Smith, Editorial Assistant

But most of all, we appreciate the hundreds of instructors and many thousands of students around the United States, Canada, and the world, who have honored us by using this textbook on their own voyages through the universe.

Andrew Fraknoi, David Morrison, and Sidney Wolff
(June 1999)

Contents Overview

Once you have visited your own planetary neighborhood, complete your *Voyages Through the Universe*. Also available from Harcourt College Publishers: *Voyages to the Stars and Galaxies.*

CONTENTS

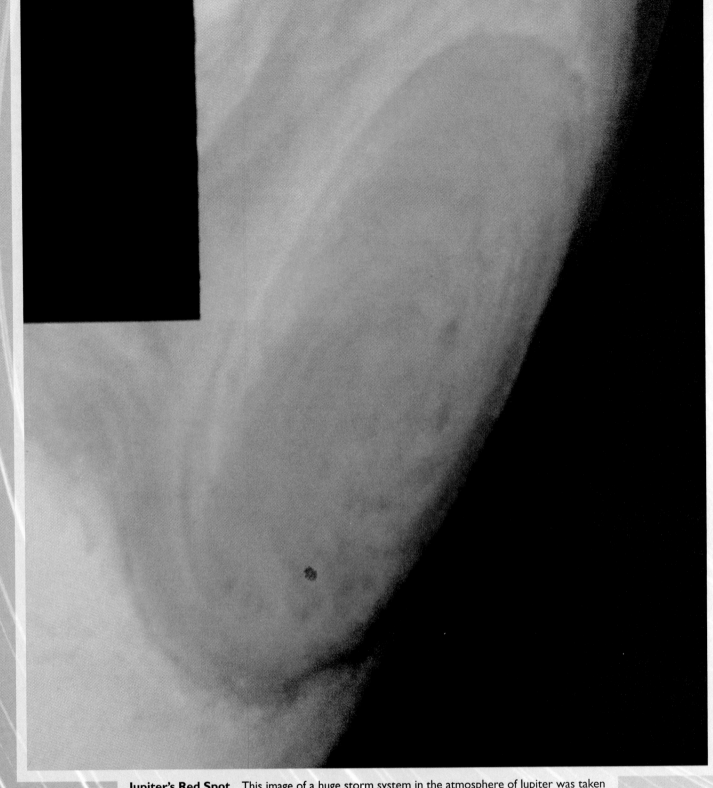

Jupiter's Red Spot This image of a huge storm system in the atmosphere of Jupiter was taken by the Galileo spacecraft on June 26, 1996. The color is somewhat heightened by computer processing—in reality the reds and oranges would be much more muted. The Red Spot is seen on the limb (edge) of the planet in this oblique view; it is thus foreshortened in by perspective in the east–west direction. In other words, it looks taller than it is wide, when in reality the Spot is wider than it is tall. The size of this giant storm changes with time, but at its greatest extent it measures almost three times the size of planet Earth. *(Galileo Imaging Team, JPL/NASA)*

Prologue and Brief Tour of the Universe

We invite you to come with us on a series of voyages to explore our cosmic neighborhood as astronomers understand it today. Out there we will find awesome and magnificent realms—worlds and worldlets quite different from anything we are used to on Earth. Nevertheless, we will learn to call the solar system our home, and we will see that its evolution is directly connected with our presence on planet Earth today. The term *solar system* refers to our Sun and the planets, moons, and assorted smaller chunks of real estate that go around it.

Along our journey we will encounter

- a "grand canyon" system so large that on Earth it would stretch from Los Angeles to Washington, D.C. (Figure P.1).
- a tiny moon whose gravity is so weak that one good throw from its surface could put a baseball into orbit.
- a storm system so vast that sometimes three Earths could fit into it side by side (see the image that opens this chapter).
- a world in so constant a state of volcanic eruption that it is literally turning itself inside out (Figure P.2).

It is these kinds of discoveries that make astronomy such an exciting field for both scientists and people in other walks of life. But we will explore more than just the objects in our "neighborhood" and the latest discoveries about them. We will pay equal attention to the *process* by which we have come to understand the realms beyond our Earth and will examine the tools we need to increase that understanding.

THE NATURE OF ASTRONOMY

Astronomy is defined as the study of the objects that lie beyond the atmosphere of our planet Earth, and of the processes by which these objects interact with one another. But, as we will see, it is much more than that. It is also humanity's attempt to organize what we learn into a clear history of the universe, from the instant of its birth in the Big Bang to the present moment in which you are finishing this sentence. For us, born as we were on planet Earth, the developments in our own solar system are crucial parts of the story.

Putting this history together is an audacious undertaking—especially for a creature of modest stature, living on a rocky ball that circles a nondescript star in an island of stars so vast that it dwarfs human understanding. We are still a long way from being finished with the story of how humanity came to exist on one of the worlds around the Sun. Throughout the book, we emphasize how much science is a "progress report" that constantly changes as new techniques and instruments allow us to probe the universe more deeply.

In considering the history of the universe, we will see again and again that the cosmos *evolves*—it changes in profound ways over long periods of time. Although you may sometimes read that the concept of evolution is controversial in science, this is simply not true. Few ideas are better established than the notion that the universe is not the same today as it was long ago. You will see, for example, that the Earth in its earliest stages could not have given rise to or supported the readers of this book. It took many millenia and much change before our planet cooled, the bombardment of its surface by cosmic projectiles reached bearable levels, and its environment became hospitable to life as we know it. Tracing the evolutionary processes that continue to shape our cosmic environment is one of the most important (and satisfying) parts of modern astronomy.

THE NATURE OF SCIENCE

Unlike religion or philosophy, science accepts nothing on faith. The ultimate judge in science is always an experiment or observation: what nature itself reveals. Science, then, is not merely a body of knowledge, but a *method* by which we attempt to understand nature and how it behaves. This method begins with many observations over a period of time. From the trends in the observations, scientists may come up with a *model* of the particular phenomenon we want to understand. Such models are always approximations of nature itself, subject to further testing.

To take a concrete astronomical example, ancient astronomers constructed a model (partly from observations, partly from philosophical beliefs) that the Earth was the center of the universe, and that everything moved around it. At first, the available observations of the Sun, Moon, and planets could be fit to this model, but eventually, better observations required the model to add circle after circle to the movements of the planets to keep the Earth at the center. As the centuries went by and improved instruments were developed for keeping track of celestial objects more precisely, the old model (even with a huge number of circles) could no longer explain all the observed facts. As we will see in Chapter 1, a new model, with the Sun at the center, fit the experimental evidence better. After a period of philosophical struggle, it became accepted as our view of the universe.

FIGURE P.1
Mars Mosaic This image of Mars was constructed at the U.S. Geological Survey in Flagstaff, Arizona, by combining many individual images taken by the Viking spacecraft starting in 1976. The view is centered on the Valles Marineris (Mariner Valley) complex of canyons, which is as long as the United States is wide. *(USGS)*

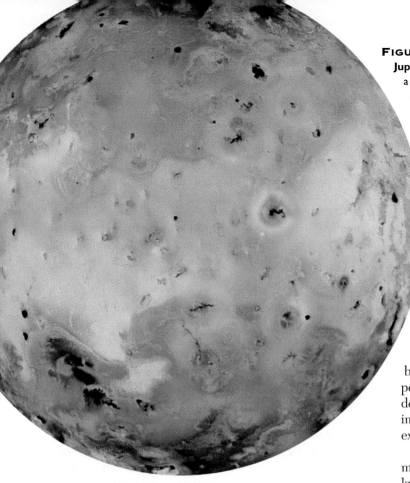

Jupiter's Moon Io   Jupiter has four large moons, whose sizes are like a small planet's. Here we see Io, the innermost one, as imaged by the Galileo spacecraft. While the colors have been computer enhanced on this picture, the surface clearly does not resemble that of any other moon with which we are familiar. Huge volcanic eruptions are seen all over Io's surface. Like an amusement park patron going on the roller coaster much too soon after lunch, Io finds the contents of its insides desperately trying to get out. The reasons for this activity are explained in Chapter 11.

(Galileo Imaging Team, JPL/NASA)

When they are first proposed, new models or ideas are sometimes called *hypotheses*. Many students today think that there can be no new hypotheses in a science such as astronomy, that everything important has already been learned. Nothing could be further from the truth. Throughout this book you will find discussions of recent, and occasionally still controversial, hypotheses in astronomy—concerning, for instance, the origin of our Moon in a huge collision with the Earth, or the presence of an ocean under the ice of one of the moons of Jupiter. Such hypotheses are based on difficult observations done at the forefront of our technology, and all require further testing before we fully incorporate them into our standard astronomical models.

This last point is crucial: If there is no possible way of *testing* a hypothesis, it does not belong in the realm of science. In some cases, theories (especially those about the very largest structures in the universe, or the very smallest structures inside the atom) may not lend themselves to testing with our current instruments. Many years may pass before the appropriate experiments or observations can be performed, but ultimately, all scientific ideas must be testable. Moreover, they must stand up to being tested again and again in as many ways as we can muster. If the experiments or observations do not support a model or hypothesis, its proponents must be ready to modify it or let it go, no matter how fond of it they may have grown.

It is this self-correcting aspect of science that sets it off from most human activities. Scientists spend a great deal of time questioning and criticizing one another. No project is funded and no report is published without extensive peer review—that is, without careful examination by other scientists in the same field. In other areas, young people are often taught to accept the authority of their elders without question, but in science (after proper training) everyone is encouraged to try new and better experiments, and to challenge any and all hypotheses.

This is one of the reasons science has made such dramatic progress. An undergraduate science major today knows more science and math than did Isaac Newton, one of the most brilliant scientists who ever lived. Even in this introductory astronomy course you will learn about objects and processes that a few generations ago no one even dreamed existed. While the domain of science is limited, within that domain its achievements have been glorious.

THE LAWS OF NATURE

Over the centuries, scientists have extracted from countless observations certain fundamental principles, called *scientific laws*. These are, in a sense, the rules of the game nature plays. One remarkable discovery about nature that underlies everything you will read in this book is that the same laws apply everywhere in the universe. The rules that govern the behavior of gravity on Earth, for example, are the same rules that determine the motion of two stars in a system so far away that your unaided eye cannot find them in the sky.

Note that without the existence of such universal laws, we could not make much headway in astronomy. If each planet or star system had completely different rules, we would have little chance of interpreting what happened in other "neighborhoods." But the consistency of the laws of nature gives us enormous power to understand distant objects without traveling to them and learning the local laws. In the same way, if every U.S. state or Canadian province had completely different laws, it would be very difficult to carry out commerce or even to understand the behavior of people in different regions. But a consistent set of laws allows us to apply what we learn or practice in one state to any other state.

You might ask whether the laws of the natural world could (under the right circumstances) be suspended. This is an enticing fantasy, but, despite many attempts, not a shred of scientific evidence has been found to support such an idea. While it would be nice if we could suspend the law of gravity and float away from our porch through effort of will alone, in real life such experiments generally result in broken bones.

This is not to say that our models cannot change. New experiments and observations can lead to new or more sophisticated models—models that can even include new phenomena and new laws about their behavior. The idea that collisions of rocks or iceballs from space could play a very significant part in the development of life on Earth is a perfect example of such a transformation, one that took place not long ago. Wishing isn't going to bring such new models into existence; only the patient process of observing nature ever more finely can reap such rewards.

One important problem about describing scientific models has to do with the limitations of language. When we try to describe complex phenomena in everyday terms, the words themselves may not be adequate to the job. For example, you may have heard the structure of the atom likened to a miniature solar system. While some aspects of our modern model of the atom do remind us of planetary orbits, many other aspects are fundamentally different.

This is why scientists often prefer to describe their theories using equations rather than words. In this book, designed to introduce the field of astronomy, we have used mainly words to discuss what scientists have learned and have avoided math beyond basic algebra. But if this course piques your interest and you go on in science, more and more of your studies will involve the language of mathematics.

NUMBERS IN ASTRONOMY

You may have heard a television reporter refer to a large number (such as the national debt) as "astronomical." In astronomy, we do have to deal with distances on a scale you may never have thought about before, and with numbers larger than any you may have encountered. Most students take a while to learn to navigate among the millions and billions that astronomers tend to throw about in their everyday discussions. With some practice, you will become just as good at it!

By the way, if you sometimes have trouble sorting out millions and billions, take heart. Even so distinguished a group as the Presidential Advisory Committee on the Future of the U.S. Space Program, in a 1990 report, listed the distance to the planet Uranus as 1.7 million miles, when in fact Uranus is 1.7 *billion* miles away. A million (1,000,000) is a thousand times less than a billion (1,000,000,000). If Uranus were that close, we would all have noticed it, because it would be bigger and brighter than the Moon in our skies.

In this text we adopt two approaches that make dealing with astronomical numbers a little bit easier. First, we use a system for writing large and small numbers called *powers-of-ten notation* (or sometimes *scientific notation*). This system is very appealing because it does away with the huge number of zeros that can seem overwhelming to the beginner.

In scientific notation, if you want to write a figure like $490,000 (which is definitely NOT the starting salary for an astronomer, but might be for a national television star), you write 4.9×10^5; the little number after the ten, called an exponent, keeps track of the number of tens you have to multiply together to get the number you want. In our example, five tens are multiplied together, so that $10 \times 10 \times 10 \times 10 \times 10$ gives 100,000. Multiply 100,000 by 4.9 and you get our astronomical starting salary. Another way to remember the basics of this notation is to note that five is the number of places you have to move the decimal point to the right to convert 4.9 to 490,000. If you are encountering this system for the first time or would like a refresher, we suggest you look at Appendix 4 for more information and examples.

Small numbers are written with negative exponents. Three millionths (0.000003) is expressed as 3.0×10^{-6}. One reason this notation is so popular among scientists— trust us, it is, even if you at first don't like it—is that it makes arithmetic a lot easier. To multiply two numbers in scientific notation, you need only add their exponents: thus $10^3 \times 10^9 = 10^{12}$. To divide numbers, just subtract exponents.

The second way we try to keep numbers simple is to use a consistent set of units—the international metric system. Unlike the British system, in which a completely arbitrary number such as 5280 feet equals a mile, metric units are related by powers of ten: A kilometer, for example, equals a thousand (10^3) meters. The metric system, which has been adopted by every major country in the world except the United States, is summarized in Appendix 5 and should become part of your vocabulary if you want to be ready for the future (when everyone will be using metric units).

LIGHT-YEARS

To give you a chance to practice scientific notation, and to set the scene for the tour of the universe in the next section, let's define a common unit astronomers use to describe distances in the universe beyond the solar system. A **light-year** (LY) is the distance that light travels in one year. Since light always travels at the same speed, and since its speed turns out to be the fastest possible speed in the universe, it makes a good standard for keeping track of distances. Some students complain about this name for a unit

The Orion Nebula This beautiful cloud of cosmic raw material (gas and dust from which new stars and planets are being made) called the Orion Nebula is about 1500 LY away. That's a distance of roughly 1.4×10^{16} km—a pretty big number! The picture we see is actually a seamless mosaic of 15 smaller images taken with the Hubble Space Telescope. The field of view is about 2.5 LY wide and shows only a small part of a vast reservoir of mostly dark material. The gas and dust in this region are illuminated by the intense light from a few extremely energetic adolescent stars in the neighborhood. *(C. R. O'Dell and NASA)*

of distance—light-years seem to imply that we are measuring time. But this mix-up of time and distance is common in everyday life as well—for example, when we tell a friend to meet us at a movie theater that's twenty minutes away.

So, how many kilometers are there in a light-year? If you are new to scientific notation, we suggest you work through all the steps of the following example for yourself on a separate sheet of paper. First, in case you are not yet a metric system fan, we should tell you that a kilometer is about 0.6 mile. Light travels at the amazing pace of 3×10^5 km per second (km/s). Think about that—light covers 300,000 km every second. In one second, it can travel seven times around the circumference of the Earth; a commercial airplane, in contrast, would take about two days to go around once, not counting time to refuel.

Now that we know how far light goes in a second, we can calculate how far it goes in a year. There are 60 (6×10^1) seconds in each minute, and 6×10^1 minutes in every hour. Thus light covers 3×10^5 km/s $\times 3.6 \times 10^3$ s/h $= 1.08 \times 10^9$ km/h.

There are 24 or 2.4×10^1 hours in a day, and 365.24 (3.65×10^2) days in a year. The product of those two numbers is 8.77×10^3 h/year. Multiplying that by 1.08×10^9 km/h gives 9.46×10^{12} km in a light-year. That's almost 10 trillion kilometers that light covers in a year. A string 1 LY long could fit around the circumference of the Earth 236 million times!

You might think that such a long unit would more than reach to the nearest star. But the stars are far more remote than our imaginations (or episodes of *Star Trek*) might lead us to believe. Even the nearest star is 4.3 LY away—more than 40 trillion km. Other stars visible to the unaided eye are hundreds or even thousands of light years away (Figure P.3). This is why astronomers are skeptical that UFOs are extraterrestrial spacecraft coming here across vast distances, briefly picking up two rural fishermen or loggers, and then going straight home. It seems like such a small reward for such a large investment.

CONSEQUENCES OF LIGHT TRAVEL TIME

There is another reason the speed of light is such a natural unit of distance for astronomers. Information about the universe comes to us almost exclusively via radiation (of which light is one example), and all such radiation travels at the speed of light—that is, 1 LY every year. This sets a limit on how quickly we can learn about events in the universe. A planet like Neptune is so far away that its light takes at least 3.2 hours to get to us; we thus see it as it was at least 3.2 hours ago. If a star is 100 LY away, the light we see from it tonight left that star 100 years ago and is just now arriving in our neighborhood. The soonest we can learn about any changes in that star—its blowing up, for example—is 100 years after the fact. For a star 500 LY away, the radiation we detect tonight left 500 years ago, and is carrying 500-year-old news.

FIGURE P.4
Telescope in Orbit The Hubble Space
Telescope, shown here being repaired aboard the
Space Shuttle Endeavour in December 1993, is an
example of the new generation of astronomical
instruments in space. *(NASA)*

Some students, accustomed to CNN and other news media known for "instant world coverage," at first find this frustrating. "You mean when I see that star up there," they ask, "I won't know what's actually happening there now for another 500 years?" But that's not really the right way to think about the situation. For astronomers, *now* is when the light reaches us here on Earth. There is no way for us to know anything about that star (or other object) until its radiation reaches us; despite the fondest dreams of science-fiction writers, instant communication through the universe is not possible.

But what at first may seem a great frustration is actually a tremendous benefit in disguise. If astronomers really want to piece together what has happened in the universe since its beginnings, they need to find evidence about each epoch of the past. Where can we find evidence today about cosmic events that occurred billions of years ago? Unfortunately, the universe doesn't leave written records (or even decent videotapes) of its main activities!

The delay in the arrival of light provides such evidence automatically. The farther out in space we look, the longer the light has taken to get here, and the longer ago it left its place of origin. By looking billions of light years out into space, astronomers are actually seeing billions of years into the past. In this way, we can reconstruct the history of the cosmos and get a sense of how it has evolved over time.

This is one reason astronomers strive to build telescopes that can collect more and more of the faint light (and other radiation) the universe sends us. The more light we collect, the fainter the objects we can make out. On average, fainter objects are farther away, and can thus tell us about periods of time even deeper in the past. New instruments, such as the Hubble Space Telescope (Figure P.4) and Hawaii's Keck Telescope (see Chapter 5), are giving astronomers views of deep space and deep time better than any we have had before.

A TOUR OF THE UNIVERSE

Let us now take a brief introductory tour of the universe as astronomers understand it today, just to get acquainted with the sorts of objects and distances astronomers deal with. While this book concerns itself mainly with the objects in our "immediate neighborhood" (cosmically speaking), we nevertheless want to show you briefly what else is out there and thus put our solar system in perspective.

We begin at home, with Earth, a nearly spherical planet about 13,000 km in diameter (Figure P.5). A space traveler entering our planetary system would easily distinguish the Earth by the large amount of liquid water that covers some two thirds of its crust. If the traveler had equipment to receive radio or television signals, or came close enough to see the lights of our cities at night, she would soon find signs that this water planet has intelligent life. (Of course, depending on what television channel the traveler tuned to, that

Humanity's Home Base Planet Earth as viewed from a perspective in space. This image was taken on May 3, 1990, by the Meteosat weather satellite high above the Earth's equator. *(ESA)*

conclusion might need to be changed to "semi-intelligent" life.)

Our nearest astronomical neighbor is the Earth's satellite, commonly called the Moon. Figure P.6 shows the Earth and the Moon to scale on the same diagram. Notice how small we have to make these bodies to fit them on the page with the right scale. The Moon's distance from Earth is about 30 times the Earth's diameter, or approximately 384,000 km, and it takes about a month for the Moon to revolve around the Earth. The Moon's diameter is 3476 km, one fourth the size of the Earth.

Light (or radio waves) takes 1.3 seconds to travel between Earth and the Moon. If you've seen videos of the Apollo flights to the Moon, you may recall that there was a delay of about 3 seconds between the time Mission Control asked a question and the time the astronauts replied. The reason is not that the astronauts were thinking slowly, but that it took the radio waves almost 3 seconds to make the round trip.

The Earth revolves around our star, the Sun, which is about 150 million km away—approximately 400 times as far as the Moon. We still call the average Earth–Sun distance an astronomical unit (AU), because in the early days of astronomy it was the most important measuring standard. Light takes a little more than 8 minutes to travel 1 AU, which means our latest news from the Sun is always 8 minutes old.

It takes Earth one year (3×10^7 s) to go around the Sun at our distance; to make it around, we must travel at approximately 110,000 km/h. (If you, like many students in the United States, still prefer miles to kilometers, you might find the following trick helpful. To convert kilometers to miles, you can multiply kilometers by 0.6. Thus 110,000 km/h becomes 66,000 mi/h—a fast clip no matter what units you use.) Since gravity holds us firmly to the Earth and there is no resistance to the Earth's motion in the vacuum of space, we participate in this breakneck journey without being aware of it day by day.

The diameter of the Sun is about 1.5 million km; our Earth could fit comfortably inside one of the minor eruptions that occur on the surface of our star. If the Sun were reduced to the size of a basketball, the Earth would be a small apple seed some 30 m from the ball.

The Earth is only one of nine planets that revolve around the Sun (Figure P.7). These planets, along with their satellites and swarms of smaller bodies, make up the *solar system,* what we might call the family of the Sun. A planet is defined as a body of significant size that orbits a star and does not produce its own light. (If a large body

FIGURE P.6
The Earth and Moon, Drawn to Scale

Our Solar Family The Sun and the planets shown to scale. Notice the size of the Earth compared with the giant planets.

consistently produces its own light, it gets to be called a star.) We are able to see the nearby planets in our skies only because they reflect the light of our local star, the Sun. If the planets were much farther away, the tiny amount of light that they manage to reflect would not be visible to us. The planets we have so far been able to discover orbiting *other* stars were found from the pull their gravity exerts on their parent stars, not from their light.

Jupiter, the largest planet in the solar system, is about 143,000 km in diameter, 11 times the size of the Earth (Figure P.8). Its distance from the Sun is five times that of the Earth, or 5 AU. On the scale where the Sun is a basketball, Jupiter is about the size of a grape, and is located about 150 m from the basketball. The orbits of the planets are shown schematically in Figure P.9. Pluto has a significantly noncircular orbit, but its *average* distance from the Sun is about 40 AU or 5.9 billion km. On our basketball scale, Pluto is a grain of sand about 1 km from the ball.

The Sun is our local star, and all the other stars are also suns—enormous balls of glowing gas that generate vast amounts of energy by nuclear reactions deep within. The other stars look faint only because they are so very far away. Continuing our analogy, Proxima Centauri, the nearest star beyond the Sun, which is 4.3 LY away, would be almost 7000 km from the basketball that represents the Sun.

When we look up at the star-studded country sky on a clear night, all the stars visible to the unaided eye turn out to be part of a single collection of stars we call the Milky Way Galaxy, or just simply the Galaxy. (When referring to

FIGURE P.8

Jupiter The largest planet in our solar system is Jupiter. We could fit almost eleven Earths side by side into its equator, and it contains as much mass as all the other planets combined. This image, which also shows one of its large satellites, was taken in February 1979 when the Voyager 1 spacecraft flew by. *(JPL/NASA)*

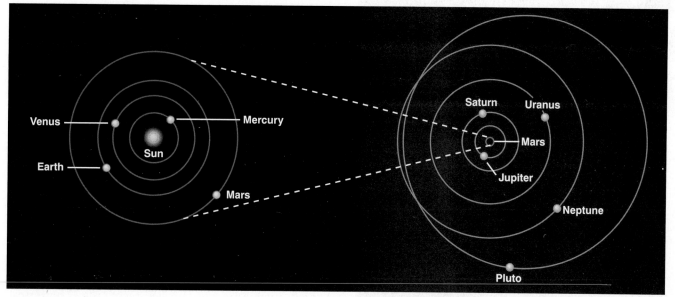

FIGURE P.9

The Orbits of the Planets in our Solar System *(Left)* The inner planets. *(Right)* The outer planets (note the change of scale).

the Milky Way, we capitalize Galaxy; when talking about other galaxies of stars, we use lowercase.) The Sun is one of hundreds of billions of stars that compose the Galaxy; its extent, as we will see, staggers the human imagination. Our Galaxy looks like a giant frisbee with a small ball in the middle. If we could move outside it and look down on the disk of the Milky Way from above, it would probably resemble the galaxy in Figure P.10, its spiral structure outlined by the blue light of hot adolescent stars.

The Sun is somewhat less than 30,000 LY from the center of the Galaxy, in a location with nothing much to distinguish it. From our position inside the Milky Way Galaxy, we cannot see through to its far rim (at least not with ordinary light) because the space between the stars is not completely empty. It contains a sparse distribution of gas (mostly the simplest element, hydrogen) intermixed with tiny solid particles that we call *interstellar dust*. This gas and dust collects into enormous clouds in many places

FIGURE P.10

A Spiral Galaxy This galaxy of billions of stars, called by its catalog number M83, is about 10 million LY away and is thought to be similar to our own Milky Way Galaxy. Here we see the giant wheel-shaped system face-on, looking down on the disk of stars. *(Cerro Tololo Interamerican Observatory/NOAO)*

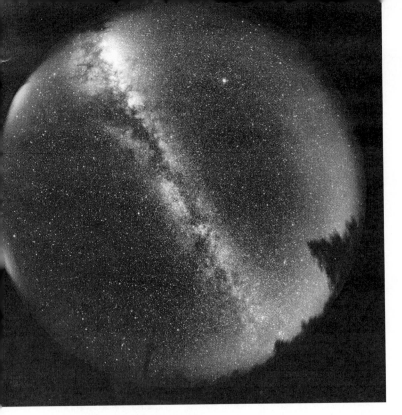

in the Galaxy, becoming the raw material for future generations of stars. Figure P.11 shows a beautiful image of the inner disk of the Galaxy as seen from our vantage point.

Typically, the interstellar material is so extremely sparse that the space between stars is a far, far better vacuum than anything we can produce in terrestrial laboratories. Yet, the dust in space, building up over thousands of light years, can block the light of more distant stars. Like the distant buildings that disappear from our view on a smoggy day in Los Angeles, the more distant regions of the Milky Way cannot be seen behind the layers of interstellar smog. Luckily, astronomers have found that stars and raw material shine with various forms of invisible radiation that do penetrate the smog, and so we have been able to build up a pretty good map of the Galaxy.

Because stars live a long time (compared to the people who like to watch them), you often hear them referred to as eternal. But in fact, no star can last forever. Since the "business" of stars is making energy, and energy production requires some sort of fuel to be used up, eventually all stars will run out of fuel and go out of business. (This news should not make you rush out to stock up on thermal underwear—our Sun still has at least 7 billion years to go.)

Ultimately, the Sun and all stars will die, and it is in their death throes that some of the most intriguing and important processes of the universe stand revealed. For example, we now know that many of the atoms in our bodies were once inside stars much larger than the Sun. These stars exploded at the ends of their lives, recycling their material back into the reservoir of the Galaxy. In this sense, all of us are literally made of "star dust."

In a very rough sense, you could think of the solar system as your house or apartment, and the Galaxy as your town, made up of many houses and buildings. In the 20th century, astronomers were able to show that just as our world is made up of many, many towns, so the universe is made up of enormous numbers of galaxies. (We define the universe to be everything that exists that is accessible to our observations.) Galaxies stretch as far in space as our telescopes can see, many billions of them within the reach of modern instruments. When galaxies were first discovered, some astronomers called them "island universes," and the term is aptly descriptive: Galaxies do look like islands of stars in the vast, dark seas of intergalactic space.

The nearest galaxy, just discovered in 1993, is a small one that lies 75,000 LY from the Sun in the direction of the constellation Sagittarius, where the smog in our own Galaxy makes it especially difficult to discern. (A *constellation*, we should note, is one of the 88 sections into which astronomers divide the sky, each named after a prominent star pattern within it.) The existence of this Sagittarius dwarf galaxy is, in fact, still controversial and will require other observations before all astronomers are convinced. Beyond it lie two other small galaxies, about 160,000 LY away. First recorded by Magellan's crew as they sailed around the world, these are called the Magellanic Clouds (Figure P.12). All three of these small galaxies are satellites

FIGURE P.13
Closest Spiral Galaxy The Andromeda Galaxy (M31) is a spiral-shaped collection of stars similar to our own Milky Way. Its flat disk of stars looks blue because its light is dominated by the vigorous light output of hot blue stars. The bulge in its center, on the other hand, contains older stars, which are predominantly yellow and red. Two smaller galaxies can be seen on either side of M31. *(Tony Hallas)*

FIGURE P.14
Fornax Cluster of Galaxies On this image, you can see a cluster of galaxies located about 60 million light-years away, in the constellation of Fornax. All the objects that are not pinpoints of light on the picture are galaxies of billions of stars. *(Image taken with the U.K. Schmidt Telescope, Anglo-Australian Observatory)*

of the Milky Way, interacting with it through the force of gravity. Ultimately, all three may even be swallowed by our much larger Galaxy.

The nearest large galaxy is a spiral quite similar to our own, located in the constellation of Andromeda and thus often called the Andromeda galaxy; it is also known by one of its catalog numbers, M31 (Figure P.13). Given the number of galaxies, no reasonable person would suggest giving all of them proper names. M31 is about 2 million LY away and, along with the Milky Way, it is part of a small cluster of over 30 galaxies that we call the Local Group.

At distances of 10 to 15 million LY we find other small galaxy groups, and then at about 50 million LY there is a more impressive system, with thousands of member galaxies, called the Virgo Cluster. We have discovered that galaxies occur mostly in clusters, both large and small (Figure P.14). Some of the clusters themselves form into larger groups called *superclusters*.

Our Local Group, as well as the Virgo Cluster, are part of such a supercluster, which stretches over a diameter of at least 60 million LY. We are just beginning to explore the structure of the universe at these enormous scales.

For now, we might just note that such ideas are far easier to state than to discover. Measurements of the properties of distant galaxies and clusters (and small remote members of our own solar system) require large telescopes, sophisticated light-amplifying devices, and painstaking labor. Every clear night, at observatories around the world, astronomers and students are at work on such mysteries as the birth of new stars and planets or the large-scale structure of the universe, mostly observing one star or one galaxy at a time, and fitting their results into the tapestry of our understanding.

THE UNIVERSE OF THE VERY SMALL

The foregoing discussion should impress on you that the universe is extraordinarily large and extraordinarily empty. The universe on average is 10,000 times emptier than our Galaxy. Yet, as we have seen, even the Galaxy is mostly empty space. The air we breathe has about 10^{19} atoms in each cubic centimeter—and we think of air as pretty empty stuff. In the interstellar gas of the Galaxy, there is about *one* atom in every cubic centimeter. Intergalactic space is so sparsely filled that to find one atom, on average, we must search through a cubic *meter* of space! Most of the universe is fantastically empty; places that are dense, such as the interiors of planets or the bodies of our readers, are tremendously rare.

Yet even the familiar solids, such as this textbook, are mostly space. If we could take such a solid apart, piece by piece, we would eventually reach the molecules of which it is formed. Molecules are the smallest particles into which matter can be divided while still retaining its chemical properties. A molecule of water (H_2O), for example,

TABLE P.1
The Cosmically Abundant Elements

Element*	Symbol	Number of Atoms per Million Hydrogen Atoms
Hydrogen	H	1,000,000
Helium	He	80,000
Carbon	C	450
Nitrogen	N	92
Oxygen	O	740
Neon	Ne	130
Magnesium	Mg	40
Silicon	Si	37
Sulfur	S	19
Iron	Fe	32

* Our list of elements is arranged in order of the atomic number, which is the number of protons in each nucleus.

consists of two hydrogen atoms and one oxygen atom, bonded together.

Molecules, in turn, are built up of atoms, which are the smallest particles of an element that can still be identified as that element. For example, an atom of gold is the smallest possible piece of gold (although one atom of gold won't impress your sweetheart very much). Nearly 100 different kinds of atoms (elements) exist in nature, but most of them are rare, and only a handful account for more than 99 percent of everything with which we come in contact. The most abundant elements in the cosmos today are listed in Table P.1; think of this table as the "greatest hits" of the universe when it comes to elements.

All atoms consist of a central, positively charged nucleus, surrounded by negatively charged electrons. The bulk of the matter in each atom is found in the nucleus which consists of positive protons and electrically neutral neutrons all tightly bound together in a very small space. Each element is defined by the number of protons in its atoms: thus any atom with 6 protons in its nucleus is called carbon, any with 50 protons is called tin, and any with 70 protons is called ytterbium. (Ytterbium, as you can probably guess, is not big on the cosmic hit parade of elements, but we like its name. For a list of the elements, see Appendix 13.)

The distance from an atomic nucleus to its electrons is typically 100,000 times the size of the nucleus itself. This is why we say that even solid matter is mostly space. The typical atom is far emptier than the solar system out to Pluto. (The distance from the Earth to the Sun, for example, is only 100 times the size of the Sun.) This is another reason atoms are not like miniature solar systems.

Remarkably, physicists have discovered that everything that happens in the universe, from the smallest atoms to the largest superclusters of galaxies, can be explained through the action of only four forces: gravity, electromagnetism (which combines the actions of electricity and magnetism),

and two forces that act at the nuclear level. We will get to know some of these forces in more detail throughout the book, but the fact that there are four forces (and not a million, or just one) has puzzled physicists and astronomers for many years, and has led to a quest for a unified picture of nature.

A CONCLUSION AND A BEGINNING

If you are typical of students new to astronomy, you have probably reached the end of our tour in this Prologue with mixed emotions. On the one hand, you may be fascinated by some of the new ideas you've read about, and eager to learn more. On the other, you may be feeling overwhelmed by the number of topics we have covered, and the number of new words and ideas we have introduced. Learning astronomy is a little like learning a new language—at first it seems there are so many new expressions that you'll never learn them all, but with practice, you soon develop facility with them.

At this point you may also feel a bit small and insignificant, dwarfed by the cosmic scales of distance and time. Such a feeling is not a bad thing from time to time—(sometimes we wish more of our politicians and film stars felt this way). And just before a difficult exam, or when you've ended a treasured relationship, it can certainly help to see your problems in a cosmic perspective. But there is another way to look at what we've learned from our first glimpses of the cosmos.

Let us consider the history of the universe from the Big Bang to today, and compress it, for easy reference, into a single year. (We have borrowed this idea from Carl Sagan's Pulitzer Prize–winning book, *The Dragons of Eden*, published in 1977 by Random House.) On this scale, the Big Bang happened at the first moment of January 1, the solar system formed around September 9, and the oldest rocks we can date on Earth go back to the third week in September (Figure P.15).

Where in this "cosmic year" does the origin of human beings fall? The answer turns out to be the evening of December 31! The invention of the alphabet doesn't occur until the 50th second of 11:59 P.M. on December 31. And the beginnings of modern astronomy are a mere fraction of a second before the New Year. Seen in a cosmic context, the amount of time we have had to study the planets and stars is minute, and our success at piecing together as much of the story as we have is remarkable.

Certainly, our attempts to understand the universe are not complete. As new instruments and new ideas allow us to gather even better data about the cosmos, our present picture of astronomy will very likely undergo many changes. In fact, we would not be surprised if your grandchildren's grandchildren found some of the contents of this book a bit primitive. But as you read our current progress report on the exploration of the universe, take a few minutes every once in a while just to savor how much we have already learned.

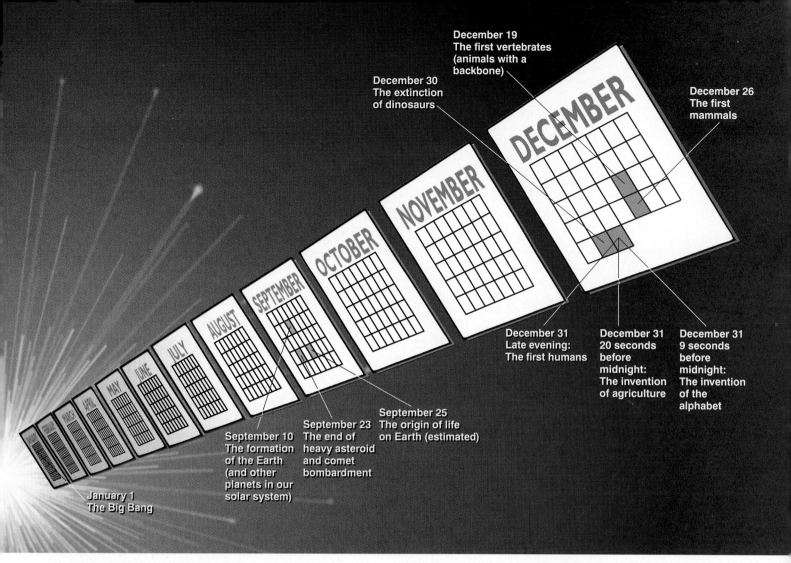

FIGURE P.15
Charting Cosmic Time On a cosmic calendar, where the time since the Big Bang is compressed into one year, creatures we would call human do not emerge on the scene until the evening of December 31.

SURFING THE WEB

If you have enjoyed some of the spectacular photographs in this prologue, and have access to the World Wide Web, we want to recommend the following Web sites to you. They will allow you to check out some of the best images astronomers have of cosmic objects, and to keep track of new images as they come in:

Planetary Photojournal [photojournal.jpl.nasa.gov]

This site, run by the Jet Propulsion Laboratory and the U.S. Geological Survey Branch of Astrogeology in Flagstaff, Arizona, is one of the most useful resources on the Web. (We may not be entirely objective, however, because two of the authors of *Voyages* served as advisors for the project to set up the site!) It currently features over 2000 of the best images from planetary exploration, with detailed captions and excellent indexing, and more are being added all the time. You can dial up images by world, feature name, date, or catalog number, and download images in a number of popular formats. The one problem with

the site is its "NASA chauvinism"; only NASA mission images are currently included.

Hubble Space Telescope [oposite.stsci.edu/pubinfo/]

All the magnificent Hubble Space Telescope images are here, with captions; many have detailed background information and new research results included. You can check the latest images, browse the ones the staff considers the Hubble's "greatest hits," or search for objects of interest to you.

National Optical Astronomy Observatories Image Gallery [www.noao.edu/noaosci.html]

NOAO, whose director is *Voyages* coauthor Sidney Wolff, includes a number of major telescopes in the United States and the Southern Hemisphere, and is a national facility, where any astronomer can apply for time. Some of the best images from NOAO instruments are collected at this site.

The constellation of Orion (a star grouping with a distinctive belt of three stars, part of the figure of the great hunter of ancient Greek mythology) can be seen above and to the left of the University of Toronto's telescope dome on Las Campanas mountain in Chile. To find our way among the stars, we rely on signposts like groups of stars that make a recognizable pattern we can point out to one another.   *(U. of Toronto)*

I   Observing the Sky: The Birth of Astronomy

THINKING AHEAD

Much to your surprise, a member of the Flat Earth Society moves in next door. He believes that the Earth is flat and that all the NASA images of a round Earth are either faked or simply show the round (but flat) disk of the Earth from above. How could you prove to this neighbor that the Earth is really round? (You can find some of our answers in the chapter and a summary in a box at the end of the chapter, but try to figure out some ways by yourself before peeking at our suggestions! You might want to discuss ideas with other students in the class.)

> "If the Lord Almighty had consulted me before embarking upon Creation, I should have recommended something simpler."
>
> Statement attributed to Alfonso X, King of Castile, after having the Ptolemaic system explained to him.

Today, few people really spend much time looking at the night sky. When we see patterns in the dark, we're much more likely to be looking at a television or movie screen than at the heavens over our heads. But in ancient days, before electric lights and television robbed so many people of the splendor of the sky, the stars and planets were an important aspect of everyone's daily life. All the records that we have—on paper and in stone—show that ancient civilizations around the world noticed, worshipped, and tried to understand the celestial lights and fit them into their own view of the world. These ancient observers found both majestic regularity and never-ending surprise in the motions of the heavens.

The Babylonians and Greeks, for example, believing the planets to be gods, studied their motions in hopes of understanding the presumed influences of those planet-gods on human affairs; thus developed the religion of astrology. Through their careful study of the planets, however, the ancient Greeks and later the Romans were also laying the foundations of the science of astronomy. As we shall see, their crowning achievement was Claudius Ptolemy's marvelously ingenious model in the 2nd century A.D.

Ptolemy's system predicted the positions of planets with reasonable precision for hundreds of years, and it was not substantially improved upon until the European Renaissance in the 16th and 17th centuries. In this chapter we shall take a look at the night sky as seen with the unaided eye and also examine some of the interesting history of how we came to understand the realm above our heads.

1.1 ## THE SKY ABOVE

Our senses suggest to us that the Earth is the center of the universe—the hub around which the heavens turn. This **geocentric** (Earth-centered) view was what almost everyone believed until the Renaissance. After all, it is simple, logical, and seemingly self-evident. Furthermore, the geocentric perspective reinforced those philosophical and religious systems that taught the unique role of human beings as the central focus of the cosmos. However, the geocentric view happens to be wrong. One of the great themes of our intellectual history is the overthrow of the geocentric

perspective. Let us therefore take a look at the steps by which we re-evaluated the place of our world in the cosmic order.

The Celestial Sphere

If you go on a camping trip or live far from city lights, your view of the sky on a clear night is pretty much identical to that seen by people all over the world before the invention of the telescope. Gazing up, you get the impression that the sky is a great hollow dome, with you at the center (Figure 1.1). The top of that dome, the point directly above your head, is called the **zenith,** and where the dome meets the Earth is called the **horizon**. It's easy to see the horizon as a circle around you from the sea or a flat prairie, but from most places where people live today, the horizon is hidden by mountains, trees, buildings, or smog.

If you lie back in an open field and observe the night sky for hours, as ancient shepherds and travelers regularly did, you will see stars rising on the eastern horizon (just as the Sun and Moon do), moving across the dome of heaven in the course of the night, and setting on the western

FIGURE 1.1
The Sky Around Us The dome of the sky, as it appears to a naive observer. The horizon is where the sky meets the ground, and the observer's zenith is the point directly overhead.

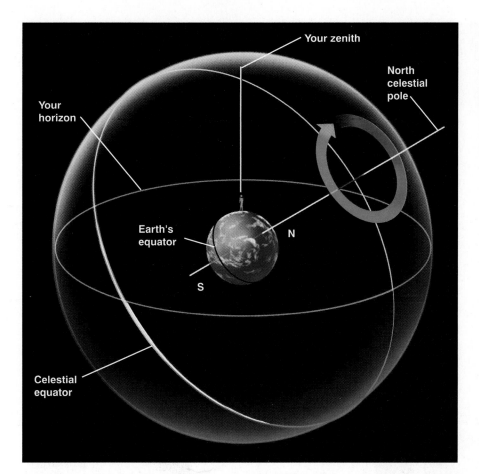

Your zenith

Your horizon

North celestial pole

Earth's equator

N

S

Celestial equator

FIGURE 1.2
Circles on the Celestial Sphere The (imaginary) celestial sphere around the Earth, on which objects in the sky can turn. Note the axis through the Earth. In reality, it is the Earth that turns around this axis, creating the illusion that the sky revolves around us. Note that the Earth in this picture has been tilted so that your location is at the top and the North Pole is where the little N is.

horizon. Watching the sky turn like this night after night, you might eventually get the idea that the dome of the sky is really part of a great sphere that is turning around you, bringing different stars into view as it turns. The early Greeks regarded the sky as just such a **celestial sphere** (Figure 1.2). Some apparently thought of it as an actual sphere of transparent crystalline material, with the stars embedded in it like tiny jewels.

Today, we know that it's not the celestial sphere that turns as the night and day proceed, but rather the planet on which we live. We can put an imaginary stick through the Earth's north and south poles; we call this stick our planet's **axis.** It is because the Earth turns on this axis every 24 hours that we see the Sun, Moon, and stars rise and set with clockwork regularity. These celestial objects are not on a dome, but at greatly varying distances from us in space. Nevertheless, it is sometimes still convenient to talk about the celestial dome or sphere to help us keep track of objects in the sky. There is even a special theater, called a *planetarium,* in which we project a simulation of the stars and planets onto a white dome.

As the celestial sphere rotates, the objects on it maintain their positions with respect to one another. A grouping of stars like the Big Dipper has the same shape during the course of the night, although it turns with the sky. During a single night, even objects that we know to have significant motions of their own, such as the nearby planets, seem fixed relative to the stars. (Only the *meteors*—brief "shooting stars" that flash into view for just a few seconds—move appreciably with respect to the celestial sphere. This is because they are not stars at all. Rather, they are small pieces of cosmic dust, burning up as they hit the Earth's atmosphere.) We can use the fact that the entire celestial sphere seems to turn together to help us set up systems for keeping track of what things are visible in the sky and where they happen to be at a given time.

Celestial Poles and Celestial Equator

To help orient us in the turning sky, astronomers use a system that extends our Earth's special points into the sky. We extend the line of the Earth's axis outward: The points where the axis meets the celestial sphere are defined as the **north celestial pole** and the **south celestial pole.** As the Earth rotates about its axis, the sky appears to turn in the opposite direction about those celestial poles (Figure 1.3). We also (in our imaginations) throw the Earth's equator onto the sky and call this the **celestial equator.** It lies halfway between the celestial poles, just as the Earth's equator lies halfway between our planet's poles.

FIGURE 1.3
Circling the North Celestial Pole Time exposure showing trails left by stars as a result of the apparent rotation of the celestial sphere. (In reality, it is the Earth that rotates.) The bright trail at top center was made by Polaris (the North Star), which is very close to the north celestial pole. *(National Optical Astronomy Observatories)*

Now let's imagine how riding on different parts of the spinning Earth affects our view of the sky. The apparent motion of the celestial sphere depends on your latitude (position north or south of the equator). Bear in mind that the Earth's axis is pointing at the celestial poles, so those two points on the sky do not appear to turn.

If you stood at the North Pole of the Earth, for example, you would see the north celestial pole overhead, at your zenith. The celestial equator, 90° from the celestial poles, lies along your horizon. As you watched them during the course of the night, the stars would all circle around the celestial pole, with none rising or setting (Figure 1.4). Only that half of the sky that is north of the celestial equator is ever visible to an observer at the North Pole. Similarly, an observer at the South Pole would see only the southern half of the sky.

If you were at the Earth's equator, on the other hand, you would see the celestial equator (which after all is just an "extension" of the Earth's equator) pass overhead through your zenith. The celestial poles, being 90° from the celestial equator, must then be at the north and south points on your horizon. As the sky turns, all stars rise and set; they move straight up from the east side of the horizon and set straight down on the west side. During a 24-hour period, all stars are above the horizon exactly half the time. (Of course, during some of those hours, the Sun is too bright for us to see them.)

What would an observer in the intermediate latitudes of the United States or Europe see? In our case, the north celestial pole is neither overhead nor on the horizon, but in between. It appears above the northern horizon at an angular height, or altitude, equal to the observer's latitude (Figure 1.4). In San Francisco, for example, where the latitude is 38° N, the north celestial pole is 38° above the northern horizon.

For an observer at 38° N latitude, the south celestial pole is 38° below the southern horizon and thus never visible. As the Earth turns, the whole sky seems to pivot about the north celestial pole. For this observer, stars within 38° of the North Pole can never set. They are always above the horizon, day and night. This part of the sky is called the north **circumpolar zone.** To observers in the continental United States, the Big and Little Dippers and Cassiopeia are examples of star groups in the north circumpolar zone. On the other hand, stars within 38° of the south celestial pole never rise. That part of the sky is the south circumpolar zone. To most U.S. observers, the Southern Cross is in that zone. (Don't worry if you are not familiar with the star groups we just mentioned. We will introduce them more formally later on.)

At this particular time in Earth history, there happens to be a star very close to the north celestial pole. It is called Polaris, the pole star, and has the distinction of being the star that moves the least amount as the northern sky turns each day. (Because it moved so little while the other stars moved much more, it played a special role in the mythology of several Native American tribes, for example. Some called it the "fastener of the sky.")

───── ☽ ◐ ◯ ◑ ◑ ☾ ─────

ASTRONOMY BASICS

What's Your Angle?

Astronomers measure how far apart objects appear in the sky by using angles. By definition, there are 360° in a circle, so a circle stretching completely around the celestial sphere contains 360°. The half-sphere or dome of the sky then contains 180° from horizon to opposite horizon. Thus, if two stars are 18° apart, their separation spans about 1/10 of the dome of the sky. To give you a sense of how big a degree is, the full Moon is about half a degree across.

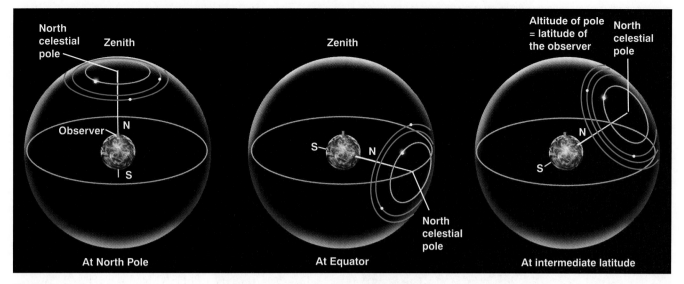

FIGURE 1.4
Star Circles at Different Latitudes The turning of the sky looks different depending on your latitude on Earth. (a) At the north pole, the stars circle the zenith and do not rise and set. (b) At the equator, the celestial poles are on the horizon, and the stars rise straight up and set straight down. (c) At intermediate latitudes, the north celestial pole is at some position between overhead and the horizon. Its angle turns out to be equal to the observer's latitude. Stars rise and set at an angle to the horizon.

Rising and Setting of the Sun

We have described the movement of stars in the night sky. The stars continue to circle during the day, but the brilliance of the Sun makes them difficult to see. (The Moon is still easily seen in the daylight, however.) On any given day, we can think of the Sun as being located at some position on the hypothetical celestial sphere. When the Sun rises—that is, when the rotation of the Earth carries the Sun above the horizon—sunlight is scattered by the molecules of our atmosphere, filling our sky with light and hiding the stars that are above the horizon.

For thousands of years, astronomers have been aware that the Sun does more than just rise and set. It gradually changes position on the celestial sphere, moving each day about 1° to the east relative to the stars. Very reasonably, the ancients thought this meant that the Sun was slowly moving around the Earth, taking a period of time we call one **year** to make a full circle. Today, of course, we know that it is the Earth that is going around the Sun, but the effect is the same: The Sun's position in our sky changes day to day. We have a similar experience when we walk around a campfire at night; we see the flames appear in front of each person seated about the fire in turn.

The path the Sun appears to take around the celestial sphere each year is called the **ecliptic** (Figure 1.5). Because of its motion on the ecliptic, the Sun rises about 4 minutes later each day with respect to the stars. The Earth must make just a bit more than one complete rotation (with respect to the stars) to bring the Sun up again. As the

months go by and we look at the Sun from different places in our orbit, we see it projected against different stars in the background, or we would, at least, if we could see the stars in the daytime. In practice, we must deduce what stars lie behind and beyond the Sun by observing the stars visible in the opposite direction at night. After a year, when the Earth has completed one trip around the Sun, the Sun will appear to have completed one circuit of the sky along the ecliptic.

The ecliptic does not lie along the celestial equator but is inclined to it at an angle of about 23°. In other words, the Sun's annual path in the sky is not lined up with the Earth's equator. This is because our planet's axis of rotation is tilted by about 23° from the plane of the ecliptic (Figure 1.6). Being tilted in this way is not at all unusual among planets; Uranus and Pluto are actually tilted so much that they orbit the Sun "on their side." The inclination of the ecliptic is the reason the Sun moves north and south in the sky as the seasons change. In Chapter 3 we will discuss the progression of the seasons in more detail.

Fixed and Wandering Stars

The Sun is not the only object that moves among the fixed stars. The Moon and each of the five planets that are visible to the unaided eye—Mercury, Venus, Mars, Jupiter, and Saturn—also slowly change their positions from day to day. During a single day, the Moon and planets all rise and set as the Earth turns, just as the Sun and stars do. But like the Sun, they have independent motions among the stars,

Constellation on the Ecliptic	Dates When the Sun Crosses It	Constellation on the Ecliptic	Dates When the Sun Crosses It
Capricornus	Jan 21–Feb 16	Leo	Aug 10–Sep 16
Aquarius	Feb 16–Mar 11	Virgo	Sept 16–Oct 31
Pisces	Mar 11–Apr 18	Libra	Oct 31–Nov 23
Aries	Apr 18–May 13	Scorpius	Nov 23–Nov 29
Taurus	May 13–June 22	Ophiuchus	Nov 29–Dec 18
Gemini	June 22–July 21	Sagittarius	Dec 18–Jan 21
Cancer	July 21–Aug 10		

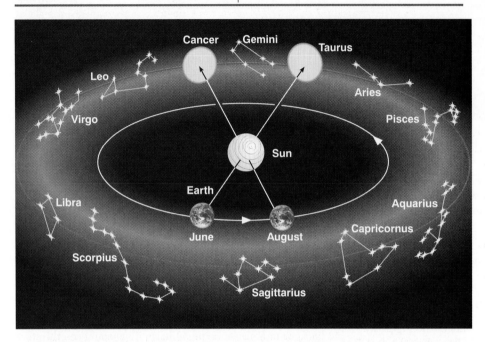

FIGURE 1.5
Constellations on the Ecliptic As the Earth revolves around the Sun, we sit on "platform Earth" and see the Sun moving around the sky. The circle in the sky that the Sun appears to make around us in the course of a year is called the ecliptic. This circle (like all circles in the sky) goes through a set of constellations; the ancients thought that these constellations, which the Sun (and the Moon and planets) visited, must be special and incorporated them into their system of astrology (see Section 1.3). Note that at any given time of the year, some of the constellations crossed by the ecliptic are visible in the night sky and others are in the day sky and thus hidden by the brilliance of the Sun. As we discuss later in the chapter, today the zodiac constellations in which we "find" the Sun each month are no longer lined up with the signs the astrologers use.

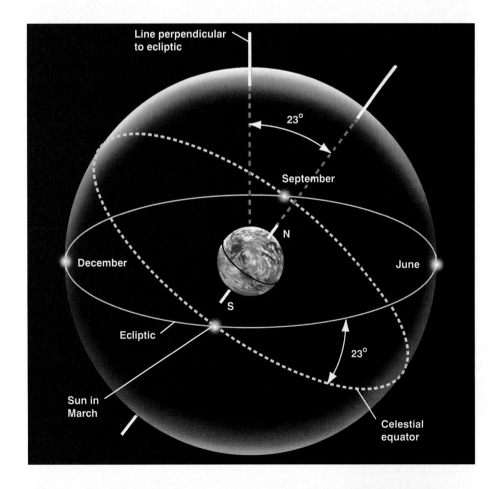

FIGURE 1.6
The Celestial Tilt The celestial equator is tilted by 23° to the ecliptic. As a result, North Americans and Europeans see the Sun north of the celestial equator and high in our sky in June, and south of the celestial equator and low in the sky in December.

superimposed on the daily rotation of the celestial sphere. Noticing these motions, the Greeks of two thousand years ago distinguished between what they called the fixed stars, those that maintain fixed patterns among themselves through many generations, and the wandering stars, or **planets.** The word *planet,* in fact, means "wanderer" in Greek.

Today we do not regard the Sun and Moon as planets, but the ancients applied the term to all seven of the moving objects in the sky. Much of ancient astronomy was devoted to observing and predicting the motions of these celestial wanderers. The Moon, being the Earth's nearest celestial neighbor, has the fastest apparent motion; it completes a trip around the sky in about one month (moonth). To do this, the Moon moves about 12°, or 24 times its own apparent width on the sky, each day.

The individual paths of the Moon and planets in the sky all lie close to the ecliptic, although not exactly on it. This is because the paths of the planets about the Sun, and of the Moon about the Earth, are all in nearly the same plane, as if they were circles on a huge sheet of paper. The planets, the Sun, and Moon are thus always found in the sky within a narrow 18°-wide belt called the **zodiac** that is centered on the ecliptic (Figure 1.5). (The root of this term is the same as that of the word *zoo* and means a collection of animals: many of the patterns of stars within the zodiac belt reminded the ancients of animals, such as a fish or a goat.)

How the planets appear to move in the sky as the months pass is a combination of their actual motions plus the motion of the Earth about the Sun; consequently, their paths are somewhat complex. As we shall see, this complexity has fascinated and challenged astronomers for centuries.

Constellations

The backdrop for the motions of the "wanderers" in the sky is the canopy of stars. If there were no clouds in the sky and we were on a flat plain with nothing to obstruct our view, we could see about 3000 stars with the unaided eye. To find their way around such a multitude, the ancients found groupings of stars that made some familiar geometric pattern or (more rarely) resembled something they knew. Each civilization found its own patterns in the stars, much like a modern Rorschach test in which you are asked to discover patterns or pictures in a set of inkblots. The ancient Chinese, Egyptians, and Greeks, among others, found their own groupings or **constellations** of stars helpful in navigating among them and in passing their star-lore on to their children.

You may be familiar with some of the star patterns we still use today, such as the Big and Little Dippers or Orion the hunter, with his distinctive belt of three stars (Figure 1.7). However, many of the stars we see are not part of a distinctive star pattern at all, and a telescope can reveal millions of stars too faint for the eye to see. Therefore, in the early decades of the 20th century, astronomers from

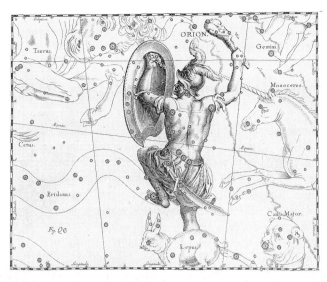

FIGURE 1.7

Orion The winter constellation of Orion, the hunter, as illustrated in the 17th-century atlas by Hevelius. *(J. M. Pasachoff and the Chapin Library)*

many countries decided to establish a worldwide system for organizing the sky.

Today, we use the term *constellation* to mean one of 88 sectors into which we divide the sky, much as the United States has been divided into 50 states. The modern boundaries between the constellations are imaginary lines in the sky running north–south and east–west, so that each point in the sky falls in a specific constellation. All the constellations are listed in Appendix 14. Whenever possible, we have named each modern constellation after the Latin translation of one of the ancient Greek star patterns that lies within it. Thus, the constellation of Orion is a kind of box on the sky, which includes, among many other objects, the stars that made up the ancient picture of the hunter. Some people use the term *asterism* to denote an especially noticeable star pattern within a constellation (or, sometimes, spanning parts of several constellations). For example, the Big Dipper is an asterism within the constellation of Ursa Major, the Big Bear.

Students are sometimes puzzled because the constellations seldom resemble the people or animals for which they were named. In all likelihood, the Greeks themselves did not name groupings of stars because they looked like actual people or objects (any more than the outline of Washington State resembles George Washington). Rather, they named sections of the sky in honor of the characters in their mythology and then fit the star configurations to the animals and people as best they could.

1.2 ANCIENT ASTRONOMY

Let us now look briefly back into history. Much of modern Western civilization is derived in one way or another from the ideas of the ancient Greeks and Romans, and this is true in astronomy as well. However, many other ancient cultures also developed sophisticated systems for observing and interpreting the sky.

Astronomy Around the World

The Babylonian, Assyrian, and ancient Egyptian astronomers knew the approximate length of the year. The Egyptians of 3000 years ago, for example, adopted a calendar based on a 365-day year. They kept careful track of the rising time of the bright star Sirius in the predawn sky, whose cycle happened to correspond with the flooding of the river Nile. The Chinese also had a working calendar and determined the length of the year at about the same time. They recorded comets, bright meteors, and dark spots on the Sun. (Many types of astronomical objects were introduced in the Prologue of this book. If you are not familiar with terms like *comets* and *meteors*, you may want to look at the Prologue.) Later, Chinese astronomers kept careful records of "guest stars"—those that are normally too faint to see, but suddenly flare up to become visible to the naked eye for a few weeks or months. We still use some of these records in studying stars that exploded a long time ago.

The Mayan culture in Central America developed a sophisticated calendar based on the planet Venus, and they made astronomical observations from sites dedicated to this purpose a thousand years ago. The Polynesians learned to navigate by the stars over hundreds of kilometers of open ocean, thus allowing them to colonize new islands far away from where they began.

In the British Isles, before the widespread use of writing, ancient people used stones to keep track of the motions of the Sun and Moon. We still find some of the great stone circles they built for this purpose, dating from as far back as 2800 B.C. The best known of these is Stonehenge, which is discussed in Chapter 3.

Early Greek and Roman Cosmology

Our concept of the cosmos—its basic structure and origin—is called **cosmology,** a Greek word. Before the invention of telescopes, humans had to depend on the simple evidence of their senses for a picture of the universe. The ancients developed cosmologies that combined their direct view of the heavens with a rich variety of philosophical and religious symbolism.

At least 2000 years before Columbus, educated people in the eastern Mediterranean region knew the Earth was round. Belief in a spherical Earth may have stemmed from the time of Pythagoras, a philosopher and mathematician who lived 2500 years ago. He believed circles and spheres to be "perfect forms" and suggested that the Earth should therefore be a sphere. As evidence that the gods liked spheres, the Greeks cited the fact that the Moon is a sphere, using evidence we will describe later.

The writings of Aristotle (384–322 B.C.), the tutor of Alexander the Great, summarize many of the ideas of his day. They describe how the progression of the Moon's *phases*—its changing shape—results from our seeing different portions of the Moon's illuminated hemisphere during the month (see Chapter 3). Aristotle also knew that the Sun has to be farther away from the Earth than is the Moon, because occasionally the Moon passes exactly between the Earth and Sun and temporarily hides the Sun from view. We call this a solar eclipse (see Chapter 3).

Aristotle cited two convincing arguments that the Earth must be round. First is the fact that as the Moon enters or emerges from the Earth's shadow during an eclipse of the Moon, the shape of the shadow seen on the Moon is always round (Figure 1.8). Only a spherical object *always* produces a round shadow. If the Earth were a disk, for example, there would be some occasions when the sunlight would be striking it edge-on and its shadow on the Moon would be a line.

FIGURE 1.8
The Earth's Round Shadow A lunar eclipse, with the Moon moving into and out of the Earth's shadow. Note the curved shape of the shadow—evidence for a spherical Earth that has been recognized since antiquity. *(Kent Wood)*

How Do We Know the Earth Is Round?

In addition to the two ways (from Aristotle's writings) discussed in this chapter, you might also reason as follows:

1. Let's watch a ship leave its port and sail into the distance on a clear day. On a flat Earth, you would just see the ship get smaller and smaller as it sailed away. But that's not what we actually observe. Instead, ships sink below the horizon, with the hull disappearing first and the mast remaining visible for a while longer. Eventually only the top of the mast can be seen, as the ship sails around the curvature of the Earth; then finally the ship disappears.

2. The Space Shuttle circles the Earth once every 90 minutes or so. Photographs taken from the Shuttle and other satellites show that the Earth is round from every perspective.

3. Suppose you made a friend in each time zone of the Earth. You could call all of them in the same hour and ask, "Where is the Sun?" On a flat Earth, each caller would give you roughly the same answer. But on a round Earth you would find that for some, the Sun would be high in the sky, while for others it would be rising, or setting, or completely out of sight (and this last group of friends would be upset with you for waking them up).

As a second argument, Aristotle explained that travelers who go south a significant distance are able to observe stars that are not visible farther north. And the height of the North Star—the star nearest the north celestial pole—decreases as a traveler moves south. On a flat Earth, everyone would see the same stars overhead. The only possible explanation is that the traveler must have moved over a curved surface on the Earth, showing stars from a different angle. (See the box "How Do We Know the Earth Is Round?" above for more ideas on proving the Earth is round.)

One brave Greek thinker, Aristarchus of Samos (310–230 B.C.), even suggested, long before Copernicus, that the Earth was moving around the Sun, but Aristotle and most of the ancient Greek scholars rejected this idea. One of the reasons for their conclusion was their understanding that if the Earth moved about the Sun, they would be observing the stars from different places along Earth's orbit. This would mean that the apparent directions of nearby stars in the sky would change during the year relative to more distant stars. (In a similar way, we see foreground objects appear to move against a more distant background whenever we are in motion. When we ride on a train, the trees in the foreground appear to shift their position relative to distant hills as the train rolls by. Unconsciously, we use this phenomenon all of the time to estimate distances around us.)

The apparent shift in the direction of an object as a result of the motion of the observer is called *parallax*. We call the shift in the apparent direction of a star due to the Earth's orbital motion *stellar parallax*. The Greeks made dedicated efforts to observe stellar parallax, even enlisting the aid of Greek soldiers with the clearest vision, but to no avail. The brighter (and presumably nearer) stars just did not seem to shift as they observed them in the spring and then again in the fall.

This meant either that the Earth was not moving or that the stars had to be so tremendously far away that the parallax shift was immeasurably small. A cosmos of such enormous extent required a leap of imagination that most ancient thinkers were not prepared to make, so they retreated to the safety of the Earth-centered view, which would dominate Western thinking for nearly two millenia.

Measurement of the Earth by Eratosthenes

The Greeks not only knew the Earth was round, they were also able to measure its size. The first fairly accurate determination of the Earth's diameter was made about 200 B.C. by Eratosthenes, a Greek living in Alexandria, Egypt. His method was a geometrical one, based on observations of the Sun.

The Sun is so distant from us that all the light rays that strike our planet approach us along essentially parallel lines. To see why, look at Figure 1.9. Take a source of light near the Earth, say at position A. Its rays strike different parts of the Earth along diverging paths. From a light source at B, or at C, still farther away, the angle between rays that strike opposite parts of the Earth is smaller. The more distant the source, the smaller is the angle between the rays. For a source infinitely distant, the rays travel along parallel lines.

Of course, the Sun is not infinitely far away, but light rays striking the Earth from a point on the Sun diverge from one another by an angle far too small to be observed with the unaided eye. As a consequence, if people all over the Earth who could see the Sun were to point at it, their fingers would all be essentially parallel to one another. (The same is also true for the planets and stars, an idea we will use in our discussion of how telescopes work.)

Eratosthenes noticed that on the first day of summer at Syene, Egypt (near modern Aswan), sunlight struck the *bottom* of a vertical well at noon. This indicated that the Sun was right over the well (that Syene was on a direct line

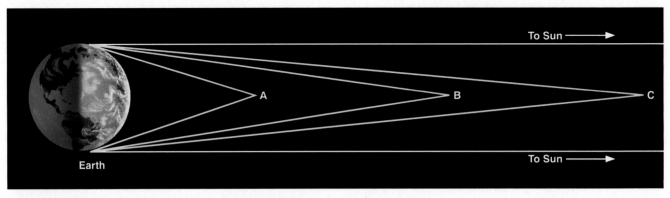

FIGURE 1.9

Light Rays from Space   The more distant an object, the more nearly parallel are the rays of light coming from it.

from the center of the Earth to the Sun). At the corresponding time and date in Alexandria, he observed that the Sun was not directly overhead but slightly south of the zenith, so that its rays made an angle with the vertical equal to about 1/50 of a circle (7°). Since the Sun's rays striking the two cities are parallel to one another, why would the two rays not make the same angle with the Earth's surface? Eratosthenes reasoned that the curvature of the round Earth meant that "straight up" was not the same in the two cities. This, he realized, could be used to measure how big the Earth is.

Alexandria, he saw, must be 1/50 of the Earth's circumference north of Syene (Figure 1.10). Alexandria had been measured to be 5000 stadia north of Syene (the *stadium* was a Greek unit of length, derived from the length of the racetrack in a stadium). Eratosthenes thus found that the Earth's circumference must be 50 × 5000, or 250,000 stadia.

It is not possible to evaluate precisely the accuracy of Eratosthenes' solution because there is doubt about which of the various kinds of Greek stadia he used as his unit of distance. If it was the common Olympic stadium, his result was about 20 percent too large. According to another interpretation, he used a stadium equal to about 1/6 km, in which case his figure was within one percent of the correct value of 40,000 km. Even if his measurement was not exact, his success at measuring the size of our planet by using only shadows, sunlight, and the power of human thought was one of the greatest intellectual achievements in history.

Hipparchus

Perhaps the greatest astronomer of pre-Christian antiquity was Hipparchus, born in Nicaea in what is present-day Turkey. He erected an observatory on the island of Rhodes in the period around 150 B.C., when the Roman Republic was increasing its influence throughout the Mediterranean region. There he measured as accurately as possible the directions of objects in the sky, compiling a pioneering star catalog with about 850 entries. He designated celestial coordinates for each star, specifying its position in the sky, just as we can specify the position of a point on the Earth by giving its latitude and longitude. He also divided the stars into **magnitudes,** according to their apparent brightness. He called the brightest ones "stars of the first magnitude," the next brightest group "stars of the second magnitude," and so forth. This system, in modified form, still remains in use today.

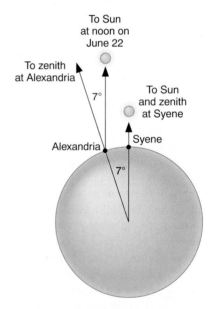

FIGURE 1.10

How Eratosthenes Measured the Size of the Earth   The Sun's rays come in parallel, but because the Earth's surface curves, a ray at Syene comes straight down, while a ray at Alexandria makes an angle of 7° with the vertical. That means, in effect, that at Alexandria the Earth's surface has curved away from Syene by 7° out of 360°, or 1/50 of a full circle. Thus the distance between the two cities must be 1/50 the circumference of the Earth.

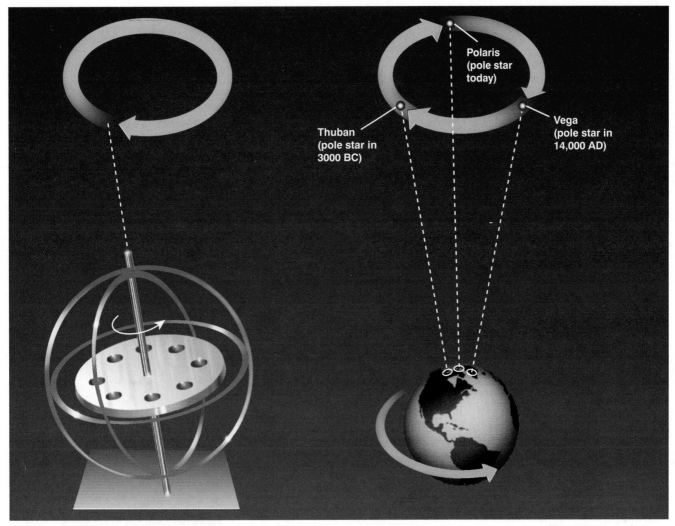

FIGURE 1.11

Precession Just as the axis of a rapidly spinning top slowly wobbles in a circle, so the axis of the Earth wobbles in a 26,000-year cycle. Today the north celestial pole is near the star Polaris, but about 5000 years ago it was close to a star called Thuban, and in 14,000 A.D. it will be closest to the star Vega.

By observing the stars and comparing his data with older observations, Hipparchus made one of his most remarkable discoveries: the position in the sky of the north celestial pole had altered over the previous century and a half. Hipparchus correctly deduced that this had happened not only during the period covered by his observations, but was in fact happening all the time: the direction around which the sky appears to rotate changes continuously.

You will recall from Section 1.1 that the north celestial pole is just the projection of the Earth's North Pole into the sky. If the north celestial pole is wobbling around, then the Earth itself must be doing the wobbling. Today we understand that the direction in which the Earth's axis points does indeed change slowly but regularly, a motion we call **precession.** If you've ever watched a child's spinning top wobble, you have observed a similar kind of motion. The top's axis describes a path in the shape of a cone, as the Earth's gravity tries to topple it over (Figure 1.11).

Because our planet is not an exact sphere, but bulges a bit at the equator, the pull of the Sun and Moon cause it to wobble like a top. It takes about 26,000 years for the Earth's axis to complete one circle of precession. As a result of this motion, the point where our axis points in the sky changes as time goes on. While Polaris is the star closest to the north celestial pole today (it will reach its very closest point around 2100 A.D.), the star Vega in the constellation of Lyra will be the north star in the year 14,000 A.D.

Ptolemy

The last great astronomer of antiquity was Claudius Ptolemy (or Ptolemaeus), who flourished in Roman

Alexandria about 140 A.D. He wrote a mammoth compilation of astronomical knowledge, which today is called by its Arabic name, the *Almagest* (meaning "The Greatest"). The *Almagest* does not deal exclusively with Ptolemy's own work, for it includes a discussion of the astronomical achievements of the past, principally those of Hipparchus. In fact, it is our main source of information about the work of Hipparchus and other Greek astronomers.

Ptolemy's most important contribution was a geometrical representation of the solar system that predicted the positions of the planets for any desired date and time. Hipparchus, not having enough data on hand to solve the problem himself, had instead amassed observational material for posterity to use. Ptolemy supplemented this material with new observations of his own and produced a cosmological model that endured more than a thousand years, until the time of Copernicus.

The complicating factor in explaining the motions of the planets is that their apparent wanderings in the sky result from the combination of their own motions with the Earth's orbital revolution. As we watch the planets move from our vantage point on the moving Earth, it's a little like watching a car race from inside one of the cars. Sometimes other cars pull ahead, while at other times we might pass opponents' cars, making them seem to move backwards for a while with respect to our car.

Figure 1.12a shows the motion of the Earth and a planet farther from the Sun, such as Mars or Jupiter. The Earth travels around the Sun in the same direction as the other planet and in nearly the same plane, but its orbital speed is faster. As a result, it periodically overtakes the planet, like a faster race car on the inside track. The figure shows where we see the planet in the sky at different times. In Figure 1.12b, we see the resulting apparent path of the planet among the stars.

Normally, planets move eastward in the sky over the weeks and months as they orbit the Sun, but from positions *B* to *D* in our figure, as the Earth passes the planet, it appears to drift backward, moving west in the sky. Even though it is actually moving to the east, the faster Earth has overtaken it and seems, from our perspective, to be leaving it behind. As the Earth rounds its orbit toward position *E*, the planet again takes up its apparent eastward motion in the sky. The temporary apparent westward motion of a planet as the Earth swings between it and the Sun is called **retrograde motion.** Such backward motion is much easier for us to understand today, now that we know the Earth is one of the moving planets and not the unmoving center of all creation. But Ptolemy was faced with the far more complex problem of explaining such motion while assuming a stationary Earth.

Furthermore, because the Greeks believed that celestial motions had to be circles, Ptolemy had to construct a model using circles alone. To do it, he needed dozens of circles, some moving around other circles, in a complex edifice that makes a modern viewer dizzy. But we must not let our modern judgment cloud our admiration for Ptolemy's achievement. In his day, a complex universe centered on

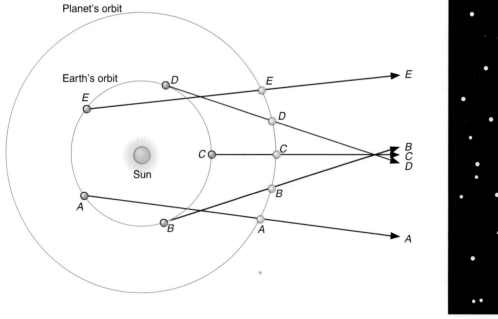

(a)   (b)

FIGURE 1.12

Retrograde Motion of a Planet Beyond the Earth's Orbit   (a) Actual positions of the planet and the Earth. (b) The apparent path of the planet as seen from the moving Earth, against the background of stars.

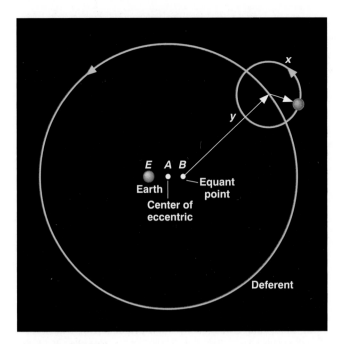

FIGURE 1.13
Ptolemy's Cosmological System Each planet orbits around a small circle called an epicycle. The system is not exactly centered on the Earth but on a point called the equant. The Greeks needed all this complexity to explain the motions in the sky, because they believed that the Earth was stationary.

the Earth was perfectly reasonable and, in its own way, quite beautiful.

Ptolemy solved the problem of explaining the observed motions of planets by having each planet revolve in a small orbit called an **epicycle.** The center of the epicycle then revolved about the Earth on a circle called a *deferent* (Figure 1.13). When the planet is at position *x* in Figure 1.13 on the epicycle orbit, it is moving in the same direction as the center of the epicycle; from Earth the planet appears to be moving eastward. When the planet is at *y*, however, its motion is in the direction opposite to the motion of the epicycle's center around the Earth. By choosing the right combination of speeds and distances, Ptolemy succeeded in having the planet moving westward at *y* at the correct speed and for the correct interval of time, thus replicating retrograde motion with his model.

However, we shall see in the next chapter that the planets, like the Earth, travel about the Sun in orbits that are ellipses, not circles. Their actual behavior cannot be represented accurately by a scheme of uniform circular motions. In order to match the observed motions of the planets, Ptolemy had to center the deferent circles not on the Earth, but at points some distance from the Earth. In addition, he introduced uniform circular motion around yet another axis, called the *equant point.* All of this considerably complicated his scheme.

It is a tribute to the genius of Ptolemy as a mathematician that he was able to develop such a complex system to account successfully for the observations of planets. It may

be that Ptolemy did not intend his cosmological model to describe reality, but merely to serve as a mathematical representation that allowed him to predict the positions of the planets at any time. Whatever his thinking, his model, with some modifications, was accepted as authoritative in the Muslim world and (later) in Christian Europe.

1.3 ASTROLOGY AND ASTRONOMY

Many ancient cultures regarded the planets and stars as representatives or symbols of the gods or other supernatural forces that controlled their lives. For them, the study of the heavens was not an abstract subject, it was directly connected to the life-and-death necessity of understanding the actions of the gods and currying favor with them. Before the time of our scientific perspective, everything that happened in nature, from the weather, to diseases and accidents, to celestial surprises like eclipses or new comets, was thought to be an expression of the whims or displeasures of the gods. Any signs that helped people understand what these gods had in mind were considered extremely important.

The movements of the seven objects that had the power to "wander" through the realm of the sky—the Sun, Moon, and five planets visible to the unaided eye—clearly must have special significance in such a system of thinking. Most ancient cultures associated these seven objects with various supernatural rulers in their pantheon and kept track of them for religious reasons. Even in the comparatively sophisticated Greece of antiquity, the planets had the names of gods and were credited with having the same powers and influences as the gods whose names they bore. From such ideas was born the ancient system called **astrology,** still practiced by some people today, in which the positions of these bodies among the stars of the zodiac are thought to hold the key to understanding what we can expect from life.

The Beginnings of Astrology

Astrology began in Babylonia about two and a half millennia ago. The Babylonians, believing that the planets and their motions influenced the fortunes of kings and nations, used their knowledge of astronomy to guide their rulers. When the Babylonian culture was absorbed by the Greeks, astrology gradually came to influence the entire Western world and eventually spread to the Orient as well.

By the 2nd century B.C., the Greeks democratized astrology by developing the idea that the planets influence every individual. In particular, they believed that the configuration of the Sun, Moon, and planets at the moment of birth affected a person's personality and fortune, the doctrine called *natal astrology.* Natal astrology reached its zenith with Ptolemy 400 years later. As famous for his astrology as for his astronomy, Ptolemy compiled the

Tetrabiblos, a treatise on astrology that remains the "bible" of the subject. It is essentially this ancient religion, older than Christianity or Islam, that is still practiced by today's astrologers.

The Horoscope

The key to natal astrology is the **horoscope,** a chart showing the positions of the planets in the sky at the moment of an individual's birth. In charting a horoscope, the planets (including the Sun and Moon, classed as wanderers or planets by the ancients) must first be located in the zodiac. When the system of astrology was set up, the zodiac was divided into 12 sectors called signs, each 30° long. Each sign was named after a constellation in the sky through which the Sun, Moon, and planets were seen to pass—the sign of Virgo after the constellation of Virgo, for example.

When someone today casually asks you your "sign," what they are asking for is your "sun-sign"—which zodiac sign the Sun was in at the moment you were born. But more than 2000 years have passed since the signs received their names from the constellations. Because of precession, the constellations of the zodiac slide westward along the ecliptic, going once around the sky in about 26,000 years. Thus today the real stars have slipped around by about 1/12 of the zodiac—about the width of one sign.

In most forms of astrology, however, the signs have remained assigned to the dates of the year they had when astrology was first set up. This means that the astrological signs and the real constellations are out of step; the sign of Aries, for example, now occupies the constellation of Pisces. When you look up your sun-sign in a newspaper astrology column, because of precession, the name of the sign associated with your birthday is no longer the name of the constellation in which the Sun was actually located when you were born. To know that constellation, you must look for the sign *before* the one that includes your birthday.

A complete horoscope shows the location of not only the sun but each planet in the sky by indicating its position in the appropriate sign of the zodiac. However, as the celestial sphere turns (owing to the rotation of the Earth), the entire zodiac moves across the sky to the west, completing a circuit of the heavens each day. Thus the position in the sky (or "house" in astrology) must also be calculated. There are more-or-less standardized rules for the interpretation of the horoscope, most of which (at least in Western schools of astrology) are derived from the *Tetrabiblos* of Ptolemy. Each sign, each house, and each planet, the last supposedly acting as a center of force, is associated with particular matters.

The detailed interpretation of a horoscope is therefore a very complicated business, and there are many schools of astrological thought on how it should be done. Although some of the rules may be standardized, how each rule is to be weighed and applied is a matter of judgment—and "art." It also means that it is very difficult to tie astrology down to specific predictions or to get the same predictions from different astrologers.

Astrology Today

Astrologers today use the same basic principles laid down by Ptolemy nearly 2000 years ago. They cast horoscopes (a process much simplified by the development of appropriate computer programs) and suggest interpretations. Sun-sign astrology (which you read in the newspapers and many magazines) is a recent simplified variant of natal astrology. Although even professional astrologers do not place much trust in such a limited scheme, which tries to fit everyone into just 12 groups, sun-sign astrology is taken seriously by many people (perhaps because it is so commonly discussed in the media). In a recent poll of teenagers in the United States, more than half said they "believed in astrology."

Today, we know much more about the nature of the planets as physical bodies, as well as about human genetics, than the ancients could. It is hard to imagine that the positions of the Sun, Moon, or planets in the sky at the moment of our birth could have anything to do with our personality or future. There are no known forces, not gravity or anything else, that could cause such effects. (For example, a simple calculation shows that the gravitational pull of the obstetrician delivering a newborn baby is greater than that of Mars.) Astrologers thus have to argue that there are unknown forces exerted by the planets that depend on their configurations with respect to one another and that do not vary according to the distance of the planet—forces for which there is no evidence.

Another curious aspect of astrology is its emphasis on planet configurations at *birth.* What about the forces that might influence us at conception? Isn't our genetic makeup more important for determining our personality than the circumstances of our birth? Would we really be a different person if we had been born a few hours earlier or later, as astrology claims? (Back when astrology was first conceived, birth was thought of as a moment of magic significance, but today we understand a lot more about the long process that precedes it.)

Actually, very few thinking people today buy the claim that our entire lives are predetermined by astrological influences at birth, but many people apparently believe that astrology has validity as an indicator of affinities and personality. A surprising number of Americans make judgments about people—whom they will hire, associate with, and even marry—on the basis of astrological information. To be sure, these are difficult decisions, and you might argue that we should use any relevant information that might help us to make the right choices. But does astrology actually provide any useful information on human personality? This is the kind of question that can be tested using the scientific method (see the box, "Testing Astrology").

The results of hundreds of tests are all the same: There is no evidence that natal astrology has any predictive power, even in a statistical sense. Why then do people often seem to have anecdotes about how well their own astrologer advised them? Effective astrologers today use the language of the zodiac and the horoscope only as the outward trappings of their craft. Mostly they work as amateur therapists, offering simple truths that clients like or need to

MAKING CONNECTIONS

Testing Astrology

In response to modern public interest in astrology, scientists have carried out a wide range of statistical tests to assess its predictive power. The simplest of these examine sun-sign astrology to determine whether—as astrologers assert—some signs are more likely than others to be associated with such objective measures of success as winning Olympic medals, earning high corporate salaries, or achieving elective office or high military rank. (You can make such a test yourself by looking up the birthdates of all members of Congress, for example, or all members of the U.S. Olympic Team.) Are our political leaders somehow selected at birth by their horoscopes and thus more likely to be Leos, say, than Scorpios?

You don't even need to be specific about your prediction in such tests. After all, many schools of astrology disagree about which signs go with which personality characteristics. To demonstrate the validity of the astrological hypothesis, it would be sufficient if the birthdays of all our leaders clustered in one or two signs in some statistically significant way. Dozens of such tests have been performed, and all have come up completely negative: the birth dates of leaders in all fields tested have been found to be randomly distributed among the signs. Sun-sign astrology does not predict anything about a person's future occupation or strong personality traits.

In a fine example of such a test, two statisticians examined the reenlistment records of the Marine Corps. We suspect our readers will agree that it takes a certain kind of personality to not only enlist but reenlist in the Marines. If sun-signs can predict strong personality traits, as astrologers claim, then the reenlistees (with similar personalities) should have been distributed preferentially in those one or few signs that matched the personality of someone who loves being a Marine. In fact, the reenlistees were distributed randomly among all the signs.

More sophisticated studies have also been done, involving full horoscopes calculated for thousands of individuals. The results of all of these studies are also negative: none of the systems of astrology has been shown to be at all effective in connecting astrological aspects to personality, success, or finding the right person to love.

Other tests show that it hardly seems to matter what a horoscope interpretation says, as long as it is vague enough, and as long as each subject feels it was personally prepared just for him or her. The French statistician Michel Gaugelin, for example, sent the horoscope interpretation for one of the worst mass-murderers in history to 150 people, but told each recipient that it was a "reading" prepared exclusively for him or her. Ninety-four percent of the readers said they recognized themselves in the interpretation of the mass-murderer's horoscope.

Geoffrey Dean, an Australian researcher, reversed the astrological readings of 22 subjects, substituting phrases that were the opposite of what the horoscope actually said. Yet his subjects said that the resulting readings applied to them just as often (95%) as the people to whom the original phrases were given. Apparently, those who seek out astrologers just want guidance, and almost any guidance will do.

---

hear. (Recent studies have shown that just about any sort of short-term therapy makes people feel a little better. This is because the very act of talking about our problems is in itself beneficial.)

The scheme of astrology has no basis in scientific fact, however. It is an interesting historical system, left over from prescientific days and best remembered for the impetus it gave people to learn the cycles and patterns of the sky. From it grew the science of astronomy, which is our main subject for discussion.

1.4 THE BIRTH OF MODERN ASTRONOMY

Astronomy made no major advances in strife-torn medieval Europe. The birth and expansion of Islam after the 7th century A.D. led to a flowering of Arabic and Jewish cultures that preserved, translated, and added to many of the astronomical ideas of the Greeks. (Many of the names of the brightest stars, for example, are today taken from the Arabic, as are such astronomical terms as zenith.)

As European culture began to emerge from its long Dark Age, trading with Arab countries led to a rediscovery of ancient texts such as the *Almagest*, and to a reawakening of interest in astronomical questions. This time of rebirth (in French, *Renaissance*) in astronomy was clearly embodied in the work of Copernicus (Figure 1.14).

Copernicus

One of the most important events of the Renaissance was the displacement of Earth from the center of the universe, an intellectual revolution initiated by a Polish cleric in the 16th century. Nicolaus Copernicus (1473–1543) was born in Torun, a mercantile town along the Vistula river. His training was in law and medicine, but his main interests were astronomy and mathematics. His great contribution to science was a critical reappraisal of the existing theories

FIGURE 1.14
Copernicus Nicolaus Copernicus (1473–1543), cleric and scientist, played a leading role in the emergence of modern science. While he could not prove that the Earth revolves about the Sun, he presented such compelling arguments for this idea that he turned the tide of cosmological thought and laid the foundations upon which Galileo and Kepler so effectively built in the following century.

(Yerkes Observatory)

One of the objections raised to the heliocentric theory was that if the Earth were moving, we would all sense or feel this motion. Solid objects would be ripped from the surface, a ball dropped from a great height would not strike the ground directly below, and so forth. But a moving person is not necessarily aware of that motion. We have all experienced seeing an adjacent train, car, or ship appear to move, only to discover that it is we who are moving.

Copernicus argued that the apparent motion of the Sun about the Earth during the course of a year could be represented equally well by a motion of the Earth about the Sun. He also reasoned that the apparent rotation of the celestial sphere could be explained by assuming that the Earth rotates while the celestial sphere is stationary. To the objection that if the Earth rotated about an axis it would fly into pieces, Copernicus answered that if such motion would tear the Earth apart, the still faster motion of the much larger celestial sphere required by the geocentric hypothesis would be even more devastating.

The Heliocentric Model

The most important idea in Copernicus' *De Revolutionibus* is that the Earth is one of six (then known) planets that revolve about the Sun. Using this concept, he was able to work out the correct general picture of the solar system. He placed the planets, starting nearest the Sun, in the correct order: Mercury, Venus, Earth, Mars, Jupiter,

of planetary motion and the development of a new Sun-centered, or **heliocentric,** model of the solar system. Copernicus concluded that the Earth is a planet and that all the planets circle the Sun. Only the Moon orbits about the Earth (Figure 1.15).

He described his ideas in detail in his book *De Revolutionibus Orbium Coelestium* (On the Revolution of Celestial Orbs), published in 1543, the year of his death. By this time, the old Ptolemaic system (like a cranky old machine) needed significant adjustments to predict the positions of the planets correctly. Copernicus wanted to develop an improved theory from which to calculate planetary positions, but in doing so, he was himself not free of all traditional prejudices.

He began with several assumptions that were common in his time, such as the idea that the motions of the heavenly bodies must be made up of combinations of uniform circular motions. But he did not assume (as most people did) that the Earth had to be in the center of the universe, and he presented a defense of the heliocentric system that was elegant and persuasive. His ideas, although not widely accepted until more than a century after his death, were much discussed among scholars and ultimately had a profound influence on the course of world history.

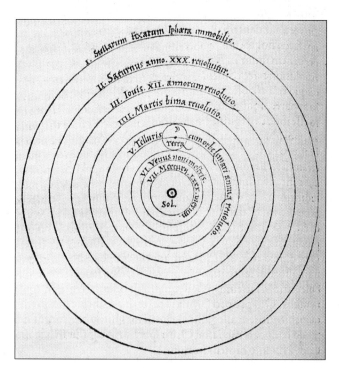

FIGURE 1.15
Copernicus' System Heliocentric plan of the solar system in the first edition of Copernicus' *De Revolutionibus.* Notice the word "Sol" for Sun in the middle. *(Crawford Collection, Royal Observatory, Edinburgh)*

and Saturn. Further, he deduced that the nearer a planet is to the Sun, the greater its orbital speed. With his theory, he was able to explain the complex retrograde motions of the planets without epicycles and to work out a roughly correct scale for the solar system.

Copernicus could not *prove* that the Earth revolves about the Sun. In fact, with some adjustments the old Ptolemaic system could have accounted as well for the motions of the planets in the sky. But Copernicus pointed out that the Ptolemaic cosmology was clumsy and lacking in the beauty and symmetry of its successor.

In Copernicus' time, in fact, few people thought there were ways to prove whether the heliocentric or the older geocentric system was correct. A long philosophical tradition, going back to the Greeks and defended by the Catholic Church, held that pure human thought combined with divine revelation represented the path to truth. Nature, as revealed by our senses, was suspect. For example, Aristotle had reasoned that heavier objects (having more of the quality that made them heavy) must fall to Earth faster than lighter ones. This is absolutely incorrect, as any simple experiment dropping two balls of different weights will show. But in Copernicus' day, experiments did not carry much weight (if you will pardon the expression); Aristotle's brilliant reasoning was more convincing.

In this environment, there was little motivation to carry out observations or experiments to distinguish between competing cosmological theories (or anything else).

It should not surprise us, therefore, that the heliocentric idea was debated for more than half a century without any tests being applied to determine its validity. (In fact, in the American colonies, the older geocentric system was still taught at Harvard University in the first years after it was founded in 1636.)

Contrast this with the situation today, when scientists rush to test each new theory and do not accept any ideas until the results are in. For example, when two researchers at the University of Utah announced in 1989 that they had discovered a way to achieve nuclear fusion (the process that powers the stars) at room temperature, other scientists at more than 25 laboratories around the United States attempted to duplicate *cold fusion* within a few weeks—without success, as it turned out.

How would we look at Copernicus' model today? When a new hypothesis or theory is proposed in science, it must first be checked for consistency with what is already known. Copernicus' heliocentric idea passes this test, for it allows planetary positions to be calculated at least as well as does the geocentric theory. The next step is to see what predictions the new theory makes that differ from those of competing ideas. In the case of Copernicus, one example is the prediction that, if Venus circles the Sun, the planet should go through the full range of phases just as the Moon does, whereas if it circles the Earth, it should not (Figure 1.16). But in those days, before the telescope, no one imagined testing this prediction.

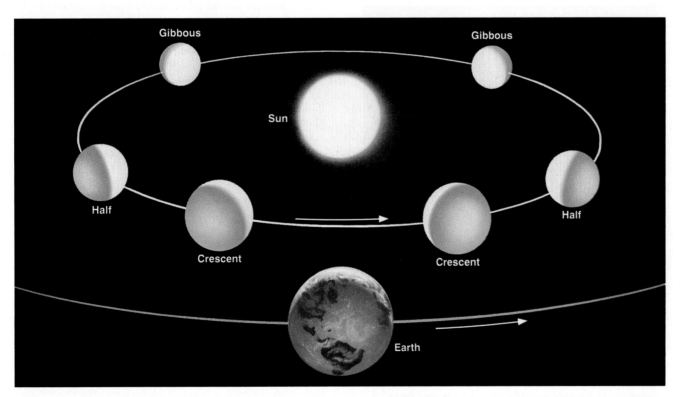

FIGURE 1.16
Venus Shows Phases   The phases of Venus according to the heliocentric theory. As Venus moves around the Sun we see changing illumination of its surface, just as we see the face of the Moon illuminated differently in the course of a month.

Galileo and the Beginning of Modern Science

Many of the modern scientific concepts of observation, experimentation, and the testing of hypotheses through careful quantitative measurements were pioneered by a man who lived nearly a century after Copernicus. Galileo Galilei (1564–1642, Figure 1.17), a contemporary of Shakespeare, was born in Pisa (Figure 1.18). Like Copernicus, he began training for a medical career, but he had little interest in the subject and later switched to mathematics. He held faculty positions at the Universities of Pisa and Padua, and eventually he became mathematician to the Grand Duke of Tuscany in Florence.

Galileo's greatest contributions were in the field of mechanics, the study of motion and the actions of forces on bodies. It was familiar to all persons then, as it is to us now, that if a body is at rest, it tends to remain at rest and requires some outside influence to start it in motion. Rest was thus generally regarded as the natural state of matter. Galileo showed, however, that rest is no more natural than motion.

If an object is slid along a rough horizontal floor, it soon comes to rest because friction between it and the floor acts as a retarding force. However, if the floor and object are both highly polished, the body, given the same initial speed, will slide farther before coming to rest. On a smooth layer of ice, it will slide farther still. Galileo reasoned that if

FIGURE 1.18
The Leaning Bell Tower of the Cathedral at Pisa   While living in Pisa, Galileo carried out experiments to show that objects of different mass fall at the same rate. According to legend, one of his experiments consisted of dropping cannonballs of different weights from this tower.   *(David Morrison)*

all resisting effects could be removed, the object would continue in a steady state of motion indefinitely. In fact, he argued, a force is required not only to start an object moving from rest but also to slow down, stop, speed up, or change the direction of a moving object. You will appreciate this if you have ever tried to stop a rolling car by leaning against it, or a moving boat by tugging on a line.

Galileo also studied the way bodies **accelerate,** or change their speed, as they fall freely or roll down a ramp. He found that such bodies accelerate uniformly; that is, in equal intervals of time they gain equal increments in speed. Galileo formulated these newly found laws in precise mathematical terms that enabled future experimenters to predict how far and how fast bodies would move in various lengths of time.

Sometime in the 1590s Galileo adopted the Copernican hypothesis of the solar system. In Roman Catholic Italy, this was not a popular philosophy, for the Church authorities still upheld the ideas of Aristotle and Ptolemy, and they had powerful political and economic reasons for

FIGURE 1.17
Galileo   Galileo Galilei (1564–1642) advocated that we perform experiments or make observations to ask nature its ways. When Galileo turned the telescope to the sky, he found that things were not the way philosophers had supposed.   *(Yerkes Observatory)*

insisting that the Earth was the center of creation. It was primarily because of Galileo and his writings that in 1616 the Church issued a prohibition decree stating that the Copernican doctrine was "false and absurd" and not to be held or defended.

Galileo's Astronomical Observations

It is not certain who first conceived of the idea of combining two or more pieces of glass to produce an instrument that enlarged images of distant objects, making them appear nearer. The first such telescopes that attracted much notice were made by the Dutch spectacle maker Hans Lippershey in 1608. Galileo heard of the discovery and, without ever having seen an assembled telescope, constructed one of his own with a three-power magnification, which made distant objects appear three times nearer and larger (Figure 1.19).

On August 25, 1609, Galileo demonstrated a telescope with a magnification of 9× to government officials of the city-state of Venice. By a magnification of 9×, we mean that the linear dimensions of the object being viewed appeared nine times larger or, alternatively, that the objects appeared nine times closer than they really were. There were obvious military advantages associated with a device for seeing distant objects. For his invention, Galileo's salary was nearly doubled, and he was granted lifetime tenure as a professor. (His university colleagues were outraged, particularly because the invention was not even original.)

Others had used the telescope before Galileo to observe things on Earth. But, in a flash of insight that changed the history of astronomy, Galileo realized that he could turn the power of the telescope toward the heavens. Before using his telescope for astronomical observations, Galileo had to devise a stable mount, and he improved the optics to provide a magnification of 30. Galileo also needed to acquire confidence in the telescope.

FIGURE 1.19
Some Telescopes Used by Galileo   The longer telescope has a wooden tube covered with paper and a lens 26 mm across.   *(Instituto e Museo di Storia della Scienza di Florenza)*

At that time, the eyes were believed to be the final arbiter of truth about sizes, shapes, and colors. Lenses, mirrors, and prisms were known to distort distant images by enlarging, reducing, or even inverting them. Galileo undertook repeated experiments to convince himself that what he saw through the telescope was identical to what he saw up close. Only then could he begin to believe that the miraculous phenomena the telescope revealed in the heavens were real.

Beginning his astronomical work late in 1609, Galileo found that many stars too faint to be seen with the naked eye became visible with his telescope. In particular, he found that some nebulous blurs resolved into many stars, and that the Milky Way—the strip of whiteness across the night sky—was also made up of a multitude of individual stars.

Examining the planets, he found four satellites, or moons, revolving about Jupiter, with periods ranging from just under 2 days to about 17 days. This discovery was particularly important because it showed that not everything has to revolve around the Earth and that there could be centers of motion that are themselves in motion. Defenders of the geocentric view had argued that if the Earth were in motion the Moon would be left behind, because it could hardly keep up with a rapidly moving planet. Yet here were Jupiter's satellites doing exactly that! (To recognize this discovery and honor his work, NASA named a spacecraft that explored the Jupiter system *Galileo*.)

With his telescope, Galileo was able to carry out the test of the Copernican theory mentioned earlier, based on the phases of Venus. Within a few months he had found that Venus goes through phases like the Moon, showing that it must revolve about the Sun, so that we see different parts of its daylight side at different times (see Figure 1.16). These observations could not be reconciled with any model in which Venus circled about the Earth. Galileo also observed the Moon and saw craters, mountain ranges, valleys, and flat dark areas that he thought might be water. These discoveries showed that the Moon might be not so dissimilar to the Earth—suggesting that the Earth, too, could belong to the realm of celestial bodies.

After Galileo's work, it became increasingly difficult to deny the Copernican view, and slowly the Earth was dethroned from its central position in the universe and given its rightful place as one of the planets attending the Sun. Galileo himself had to appear before the Inquisition to answer charges that his work was heretical, and he was condemned to house arrest. His books were on the Church's forbidden list until 1836, although in countries where the Roman Catholic Church held less sway, they were widely read and discussed. Not until 1992 did the Catholic Church publicly admit that it had erred in the matter of censoring Galileo's ideas.

The new ideas of Copernicus and Galileo began a revolution in our conception of the cosmos. It eventually became evident that the universe is a vast place, and that the

Earth's role in it is relatively unimportant. The idea that the Earth moves around the Sun like the other planets raised the possibility that they might be worlds themselves, perhaps even supporting life. And as the Earth was demoted from its position at the center of the universe, so too was humanity. The universe, despite what we may wish, does not revolve around us, and we must find a more modest place for ourselves in the scheme of the cosmos.

Most of us take these things for granted today, but four centuries ago such concepts were frightening and heretical for some, immensely stimulating for others. The pioneers of the Renaissance started the European world along the path toward science and technology that we still tread today. For them, nature was rational and ultimately knowable, and experiments and observations provided the means to reveal its secrets.

SEEING FOR YOURSELF

Observing the Planets

At almost any time of the night, and at any season, you can spot one or more bright planets in the sky. All five of the planets known to the ancients—Mercury, Venus, Mars, Jupiter, and Saturn—are more prominent than any but the brightest stars, and they can be seen even from urban locations if you know where and when to look. One way to tell planets from bright stars is that planets twinkle less.

> One way to tell planets from bright stars is that planets twinkle less.

Venus, which stays close to the Sun from our perspective, appears either as an "evening star" in the west after sunset, or as a "morning star" in the east before sunrise. It is the brightest object in the sky after the Sun and Moon. It far outshines any real star, and under the most favorable circumstances it can even cast a visible shadow. Some young military recruits have tried to shoot Venus down as an approaching enemy craft or UFO!

Mars, with its distinctive red color, can be nearly as bright as Venus is when close to the Earth, but normally it remains much less conspicuous. Jupiter is most often the second-brightest planet, approximately equaling in brilliance the brightest stars. Saturn is dimmer, and it varies considerably in brightness, depending on whether its rings are seen nearly edge-on (faint) or more widely opened (bright).

Mercury is quite bright, but few people ever notice it because it never moves very far from the Sun (it's never more than 28° away in the sky) and is always seen against bright twilight skies. There is a story (probably apocryphal) that even the great Copernicus never saw the planet Mercury.

True to their name, the planets "wander" against the background of the "fixed" stars. Although their apparent motions are complex, they reflect an underlying order, upon which the heliocentric model of the solar system, as described in this chapter, was based. The positions of the planets are often listed in newspapers (sometimes on the weather page), and clear maps and guides to their locations can be found each month in such magazines as *Sky & Telescope* and *Astronomy* (available at most libraries). There are also a number of computer programs (including shareware and commercial programs) that allow you to calculate and display where the planets are on any night (including Red-Shift™, a package that can be purchased with our textbook).

Activities:

1. If either Venus or Mercury is visible, plot its position with respect to the background stars and note the date. Use a map of the relevant portion of the sky, or sketch the location of the brighter stars. Wait a few days and repeat the observation. Estimate how many degrees the planet moves per day. (The width of your thumb, held straight out at arm's length, covers about 1° of sky; your fist at arm's length covers about 10°.) If you make enough observations, you can estimate each planet's maximum distance in degrees from the Sun.

2. Use binoculars to see if you can determine the shape of Venus. This experiment will be easier when Venus is on the same side of the Sun as the Earth.

3. Determine which of the outer planets—Mars, Jupiter, and Saturn—are visible. For each visible planet, draw its position with respect to the background stars and note the date. Repeat this experiment every few days for two months or as long as each planet is visible. Is the planet moving east or west? Did it appear to change direction? Did you observe retrograde motion for any of the planets?

The Constellations and Their Stars [www.astro.wisc.edu/~dolan/constellations/]

Graduate student Chris Dolan has assembled handy information about all the constellations and the most important stars, nebulae, clusters, and galaxies that can be found in them. You can access things by name or by month of visibility, and there are good links to related sites for constellation fans.

The Galileo Project [es.rice.edu:80/ES/humsoc/Galileo/]

Noted historian of astronomy Albert VanHelden of Rice University maintains this rich lode of information about Galileo's life, work, and times, with text, images, maps, timelines, and a very creative organizational scheme.

Ancient Astronomy Site [arcturus.pomona.edu]

Brian Penprase of Pomona College and his students have assembled this introduction to the astronomical lore and knowledge of ancient civilizations. It features an interactive atlas of world astronomy, timelines for each continent, tours and images, course materials, and links to other sites the world over.

The History of Astronomy Site [www.astro.uni-bonn.de/~pbrosche/astoria.html]

Professor Wolfgang Dick maintains this comprehensive Web site on behalf of the German Working Group for the History of Astronomy; it contains a huge number of links to sites about all aspects of astronomical history. Among its best features is the ability to select the name of a historical astronomer and find links that provide instant biographical information in most cases.

Activities about Astrology [www.aspsky.org/html/astro/act3/astrology.html]

This site, designed for school teachers and astronomers who volunteer to work with them, nevertheless has a number of items of interest to college students and instructors. There are suggested activities for "testing" the hypothesis of astrology, an article by *Voyages* co-author Andrew Fraknoi suggesting some embarrassing questions to ask astrology believers, and a bibliography of astrology debunking resources.

SUMMARY

1.1 The direct evidence of our senses supports a **geocentric** perspective, with the **celestial sphere** pivoting on the **celestial poles** and rotating about a stationary Earth. We see only half of this sphere at one time, limited by the **horizon;** the point directly overhead is our **zenith.** The Sun's annual path on the celestial sphere is the **ecliptic,** a line that runs through the center of the **zodiac,** the 18°-wide strip of sky within which we always find the Moon and planets. The celestial sphere is organized into 88 **constellations,** or sectors.

1.2 Ancient Greeks such as Aristotle recognized that the Earth and Moon are spheres, and understood the phases of the Moon, but because of their inability to detect stellar **parallax,** they rejected the idea that the Earth moves. Eratosthenes measured the size of the Earth with surprising precision. Hipparchus carried out many astronomical observations, making a star catalog, defining the system of stellar magnitudes, and discovering **precession** from the apparent shift in the position of the north celestial pole. Ptolemy of Alexandria summarized classical astronomy in his *Almagest;* he explained planetary motions, including **retrograde motion,** with remarkably good accuracy using a model centered on the Earth. This geocentric model, based on combinations of uniform circular motion using the **epicycles,** was accepted as authority for more than a thousand years.

1.3 The ancient religion of **astrology,** whose main contribution to civilization was a heightened interest in the heavens, began in Mesopotamia. It reached its peak in the Greco-Roman world, especially as recorded in the *Tetrabiblos* of Ptolemy. Natal astrology is based on the assumption that the positions of the planets at the time of our birth, as described by a **horoscope,** determines our future. However, modern tests clearly show that there is no evidence for this, even in a broad statistical sense.

1.4 Nicolaus Copernicus introduced the **heliocentric cosmology** to Renaissance Europe in his book *De Revolutionibus.* Although he retained the Aristotelian idea of uniform circular motion, Copernicus suggested that the Earth is a planet and that the planets all circle about the Sun, dethroning the Earth from its position at the center of the universe. Galileo Galilei was the father of both modern experimental physics and telescopic astronomy. He studied the **acceleration** of moving objects, and in 1610 began telescopic observations, discovering the nature of the Milky Way, the large-scale features of the Moon, the phases of Venus, and four satellites of Jupiter. Although he was accused of heresy for his support of the heliocentric cosmology, Galileo's observations and brilliant writings convinced most of his scientific contemporaries of the reality of the Copernican theory.

(This section in each chapter will be devoted to activities that small groups of students can collaborate on, either during class, in a discussion section, or as independent projects. Your instructor may assign some of these or you can use them in a study group to extend your understanding of astronomy.)

A You can begin by considering the question with which we began this chapter. How many ways can you think of to prove to a rabid member of the "Flat Earth Society" that our planet is indeed round?

B Have your group make a list of ways in which a belief in astrology (the notion that your life path or personality is controlled by the position of the Sun, Moon, and planets at the time of your birth) might be harmful to an individual or to society at large.

C Members of the group should compare their experiences with the night sky. Have you seen the Milky Way? Can you identify any constellations? Why do you think so many fewer people know the night sky today than at the time of the ancient Greeks? What reasons can you think of why a person may want to be acquainted with the night sky?

D Constellations commemorate great heroes, dangers, or events in the legends of the people who name them. Suppose we had to start from scratch today, naming the patterns of stars in the sky. What would you choose to commemorate by naming a constellation after it/him/her and *why*? Can the members of your group agree on any choices?

E Although astronomical mythology no longer holds a powerful sway over the modern imagination, we still find proof of the power of astronomical images in the number of products in the marketplace that have astronomical names. How many can your group come up with? (Think of things like Milky Way™ candy bars, Saturn™ cars, or Comet™ cleanser.)

REVIEW QUESTIONS

1. From where on Earth could you observe all of the stars during the course of a year? What fraction of the sky can be seen from the North Pole?

2. Describe a practical way to determine in which constellation the Sun is found at any time of the year.

3. What is a constellation as astronomers define it today? What does it mean when an astronomer says, "I saw a comet in Orion last night"?

4. Give four ways to demonstrate that the Earth is round.

5. Explain why we see retrograde motion of the planets, according to both geocentric and heliocentric cosmologies.

6. Draw a picture that explains why Venus goes through phases the way the Moon does, according to the heliocentric cosmology. Does Jupiter also go through phases as seen from the Earth? Why?

7. In what ways did the work of Copernicus and Galileo differ from the traditional views of the ancient Greeks and of the Catholic Church?

THOUGHT QUESTIONS

8. Show with a simple diagram how the lower parts of a ship disappear first as it sails away from you on a spherical Earth. Use the same diagram to show why lookouts on old sailing ships could see farther from the masthead than from the deck. Would there be any advantage to posting lookouts on the mast if the Earth were flat? (Note that these nautical arguments for a spherical Earth were quite familiar to Columbus and other mariners of his time.)

9. Parallaxes of stars were not observed by ancient astronomers. How can this fact be reconciled with the heliocentric hypothesis?

10. Design an experiment to test whether or not the planets and their motions influence human behavior.

11. Why do you think so many people believe in astrology? What psychological needs does such a belief system satisfy?

12. Consider three cosmological perspectives: (1) the geocentric perspective, (2) the heliocentric perspective, and (3) the modern perspective in which the Sun is a minor star on the outskirts of one galaxy among billions. Discuss some of the cultural and philosophical implications of each point of view.

13. The Moon moves relative to the background stars. Go outside at night and note the position of the Moon relative to nearby stars. Repeat the observation a few hours later. How far has the Moon moved? (For reference, the diameter of the Moon is about ½°.) Based on your estimate of its motion, how long will it take for the Moon to return to the position relative to the stars in which you first observed it?

14. The north celestial pole appears at an altitude above the horizon that is equal to the observer's latitude. Identify Polaris, the North Star, which lies very close to the north celestial pole. Measure its altitude. (This can be done with a protractor. Alternatively, your fist, extended at arm's length, spans a distance approximately equal to 10°.) Compare this estimate with your latitude. (Note that this experiment cannot be easily performed in the Southern Hemisphere because Polaris itself is not visible in the south and no bright star is located near the south celestial pole.)

15. Suppose Eratosthenes had found that in Alexandria at noon on the first day of summer, the line to the Sun makes an angle of 30° with the vertical. What then would he have found for the Earth's circumference?

16. Suppose Eratosthenes' results for the Earth's circumference were quite accurate. If the diameter of the Earth is 12,740 km, evaluate the length of his stadium in kilometers.

17. Suppose you are on a strange planet and observe, at night, that the stars do not rise and set but circle parallel to the horizon. Now you walk in a constant direction for 8000 miles, and at your new location on the planet you find that all stars rise straight up in the east and set straight down in the west, perpendicular to the horizon.
   a. How could you determine the circumference of the planet without any further observations?
   b. What evidence is there that the Greeks could have done what you suggest?
   c. What is the circumference, in miles, of that planet?

SUGGESTIONS FOR FURTHER READING

Culver, B. and Ianna, P. *Astrology: True or False.* 1988, Prometheus Books. The best skeptical book about astrology.

Ferris, T. *Coming of Age in the Milky Way.* 1988, Morrow. A general history of our understanding of the organization of the universe.

Fraknoi, A. "Your Astrology Defense Kit" in *Sky & Telescope,* Aug. 1989, p. 146. A review of the tenets and tests of astrology, along with some skeptical questions.

Gingerich, O. "From Aristarchus to Copernicus" in *Sky & Telescope,* Nov. 1983, p. 410.

Gingerich, O. "How Galileo Changed the Rules of Science" in *Sky & Telescope,* Mar. 1993, p. 32.

Gurshtein, A. "In Search of the First Constellations" in *Sky & Telescope,* June 1997, p. 47. This and an earlier article in the Oct. 1995 issue of *Sky & Telescope* present intriguing ideas about the origins of the ancient star groupings.

Krupp, E. *Beyond the Blue Horizon: Myths and Legends of the Sun, Moon, Stars, and Planets.* 1991, HarperCollins. Superb introductory book on the sky stories of ancient cultures.

Krupp, E. *Skywatchers, Shamans, and Kings.* 1997, Wiley. An excellent primer on ancient monuments and astronomical systems, and the terrestrial purposes they served.

Kuhn T. *The Copernican Revolution.* 1957, Harvard U. Press. Classic study of the changes wrought by Copernicus' work.

Reston, J. *Galileo: A Life.* 1994, HarperCollins. A well-written, popular-level biography by a journalist.

Sagan, C. *Cosmos.* 1980, Random House. The chapter titled "Backbone of Night" focuses on the ancient Greek astronomers.

In February 1984, astronaut Bruce McCandless, having let go of the tether to the Space Shuttle, became the first human satellite of planet Earth. Here he is shown riding the Manned Maneuvering Unit, whose nitrogen gas jets enabled him to move around in orbit and ultimately return to the Shuttle. As he circled our planet, he was subject to the same basic laws of motion that govern all other satellites. *(NASA)*

2 Orbits and Gravity

THINKING AHEAD

To reach a scientific understanding of the natural world, scientists had to figure out the basic rules or *laws* of nature. Thinking back over the different scientific laws you may have studied in your earlier classes, which do you think are the most fundamental for describing how the physical world works?

If we could look down on the solar system from somewhere out in space, far above the plane of the planets' orbits, interpreting planetary motions would be much simpler. But the fact is, we must observe the positions of all the other planets from our own moving planet. Scientists of the Renaissance did not know the details of the Earth's orbit any better than the orbits of the other planets. Their problem, as we saw in Chapter 1, was that they had to deduce the nature of all planetary motion using only their observations of the other planets' positions in the sky. To solve this complex problem more fully, better observations and better conceptual models were needed.

FIGURE 2.1

Tycho at Hven   A stylized engraving showing Tycho Brahe using his instruments to measure the altitude of a celestial object above the horizon. Note that the scene includes hints of the grandeur of Tycho's observatory.   *(Granger Collection, New York)*

2.1   THE LAWS OF PLANETARY MOTION

At about the time that Galileo was beginning his experiments with falling bodies, the efforts of two other scientists dramatically advanced our understanding of the motions of the planets. These two astronomers were the observer Tycho Brahe and the mathematician Johannes Kepler. Together, they placed the speculations of Copernicus on a sound mathematical basis and paved the way for the work of Isaac Newton in the next century.

Tycho Brahe

Three years after the publication of Copernicus' *De Revolutionibus*, Tycho Brahe (1546–1601) was born to a family of Danish nobility. He developed an early interest in astronomy and as a young man made significant astronomical observations. Among these was a careful study of what we now know was an exploding star that flared up to great brilliance in the night sky. His growing reputation gained him the patronage of the Danish King Frederick II, and at the

age of 30 Tycho was able to establish a fine astronomical observatory on the North Sea island of Hven (Figure 2.1). Tycho was the last and greatest of the pretelescopic observers in Europe.

At Hven, Tycho made a continuous record of the positions of the Sun, Moon, and planets for almost 20 years. His extensive and precise observations enabled him to note that the positions of the planets varied from those given in published tables, which were based on the work of Ptolemy. But Tycho was an extravagant and cantankerous fellow, and he accumulated enemies among government officials. When his patron, Frederick II, died in 1597, Tycho lost his political base and decided to leave Denmark. He took up residence near Prague, where he became court astronomer to the Emperor Rudolf of Bohemia. There, in the year before his death, Tycho found a most able young mathematician, Johannes Kepler, to assist him in analyzing his extensive planetary data.

Kepler

Kepler (1571–1630) (Figure 2.2) was born into a poor family in the German province of Württemberg and lived much of his life amid the turmoil of the Thirty Years' War. He attended university at Tübingen and studied for a theological career. There he learned the principles of the Copernican system and became converted to the heliocentric hypothesis. Eventually, Kepler went to Prague to serve as an assistant to Tycho, who set him to work trying to find a satisfactory theory of planetary motion—one that

FIGURE 2.2

Kepler   Johannes Kepler (1571–1630) was a German mathematician and astronomer. His discovery of the basic quantitative laws that describe planetary motion placed the heliocentric cosmology of Copernicus on a firm mathematical basis.
(AIP/Niels Bohr Library)

was compatible with the long series of observations made at Hven.

Tycho was reluctant to provide Kepler with much material at any one time, for fear that Kepler would discover the secrets of the universal motions by himself, thereby robbing Tycho of some of the glory. Only after Tycho's death in 1601 did Kepler get possession of the priceless records. Their study occupied most of Kepler's time for more than 20 years.

Kepler's most detailed study was of Mars, for which the observational data were the most extensive. In 1609 he published the first results of his work in *The New Astronomy*, where we find his first two laws of planetary motion. Their discovery was a profound step in the development of modern science.

The Orbit of Mars

Kepler began his research under the assumption that the orbits of planets were circles, but the observations contradicted this idea. Working with data for Mars, he eventually discovered that the orbit of that planet had the shape of a somewhat flattened circle, or **ellipse.** Next to the circle, the ellipse is the simplest kind of closed curve, belonging to a family of curves known as *conic sections* (Figure 2.3).

You might recall from math classes that in a circle, the distance from a special point called the center to anywhere on the circle is exactly the same. In an ellipse, the *sum* of the distances from two special points inside the ellipse to any point on the ellipse is always the same. These two points inside the ellipse are called its **foci** (singular: **focus**), a word invented for this purpose by Kepler.

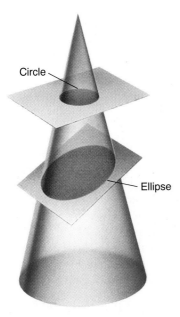

FIGURE 2.3
Conic Sections   The circle and the ellipse are both formed by the intersection of a plane with a cone. This is why both curves are called conic sections.

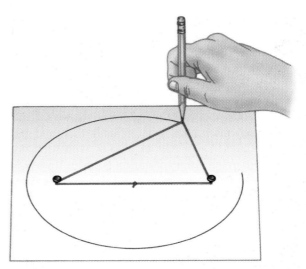

FIGURE 2.4
Drawing an Ellipse   We can construct an ellipse with two tacks and a string, where the tacks are the foci of the ellipse. Note that the length of the string remains the same, so that the sum of the distances from any point on the ellipse to the foci is always constant.

This property suggests a simple way to draw an ellipse (Figure 2.4). We tie the ends of a length of string to two tacks pushed through a sheet of paper into a drawing board, so that the string is slack. If we push a pencil against the string, making the string taut, and then slide the pencil against the string all around the tacks, the curve that results is an ellipse. At any point where the pencil may be, the sum of the distances from the pencil to the two tacks is a constant length—the length of the string. The tacks are at the two foci of the ellipse.

The widest diameter of the ellipse is called its **major axis.** Half this distance—that is, the distance from the center of the ellipse to one end—is the **semimajor axis,** which is usually used to specify the size of the ellipse. For example, the semimajor axis of the orbit of Mars, which also turns out to be the planet's average distance from the Sun, is 228 million km.

The shape (roundness) of an ellipse depends on how close together the two foci are, compared with the major axis. The ratio of the distance between the foci to the length of the major axis is called the **eccentricity** of the ellipse. If an ellipse is drawn using strings and tacks, the length of the major axis is the length of the string and the eccentricity is the distance between the tacks divided by the length of the string.

If the foci (or tacks) are in the same place, the eccentricity is zero and the ellipse is just a circle; thus, a circle is an ellipse of zero eccentricity. We can make ellipses of various shapes by varying the spacing of the tacks (as long as they are not farther apart than the length of the string). The greater the eccentricity, the more elongated is the ellipse, up to a maximum eccentricity of 1.0.

The size and shape of an ellipse are completely specified by its semimajor axis and its eccentricity. Kepler found that Mars has an elliptical orbit, with the Sun at one focus (the other focus is empty). The eccentricity of the orbit of Mars is only about 0.1; its orbit, drawn to scale, would be practically indistinguishable from a circle. Yet the difference is critical for understanding planetary motions.

Kepler generalized this result in his first law and said that the orbits of all the planets are ellipses. Here was a decisive moment in the history of human thought: It was not necessary to have only circles in order to have an acceptable cosmos. The universe could be a bit more complex than the Greek philosophers had wanted it to be.

Kepler's second law deals with the speed with which each planet moves along the ellipse. Working with Tycho's observations of Mars, Kepler discovered that the planet speeds up as it comes closer to the Sun and slows down as it pulls away from the Sun. He expressed the precise form of this relation by imagining that the Sun and Mars are connected by a straight, elastic line (Figure 2.5).

When Mars is closer to the Sun (positions 1 and 2 in Figure 2.5), the elastic line is not stretched as much, and the planet moves rapidly. Further from the Sun, as in positions 3 and 4, the line is stretched a lot, and the planet does not move as fast. As Mars travels in its elliptical orbit around the Sun, the elastic line sweeps out areas of the ellipse (the colored regions in our figure). Kepler found that in equal intervals of time, the areas swept out in space by this imaginary line are always equal; that is, the area of the region from 1 to 2 is the same as that from 3 to 4.

This is also a general property of the orbits of all planets. A planet in a circular orbit always moves at a constant speed, but in an elliptical orbit the planet's speed varies considerably, following the rule Kepler found.

Laws of Planetary Motion

Kepler's first two laws of planetary motion describe the shape of a planet's orbit and allow us to calculate the speed of its motion at any point in the orbit. Kepler was pleased to have discovered such fundamental rules, but they did not satisfy his quest to understand planetary motions. He wanted to know why the orbits of the planets were spaced as they are, and to find a mathematical pattern in their orbital periods—a "harmony of the spheres," as he called it. For many years he worked to discover mathematical relationships governing planetary spacings and the time each planet took to go around the Sun.

In 1619, Kepler succeeded in finding a simple algebraic relation that links the semimajor axes of the planets' orbits and their periods of revolution. The relation is now known as Kepler's third law. It applies to all of the planets, including the Earth, and it provides a means for calculating their relative distances from the Sun.

For the solar system, Kepler's third law takes its simplest form when the period is expressed in years (the

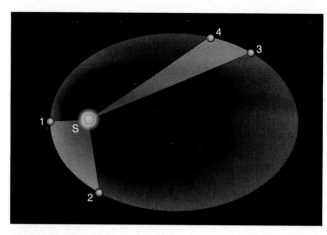

FIGURE 2.5

Kepler's Second Law: The Law of Equal Areas A planet moves most rapidly on its elliptical orbit when at position 1, nearest the Sun, which is at one focus of the ellipse. The orbital speed of the planet varies in such a way that in equal intervals of time a line between the Sun and the planet sweeps out equal areas. Thus, the area swept out from 1 to 2 is the same as that from 3 to 4. Note that the eccentricities of the planets' orbits are substantially less than shown here.

revolution period of the Earth) and the semimajor axis of the orbit is expressed in terms of the Earth's average distance from the Sun, called the **astronomical unit (AU)**. One astronomical unit is equal to 1.5×10^8 km. Kepler's third law is, then,

$$(\text{distance})^3 = (\text{period})^2$$

For example, this relationship tells us how to calculate Mars' average distance from the Sun (the semimajor axis of its orbit) from its period of 1.88 years. The period squared is $1.88 \times 1.88 = 3.53$, and this equals 1.52 cubed ($1.52 \times 1.52 \times 1.52$). Thus, the planet's semimajor axis in astronomical units must be 1.52. In other words, to go around the Sun in a little less than two years, Mars must be about 50% farther from the Sun than Earth is.

Kepler's three laws of planetary motion can be summarized as follows:

Kepler's First Law Each planet moves about the Sun in an orbit that is an ellipse, with the Sun at one focus of the ellipse.

Kepler's Second Law The straight line joining a planet and the Sun sweeps out equal areas in space in equal intervals of time.

Kepler's Third Law The squares of the planets' periods of revolution are in direct proportion to the cubes of the semimajor axes of their orbits.

These three laws provided a precise geometric description of planetary motion within the framework of the Copernican system. With these tools, it was possible to calculate planetary positions with undreamed-of precision. But Kepler's laws are purely descriptive; they do not help us understand what forces of nature constrain the planets to follow this particular set of rules. That step was left to Newton.

2.2 NEWTON'S GREAT SYNTHESIS

It was the genius of Isaac Newton (1643–1727) that found a conceptual framework that completely explained the observations and rules assembled by Galileo, Tycho, Kepler, and others. Newton was born in Lincolnshire, England, in the year after Galileo's death (Figure 2.6). Against the advice of his mother, who wanted him to stay home and help with the family farm, he entered Trinity College at Cambridge in 1661 and eight years later was appointed Professor of Mathematics there. Among Newton's contemporaries in England were architect Christopher Wren, authors Samuel Pepys and Daniel Defoe, and composer G. F. Handel.

Newton's Laws of Motion

As a young man in college, Newton became interested in natural philosophy, as science was then called. He worked out some of his first ideas on mechanics and optics during the plague years of 1665 and 1666, when students were sent home from college. Newton, a moody and often difficult man, continued to work on his ideas in private, even inventing new mathematical tools to help him deal with the complexities involved. Eventually, his friend Edmund Halley prevailed on him to collect and publish the results of his remarkable investigations on motion and gravity. The result was a volume that set out the underlying system of the physical world, *Philosophiae Naturalis Principia Mathematica.* The *Principia,* as the book is generally known, was published at Halley's expense in 1687.

At the very beginning of the *Principia,* Newton states three laws that he presumes to govern the motions of all objects:

Newton's First Law Every body continues doing what it is already doing—being in a state of rest, or moving uniformly in a straight line—unless it is compelled to change by an outside force.

Newton's Second Law The change of motion of a body is proportional to the force acting on it, and is made in the direction in which that force is acting.

Newton's Third Law To every action there is an equal and opposite reaction (or, the mutual actions of two bodies upon each other are always equal and act in opposite directions).

In the original Latin, the three laws contain only 59 words, but those few words set the stage for modern science. Let us examine them more carefully.

Interpretation of Newton's Laws

Newton's first law is a restatement of one of Galileo's discoveries, called the **conservation of momentum**—where momentum is a measure of a body's motion. The law states that in the absence of any outside influence, a body's momentum remains unchanged. We use this word in everyday expressions as well, as in "This bill in Congress has a lot of momentum: It seems unstoppable." In other words, a stationary object stays put, and a moving object keeps moving. Momentum depends on three factors. The first is *speed*—how fast a body moves (zero if it is stationary). The second is the *direction* in which the body is moving. Scientists use the term velocity to describe both speed and direction. For example, 20 kilometers per hour (km/h) due south is velocity; 20 km/h is speed. The third factor in momentum is what Newton called *mass.* Mass is a measure of the amount of matter in a body, as we will discuss later.

As mentioned in Chapter 1, it is not so easy to see this rule in action in the everyday world. Objects in motion do not remain in motion, because outside forces are always acting. One important force is friction, which tends to slow things down. If you roll a ball down the sidewalk, it eventually comes to a stop because the rubbing of the ball against the sidewalk exerts a force. But out in space between the stars, where friction is negligible, objects could in fact continue to move (coast) indefinitely. Newton's first law is sometimes called the *law of inertia,* inertia being the tendency of objects (and legislatures) to keep doing what they are already doing.

The momentum of a body can change only under the action of an outside influence. Newton's second law defines *force* in terms of its ability to *change momentum.* A force

FIGURE 2.6
Newton Isaac Newton (1643–1727), whose work on the laws of motion, gravity, optics, and mathematics laid the foundations of much of physical science. *(AIP/Niels Bohr Library)*

exam, the faster it will fly across the room. How much a force will accelerate an object is also determined by the object's mass. If you kick a tennis ball with the same force that you kick the textbook, the tennis ball—having less mass—can be accelerated to a greater speed.

Newton's third law is the most profound. Basically, it is a generalization of the first law, but it also gives us a way to define mass. If we consider a system of two or more objects isolated from outside influences, Newton's first law says that the total momentum of the system of objects should remain constant. Therefore, any change of momentum within the system must be balanced by another change that is equal and opposite to it, so that the momentum of the entire system is not changed.

This means that forces in nature do not occur alone: We find that in each situation there is always a pair of forces that are equal to and opposite each other. If a force is exerted on an object, it must be exerted by something else, and the object will exert an equal and opposite force back on that something. Suppose that during that same difficult midterm, another student runs screaming from the room and jumps out a (not very high) window. The force pulling him down after he jumps (as we will see in the next section) is the gravitational force between him and the Earth. Both he and the Earth must suffer the same total change of momentum because of the influence of this mutual force. So, both the student and the Earth are accelerated by each other's pull. However, the student does much more of the moving. Because the Earth has enormously greater mass, it can experience the same change of momentum by accelerating only a very small amount. Things fall toward the Earth all the time, but the acceleration of our planet as a result is far too small to be noticed or measured.

A more obvious example of the mutual nature of forces between objects is familiar to all who have batted a baseball. The recoil you feel as you swing your bat shows that the ball exerts a force on it during the impact, just as the bat does on the ball. Similarly, when a rifle is discharged, the force pushing the bullet out of the muzzle is equal to that pushing backward upon the gun and the shoulder of the person shooting it.

This, in fact, is the principle behind jet engines and rockets: the force that discharges the exhaust gases from the rear of the rocket is accompanied by force that shoves the rocket forward. The exhaust gases need not push against air or the Earth; a rocket actually operates best in a vacuum (Figure 2.7).

(which you can think of as a push or a pull) has both size and direction. When force is applied to a body, the momentum changes in the direction of the applied force. This means that a force is required to change either the speed or the direction of a body, or both—that is, to start it moving, to speed it up, to slow it down, to stop it, or to change its direction.

Any such change in an object's state of motion is called **acceleration.** Newton showed that the acceleration of a body was proportional to the force being applied to it. The harder you kick your astronomy textbook after a difficult

Mass, Volume, and Density

Before we go on to discuss Newton's other work, we want to take a brief look at some terms that will be important to sort out clearly. We begin with *mass,* which is a measure of the amount of material in an object. Today, scientists often think of it in terms of the numbers of subatomic particles (such as protons and neutrons) that make up an object.

Volume is a measure of the physical space occupied by a body, say in cubic centimeters or liters. In short, the volume is the "size" of an object; it has nothing to do with its mass. A penny and an inflated balloon may both have the same mass, but they have very different volumes.

The penny and balloon are also very different in **density,** which is a measure of how much mass we have per unit volume. Specifically, density is the ratio of mass to volume. Note that often in everyday language we use "heavy" and "light" as indications of density (rather than weight), as, for instance, when we say that iron is heavy or that puff pastry is light.

The units of density that will be used in this book are grams per cubic centimeter (g/cm3).[1] If a block of some material has a mass of 300 g and a volume of 100 cm^3, its density is 3 g/cm^3. Familiar materials span a considerable range in density, from artificial materials such as plastic insulating foam (less than 0.1 g/cm^3) to gold (19 g/cm^3) (Table 2.1). In the astronomical universe, much more remarkable densities can be found, all the way from a comet's tail (10^{-16} g/cm^3) to a neutron star (10^{15} g/cm^3).

To sum up, then, mass is "how much," volume is "how big," and density is "how tightly packed."

Angular Momentum

The concept of **angular momentum** is a bit more complex, but it is important for understanding many astronomical objects. Angular momentum is a measure of the momentum of an object as it rotates or revolves about some fixed point. Whenever we deal with the revolution of spinning objects, from planets to galaxies, we have to consider angular momentum. The angular momentum of an object is defined as the product of three quantities: its

FIGURE 2.8
The Conservation of Angular Momentum When a spinning figure skater brings her arms in, their distance from the spin center is small, so her speed increases. When her arms are out, their distance from the spin center is greater, so she slows down.

mass, its velocity, and its distance from the fixed point around which it turns.

If these three quantities remain constant—that is, if the motion takes place at a constant speed and at a fixed distance from the spin center—then the angular momentum is also a constant. So, ignoring air resistance, if you tie a string to your astronomy textbook and twirl it around your head at constant speed, you will have a system with constant angular momentum.

More generally, angular momentum is constant, or conserved, in any rotating system in which no external forces act, or in which the only forces are directed toward or away from the spin center. An example of such a system is a planet orbiting the Sun. Kepler's second law is an example of the conservation of angular momentum. When a planet approaches the Sun on its elliptical orbit, the distance to the spin center decreases; the planet speeds up to keep the angular momentum the same. Similarly, when the planet is further from the Sun, it revolves more slowly.

Just as a planet speeds up when it approaches the Sun, a shrinking cloud of dust or a star collapsing on itself (both situations you will encounter as you read on) increases its spin rate as it contracts. There is less distance to the spin center, so the speed goes up to keep angular momentum the same. The concept is also illustrated by figure skaters, who bring their arms and legs in to spin more rapidly, and extend their arms and legs to slow down (Figure 2.8). (You can duplicate this yourself on a well-oiled piano stool, by

TABLE 2.1
Densities of Materials

Material	Density (g/cm^3)
Gold	19.3
Lead	11.4
Iron	7.9
Earth (bulk)	5.6
Rock (typical)	2.5
Water	1.0
Wood (typical)	0.8
Insulating foam	0.1
Silica gel	0.02

[1]Generally we use the standard metric (or SI) units in this book. The proper metric unit of density is kg/m^3. But to most people g/cm^3 provides a more meaningful unit because the density of water is exactly 1 g/cm^3. Density expressed in g/cm^3 is sometimes called specific density or specific weight.

starting yourself spinning slowly with your arms extended, and then pulling your arms in.)

 ## 2.3 UNIVERSAL GRAVITY

The Law of Gravity

Newton's laws of motion showed that, left to themselves, objects at rest stay at rest, and those in motion continue moving uniformly in a straight line. Thus, it is the straight line, not the circle, that defines the most natural state of motion. In the case of the planets, then, some force must be bending their paths from straight lines into ellipses. That force, Newton proposed, is gravity.

Although that may sound obvious today, in Newton's time gravity was something associated with the Earth alone. Everyday experience shows us that the Earth exerts a gravitational force upon objects at its surface. If you drop something off a leaning tower in Pisa, it falls toward the Earth, accelerating as it falls. Newton's insight was that the Earth's gravity might extend as far as the Moon and produce the acceleration required to curve the Moon's path from a straight line and keep it in its orbit. He further speculated that gravity is not limited to the Earth, but that there is a general force of attraction between all material bodies. If so, the attractive force between the Sun and each of the planets could keep each in its orbit.

Once Newton boldly hypothesized that there is a *universal attraction* among all bodies everywhere in space, he had to determine the exact nature of the attraction. The precise mathematical description of that gravitational force had to dictate that the planets move exactly as Kepler had observed them to (as expressed in Kepler's three laws). Also, the law of gravity had to predict the correct behavior of falling bodies on the Earth, as observed by Galileo. How must gravitational force depend on distance in order for these conditions to be met?

The answer to this question required mathematical tools that had not yet been developed. But this did not deter Isaac Newton, who invented what we today call *calculus* to deal with this problem. Eventually he was able to conclude that the force of gravity must drop off with increasing distance between the Sun and a planet (or between any two objects) in proportion to the *inverse square* of their separation. In other words, if a planet were twice as far from the Sun, the force would be $1/2^2$ or 1/4 as large. Put the planet three times farther away, and the force is $1/3^2$ or 1/9 as strong.

Newton also concluded that the gravitational attraction between two bodies must be proportional to their masses. The more mass an object has, the stronger the pull of its gravity. Expressed as a formula, the gravitational attraction between any two objects is given by one of the most famous formulas in all of science:

$$\text{force} = GM_1M_2/R^2$$

where M_1 and M_2 are the masses of the two objects and R is their separation. The number represented by G is called the constant of gravitation. With such a force and the laws of motion, Newton was able to show mathematically that the only orbits permitted were exactly those described by Kepler's laws.

Newton's law of gravity works for the planets, but is it really universal? The gravitational theory should also predict the observed acceleration of the Moon toward the Earth, falling around the Earth at a distance of about 60 times the radius of the Earth, as well as of an object (say an apple) dropped near the Earth's surface. The falling of an apple is something we can measure quite easily; can we use it to predict the motions of the Moon?

Newton's theory says that the force on (and therefore the acceleration of) an object toward the Earth should be inversely proportional to the square of its distance from the center of the Earth. Objects like apples at the surface of the Earth (R = 1 Earth radius, the distance from its center) are observed to accelerate downward at 9.8 meters per second per second (9.8 m/s^2).

Therefore, if the law holds, the Moon, 60 Earth radii from its center, should experience an acceleration toward the Earth that is $1/60^2$, or 3600 times less—that is, about 0.00272 m/s^2. This is precisely the observed acceleration of the Moon in its orbit. What a triumph! Imagine the thrill Newton must have felt to realize he had discovered and verified a law that holds for the Earth, apples, the Moon, and, as far as we know, everything in the universe!

Putting the math aside, what Newton's law suggests is that gravity is a "built-in" property of mass. Wherever masses occur, they will interact via the force of gravity. The more mass there is, the greater the force it can exert. Here on Earth, the largest concentration of mass is, of course, the planet we stand on, and its pull dominates the gravitational interactions we experience. But everything with mass attracts everything else with mass anywhere in the universe. For example, you and this textbook (each having mass) attract each other via gravity. (We certainly hope it's not the *only* way you find our book attractive.) If you let the book fall, however, it is pulled much more strongly by the Earth, so it falls to the Earth, not toward you. But if you and the textbook somehow found yourselves out in space, far from any stars or planets, you would discover that gravitational attraction would start to pull you together.

Newton's law also suggests that gravity never becomes zero. It quickly gets weaker with distance, but it continues to act to some degree no matter how far away you get. The pull of the Sun is stronger at Mercury than at Pluto, but it can be felt far beyond Pluto, where (as we shall see) we have good evidence that it continues to make enormous numbers of smaller bodies move around huge orbits. And its pull joins with the pull of billions of other stars to create the gravitational pull of the Milky Way Galaxy in which we live. That force, in turn, can make other smaller galaxies orbit around the Milky Way, and so forth.

Why is it then, you may ask, that the astronauts aboard the Space Shuttle appear "weightless" and we see images on television of people and objects floating in the spacecraft? After all, the astronauts in the Shuttle are only a few

FIGURE 2.9
Astronaut in Free Fall Aboard the Space Shuttle
Columbia in July 1997, astronaut Susan Still floats
before the Spacelab control center. Both she and
the Shuttle are falling freely around the Earth, and
so she experiences no force relative to the Shuttle
(just as people in a freely falling elevator would feel
weightless for the brief time they had left to live.)
(NASA)

hundred kilometers above the surface of the Earth, not really a significant distance compared to the size of the Earth. Gravity is certainly not a great deal weaker that much farther away. The astronauts feel weightless for the same reason that passengers in an elevator whose cable has broken or in an airplane whose engines no longer work feel weightless[2] (Figure 2.9). They are *falling,* and in free fall, they accelerate at the same rate as everything around them, including their spacecraft and a sandwich that got away from their grip. Thus, *relative* to the Shuttle, the astronauts experience no additional force and so feel "weightless." Unlike the falling elevator passengers, however, the astronauts are falling *around* the Earth, not *to* the Earth (and will live to tell their story).

Orbital Motion and Mass

Kepler's laws are descriptions of the orbits of objects moving according to Newton's laws of motion and the law of gravity. Knowing that gravity is the force that attracts planets toward the Sun, however, allowed Newton to rethink Kepler's third law. Recall that Kepler had found a relationship between the period of a planet's revolution and its distance from the Sun. Newton's law of gravity can be used to show mathematically that this relationship is actually

$$D^3 = (M_1 + M_2) \times P^2$$

As explained in Section 2.1, we express distances in the solar system in astronomical units and periods in years. But Newton's formulation introduces the additional factor of the masses of the Sun (M_1) and planet (M_2), both expressed in units of the Sun's mass.

How did Kepler miss this factor? In units of the Sun's mass, the mass of the Sun is 1; in units of the Sun's mass, the mass of a typical planet is a negligibly small fraction, and so $(M_1 + M_2)$ is very, very close to 1. This makes Newton's formula look almost the same as Kepler's. The tiny mass of the planets is the reason that Kepler did not realize that both masses had to be included in the calculation. There are many cases in astronomy, however, in which we do need to include the two mass terms—for example, when two stars or two galaxies orbit one another.

Including the mass term allows us to use this formula in a new way. If we can measure the motions (distances and periods) of objects acting under their mutual gravity, the formula will permit us to deduce their masses. For example, we can calculate the mass of the Sun by using the distances and periods of the planets, or the mass of Jupiter by noting the motions of its moons. Indeed, Newton's reformulation of Kepler's third law is one of the most powerful concepts in astronomy. Our ability to deduce the masses of objects from their motions is key to understanding the nature and evolution of many astronomical bodies. We will use this law repeatedly throughout this text in calculations that range from the orbits of comets to the interactions of galaxies.

2.4 ORBITS IN THE SOLAR SYSTEM

Description of an Orbit

Celestial mechanics is the study of the motions of astronomical objects, using gravitational theory. The path of an object through space is called its orbit, whether that object is a spacecraft, planet, star, or galaxy. An orbit, once determined, allows the future positions of the object to be calculated. When only two objects are involved (such as a planet in orbit about the Sun), three quantities

---

[2]In the film *Apollo 13,* the scenes where the astronauts were "weightless" were actually filmed in a falling airplane. As you might imagine, the plane only fell for short periods before the engines cut in again.

TABLE 2.2
Orbital Data for the Planets

Planet	Semimajor Axis (AU)	Period (yr)	Eccentricity
Mercury	0.39	0.24	0.21
Venus	0.72	0.62	0.01
Earth	1.00	1.00	0.02
Mars	1.52	1.88	0.09
(Ceres)	2.77	4.60	0.08
Jupiter	5.20	11.86	0.05
Saturn	9.54	29.46	0.06
Uranus	19.19	84.07	0.05
Neptune	30.06	164.80	0.01
Pluto	39.60	248.60	0.25

are required to describe the orbit. These are the *size* (the semimajor axis), the *shape* (the eccentricity), and the *period* of revolution.

Two points in any orbit have been given special names. The place where the planet is closest to the Sun (*Helios* in Greek) is called the **perihelion** of its orbit, and the place where it is farthest away and moves the most slowly is the **aphelion.** For a satellite orbiting the Earth (*Geos* in Greek), the corresponding terms are **perigee** and **apogee.**

Orbits of the Planets

Today, Newton's work enables us to calculate and predict the orbits of the planets with marvelous precision. We know nine planets, beginning with Mercury closest to the Sun and extending outward to Pluto. The average orbital data for the planets are summarized in Table 2.2. (In case you don't recognize it, Ceres is the largest of the *asteroids;* see the next section.)

According to Kepler's laws, Mercury must have the shortest period of revolution (88 Earth days); thus it has the highest orbital speed (averaging 48 km/s). At the opposite extreme, Pluto has a period of 249 years and an average orbital speed of just 5 km/s.

All of the planets have orbits of rather low eccentricity. The most eccentric orbits are those of Mercury (0.21) and Pluto (0.25); all of the rest have eccentricities of less than 0.1. It is fortunate for the development of science that Mars has an eccentricity greater than many of the other planets, for otherwise the pretelescopic observations of Tycho would not have been sufficient for Kepler to deduce that its orbit had the shape of an ellipse rather than a circle.

The planetary orbits are also confined close to a common plane, which is near the plane of the Earth's orbit (the ecliptic). The strange orbit of Pluto is inclined about 17° to

the average, but all of the other planets lie within 10° of the common plane of the solar system.

Orbits of Asteroids and Comets

In addition to the nine planets, there are many smaller objects in the solar system. Some of these are natural satellites that orbit all of the planets except Mercury and Venus. In addition, there are two classes of smaller objects in heliocentric orbits: the asteroids and the comets. Both asteroids and comets are believed to be small chunks of material left over from the formation process of the solar system.

Asteroids and comets differ from each other in composition and in the natures of their orbits. In general, the asteroids have orbits with smaller semimajor axes than do the comets (Figure 2.10). The great majority of them lie between 2.2 and 3.3 AU, in the region known as the **asteroid belt.** As you can see in Table 2.2, the asteroid belt (represented by its largest member, Ceres) is in the middle of a gap between the orbits of Mars and Jupiter. It is because these two planets are so far apart that stable orbits of small bodies can exist in the region between them.

Comets generally have orbits of larger size and greater eccentricity than those of the asteroids. Typically, the eccentricities of their orbits are 0.8 or higher. According to Kepler's second law, therefore, they spend most of their time far from the Sun, moving very slowly. As they approach perihelion, the comets speed up and whip through the inner parts of their orbits more rapidly.

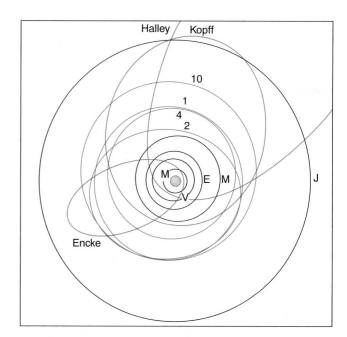

FIGURE 2.10
Solar System Orbits We see the orbits of typical comets and asteroids compared with those of the planets Mercury, Venus, Earth, Mars, and Jupiter (black circles). Shown in red are three comets: Halley, Kopff, and Encke. In blue are the four largest asteroids: 1 Ceres, 2 Pallas, 4 Vesta, and 10 Hygeia.

MAKING CONNECTIONS

Astronomy and the Poets

When Copernicus, Kepler, Galileo, and Newton formulated the fundamental rules that underlie everything in the physical world, they changed much more than the face of science. For some, they gave humanity the courage to let go of old superstitions and see the world as rational and manageable; for others, they upset comforting, ordered ways that had served humanity for centuries, leaving only a dry, mechanical clockwork universe in their wake.

Poets of the time reacted to such changes in their own work and debated whether the new world picture was an appealing or frightening one. John Donne (1573–1631), in a poem called *Anatomy of the World,* laments the passing of the old certainties:

> The new philosophy [science] calls all in doubt,
> The element of fire is quite put out;
> The Sun is lost, and th' earth, and no man's wit
> Can well direct him where to look for it.

(Here the "element of fire" refers also to the sphere of fire, which medieval thought placed between the Earth and the Moon.)

By the next century, poets like Alexander Pope were celebrating Newton and the Newtonian world view. Pope's famous couplet, written upon Newton's death, goes:

> Nature, and nature's laws lay hid in night.
> God said, Let Newton be! and all was light.

In his 1733 poem, *An Essay on Man,* Pope revels in the complexity of the new views of the world, incomplete though they are:

> Of man, what see we, but his station here,
> From which to reason, to which refer?...
> He, who thro' vast immensity can pierce,
> See worlds on worlds compose one universe,

> Observe how system into system runs,
> What other planets circle other suns,
> What vary'd being peoples every star,
> May tell why Heav'n has made us as
> we are...
> All nature is but art, unknown to thee;
> All chance, direction, which thou canst
> not see;
> All discord, harmony not understood;
> All partial evil, universal good:
> And, in spite of pride, in erring reason's spite,
> One truth is clear, whatever is, is right.

Poets and philosophers continued to debate whether humanity was exalted or debased by the new views of science. The 19th-century poet Arthur Hugh Clough (1819–1861) cries out in his poem *The New Sinai*:

> And as of old from Sinai's top God said that God
> is one,
> By science strict so speaks He now to tell us, there
> is None!
> Earth goes by chemic forces; Heaven's a
> Mécanique Celeste!
> And heart and mind of humankind a watchwork as
> the rest!

(A "mécanique celeste" is a clockwork model to demonstrate celestial motions.)

The 20th-century poet Robinson Jeffers (whose brother was an astronomer) saw it differently in a poem called *Star Swirls*:

> There is nothing like astronomy to pull the stuff
> out of man.
> His stupid dreams and red-rooster importance:
> Let him count the star-swirls.

2.5 MOTIONS OF SATELLITES AND SPACECRAFT

Space Flight and Satellite Orbits

The law of gravity and Kepler's laws describe the motions of Earth satellites and interplanetary spacecraft as well as the planets. Sputnik, the first artificial Earth satellite, was launched by what was then called the Soviet Union on October 4, 1957. Since that time, thousands of satellites have been placed into orbit around the Earth, and spacecraft have also orbited the Moon, Venus, Mars, Jupiter, and the asteroid Eros.

Once an artificial satellite is in orbit, its behavior is no different from that of a natural satellite, such as our Moon. If the satellite is high enough to be free of atmospheric friction, it will remain in orbit forever, following Kepler's laws in a perfectly respectable way. However, although there is no difficulty in maintaining a satellite once it is in orbit, a great deal of energy is required to lift the spacecraft off the Earth and accelerate it to orbital speed.

To illustrate how a satellite is launched, imagine a gun firing a bullet horizontally from the top of a high mountain,

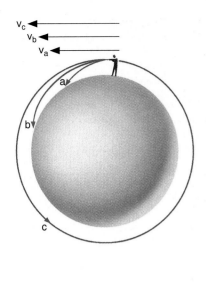

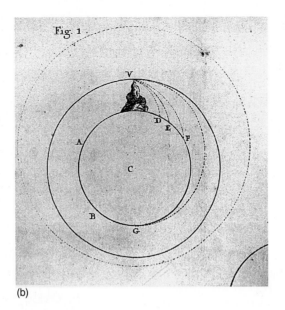

FIGURE 2.11
Firing a Bullet into Orbit (a) For paths *a* and *b*, the velocity is not enough to prevent gravity from pulling the bullet back to Earth; in case *c*, the velocity allows the bullet to fall completely around the Earth. (b) A diagram by Newton in his *De Mundi Systematic*, 1731 edition, illustrating the same concept shown in panel (a). *(Crawford Collection, Royal Observatory, Edinburgh)*

(a)   (b)

(Figure 2.11a—adapted from a similar diagram by Newton, shown in Figure 2.11b). Imagine, further, that the friction of the air could be removed and that nothing can get in the bullet's way; other mountains, buildings, and so on, are all absent. Then the only force that acts on the bullet after it leaves the muzzle is the gravitational force between the bullet and Earth.

If the bullet is fired with some velocity we can call v_a, it continues to have that forward speed, but meanwhile the gravitational force acting upon it pulls it downward toward the Earth, where it strikes the ground at point *a*. However, if it is given a higher muzzle velocity, v_b, its higher forward speed carries it farther before it hits the ground. This is because, regardless of its forward speed, the downward gravitational force is the same. Thus, this faster-moving bullet strikes the ground at point *b*.

If our bullet is given a high enough muzzle velocity, v_c, the curved surface of the Earth causes the ground to tip out from under it so that it remains the same distance above the ground and falls around the Earth in a complete circle. The speed needed to do this—called the **circular satellite velocity**—is about 8 km/s or about 17,500 miles per hour (mph) in more familiar units.

Each year more than 50 new satellites are launched into orbit by such nations as Russia, the United States, China, Japan, India, and Israel, as well as by the European Space Agency (ESA), a consortium of European nations (Figure 2.12). Most satellites are launched into low Earth orbit, since this requires the minimum launch energy. At the orbital speed of 8 km/s, they circle the planet in about 90 min.

Low Earth orbits are not stable indefinitely, because the drag generated by friction with the thin upper atmosphere eventually leads to a loss of energy and "decay" of the orbit. (In the terms of Newton's work, we can say that friction is another outside force acting on the satellite.) Upon re-entering the denser parts of the atmosphere, most satellites are burned up by atmospheric friction, although some solid parts may reach the surface. Of course, such piloted rockets as the Space Shuttle and other recoverable payloads are designed to survive reentry intact.

Interplanetary Spacecraft

The exploration of the solar system has been carried out largely by robot spacecraft sent to the other planets. To escape Earth, these craft must achieve **escape velocity,** the speed needed to move away from the Earth forever, which is about 11 km/s (about 25,000 mph). After this they coast to their targets, subject only to minor trajectory adjustments provided by small thruster rockets on board. In interplanetary flight, these spacecraft follow Keplerian orbits around the Sun, modified only when they pass near one of the planets.

While it is close to its target, a spacecraft is deflected by the planet's gravitational force into a modified orbit, either gaining or losing energy in the process. By carefully choosing the aim point in a planetary encounter, controllers have actually been able to use a planet's gravity to redirect a flyby spacecraft to a second target. *Voyager 2* (Figure 2.13) used a series of gravity-assisted encounters to yield successive flybys of Jupiter (1979), Saturn (1980), Uranus (1986), and Neptune (1989). The *Galileo* spacecraft, launched in 1989, flew past Venus once and Earth twice to gain the energy required to reach its ultimate goal at Jupiter.

If we wish to orbit a planet, we must slow the spacecraft with a rocket when the spacecraft is near its destination, allowing it to be captured into an elliptical orbit. Additional rocket thrust is required to bring a vehicle down from orbit for a landing on the surface. Finally, if a return trip to Earth is planned, the landed payload must include enough propulsive power to repeat the entire process in reverse.

FIGURE 2.12
Satellites in Earth Orbit A plot of all known unclassified satellites and satellite debris in Earth orbit larger than about the size of softball as of 1995. *(U.S. Space Command, NORAD)*

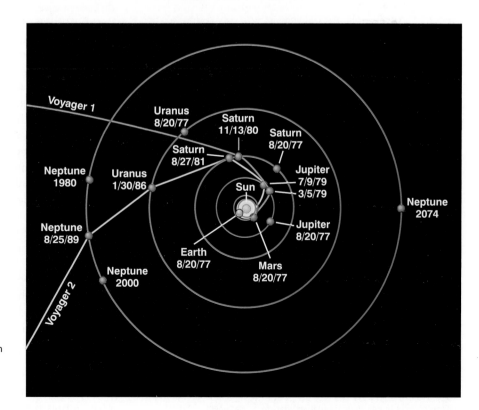

FIGURE 2.13
Using Gravity Assists to Get to the Outer Planets The flight paths of *Voyager 1* (red) and *Voyager 2* (yellow) through the outer solar system, taking advantage of the gravitation of each planet to adjust the trajectory toward the next target. *(NASA/JPL)*

2.6 GRAVITY WITH MORE THAN TWO BODIES

Until now we have considered the Sun and a planet (or a planet and one of its satellites) as nothing more than a pair of bodies revolving around each other. In fact, all the planets exert gravitational forces upon one another as well. These interplanetary attractions cause slight variations from the orbits that would be expected if the gravitational forces between planets were neglected. Unfortunately, the problem of treating the motion of a body that is under the gravitational influence of two or more other bodies is very complicated and can be handled properly only with large computers.

The Interactions of Many Bodies

Here's an example of what we mean: Suppose you have a cluster of a thousand stars all orbiting a common center (such clusters are quite common, as we shall see). If we know the exact position of each star at any given instant, we can calculate the combined gravitational force of the entire group on any one member of the cluster. Knowing the force on the star in question, we can therefore find how it will accelerate. If we know how it was moving to begin with, we can then calculate how it will move in the next instant of time, thus tracking its motion.

However, the problem is complicated by the fact that the other stars are also moving and thus changing the effect they will have on our star. Therefore, we must simultaneously calculate the acceleration of each body produced by the combination of the gravitational attractions of all the others in order to track the motions of all of them, and hence of any one. Such complex calculations have been carried out with modern computers to track the evolution of hypothetical clusters of stars with up to a million members (Figure 2.14).

Within the solar system, the problem of computing the orbits of planets and spacecraft is somewhat simpler. We have seen that Kepler's laws, which do not take into account the gravitational effects of the other planets on an orbit, really work quite well. This is because these additional influences are very small in comparison to the dominant gravitational attraction of the Sun. Under such circumstances, it is possible to treat the effects of other bodies as small *perturbations* (or disturbances) on the force exerted by the Sun. During the 18th and 19th centuries, mathematicians developed many elegant techniques for calculating perturbations, permitting them to predict very precisely the positions of the planets. Such calculations eventually led to the discovery of a new planet in 1846.

Discovery of Neptune

The discovery of the eighth planet, Neptune, was one of the high points in the development of gravitational theory. In 1781, William Herschel, a musician and unpaid astronomer, accidently discovered the seventh planet, Uranus. It happens that Uranus had been observed a century before, but in none of those earlier sightings was it recognized as a planet; rather, it was simply recorded as a star.

By 1790, an orbit had been calculated for Uranus using observations of its motion in the decade following its discovery. Even after allowance was made for the perturbing effects of Jupiter and Saturn, however, it was found that Uranus did not move on an orbit that exactly fit the earlier observations of it made since 1690. By 1840, the discrepancy between the positions observed for Uranus and those predicted from its computed orbit amounted to about 0.03°—an angle barely discernible to the unaided eye but still larger than the probable errors in the orbital calculations. In other words, Uranus just did not seem to move on the orbit predicted from Newtonian theory.

In 1843, John Couch Adams (Figure 2.15), a young Englishman who had just completed his studies at Cambridge, began a detailed mathematical analysis of the irregularities in the motion of Uranus to see if they might be produced by the pull of an unknown planet. His calculations indicated the existence of a planet more distant than Uranus from the Sun. In October 1845, Adams delivered his results to George Airy, the British Astronomer Royal, informing him where in the sky to find the new planet. We now know that Adams' predicted position for the new body was correct to within 2°, but for a variety of reasons, Airy did not follow up right away.

Meanwhile, French mathematician Urbain Jean Joseph Leverrier, unaware of Adams or his work, attacked the same problem and published its solution in June 1846. Airy, noting that Leverrier's predicted position for the

FIGURE 2.14
Modern Computing Power Supercomputers at NASA Ames Research Center are capable of tracking the motions of more than a million objects under their mutual gravitation. *(NASA/ARC)*

Mathematicians Who Discovered a Planet John Couch Adams (1819–1892) and Urbain J. J. Leverrier (1811–1877) share the credit for discovering the planet Neptune. *(a, Yerkes Observatory b, Corbis/Bettmann)*

unknown planet agreed to within 1° with that of Adams, suggested to James Challis, director of the Cambridge Observatory, that he begin a search for the new object. The Cambridge astronomer, having no up-to-date star charts of the Aquarius region of the sky where the planet was predicted to be, proceeded by recording the positions of all the faint stars he could observe with his telescope in that location. It was Challis' plan to repeat such plots at intervals of several days, in the hope that the planet would distinguish itself from a star by its motion. Unfortunately, he was negligent in examining his observations; although he had actually seen the planet, he did not recognize it.

About a month later, Leverrier suggested to Johann Galle, an astronomer at the Berlin Observatory, that he look for the planet. Galle received Leverrier's letter on September 23, 1846, and, possessing new charts of the Aquarius region, found and identified the planet that very night. It was less than a degree from the position Leverrier predicted. The discovery of the eighth planet, now known as Neptune (the Latin name for the god of the sea), was a major triumph for gravitational theory, for it dramatically confirmed the generality of Newton's laws. The honor for the discovery is properly shared by the two mathematicians, Adams and Leverrier.

We should note that the discovery of Neptune was not a complete surprise to astronomers, who had long suspected the existence of the planet based on the "disobedient" motion of Uranus. On September 10, 1846, two weeks before Neptune was actually found, John Herschel, son of the discoverer of Uranus, remarked in a speech before the British Association, "We see [the new planet] as Columbus saw America from the shores of Spain. Its movements have been felt trembling along the far-reaching line of our analysis with a certainty hardly inferior to ocular demonstration."

SURFING THE WEB

🖥 Images of Tycho Brahe
[www.mhs.ox.ac.uk/tycho/]
Virtual exhibit from the Museum of the History of Science in Oxford, England, with images of and information about Tycho and his contemporaries.

🖥 Isaac Newton Home Page
[newton.gws.uky.edu/cover.html]
James Force and Charles Krumpelman at the University of Kentucky have assembled this site, which features a nice biography of Newton, links to other sites, and information about some of Newton's manuscripts. (Other lists of Newton links can be found at the Newtonia site [www.info.cern.ch/m/mcnab/www/n].)

🖥 Sites for Observing Earth Satellites
With thanks to Stuart Goldman's article in *Sky & Telescope* magazine, August 1996, p. 86, here are some sites for those who wish to track Earth orbiting satellites in the sky:

• Visual Satellite Observer's Home Page [www2.satellite.eu. org/sat/vsohp/satintro.html] has lots of instructions and links.

• The Satellite Observing Resources Page [www.accesscom.com/~iburrell/sat/sattrack.html] also has links and tutorials.

Two on-line calculators to tell you when a satellite will be visible over your location are Satellite Tracking Pass Predictions [acsprod1.acs.ncsu.edu/scripts/HamRadio/sattrack] and, Earth Satellite Ephemeris Service [www.chara.gsu.edu/sat.html]

🖥 Mathematical Discovery of Planets Site
[www-groups.dcs.st-and.ac.uk/~history/HistTopics/ Neptune_and_Pluto.html]
Has the fuller story of how Neptune was predicted and discovered, and who did what to whom. A spirited defense of George B. Airy (who comes across as the "villain" in many versions of the story), written by historian Alan Chapman, can be found at www.ast.cam.ac.uk/~naw96/lassell/adams-airy.html

2.1 Tycho Brahe was the most skillful of the pretelescopic astronomical observers. His accurate observations of planetary positions provided the data used by Johannes Kepler to derive the three fundamental laws of planetary motion that bear his name: (1) planetary orbits are **ellipses** (a figure described by its **semimajor axis** and **eccentricity**) with the Sun at one focus; (2) in equal intervals, a planet's orbit sweeps out equal areas; and (3) if times are expressed in years, and distances in **astronomical units,** the relationship between period (P) and semimajor axis (D) of an orbit is given by $P^2 = D^3$.

2.2 In his *Principia*, Isaac Newton established the three laws that govern the motion of objects: (1) Bodies continue at rest or in uniform motion unless acted upon by an outside force; (2) an outside force causes an acceleration (and changes the **momentum**) of an object; and (3) for each action there is an equal and opposite reaction. Momentum is a measure of the motion of an object and depends on both its mass and its velocity. **Angular momentum** is a measure of motion of a spinning or revolving object. The **density** of an object is its mass divided by its volume.

2.3 Gravity, the attraction of all mass for all other mass, is the force that keeps the planets in orbit. Newton's law of gravity relates gravitational force to mass and distance ($F = GM_1M_2/R^2$). Newton was able to show the equivalence of gravitational force (weight) on Earth to the gravitational force between objects in space. When Kepler's laws are re-examined in the light of gravitational theory, it becomes clear that the masses of both Sun and planet are important for the third law, which becomes $(M_1 + M_2) \times P^2 = D^3$. Mutual gravitational effects permit us to calculate the masses of astronomical objects, from comets to galaxies.

2.4 The lowest point in a satellite orbit is its **perigee,** and the highest point is its **apogee** (corresponding to **perihelion** and **aphelion** for an orbit about the Sun). The planets all follow orbits about the Sun that are nearly circular and in the same plane. Most asteroids are found between Mars and Jupiter in the **asteroid belt,** whereas comets generally follow orbits of high eccentricity.

2.5 The orbit of an artificial satellite depends on the circumstances of its launch. The **circular satellite velocity** at the Earth's surface is 8 km/s, and the **escape velocity** is 11 km/s. There are many possible interplanetary trajectories, including those that use gravity-assisted flybys of one object to redirect the spacecraft toward its next target.

2.6 Gravitational problems that involve more than two interacting bodies are much more difficult to deal with than two-body problems. They require large computers for accurate solutions. If one object dominates gravitationally, it is possible to calculate the effects of a second object in terms of small perturbations. This approach was used by Adams and Leverrier to predict the position of Neptune from its perturbations of the orbit of Uranus and thus discover a new planet mathematically.

A An eccentric, but very rich, alumnus of your college makes a bet with the Dean that if you drop a marble and a bowling ball from the top of a tall water tower, the bowling ball would hit the ground first. Have your group discuss if you would make a side-bet on the fact that the alum is right. How would you decide who is right?

B Suppose a member of your group felt unhappy about his or her weight. Where could he or she go to weigh one fourth as much as they do now? Where would the person weigh even less? Would changing the unhappy person's weight have any effect on his or her mass?

C When the Apollo astronauts landed on the Moon, some commentators complained that it ruined the mystery and "poetry" of the Moon forever (and that lovers could never gaze at the full Moon in the same way again). Others felt that knowing more about the Moon could only enhance its interest to us as we see it from Earth. How does your group feel? Why?

D Figure 2.12 shows an impressive swarm of satellites in orbit around the Earth. What do you think all these satellites do? How many categories of functions for Earth satellites can your group come up with?

1. State Kepler's three laws in your own words.

2. Why did Kepler need Tycho Brahe's data to formulate his laws?

3. Which has more mass, an armful of feathers or an armful of lead? Which has more volume, a kilogram of feathers or a kilogram of lead? Which has more density, a kilogram of feathers or a kilogram of lead?

4. Explain how Kepler was able to find a relationship (his third law) between the periods and distances of the planets that did not depend on the masses of the planets or the Sun.

5. Write out Newton's three laws of motion in terms of what happens with the momentum of objects.

6. What planet has the largest
  a. semimajor axis **b.** speed of revolution around the Sun
  c. period of revolution around the Sun **d.** eccentricity

7. Why do we say that Neptune was the first planet to be discovered through the use of mathematics?

THOUGHT QUESTIONS

8. Is it possible to escape the force of gravity by going into orbit about the Earth? How does the force of gravity in the Russian space station *Mir* (orbiting 500 km above the Earth's surface) compare with that on the ground? (*Hint:* The Earth's gravity acts as if all the mass were concentrated at the center of the Earth. Is *Mir* significantly farther from the Earth's center than the Earth's surface is?)

9. What is the momentum of a body whose velocity is zero? How does Newton's first law of motion include the case of a body at rest?

10. Evil space aliens drop you and your astronomy instructor 1 km apart out in space, far from any star. Discuss the effects of gravity on each of you.

11. A body moves in a perfectly circular path at constant speed. Are there forces acting in such a system? How do you know?

12. As air friction causes a satellite to spiral inward closer to the Earth, its orbital speed increases. Why?

13. Use a history book or an encyclopedia to find out what else was happening in England during Newton's lifetime, and discuss what trends of the time might have contributed to his accomplishments and the rapid acceptance of his work.

PROBLEMS

14. What is the semimajor axis of a circle of diameter 24 cm? What is its eccentricity?

15. If 24 g of material fills a cube 2 cm on a side, what is the density of the material?

16. Draw an ellipse by the procedure described in the text, using a string and two tacks. Arrange the tacks so that they are separated by one-tenth the length of the string. Comment on the appearance of your ellipse. This (if you have been careful in your construction) is approximately the shape of the orbit of Mars.

17. The Earth's distance from the Sun varies from 147.2 million to 152.1 million km. What is the eccentricity of its orbit? (*Hint:* The distance between the foci of the ellipse is twice the distance between the Sun and the center of the ellipse. To find this, you need to first find the center of the ellipse by dividing the major axis in two. It helps to draw a diagram for yourself.)

18. Look up the revolution periods and distances from the Sun for Venus, Earth, Mars, and Jupiter. Calculate D^3 and P^2 (in the units specified in the text), and verify that they obey Kepler's third law.

19. What would be the period of a planet whose orbit has a semimajor axis of 4 AU? Of an asteroid with a semimajor axis of 10 AU?

20. What is the distance from the Sun (in astronomical units) of an asteroid with a period of revolution of eight years? What is the distance of a planet whose period is 45.66 days?

21. Newton showed that the periods and distances in Kepler's third law depend on the masses of the objects. What would be the period of revolution of the Earth (at 1 AU from the Sun) if the Sun had twice its present mass?

22. By what factor would a person's weight at the surface of the Earth be reduced if the Earth had its present mass but eight times its present volume? Its present size but only one-third its present mass?

SUGGESTIONS FOR FURTHER READING

Casper, M. *Kepler.* 1959, Collier. Biography by a respected historian.

Christianson, G. "Newton's *Principia:* A Retrospective" in *Sky & Telescope,* July 1987, p. 18.

Christianson, G. "The Celestial Palace of Tycho Brahe" in *Scientific American,* Feb. 1961, p. 118.

Cohen, I. "Newton's Discovery of Gravity" in *Scientific American,* Mar. 1981, p. 166.

Gingerich, O. *The Eye of Heaven: Ptolemy, Copernicus and Kepler.* 1993, American Institute of Physics Press.

King-Hele, D. and Eberst, R. "Observing Artificial Satellites" in *Sky & Telescope,* May 1986, p. 457.

Koestler, A. *The Sleepwalkers: A History of Man's Changing Vision of the Universe.* 1959, Macmillan. A journalist's recreation of Renaissance astronomical developments.

Thoren, V. *The Lord of Uraniborg.* 1990, Cambridge U. Press. Definitive modern study of Brahe's life and work.

Wilson, C. "How Did Kepler Discover His First Two Laws?" in *Scientific American,* Mar. 1972.

As captured with a fish-eye lens aboard the Space Shuttle on December 9, 1993, the Earth hangs above the Hubble Space Telescope as it is repaired. The reddish continent is Australia, its size and shape distorted by the special lens. Because the seasons in the Southern Hemisphere are opposite ours, it is summer in Australia on this December day. *(NASA)*

3 Earth, Moon, and Sky

THINKING AHEAD

If the Earth's orbit is nearly a perfect circle (as we saw in the previous chapters), why is it hotter in summer and colder in winter in many places around the globe? And why are the seasons in Australia or Peru the opposite of those in the United States or Europe?

The story is told that Galileo, as he left the Hall of the Inquisition following his retraction of the doctrine that the Earth rotates and revolves about the Sun, said under his breath, "But nevertheless it moves." The story is perhaps apocryphal, but certainly Galileo knew that the Earth was in motion, whatever church authorities said.

It is the motions of the Earth that produce the seasons and give us our measures of time and date. The Moon's motions around us provide the concept of the month and the cycle of lunar phases. In this chapter we will examine some of the basic phenomena of our everyday world in their astronomical context.

Locating Places on the Earth

Let's begin by fixing our position on the surface of the Earth. As we discussed in Chapter 1, the Earth's axis of rotation defines the locations of its North and South Poles, and of its equator, halfway between. Two other directions are also defined by the Earth's motions: east is the direction toward which the Earth rotates, and west is its opposite. At almost any point on the Earth, the four directions—north, south, east, and west—are well defined, despite the fact that our planet is round rather than flat. The only exceptions are exactly at the North and South Poles, where the directions east and west are ambiguous (because the poles do not turn).

We can use these ideas to define a system of coordinates attached to our planet. Such a system, like the layout of streets and avenues in Manhattan or Salt Lake City, helps us find where we are or want to go. Coordinates on a sphere, however, are a little more complicated than those on a flat surface. We must define circles on the sphere that are equivalent to the rectangular grid that specifies position on a plane.

A **great circle** is any circle on the surface of a sphere whose center is at the center of the sphere. For example, the Earth's equator is a great circle on the Earth's surface, halfway between the North and South Poles. We can also imagine a series of great circles that pass through the North and South Poles. These circles are called **meridians;** they each cross the equator at right angles.

Any point on the surface of the Earth will have a meridian passing through it (Figure 3.1). This meridian specifies the east–west location, or *longitude*, of that place.

FIGURE 3.2
The Royal Greenwich Observatory in England At the internationally agreed-upon zero point of longitude, tourists can stand and straddle the exact line where longitude "begins." *(© 1993 Hal Berol/Visuals Unlimited)*

By international agreement (and it took many meetings for the world's countries to agree), your longitude is defined as the number of degrees of arc along the equator between your meridian and the one passing through Greenwich, England.

Why Greenwich, you might ask? Every country wanted 0° longitude to pass through its own capital. Greenwich, the site of the old Royal Observatory (Figure 3.2), was selected because it was between continental Europe and the United States, and because it was the site for much of the development of a way to measure longitude at sea. Longitudes are measured either to the east or the west of the Greenwich meridian from 0° to 180°. As an example, the longitude of the clock-house benchmark of the U.S. Naval Observatory in Washington, D.C., is 77.066° W.

Your *latitude* (or north–south location) is the number of degrees of arc you are away from the equator along your meridian. Latitudes are measured either north or south of the equator from 0° to 90°. As an example, the latitude of the previously mentioned Naval Observatory benchmark is 38.921° N. The latitude of the South Pole is 90° S.

Locating Places in the Sky

Positions in the sky are measured in a way that is very similar to the one we use on the surface of the Earth. Instead of latitude and longitude, astronomers use coordinates called declination and right ascension. In denoting positions of objects in the sky, it is often convenient to make use of the fictitious celestial sphere. We saw in Chapter 1

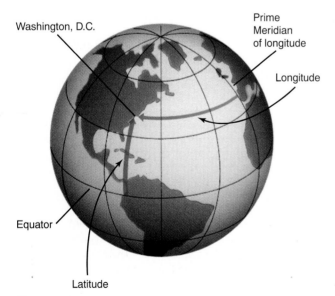

Washington, D.C.

Prime Meridian of longitude

Longitude

Equator

Latitude

FIGURE 3.1
The Latitude and Longitude of Washington, D.C.

that the sky appears to rotate about points above the North and South Poles of the Earth—the north celestial pole and the south celestial pole. Halfway between the celestial poles, and thus 90° from each, is the celestial equator, a great circle on the celestial sphere that is in the same plane as the Earth's equator. We can use these markers in the sky to set up a system of celestial coordinates.

Declination on the celestial sphere is measured the same way that latitude is measured on the sphere of the Earth: from the celestial equator toward the north (positive) or south (negative). So Polaris, the star near the north celestial pole, has a declination of almost +90°.

Right ascension (RA) is like longitude, except that instead of Greenwich, its arbitrarily chosen point of origin is the *vernal equinox,* a point in the sky where the ecliptic (the Sun's path) crosses the celestial equator. RA can be expressed in units of angle (degrees) or in units of time. This is because the celestial sphere seems to turn around the Earth once a day, and so the 360° of RA that it takes to go once around the celestial sphere can just as well be set equal to 24 hours. Then each 15° of arc is equal to 1 hour of time. The hours can be further subdivided into minutes. For example, the celestial coordinates of the bright star Vega (Appendix 11) are RA 18 h 36.2 m (= 279.05°) and declination +38.77°.

One way to visualize these circles in the sky is to imagine the Earth as a transparent sphere with the terrestrial coordinates (latitude and longitude) painted on it with dark paint. Imagine the celestial sphere around us as a giant ball, painted white on the inside. Then imagine yourself at the center of the Earth, with a bright light bulb in the middle, looking out through its transparent surface to the sky. The terrestrial poles, equator, and meridians will be projected as dark shadows on the celestial sphere, giving us the system of coordinates in the sky.

The Turning Earth

We have seen that the apparent rotation of the celestial sphere could be accounted for either by a daily rotation of the sky around a stationary Earth, or by the rotation of the Earth itself. Since the 17th century, it has been generally accepted that it is the Earth that turns, but not until the 19th century did the French physicist Jean Foucault provide a direct and unambiguous demonstration of this rotation. In 1851, he suspended a 60-m pendulum weighing about 25 kg from the domed ceiling of the Pantheon in Paris and then started the pendulum swinging evenly. If the Earth had not been turning, there would have been no force to alter the pendulum's plane of oscillation and so it would have continued tracing the same path. Yet after a few minutes it was apparent that the pendulum's plane of motion was turning. Foucault explained that it was not the pendulum that was shifting, but rather the Earth that was turning underneath it (Figure 3.3). You can now find such pendulums in many science centers and planetaria around the world.

FIGURE 3.3
The Foucault Pendulum *(© Bob Emott, Photographer)*

Can you think of other ways to prove that it is the Earth and not the sky that is turning?

 THE SEASONS

One of the fundamental facts of life at the mid-latitudes, where most of this book's readers live, is that there are significant variations in the heat we receive from the Sun over the course of the year. We thus divide the year into *seasons,* each with its different amount of sunlight. The difference between seasons gets more pronounced the farther north or south from the equator we travel, and the seasons in the Southern Hemisphere are the reverse of what we find in the northern half of the Earth. With these observed facts in mind, let us ask what causes the seasons.

According to recent surveys, most people believe that the seasons are the result of the changing distance between the Earth and the Sun. This sounds reasonable at first: It should be colder when the Earth is farther from the Sun. But the facts don't bear out this hypothesis. Although the Earth's orbit around the Sun is an ellipse, its distance from the Sun varies by only about three percent. That's not enough to cause significant variations in the Sun's heating.

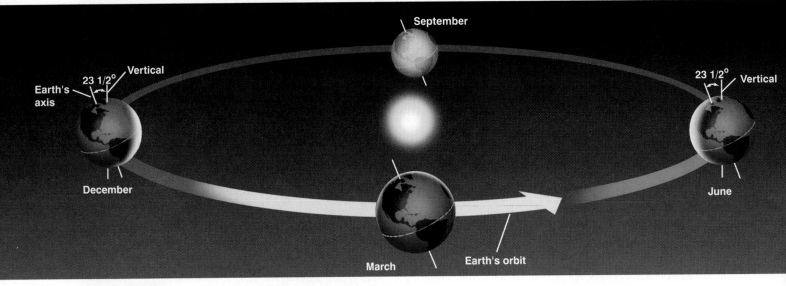

FIGURE 3.4

Seasons We see the Earth at different seasons as it circles the Sun. During our winter in the north, the Southern Hemisphere "leans into" the Sun and is illuminated more directly. In summer, it is the Northern Hemisphere that is leaning into the Sun and has longer days. In spring and autumn, the two hemispheres receive more equal shares of sunlight.

To make matters worse for people in North America who hold this hypothesis, the Earth is actually closest to the Sun in January, when the Northern Hemisphere is in the middle of winter. And if distance were the governing factor, why would the two hemispheres have opposite seasons? As we shall show, the seasons are in fact caused by the 23° tilt of the Earth's axis relative to the plane in which we circle the Sun.

The Seasons and Sunshine

Figure 3.4 shows the Earth's annual path around the Sun, with the Earth's axis tilted by 23°. Note that the axis continues to point in the same direction in the sky in the course of the year. As the Earth travels around the Sun, in June the Northern Hemisphere "leans into" the Sun and is more directly illuminated. In December, the situation is reversed: The Southern Hemisphere leans into the Sun, and the Northern Hemisphere leans away. In September and March, the Earth leans "sideways"—neither into the Sun nor away from it—and so the two hemispheres are equally favored with sunshine.

How does the Sun's favoring one hemisphere translate into making it warmer for us down on the surface of the Earth? There are two effects we need to consider. When we lean into the Sun, sunlight hits us at a more direct angle and is more effective at heating the Earth's surface (Figure 3.5). You can get a similar effect by shining a flashlight onto a wall. If you shine the flashlight straight on (if the wall "leans into" the flashlight, so to speak), you get an intense spot of light on the wall. But if you hold the flashlight at an angle (if the wall "leans out" of the beam), then the spot of

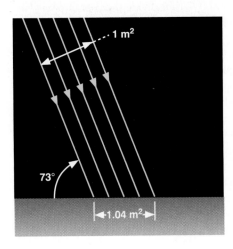
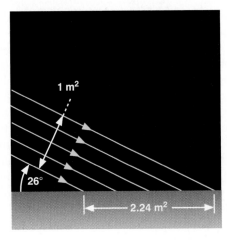

FIGURE 3.5

The Sun's Rays in Summer and Winter In summer the Sun appears high in the sky and its rays hit the Earth more directly, spreading out less. In winter the Sun is low in the sky and its rays spread out over a much wider area, becoming less effective at heating the ground.

light will be more spread out. In the same way, the sunlight in June is more direct and intense on the Northern Hemisphere, and hence more effective at heating.

The second effect has to do with the length of time the Sun spends above the horizon (Figure 3.6). Even if you've never thought about astronomy before, we're sure you have observed that the days get longer in summer and shorter in winter. Let's see why this happens.

As we saw in Chapter 1, an equivalent way to look at our path around the Sun each year is to pretend that the Sun moves around the Earth (on a circle called the ecliptic). Because the Earth's axis is tilted, the ecliptic is tilted by 23° relative to the celestial equator (review Figure 1.6), and where we see the Sun in the sky changes as the year wears on. In June, the Sun is north of the celestial equator and spends more time with those who live in the Northern Hemisphere. It rises high in the sky and is above the horizon in the United States for as much as 15 hours. Thus, the Sun not only heats us with more direct rays, it has more time to do it each day. (Notice in Figure 3.6 that our gain is the Southern Hemisphere's loss. There, the June Sun is low in the sky, meaning fewer daylight hours. In Chile, for example, June is a colder, darker time of year.) In December, when the Sun is south of the celestial equator, the situation is reversed.

Let's look at what the Sun's illumination on Earth looks like at some specific dates of the year, when these effects are at their maximum. On about June 22 (the date we who live in the Northern Hemisphere call the *summer solstice*

or sometimes the first day of summer), the Sun shines down most directly upon the Northern Hemisphere of the Earth. It appears 23° north of the equator and thus on that date passes through the zenith of places on the Earth that are at 23° N latitude. The situation is shown in detail in Figure 3.7. To a person at latitude 23° N (near Hawaii, for example), the Sun is directly overhead at noon. This latitude, where the Sun can appear at the zenith at noon on the first day of summer, is called the *Tropic of Cancer.*

We also see in Figure 3.7 that the Sun's rays shine down all around the North Pole. As the Earth turns on its axis, the North Pole will always be illuminated by the Sun; all places within 23° of the pole have sunshine for 24 hours on the first day of summer. The Sun is as far north on this date as it can get; thus, 90° − 23° (or 67° N) is the southernmost latitude where the Sun can be seen for a full 24-hour period (the "land of the midnight Sun"). That circle of latitude is called the *Arctic Circle.*

Many early cultures scheduled special events around the summer solstice to celebrate the longest days and thank their gods for making the weather warm. This required them to keep track of the lengths of the days and the northward trek of the Sun in order to know the right day for the "party." (You can do the same thing by watching for several weeks, from the same observation point, where the Sun rises or sets relative to a fixed landmark. In summer, the Sun will rise farther and farther north of east, and set farther and farther north of west, reaching the maximum around the summer solstice.)

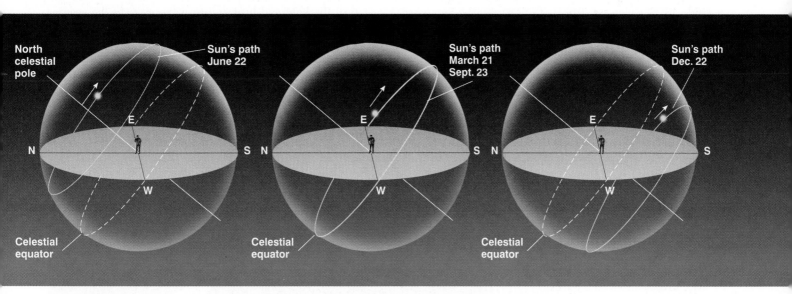

FIGURE 3.6
The Sun's Path in the Sky for Different Seasons   On June 22, the Sun rises north of east and sets north of west; for observers in the Northern Hemisphere of the Earth, it spends about 15 hours above the horizon. On December 22, the Sun rises south of east and sets south of west; it spends only 9 hours above the horizon, which means short days and long nights in northern lands (and a strong need for people to hold celebrations to cheer themselves up). On March 21 and September 21, the Sun spends an equal amount of time above and below the horizon.

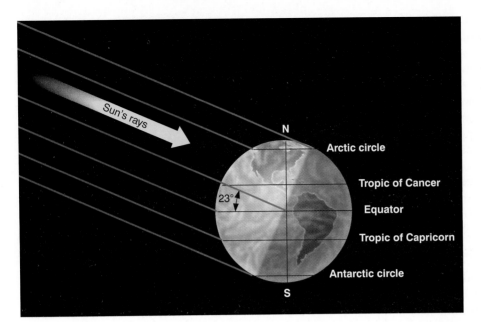

FIGURE 3.7
The Earth on June 22 This is the date of the summer solstice in the Northern Hemisphere. Note that as the Earth turns on its axis (the line connecting the North and South Poles), the North pole is in constant sunlight while the South Pole is veiled in 24-hour darkness. The Sun is at the zenith for observers on the Tropic of Cancer.

Now look at the South Pole in Figure 3.7. On June 22, all places within 23° of the South Pole—that is, south of what we call the *Antarctic Circle*—do not see the Sun at all for 24 hours. Day and night are both dark.

The situation is reversed six months later, about December 22 (the date of the *winter solstice* or the first day of winter in the Northern Hemisphere), as shown in Figure 3.8. Now it is the Arctic Circle that has the 24-hour night and the Antarctic Circle that has the midnight Sun. At latitude 23° S, called the *Tropic of Capricorn*, the Sun passes through the zenith at noon. Days are longer in the Southern Hemisphere and shorter in the north. In the United States

and southern Europe, we might get only 9 or 10 hours of sunshine during the day. It is winter in the Northern Hemisphere and summer in the Southern Hemisphere.

Every culture that developed some distance north of the equator has a celebration around December 22, to help people deal with the depressing lack of sunlight and the often dangerously cold temperatures. Originally, this was often a time for huddling with family and friends, for sharing the reserves of food and drink, for rituals asking the gods to return the light and heat and turn the cycle of the seasons around. Many cultures constructed elaborate devices for anticipating when the shortest day of the year was coming.

FIGURE 3.8
The Earth on December 22 This is the date of the winter solstice in the Northern Hemisphere. Now the North Pole is in darkness for 24 hours and the South Pole is illuminated. The Sun is at the zenith for observers on the Tropic of Capricorn and thus is low in the sky for the residents of the United States and Canada (just the right time to sacrifice a ritual turkey and huddle with your loved ones).

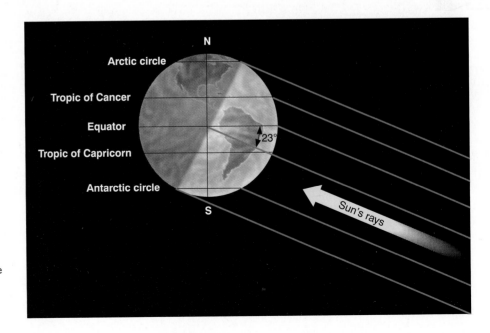

Stonehenge in England (see Section 3.4), built long before the invention of writing, is probably one such device. In our own time, we continue the winter solstice tradition with Christmas and Chanukah celebrations (although few of us now feel the need to pause while eating the sacrificial turkey to ask the sun god to make the days longer).

Halfway between the solstices, on about March 21 and September 23, the Sun is on the celestial equator. From Earth, it appears above our planet's equator, and favors neither hemisphere. Every place on the Earth then receives exactly 12 hours of sunshine and 12 hours of night. The points where the Sun crosses the celestial equator are called the *vernal* (spring) and *autumnal* (fall) *equinoxes.*

The Seasons at Different Latitudes

The seasonal effects are different at different latitudes on Earth. At the equator, for instance, all seasons are much the same. Every day of the year, the Sun is up half the time, so there are always 12 hours of sunshine and 12 hours of night. Local residents define the seasons by the amount of rain rather than by the amount of sunlight. As we travel north or south, the seasons become more pronounced, until we reach extreme cases in the Arctic and Antarctic.

At the North Pole, all celestial objects that are north of the celestial equator are always above the horizon and, as the Earth turns, circle around parallel to it. The Sun is north of the celestial equator from about March 21 to September 23, so at the North Pole, the Sun rises when it reaches the vernal equinox and sets when it reaches the autumnal equinox. Each year there are six months of sunshine at each pole, followed by six months of darkness.

Clarifications about the Real World

In our discussions so far, we have been describing the rising and setting of the Sun and stars as they would appear if the Earth had little or no atmosphere. In reality, however, the atmosphere has the curious effect of allowing us to see a little way "over the horizon." This effect is a result of *refraction,* the bending of light in a medium such as air or water, something we will discuss in Chapter 5. Because of this atmospheric refraction, the Sun appears to rise earlier and to set later than it would if no atmosphere were present.

In addition, the atmosphere scatters light and provides some twilight illumination even when the Sun is below the horizon. Astronomers define morning twilight as beginning when the Sun is 18° below the horizon, and evening twilight extends until the Sun sinks more than 18° below the horizon.

These atmospheric effects require small corrections in many of our statements about the seasons. At the equinoxes, for example, the Sun appears to be above the horizon for a few minutes longer than 12 hours, and below the horizon for less than 12 hours. These effects are most dramatic at the Earth's poles, where the Sun actually rises more than a week before it reaches the celestial equator. As

a consequence of twilight, the period of real darkness at each pole lasts for only about three months of each year, rather than six.

Also, you probably know that the summer solstice (June 22) is not the warmest day of the year, even if it is the longest. The hottest months in the Northern Hemisphere are July and August. This is because our weather involves the air and water covering the Earth's surface, and these large reservoirs do not heat up instantaneously. After all, a swimming pool does not get warm the moment the Sun rises but is warmest late in the afternoon, after it has had time to absorb the Sun's heat. In the same way, the Earth gets warmer after it has had a chance to absorb the extra sunlight that is the Sun's summer gift to us. And the coldest times of winter are a month or more after the winter solstice.

3.3 KEEPING TIME

The measurement of time is based on the rotation of the Earth. Throughout most of human history, time has been reckoned by the positions of the Sun and stars in the sky. Only recently have mechanical and electronic clocks taken over this function in regulating our lives.

The Length of the Day

The most fundamental astronomical unit of time is the day, measured in terms of the rotation of the Earth. There is, however, more than one way to define the day. Usually, it is the rotation period of the Earth with respect to the Sun, called the **solar day.** After all, for most people sunrise is more important than the rising time of Arcturus or some other star, so we set our clocks to some version of sun-time. However, astronomers also use a **sidereal day,** which is defined in terms of the rotation period of the Earth with respect to the stars.

A solar day is slightly longer than a sidereal day, because (as you can see from Figure 3.9) the Earth moves a significant distance along its path around the Sun in a day. Suppose we start the day when the Earth's orbital position is at A, with both the Sun and some distant star (located in direction C) being above an observer at point O on the Earth. When the Earth has completed one rotation with respect to the distant star, C is again above O. However, notice that because of the movement of the Earth along its orbit from A to B, the Sun has not yet reached a position above O. To complete a solar day, the Earth must rotate an additional amount, equal to 1/365 of a full turn. The time required for this extra rotation is 1/365 of a day, or about 4 min, so the solar day is about 4 minutes longer than the sidereal day.

Because our ordinary clocks are set to solar time, stars rise 4 minutes earlier each day. Astronomers prefer sidereal time for planning their observations because in that system, a star rises at the same time every day.

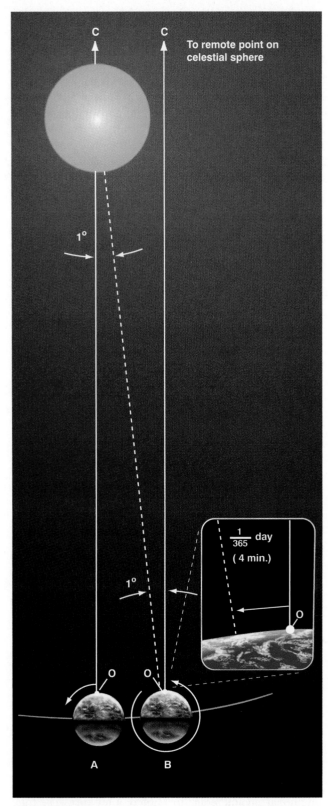

FIGURE 3.9
The Difference Between a Sidereal and a Solar Day This is a top view, looking down as the Earth orbits the Sun. Because the Earth moves around the Sun (roughly 1° per day), after one complete rotation of the Earth relative to the stars, we do not see the Sun in the same position.

Apparent Solar Time

Apparent solar time is reckoned by the actual position of the Sun in the sky (or, during the night, its position below the horizon). This is the kind of time indicated by sundials, and it probably represents the earliest measure of time used by ancient civilizations. Today we adopt the middle of the night as the starting point of the day and measure time in hours elapsed since midnight.

During the first half of the day, the Sun has not yet reached the meridian (the great circle in the sky that passes through our zenith). We designate those hours as before midday (*ante meridiem*, or A.M.). We customarily start numbering the hours after noon over again, and designate them by P.M. (*post meridiem*) to distinguish them from the morning hours.

Although apparent solar time seems simple, it is not really very convenient to use. The exact length of an apparent solar day varies slightly during the year. The eastward progress of the Sun in its annual journey around the sky is not uniform because the speed of the Earth varies slightly in its elliptical orbit. Another reason is that the Earth's axis of rotation is not perpendicular to the plane of its revolution. Thus, apparent solar time does not advance at a uniform rate. After the invention of clocks that ran at a uniform rate, it became necessary to abandon the apparent solar day as the fundamental unit of time.

Mean Solar Time and Standard Time

Mean solar time is based on the average value of the solar day over the course of the year. A mean solar day contains exactly 24 hours, and it is what we use in our everyday timekeeping. Although mean solar time has the advantage of progressing at a uniform rate, it is still inconvenient for practical use because it is determined by the position of the Sun. For example, noon occurs when the Sun is overhead. But because we live on a round Earth, the exact time of noon is different as you change your longitude by moving east or west.

If mean solar time were strictly observed, people traveling east or west would have to reset their watches continually as the longitude changed, just to read the local mean time correctly. For instance, a commuter traveling from Oyster Bay on Long Island to New York City would have to adjust the time on the trip through the East River tunnel, because Oyster Bay time is actually about 1.6 min more advanced than that of Manhattan. (Imagine an airplane trip in which the flight attendant gets on the intercom every minute, saying, "Please reset your watch for local mean time.")

Until near the end of the last century, every city and town in the United States did keep its own local mean time. With the development of railroads and the telegraph, however, the need for some kind of standardization became evident. In 1883, the nation was divided into four standard time zones (now five, including Hawaii and Alaska). Within each zone, all places keep the same *standard time,* with the local mean solar time of a standard line of longitude running more or less through the middle of each zone. Now travelers reset their watches only when the time change has amounted to a full hour. Pacific standard time is 3 hours earlier than eastern standard time, a fact that becomes painfully obvious in California when someone on the East Coast forgets and calls you at 5 A.M.

For local convenience, the boundaries between the U.S. time zones are chosen to correspond to divisions between states. Since 1884, standard time has been in use around the world by international agreement (with 24 time zones circling the globe). Almost all countries have adopted one or more standard time zones, although one of the largest nations, India, has settled on a half-zone, being 5½ from Greenwich standard.

Daylight saving time is simply the local standard time of the place plus one hour. It has been adopted for spring and summer use in most states in the United States, as well as in many other countries, to prolong the sunlight into evening hours, on the apparent theory that it is easier to change the time by government action than it would be for individuals or businesses to adjust their own schedules to produce the same effect. It does not, of course, "save" any daylight at all—because the amount of sunlight is determined by the seasonal effects we discussed in Section 3.2 and not by what we do with our clocks.

The International Date Line

The fact that time is always more advanced to the east presents a problem. Suppose you travel eastward around the world. You pass into a new time zone, on the average, about every 15° of longitude you travel, and each time you dutifully set your watch ahead an hour. By the time you have completed your trip, you have set your watch ahead through a full 24 hours and thus gained a day over those who stayed at home.

The solution to this dilemma is the international date line, set by international agreement to run approximately along the 180° meridian of longitude. The date line runs about down the middle of the Pacific Ocean, although it jogs a bit in a few places to avoid cutting through groups of islands and through Alaska (Figure 3.10). By convention, at the date line the date of the calendar is changed by one day. Crossing the date line from west to east, thus advancing your time, you compensate by decreasing the date; crossing from east to west, you increase the date by one day. To maintain our planet on a rational system of time keeping, we simply must accept that the date will differ in different cities at the same time. A well-known example is the date when the Imperial Japanese Navy bombed Pearl Harbor in Hawaii, known in the United States as Sunday, December 7, 1941, but taught to Japanese students as Monday, December 8.

3.4 THE CALENDAR

The Challenge of the Calendar

"What's today's date?" is one of the most common questions you can ask (usually when writing a check, or worrying about the next exam). Long before the era of digital watches that tell the date, people used calendars to help measure the passage of time.

There are two traditional functions of any calendar. First, it must keep track of time over the course of longer spans, allowing people to anticipate the cycle of the seasons and to honor special religious or personal anniversaries. Second, to be useful to a large number of people, a calendar must use natural time intervals that everyone can agree on—those defined by the motions of the Earth, Moon, and sometimes even the planets. The natural units of our calendar are the *day,* based on the period of rotation of the Earth; the *month,* based on the period of revolution of

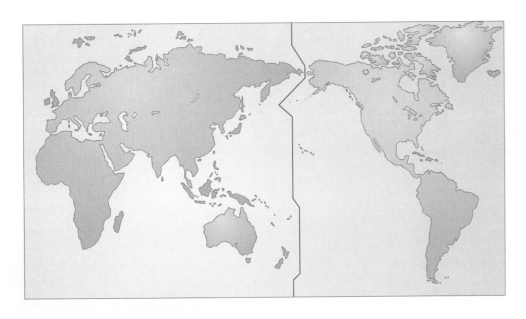

FIGURE 3.10
Where the Date Changes
The international date line is an arbitrarily drawn line on the Earth where the date changes. So that neighbors do not have different days, the line is located where the Earth's surface is mostly water.

the Moon about the Earth; and the *year*, based on the period of revolution of the Earth about the Sun. Difficulties have resulted from the fact that these three periods are not commensurable—that is, one does not divide evenly into any of the others.

The rotation period of the Earth is, by definition, 1.0000 day. The period required by the Moon to complete its cycle of phases, called the lunar month, is 29.5306 days. The basic period of revolution of the Earth, called the tropical year, is 365.2422 days. The ratios of these numbers are not very convenient for calculations. In other words, it turns out that our celestial clocks are pretty incompatible. This is the historic challenge of the calendar, dealt with in various ways by different cultures.

Early Calendars

Even the earliest cultures were concerned with the keeping of time and the calendar. Particularly interesting are monuments left by Bronze Age people in northwestern Europe, especially the British Isles. The best preserved of the monuments is Stonehenge (Figure 3.11), about 13 km from Salisbury in southwest England. It is a complex array of stones, ditches, and holes arranged in concentric circles. Carbon dating and other studies show that Stonehenge was built during three periods ranging from about 2800 B.C. to 1500 B.C. Some of the stones are aligned with the directions of the Sun and Moon during their risings and settings at critical times of the year (such as the summer and winter solstices), and it is generally believed that at least one function of the monument was connected with the keeping of a calendar.

The Maya in Central America, who thrived between the 2nd and 10th centuries A.D., were also concerned with the keeping of time. Their calendar was as sophisticated as, and perhaps more complex than, contemporary calendars in Europe. The Maya did not attempt to correlate their calendar accurately with the length of the year or lunar month. Rather, their calendar was a system for keeping track of the passage of days and for counting time far into the past or future. Among other purposes, it was useful for predicting astronomical events—for example, the positions of Venus in the sky (Figure 3.12).

The ancient Chinese developed an especially complex calendar, largely limited to a few privileged hereditary court astronomer–astrologers. In addition to the motions of the Earth and Moon, they were able to fit in the approximately 12-year cycle of Jupiter, which was central to their system of astrology. The Chinese still preserve some aspects of this system in their cycle of 12 "years"—the Year of the Dragon, the Year of the Pig, and so on—that are defined by the position of Jupiter in the zodiac.

Our Western calendar derives from Greek calendars dating from at least the 8th century B.C. They led, eventually, to the *Julian calendar*, introduced by Julius Caesar, which approximated the year at 365.25 days, fairly close to the actual value of 365.2422. The Romans achieved this approximation by declaring years to have 365 days each, with the exception of every fourth year. The *leap year* was to have one extra day, bringing its length to 366 days and thus making the average length of the year in the Julian calendar 365.25 days.

The Romans had dropped the almost impossible task of trying to base their calendar on the Moon as well as the Sun, although a vestige of older lunar systems can be seen

FIGURE 3.11

Part of Stonehenge This ancient monument was built between 2800 and 1500 B.C., and used to keep track of the motions of the Sun and Moon. Today, heedless tourists and vandals have disturbed and chipped away at the stones to such a degree that the site is now fenced in and entry is restricted.
(David Morrison)

in the fact that our months have an average length of about 30 days. However, lunar calendars remained in use in other cultures, and Islamic calendars are still primarily lunar rather than solar.

The Gregorian Calendar

Although the Julian calendar (which was adopted by the early Christian Church) represented a great advance, its average year still differed from the true year by about 11 minutes, an amount that accumulates over the centuries to an appreciable error. By 1582, that 11 minutes per year had added up to the point where the first day of spring was occurring on March 11, instead of March 21. If the trend were allowed to continue, eventually the Christian celebration of Easter would be occurring in early winter. Pope Gregory XIII, a contemporary of Galileo, felt it necessary to institute further calendar reform.

The Gregorian calendar reform consisted of two steps. First, ten days had to be dropped out of the calendar to bring the vernal equinox back to March 21; by proclamation, the day following October 4, 1582, became October 15. The second feature of the new Gregorian calendar was a change in the rule for leap year, making the average length of the year more closely approximate the tropical year. Gregory decreed that three of every four century years, all leap years under the Julian calendar, would be common years henceforth. The rule was that only century years divisible by 400 would be leap years. Thus, 1700, 1800, and 1900—all divisible by 4 but not by 400—were not leap years in the Gregorian calendar. On the other hand, the years 1600 and 2000, both divisible by 400, were leap years. The average length of this Gregorian year, 365.2425 mean solar days, is correct to about one day in 3300 years.

The Catholic countries immediately put the Gregorian reform into effect, but countries under control of the Eastern Church and most Protestant countries did not adopt it until much later. It was 1752 when England and the American colonies finally made the change. By parliamentary decree, September 2, 1752, was followed by September 14. Although special laws were passed to prevent such abuses as landlords collecting a full month's rent for September, there were still riots, and people demanded their 12 days back. Russia did not abandon the Julian calendar until the time of the Bolshevik revolution. The Russians then had to omit 13 days to come into step with the rest of the world.

3.5 PHASES AND MOTIONS OF THE MOON

After the Sun, the Moon is the brightest and most obvious object in the sky. Unlike the Sun, it does not shine under its own power, but merely glows with reflected sunlight. If you

FIGURE 3.12
Ruins of the Caracol This Mayan observatory at Chichén Itza in the Yucatan, Mexico, dates from around 1000 A.D. *(David Morrison)*

were to follow its progress in the sky for a month, you would observe a cycle of *phases*, with the Moon starting out dark and getting more and more illuminated by sunlight over the course of about two weeks. After the Moon's disk becomes fully visible, it then begins to fade, returning to dark about two weeks later. These changes fascinated and mystified many early cultures, which came up with marvelous stories and legends to explain the cycle of the Moon. Even in the modern world, many people don't understand what causes the phases, thinking that they are somehow related to the shadow of the Earth. Let us see how the phases can be explained by the motion of the Moon relative to the bright light source in the solar system, the Sun.

Lunar Phases

Although we know the Sun moves 1/12 of its path around the sky each month, for purposes of explaining the phases, we can assume that the Sun's light comes from roughly the same direction during the course of a four-week lunar cycle. The Moon, on the other hand, moves completely around the Earth in that time. As we watch the Moon from our vantage point on Earth, how much of its face we see illuminated by sunlight depends on the angle the Sun makes with the Moon.

Here is a simple experiment to show you what we mean: Stand about 6 ft in front of a bright electric light in a completely dark room (or outdoors at night) and hold in your hand a small round object such as a tennis ball or an orange. Your head can then represent the Earth, the light the Sun, and the ball the Moon. Move the ball around your head (making sure you don't cause an eclipse by blocking the light with your head). You will see phases just like

Astronomy and the Days of the Week

The week seems independent of celestial motions, although its length may have been based on the time between quarter phases of the Moon. In Western culture, the seven days of the week are named after the seven "wanderers" that the ancients saw in the sky: the Sun, Moon, and five planets visible to the unaided eye (Mercury, Venus, Mars, Jupiter, and Saturn).

In English, we can easily recognize Sun-day, Moon-day, and Saturn-day, but the other days are named after the Norse equivalents of the Roman gods that gave their names to the planets. In languages more directly related to Latin, the correspondences are clearer. Wednesday, Mercury's day, for example, is Mercoledi in Italian, Mercredi in French, and Miercoles in Spanish. Mars gives its name to Tuesday (Martes in Spanish), Jupiter or Jove to Thursday (Giovedi in Italian), and Venus to Friday (Vendredi in French).

There is no reason that the week has to have seven days rather than five or eight. It is interesting to speculate that if we had lived in a planetary system where fewer planets were visible, we might well have had a shorter week.

---

those of the Moon on the ball. (Another good way to get acquainted with the phases and motions of the Moon is to follow our satellite in the sky for a month or two, recording its shape, its direction from the Sun, and when it rises and sets.)

Let's examine the Moon's cycle of phases using Figure 3.13, which depicts the Moon's behavior for the entire "moonth" (month). The trick to this figure is that you must imagine yourself standing on the Earth, *facing* the Moon in each of its phases. So for the position labeled *A,* you are on the right side of the Earth and it's the middle of the day; for position *E,* you are on the left side of the Earth in the middle of the night. Note that in every position on Figure 3.13, the Moon is half illuminated and half dark (as a ball in sunlight should be). The difference at each position has to do with what part of the Moon faces Earth.

The Moon is said to be *new* when it is in the same general direction in the sky as the Sun (position *A*). Here its illuminated (bright) side is turned away from us and its dark side is turned toward us. In this phase it is invisible; its dark, rocky surface does not give off any light of its own. Because the new moon is in the same part of the sky as the Sun, it rises at sunrise and sets at sunset.

But the Moon does not remain in this phase long, because it moves 12° eastward each day in its monthly path around us. A day or two after the new phase, the thin *crescent* first appears, as we begin to see a small part of the Moon's illuminated hemisphere. It has moved into a position where it now reflects a little sunlight toward us on one side. The bright crescent increases in size on successive days as the Moon moves farther and farther around the sky away from the direction of the Sun (position *B*). Because the Moon is moving eastward away from the Sun, it rises later and later each day, like a student during summer vacation.

After about one week, the Moon is one-quarter of the way around its orbit (position *C*) and so we say it is at the *first quarter* phase. Now half of the Moon's illuminated side is visible to Earth observers. Because of its eastward motion, the Moon now lags about one quarter of the day behind the Sun, rising roughly around noon and setting roughly around midnight.

During the week after the first quarter phase, we see more and more of the Moon's illuminated hemisphere (position *D*), a phase that is called *waxing* (or growing) *gibbous*. Eventually, the Moon arrives at position *E* in our figure, where it and the Sun are opposite each other in the sky. The side of the Moon turned toward the Sun is turned toward the Earth, and we have *full* phase.

When the Moon is full, it lags half an orbit and half a day behind the Sun (and so they are opposite each other in the sky). This means the Moon does the opposite of what the Sun does, rising at sunset and setting at sunrise. Note what that means in practice. The completely illuminated (and thus very noticeable) Moon rises just as it gets dark, remains in the sky all night long and sets as the Sun's first rays are seen at dawn. And when is it highest in the sky and most noticeable? At midnight, a time made famous in generations of horror novels and films. (Note how the behavior of a vampire like Dracula parallels the behavior of the full Moon: Dracula rises at sunset, does his worst mischief at midnight, and must be in his coffin by sunrise. The old legends were a way of personifying the behavior of the full Moon, which was a much more dramatic part of people's lives in the days before city lights and television.)

Folk wisdom has it that more crazy behavior is seen during the time of the full moon (the Moon even gives a name to crazy behavior—"lunacy"). But, in fact, statistical tests of this "hypothesis," involving thousands of records from hospital emergency rooms and police files, do not reveal any correlation with the phases of the Moon. As many homicides occur during the new moon or the crescent moon as during the full moon. Most investigators believe that the real story is not that more crazy behavior happens

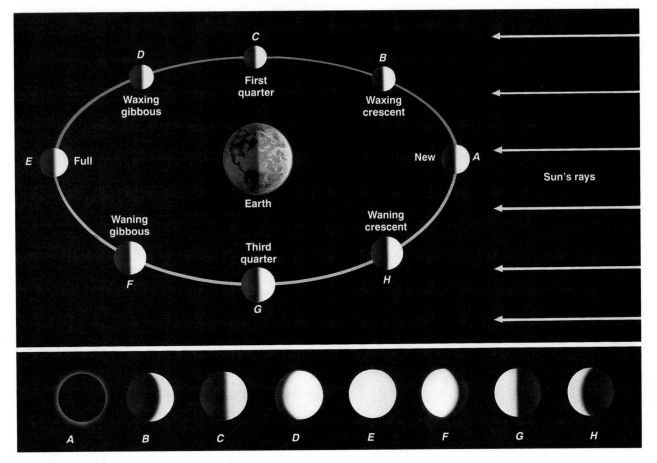

FIGURE 3.13

The Phases of the Moon   The appearance of the Moon during the course of a complete monthly cycle. The upper part shows a perspective from space, with the Sun off to the right in a fixed position. Imagine yourself standing on the Earth, facing the Moon in each part of its orbit around the Earth. In position *A*, for example, you would face the Moon from the right side of the Earth, in the middle of the day. The strip below shows the appearance of the Moon from Earth as you would see it from each lettered position. (Please note that the distance of the Moon from the Earth is not to scale in this diagram: The Moon is roughly 30 Earth diameters away from us, but we would have needed a big expensive foldout to show you the diagram to scale!)

on nights with a full moon, but rather that we are more likely to notice such behavior with the aid of a bright celestial light that is up all night long.

During the two weeks following the full moon, the Moon goes through the same phases again in reverse order (points *F*, *G*, and *H* in Figure 3.13), returning to new phase after about 29.5 days. About a week after full moon, for example, the Moon is now at *third quarter*—meaning that it is three quarters of the way around, not that it is three-fourths illuminated. In fact, half of the visible side of the Moon is again dark. At this phase, the Moon is now rising around midnight and setting around noon.

Note that there is one thing quite misleading about Figure 3.13. If you look at the Moon in position *E*, although it is full in theory, it appears as if its illumination would in fact be blocked by a big fat Earth, and hence we

would not see anything on the Moon except the Earth's shadow. In reality, the Moon is nowhere near as close to the Earth (nor is its path so identical with the Sun's in the sky) as this diagram (and the diagrams in every school textbook) might lead you to believe. The Moon is actually *30 Earth diameters* away from us; in the Prologue you will find a diagram that shows the two bodies to scale. In reality, the Earth's shadow misses the Moon most months, and we regularly get treated to a full moon. The times when the Earth's shadow does fall on the Moon are called lunar eclipses and will be discussed in Section 3.7.

You can see from this tour of the Moon's phases and times of rising and setting that the writer of the old song that went "I've got the Sun in the morning, and the Moon at night" did not quite remember his college astronomy course. It is just during the full phase that the Moon is up only at night. At other times of the month, it may be visible all morning (third quarter) or all afternoon (first quarter) in the daytime sky.

The Moon's Revolution and Rotation

The Moon's sidereal period—that is, the period of its revolution about the Earth measured with respect to the stars—is a little over 27 days: 27.3217 days to be exact. The time interval in which the phases repeat—say from full to full—is 29.5306 days. The difference is again the fault of the Earth's motion around the Sun. The Moon must make more than a complete turn around the moving Earth to get back to the same phase with respect to the Sun. As we saw, the Moon changes its position on the celestial sphere rather rapidly: even during a single evening, the Moon creeps visibly eastward among the stars, traveling its own width in a little less than 1 hour. The delay in moonrise from one day to the next, caused by this eastward motion, averages about 50 minutes.

The Moon *rotates* on its axis in exactly the same time that it takes to *revolve* about the Earth. As a consequence (as shown in Figure 3.14), the Moon always keeps the same face turned toward the Earth. You can try this yourself by "orbiting" your roommate or another volunteer. Start by facing your roommate. If you make one rotation (spin) with your shoulders in the exact same time that you revolve

around him or her, you will continue to face your roommate during the whole "orbit."

The differences in the Moon's appearance from one night to the next are due to changing illumination by the Sun, not to its own rotation. You sometimes hear the back side of the Moon (the side we never see) called the "dark side." This is a misunderstanding of the real situation: Which side is light and which is dark changes as the Moon moves around the Earth. The back side is dark no more frequently than the front side. Since the Moon rotates, the Sun rises and sets on all sides of the Moon.

3.6 OCEAN TIDES AND THE MOON

Anyone living near the sea is familiar with the twice-daily rising and falling of the tides. Early in history it was clear that tides must be related to the Moon, because the daily delay in high tide is the same as the daily delay in the Moon's rising. A satisfactory explanation of the tides, however, awaited the theory of gravity, supplied by Newton.

The Pull of the Moon on the Earth

The gravitational forces exerted by the Moon at several points on the Earth are illustrated in Figure 3.15. These

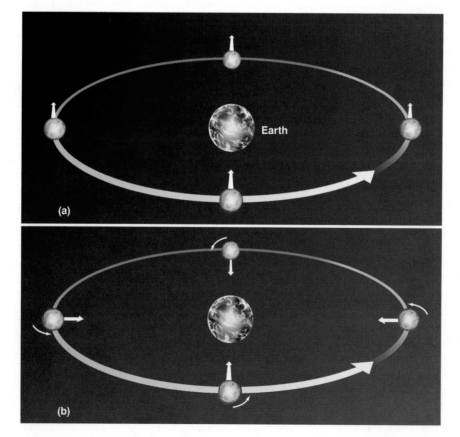

FIGURE 3.14
The Moon Without and with Rotation
In this figure we stuck a white arrow into a fixed point on the Moon to keep track of its sides. (a) If the Moon did not rotate as it orbited the Earth, it would present all of its sides to our view; hence the white arrow would only point directly toward the Earth in the bottom position on the diagram. (b) Actually, the Moon rotates in the same period that it revolves, so we always see the same side (the white arrow keeps pointing to the Earth).

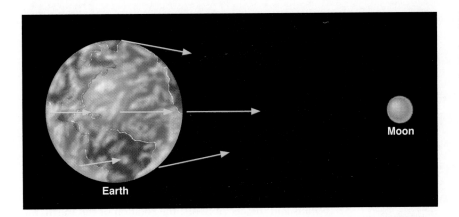

forces differ slightly from one another because the Earth is not a point, but has a certain size: All parts are not equally distant from the Moon, nor are they all in exactly the same direction from the Moon.

Moreover, the Earth is not perfectly rigid. As a result, the differences among the forces of the Moon's attraction on different parts of the Earth (called differential forces) cause the Earth to distort slightly. The side of the Earth nearest the Moon is attracted toward the Moon more strongly than is the center of the Earth, which in turn is attracted more strongly than is the side opposite the Moon. Thus, the differential forces tend to stretch the Earth slightly into a *prolate spheroid* (a football shape), with its long diameter pointed toward the Moon.

If the Earth were made of water, it would distort until the Moon's differential forces over different parts of its surface came into balance with the Earth's own gravitational forces pulling it together. Calculations show that in this case the Earth would distort from a sphere by amounts ranging up to nearly 1 m. Measurements of the actual deformation of the Earth show that the solid Earth *does* distort, but only about one third as much as water would, because of the great rigidity of the Earth's interior.

Because the tidal distortion of the solid Earth amounts at its greatest to only about 20 cm, the Earth does not distort enough to balance the Moon's differential forces with its own gravity. Hence, objects at the Earth's surface experience tiny horizontal tugs, tending to make them slide about. These *tide-raising forces* are too insignificant to affect solid objects like astronomy students or rocks in the Earth's crust, but they do affect the waters in the oceans.

The Formation of Tides

The tide-raising forces, acting over a number of hours, produce motions of the water that result in measurable tidal bulges in the oceans. Water on the side of the Earth facing the Moon is drawn toward it, piling up to greater depths there, with the greatest depths at the point just below the Moon. On the side of the Earth opposite the Moon, water flows to produce a tidal bulge there as well (Figure 3.16).

Note that the tidal bulges in the oceans do not result from the Moon's compressing or expanding the water, nor from the Moon's lifting the water "away from the Earth." Rather, they result from an actual flow of water over the Earth's surface toward the regions below and opposite the Moon, causing the water to pile up to greater depths at those places (Figure 3.17).

In the idealized (and, as we shall see, oversimplified) model just described, the height of the tides would be only a few feet. The rotation of the Earth would carry an observer at any given place alternately into regions of deeper and shallower water. An observer being carried toward the regions under or opposite the Moon, where the water was deepest, would say, "The tide is coming in"; when carried away from those regions, the observer would say, "The tide is going out." During a day, the observer would be carried through two tidal bulges (one on each side of the Earth) and so would experience two high tides and two low tides.

The Sun also produces tides on the Earth, although it is less than half as effective as the Moon at tide raising. The actual tides we experience are a combination of the larger effect of the Moon and the smaller effect of the Sun. When the Sun and Moon are lined up (at new moon or full

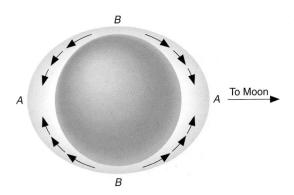

FIGURE 3.16
Tidal Bulges in an "Ideal" Ocean

FIGURE 3.17
Minas Basin Seen at Low and High Tide *(Courtesy Nova Scotia Tourism)*

moon), the tides produced reinforce each other and so are greater than normal (Figure 3.18). These are called *spring tides* (the name is connected not to the season, but to the idea that higher tides "spring up"). Spring tides are

approximately the same, whether the Sun and Moon are on the same or opposite sides of the Earth, because tidal bulges occur on both sides. When the Moon is at first quarter or last quarter (at right angles to the Sun's direction), the tides produced by the Sun partially cancel the tides of the Moon, making them lower than usual. These are called *neap tides*.

The "simple" theory of tides, described in the preceding paragraphs, would be sufficient if the Earth rotated very slowly and were completely surrounded by very deep oceans. However, the presence of land masses stopping the flow of water, the friction in the oceans and between oceans and the ocean floors, the rotation of the Earth, the wind, the variable depth of the ocean, and so on, all complicate the picture. This is why in the real world, some places have very small tides while in other places huge tides become tourist attractions.

3.7 ECLIPSES OF THE SUN AND MOON

One of the fortunate coincidences of living on Earth at the present time is that the two most prominent astronomical objects, the Sun and the Moon, have nearly the same apparent size in the sky. Although the Sun is about 400 times larger in diameter than is the Moon, it is also about 400 times farther away, so both the Sun and the Moon have the same angular size—about $1/2°$. As a result, the Moon, as seen from the Earth, can appear to cover the Sun, producing one of the most impressive events in nature.

Any solid object in the solar system casts a shadow by blocking the light of the Sun from a region behind it. This shadow becomes apparent whenever another object moves into it. In general, an **eclipse** occurs whenever any part of either the Earth or the Moon enters the shadow of the

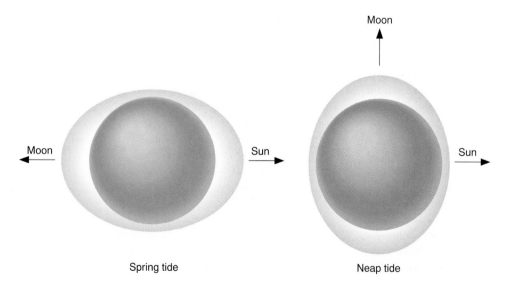

Moon ← | Sun → | Sun →

Spring tide   Neap tide

FIGURE 3.18
Tides Caused by Different Alignments of the Sun and Moon (a) In spring tides, the Sun's and Moon's pulls reinforce each other. (b) In neap tides, the Sun and the Moon pull at right angles to each other and the resulting tides are lower than usual.

George Darwin and the Slowing of the Earth

The rubbing of water over the face of the Earth involves an enormous amount of energy. Over long periods of time, the friction of the tides is actually slowing down the rotation of the Earth. Our day gets longer by about 0.002 seconds each century. That seems very small, but such tiny changes can add up over millions and billions of years.

Although the Earth's spin is slowing down, the angular momentum

G. Darwin   *(© Royal Society)*

(see Chapter 2) in a system such as the Earth–Moon system cannot change. Thus some other spin motion must speed up to take the extra angular momentum. The details of what happens were worked out over a century ago by George Darwin (1845–1912), the son

———————————————

"Darwin calculated that the Moon will slowly spiral outward, away from the Earth."

———————————————

of naturalist Charles Darwin. George Darwin had a strong interest in science but studied law for six years and was admitted to the bar. However, he never practiced law, returning to science instead and eventually becoming a professor at Cambridge University. He was a protégé of Lord Kelvin, one of the

great physicists of the 19th century, and became interested in the long-term evolution of the solar system. He specialized in making detailed (and difficult) mathematical calculations of how orbits and motions will change over geologic time.

What Darwin calculated for the Earth–Moon system was that the Moon will slowly spiral outward, away from the Earth. As it moves farther away, it will orbit less quickly (just as planets farther from the Sun move more slowly in their orbits). Thus, the month will get longer. Also, because the Moon will be more distant, total eclipses of the Sun (see Section 3.7) will no longer be visible from Earth.

The day and the month will both continue to get longer, although bear in mind that the effects are very gradual. The calculations show that ultimately— billions of years in the future—the day and the month will be the same length (about 47 of our present days), and the Moon will be stationary in the sky over the same spot on the Earth.

---

other. When the Moon's shadow strikes the Earth, people within that shadow see the Sun at least partially covered by the Moon; that is, they witness a **solar eclipse.** When the Moon passes into the shadow of the Earth, people on the night side of the Earth see the Moon darken in what is called a **lunar eclipse.**

The shadows of the Earth and the Moon consist of two parts: a dark cone where the shadow is darkest, called the *umbra,* and a lighter, more diffuse region of darkness called the *penumbra.* As you can imagine, the most spectacular eclipses occur when a body enters the umbra. Figure 3.19 illustrates the appearance of the Moon's shadow and what the Sun and Moon would look like from different points within the shadow.

If the path of the Moon in the sky were identical to the path of the Sun (the ecliptic), we might expect to see an eclipse of the Sun and the Moon each month, whenever the Moon got in front of the Sun or into the shadow of the Earth. However, the Moon's orbit is tilted relative to the plane of the Sun's orbit by about 5° (imagine two hula hoops with a common center, but tilted a bit). As a result, most months the Moon is sufficiently above or below the

Sun to avoid an eclipse. But when the two paths cross (twice a year), it is then "eclipse season" and eclipses are possible.

Eclipses of the Sun

The apparent or angular sizes of both Sun and Moon vary slightly from time to time, as their distances from the Earth vary. (Figure 3.19 shows the distance of the observer varying at points A through D, but the idea is the same.) Much of the time, the Moon looks slightly smaller than the Sun and cannot cover it completely, even if the two are perfectly aligned. However, if an eclipse of the Sun occurs when the Moon is somewhat nearer than its average distance, the Moon can completely hide the Sun, producing a total solar eclipse. A total eclipse of the Sun occurs at those times when the umbra of the Moon's shadow reaches the surface of the Earth.

The geometry of a total solar eclipse is illustrated in Figure 3.20. If Sun and Moon are properly aligned, then the Moon's darkest shadow intersects the ground at a small

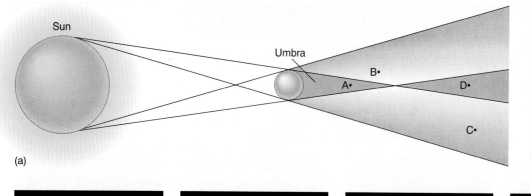

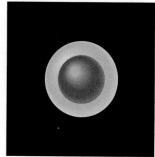

(a)

Umbra

B•

A•

D•

C•

FIGURE 3.19
Explaining Solar Eclipses
(a) The shadow cast by a spherical body (the Moon, for example). Notice the dark umbra and the lighter penumbra. Four points in the shadow are labeled with letters. (b) What the Sun and Moon would look like in the sky at the four labeled points.

(b)

point on the Earth's surface. Anyone on the Earth within the small area covered by the tip of the Moon's shadow will, for a few minutes, be unable to see the Sun and will witness a total eclipse. At the same time, observers on a larger area of the Earth's surface who are in the penumbra will see only a part of the Sun eclipsed by the Moon: We call this a partial solar eclipse.

As the Moon moves eastward in its orbit, the tip of its shadow sweeps eastward at about 1500 km/h along a thin band across the surface of the Earth; the total solar eclipse progresses along this band. The thin zone across the Earth within which a total solar eclipse is visible (weather permitting) is called the *eclipse path*. Within a region about 3000 km on either side of the eclipse path, a partial solar eclipse is visible. It does not take long for the Moon's shadow to sweep past a given point on Earth. The duration of totality may be only a brief instant; it can never exceed about 7 minutes.

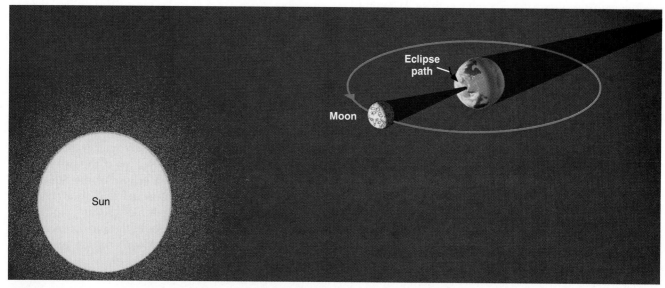

Eclipse path

Moon

Sun

FIGURE 3.20
Geometry of a Total Solar Eclipse Note that our diagram is not to scale. The Moon blocks the Sun as seen from some parts of the Earth and casts a shadow on our planet.

Because a total eclipse of the Sun is so spectacular, it is well worth trying to see one if you can. There are people whose hobby is "eclipse chasing" and who brag about how many they have seen in their lifetimes. Because much of the Earth's surface is water, eclipse chasing can involve lengthy boat trips (and often requires air travel as well). As a result, eclipse chasing is rarely within the budget of a typical college student. Nevertheless, a list of future eclipses is given for your reference in Appendix 9, just in case you hit it rich early.

Appearance of a Total Eclipse

What can you see if you are lucky enough to catch a total eclipse? A solar eclipse starts when the Moon just begins to silhouette itself against the edge of the Sun's disk. A partial phase follows, during which more and more of the Sun is covered by the Moon. About an hour after the eclipse begins, the Sun becomes completely hidden behind the Moon. In the few minutes immediately before this period of totality begins, the sky noticeably darkens, some flowers close up, and chickens may go to roost. As an eerie twilight suddenly descends during the day; other animals (and people) may get disoriented.

In the last instant before totality, the only parts of the Sun that are visible are those that shine through the lower valleys in the Moon's irregular profile and line up along the periphery of the advancing edge of the Moon—a phenomenon called *Baily's beads*, after the British astronomer who first described them. A final flash of sunlight through a lunar valley produces a brilliant flare on the disappearing crescent of the Sun: Eclipse watchers call this "the diamond ring." During totality the sky is dark enough that planets become visible in the sky, and usually the brighter stars do as well.

As Baily's beads disappear and the bright disk of the Sun becomes entirely hidden behind the Moon, the Sun's remarkable **corona** flashes into view (Figure 3.21). The corona is the Sun's outer atmosphere, consisting of sparse gases that extend for millions of miles in all directions from the apparent surface of the Sun. It is ordinarily not visible because the light of the corona is feeble compared with that from the underlying layers of the Sun. Only when the brilliant glare from the Sun's visible disk is blotted out by

SEEING FOR YOURSELF

How to Observe Solar Eclipses

It is extremely dangerous to look at the Sun: Even a brief exposure can damage your eyes. Normally, few sane people are tempted to do this, because it is painful (and something your mother told you never to do!). But during the course of an eclipse, the temptation to take a look is strong. Don't give in. The fact that the Moon is covering part of the Sun doesn't make the uncovered part any less dangerous to look at. Still, there are perfectly safe ways to follow the course of a solar eclipse, if you are lucky enough to be in the path of the shadow.

The easiest technique is to make a pinhole projector. Take a piece of cardboard with a small (1-mm) hole punched in it and hold it several feet above a light surface, such as a concrete sidewalk or a white sheet of paper, so that the hole is "aimed" at the Sun. The hole produces a fuzzy but adequate image of the eclipsed Sun (see diagram).

You can also cover up all but about a dime's worth of a hand mirror and use the small mirror surface that remains to project an image of the eclipsed Sun onto a wall. This takes a little more practice.

Although there are safe filters for looking at the Sun directly, people have suffered eye damage by looking through improper filters (or no filter at all). For example, neutral-density photographic filters are not safe, for they transmit infrared radiation that can cause severe damage to the retina. Also unsafe are smoked glass, completely exposed color film, and many other homemade filters.

It is safe to look at the Sun directly when it is *totally* eclipsed, even through binoculars or telescopes. Unfortunately the total phase, as we discussed, is all too brief. But if you know when it is coming (and going), be sure you look, for it's an unforgettably beautiful sight. And, despite the ancient folklore that presents them as dangerous times to be outdoors, the partial phases of eclipses—as long as you are not looking directly at the Sun—are certainly not any more dangerous than being out in sunlight.

During past eclipses, unnecessary panic has been created by uninformed public officials acting with the best intentions. There were two marvelous total eclipses in Australia in the 20th century during which townspeople held newspapers over their heads for protection and schoolchildren cowered indoors with their heads under their desks. What a pity that all those people missed what would have been one of the most memorable experiences of their lifetimes!

> "It is safe to look at the Sun directly when it is *totally* eclipsed, even through binoculars or telescopes."

How to watch a solar eclipse safely during its partial phases using a pinhole projector.

FIGURE 3.21

Sun's Corona The corona (thin outer atmosphere) of the Sun, visible during the July 11, 1991, total solar eclipse, photographed from near La Paz, Mexico, by a dedicated eclipse chaser. *(©1991 Stephen J. Edberg)*

the Moon during a total eclipse is the pearly white corona visible.

The total phase of the eclipse ends, as abruptly as it began, when the Moon begins to uncover the Sun. Gradually the partial phases of the eclipse repeat themselves, in reverse order, until the Moon has completely uncovered the Sun.

Eclipses of the Moon

A lunar eclipse occurs when the Moon enters the shadow of the Earth. The geometry of a lunar eclipse is shown in Figure 3.22. The Earth's dark shadow is about 1.4 million km long, so at the Moon's distance (an average of 384,000 km) it could cover about four full moons. Unlike a solar eclipse, which is visible only in certain local areas on the Earth, a lunar eclipse is visible to everyone who can see the Moon. Because a lunar eclipse can be seen (weather permitting) from the entire night side of the Earth, lunar eclipses are observed far more frequently from a given place on Earth than are solar eclipses.

An eclipse of the Moon is total only if the Moon's path carries it through the Earth's umbra. If the Moon does not enter the umbra completely, we have a partial eclipse of the Moon. Sometimes the Moon can miss the umbra altogether and just enter the larger penumbra of the Earth: In that case we have a penumbral eclipse of the Moon. The Moon's light does not decrease very much in the penumbra, so these eclipses are often hard to notice.

A lunar eclipse can take place only when the Sun, Earth, and Moon are in a line, which means the Moon will be in full phase before the eclipse, making the darkening even more dramatic. About 20 minutes or so before the Moon reaches the dark shadow, it dims somewhat as the Earth partly blocks the sunlight. As the Moon begins to dip into the shadow, the curved shape of the Earth's shadow upon it soon becomes apparent.

Even when totally eclipsed, the Moon is still faintly visible, usually appearing a dull coppery red. The illumination on the eclipsed Moon is sunlight that has been bent into the Earth's shadow by passing through the Earth's atmosphere.

After totality the Moon moves out of the shadow and the sequence of events is reversed. The total duration of the eclipse depends on how closely the Moon's path approaches the axis of the shadow. For an eclipse where the Moon goes through the center of the Earth's shadow, each partial phase consumes at least 1 hour, and totality can last as long as 1 hour and 40 minutes. Total eclipses of the Moon occur, on the average, about once every two or three years.

FIGURE 3.22

Geometry of a Lunar Eclipse The Moon is full as seen from Earth at A; at position B, the Moon has begun to move into the Earth's shadow; at position C, the eclipse is over and the Moon is moving out of the Earth's shadow. (Note that this diagram is also not to scale and that the distance the Moon moves in its orbit during the eclipse has been exaggerated here for educational purposes.)

Calendar Zone [www.calendarzone.com/]

This rich, sometimes strange, but always enjoyable site has everything you've ever wanted to ask or know about calendars and timekeeping—sometimes as part of the site, and sometimes through links around the world. In addition to astronomical ideas, the site features the cultural and political aspects of time systems as well as software for keeping track of time.

Calendars and Their History [astro.nmsu.edu/~lhuber/leaphist.html]

The late historian of astronomy Leroy Doggett wrote this introduction to calendars for *The Explanatory Supplement to the Astronomical Almanac,* and it still remains one of the best guides to different calendar systems and how they are based on the motions of celestial objects. Includes sections on the Hebrew, Islamic, Indian, and Chinese calendars. (Note that this is just a plain text site, with no graphics or links.)

The Reasons for the Seasons [www.aspsky.org/html/tnl/29/29.html]

An issue of a newsletter for astronomy teachers that goes into more depth on what causes the seasons and how the different effects combine.

Eclipse Home Page [sunearth.gsfc.nasa.gov/eclipse/eclipse.html]

Fred Espenak, of NASA's Goddard Space Flight Center, has published eclipse calculations and guidebooks for many years, and his site contains a wealth of information on the details of lunar and solar eclipses, past and future, as well as observing and photography hints.

The Sky Online Eclipse Page [www.skypub.com/eclipses/eclipses.shtml]

Sky & Telescope magazine's sites for eclipses features good tourist information about eclipse tours, observing and photography hints, and lots of links. Good starting site for eclipse chasers or fans.

Astronomical Data Services [aa.usno.navy.mil/AA/data]

This rich site from the U.S. Naval Observatory allows you to ask many questions about the Earth, Moon, and sky. Want to know when the full moon or new moon will fall in August of 2004? Or when the next lunar eclipse will occur? Or when the Moon or Sun will rise and set for any day of the year? It's all here and much more besides, in tables and as on-screen calculators.

Sundials on the Internet [www.sundials.co.uk/]

Before modern clocks, shadows cast by the Sun were a primary way of telling time. This site, maintained by sun-dial devotees has information about sundials around the world, the theory of sundials, references, and many links.

SUMMARY

3.1 The terrestrial system of latitude and longitude makes use of the **great circles** called **meridians.** An analogous celestial coordinate system is called **right ascension (RA)** and **declination,** with the vernal equinox serving as reference point (like the prime meridian at Greenwich on the Earth). These coordinate systems help us locate any object on the celestial sphere. The Foucault pendulum is a way to demonstrate that the Earth rather than the sky is turning.

3.2 The familiar cycle of the seasons results from the 23° tilt of the Earth's axis of rotation. At the *summer solstice,* the Sun is higher in the sky and its rays strike the Earth more directly. The Sun is in the sky for more than half the day and can heat the Earth longer. At the *winter solstice,* the Sun is low in the sky and up for fewer than 12 hours. At the *vernal* and *autumnal equinoxes,* the Sun is on the celestial equator, and we get 12 hours of day and night. The seasons are different at different latitudes.

3.3 The basic unit of astronomical time is the day (either the **solar day** or the **sidereal day**). **Apparent solar time** is based on the position of the Sun in the sky, and **mean solar time** is based on the average value of a solar day during the year. By international agreement, we define 24 time zones around the world, each with its own standard time. The convention of the **international date line** is necessary to reconcile times in different parts of the Earth.

3.4 The fundamental problem of the calendar is to reconcile the incommensurable lengths of the day, the month, and the year. Most modern calendars, beginning with the Roman (Julian) calendar of the 1st century B.C., neglect the problem of the month and concentrate on achieving the correct number of days in a year by using such conventions as the leap year. Today, most of the world has adopted the Gregorian calendar established in 1582.

3.5 The Moon's monthly cycle of *phases* results from the changing angle of its illumination by the Sun. The full moon is visible in the sky only during the night; other phases are visible during the day. Because its period of revolution is the same as its period of rotation, the Moon always keeps the same face toward the Earth.

3.6 The twice-daily ocean *tides* are primarily the result of the Moon's differential gravitational force on the material of the Earth's crust and ocean. These tidal forces cause ocean water to flow into two tidal bulges on opposite sides of the Earth; each day the Earth rotates through these bulges. Actual ocean tides are complicated by the additional effects of the Sun, and by the shape of the coasts and ocean basins.

3.7 The Sun and Moon have nearly the same angular size (about 0.5°). A **solar eclipse** occurs when the Moon moves between the Sun and the Earth, casting its shadow on a part of the Earth's surface. If the eclipse is total, the observer is in the Moon's *umbra,* the light from the bright disk of the Sun is completely blocked, and the solar atmosphere (the **corona**) comes into view. Solar eclipses take place rarely in any one location, but they are among the most spectacular sights in nature. A **lunar eclipse** takes place when the Moon moves into the Earth's shadow; it is visible (weather permitting) from the entire night hemisphere of the Earth.

INTER-ACTIVITY

A Have your group brainstorm about other ways (besides the Foucault Pendulum) that you could prove that it is our Earth that is turning once a day, and not the sky turning around us. (*Hint:* How does the spinning of the Earth affect the oceans and the atmosphere?)

B What would the seasons on Earth be like if the Earth's axis were not tilted? How many things about life on Earth can you think of that would be different in this case?

C After college and graduate training, members of your group are asked to set up a school in New Zealand. Describe some ways your school schedule in the Southern Hemisphere would have to differ from what we are used to in the North.

D During the traditional U.S. Christmas vacation weeks, you are sent to the vicinity of the South Pole on a research expedition (depending on how well you did on your astronomy midterm, either as a research assistant or as a short-order cook!) Have your group discuss how the days and nights will be different there and how these differences might affect you.

E Discuss with your group all the stories you have heard about the Moon and crazy behavior. Why do you think people associate crazy behavior with the full Moon? What other legends besides vampire stories are connected with the phases of the Moon?

F Your college town becomes the founding site for a strange new cult that worships the Moon. These true believers gather regularly around sunset and do a dance where they must extend their arms in the direction of the Moon. Have your group discuss which way their arms will be pointing *at sunset* when the Moon is new, first quarter, full, and third quarter.

REVIEW QUESTIONS

1. Discuss how latitude and longitude on Earth are similar to declination and right ascension in the sky.

2. What is the latitude of the North Pole? The South Pole? Why does longitude have no meaning at the North and South Poles?

3. Make a table showing each main phase of the Moon and roughly when the Moon rises and sets for each phase. During which phase can you see the Moon in the middle of the morning? In the middle of the afternoon?

4. What are the advantages and disadvantages of apparent solar time? How is the situation improved by introducing mean solar time and standard time?

5. What are the two ways that the tilt of the Earth's axis causes the summers in the United States to be warmer than the winters?

6. Why is it difficult to construct a practical calendar based on the Moon's cycle of phases?

7. Explain why there are two high tides and two low tides every day. Strictly speaking, should the period during which there are two high tides be 24 hours? If not, what should the interval be?

8. What is the phase of the Moon during a total solar eclipse? During a total lunar eclipse?

THOUGHT QUESTIONS

9. Where are you on the Earth according to the following descriptions? (Refer back to Chapter 1 as well as this chapter.)
  a. The stars rise and set perpendicular to the horizon.
  b. The stars circle the sky parallel to the horizon.
  c. The celestial equator passes through the zenith.
  d. In the course of a year, all stars are visible.
  e. The Sun rises on September 23 and does not set until March 21 (ideally).

10. In countries at far northern latitudes, the winter months tend to be so cloudy that astronomical observations are nearly impossible. Why can't good observations of the stars be made at those places during the summer months?

11. What is the phase of the Moon if it
  a. rises at 3:00 P.M.?
  b. is highest in the sky at 7:00 A.M.?
  c. sets at 10:00 A.M.?

12. A car accident occurs around midnight on the night of a full moon. The driver at fault claims he was blinded momentarily by the Moon rising on the eastern horizon. Should the police believe him?

13. The secret recipe to the ever-popular veggie burgers in the college cafeteria is hidden in a drawer in the director's office. Two students decide to break in and get their hands on it, but they want to do it a few hours before dawn on a night when there is no Moon, so they are less likely to be caught. What phases of the Moon would suit their plans?

14. Your granduncle, who often exaggerates events in his own life, tells you about a terrific adventure he had on February 29, 1900. Why would this story make you suspicious?

15. One year, when money is no object, you enjoy your birthday so much that you want to have another one right away. You get into your supersonic jet. Where should you and the people celebrating with you travel? From what direction should you approach? Explain.

16. Suppose you lived in the crater Copernicus on the side of the Moon facing the Earth.
  a. How often would the Sun rise?
  b. How often would the Earth set?
  c. During what fraction of the time would you be able to see the stars?

17. In a lunar eclipse, does the Moon enter the shadow of the Earth from the east or west side? Explain why.

18. Describe what an observer at the crater Copernicus would see while the Moon is eclipsed. What would the same observer see during what would be a total solar eclipse as viewed from the Earth?

19. The day on Mars is 1.026 Earth days long. The martian year lasts 686.98 Earth days. The two moons of Mars take 0.32 Earth days (for Phobos) and 1.26 Earth days (for Deimos) to circle the planet. You are given the task of coming up with a martian calendar for a new Mars colony. What might you do?

20. a. If a star rises at 8:30 P.M. tonight, approximately what time will it rise two months from now?
  b. What is the altitude of the Sun at noon on December 22, as seen from a place on the Tropic of Cancer?

21. Suppose the tilt of the Earth's axis were only 16°. What, then, would be the difference in latitude between the Arctic Circle and the Tropic of Cancer? What would be the effect on the seasons compared with that produced by the actual tilt of 23°?

22. Consider a calendar based entirely on the day and the month (the Moon's period from full phase to full phase). How many days are there in a month? Can you figure out a scheme analogous to leap year to make this calendar work? Can you also incorporate the idea of a week into your lunar calendar?

23. Show that the Gregorian calendar will be in error by one day in about 3300 years.

Aveni, A. *Empires of Time: Calendars, Clocks, and Cultures.* 1989, Basic Books.

Bartky, I. and Harrison, E. "Standard and Daylight Saving Time" in *Scientific American*, May 1979.

Brown, H. *Man and the Stars.* 1978, Oxford U. Press. Contains good sections on the history of astronomical timekeeping.

Coco, M. "Not Just Another Pretty Phase" in *Astronomy*, July 1994, p. 76. Moon phases explained.

DeVorkin, D. *Practical Astronomy.* 1986, Smithsonian Institution Press.

Gingerich, O. "Notes on the Gregorian Calendar Reform" in *Sky & Telescope*, Dec. 1982, p. 530.

Harris, J. and Talcott, R. *Chasing the Shadow: An Observer's Guide to Eclipses.* 1994, Kalmbach.

Kluepfel, C. "How Accurate Is the Gregorian Calendar?" in *Sky & Telescope*, Nov. 1982, p. 417.

Littmann, M. and Willcox, K. *Totality.* 1991, U. of Hawaii Press. Fine introduction to science and lore of eclipses.

Olson, D. and Doescher, R. "Lincoln and the Almanac Trial" in *Sky & Telescope*, Aug. 1990, p. 184. How he used the Moon.

Pasachoff, J. and Ressmeyer, R. "The Great Eclipse" in *National Geographic*, May 1992, p. 30. About the July 1991, solar eclipse, with spectacular photographs.

Rey, H. *The Stars: A New Way to See Them.* 1976, Houghton Mifflin. Good introduction to time, the seasons, and celestial coordinates by the author of the "Curious George" children's stories.

The Very Large Array of radio telescopes in New Mexico seen with a rainbow. Just as water droplets in our atmosphere can disperse sunlight into a rainbow of colors, so instruments called spectrometers allow astronomers to break the light from planets, stars, or galaxies into their component colors. The details of these colors can help us decipher what conditions are like in the objects from which they originate. *(Photo by Martha Haynes and Riccardo Giovanelli, Cornell University)*

4 Radiation and Spectra

THINKING AHEAD

The nearest star is so far away that the fastest spacecraft humans have built would take almost 100,000 years to get there. Yet we very much want to know what material this neighbor star is composed of and how it differs from our own Sun. How can we learn about the chemical makeup of stars that we cannot hope to visit or sample?

"In the last hundred years the power of interpreting the messages brought by light has increased greatly. We are in touch with the stars."

Sir William Bragg in *The Universe of Light* (1933, G. Bell and Sons).

In astronomy, most of the objects that we study are completely beyond our reach. The temperature of the Sun is so high that a spacecraft would be fried long before reaching the solar surface, and the stars are much too far away to visit in our lifetimes with the technology now available. Even light, which travels at a speed of 300,000 km/s, takes more than four years to reach us from the *nearest* star. If we want to learn about the Sun and stars, or even about most members of our solar system, we must rely on techniques that allow us to analyze them from a distance.

And here we are in luck. Coded into the light and other kinds of radiation that reach us from objects in the universe is a wide range of information about what those objects are like and how they work. If we can decipher this "cosmic code" and read the messages it contains, we can learn an enormous amount about the cosmos without ever having to leave the Earth or its immediate environment.

The light and other radiation we receive from the stars and planets is generated by processes at the atomic level—by changes in the way the parts of the atom interact and move. Thus, to appreciate how light is generated we must first explore how atoms work. There is a bit of irony in the fact that in order to understand some of the largest structures in the universe, we must become acquainted with some of the smallest.

Notice that we have twice used the term *light and other radiation.* One of the key ideas explored in this chapter is that visible light is not unique—it is merely the most familiar example of a much larger family of radiation that can carry information to us.

4.1 THE NATURE OF LIGHT

As we saw in earlier chapters, Newton's theory of gravity accounts for the motions of the planets as well as the objects on the Earth. Application of this theory to a variety of problems dominated the work of scientists for nearly two centuries. In the 19th century, many physicists turned to the study of electricity and magnetism, which—as we shall see in a moment—are intimately connected with the production of light.

The scientist who played a role in this field analogous to Newton's role in the study of gravity was physicist James Clerk Maxwell (1831–1879), born and educated in Scotland (Figure 4.1). Inspired by a number of ingenious experiments that showed an intimate relationship between electricity and magnetism, Maxwell developed a single theory that describes both with just a small number of elegant equations. It is this theory that allows us to understand the nature and behavior of light.

4.1_1

Maxwell's Theory of Electromagnetism

We will look at the structure of the atom in more detail in Section 4.4, but we begin by noting that the typical atom consists of several types of particles, a number of which have not only mass, but an additional property called electric charge. In the nucleus (central part) of every atom are *protons,* which are positively charged; outside the nucleus are *electrons,* which have a negative charge.

Maxwell's theory deals with these electric charges and their effects, especially when they are moving. In the vicinity of an electric charge, another charge feels a force of attraction or repulsion: Opposite charges attract; like charges repel. When charges are not in motion, we observe only this electric attraction or repulsion. If charges are in motion, however, (as they are inside every atom and in a wire carrying a current), then we measure another force, called *magnetism.*

Magnetism was well known for much of recorded human history, but its cause was not understood until the 19th century. Experiments with electric charges demonstrated that magnetism was the result of moving charged particles. Sometimes the motion is clear, as in the coils of heavy wire that make an industrial electromagnet. Other times it is more subtle, as in the kind of magnet you buy in a hardware store, in which many of the electrons inside the atoms are spinning in roughly the same direction—it is the alignment of their motion that causes the material to become magnetic.

FIGURE 4.1
James Clerk Maxwell Maxwell (1831–1879) unified the rules governing electricity and magnetism into a coherent theory.
(American Institute of Physics, Niels Bohr Library)

We use the word *field* in physics to describe the action of forces that one object exerts on other distant objects. For example, we say the Sun produces a *gravitational field* that controls the Earth's orbit, even though the Sun and the Earth do not come directly into contact. Using this terminology, we can say that stationary electric charges produce *electric fields,* and moving electric charges produce *magnetic fields.*

Actually, the relationship between electric and magnetic phenomena is even more profound. Experiments showed that changing magnetic fields could produce electric currents (and thus changing electric fields), and changing electric currents could (in turn) produce changing magnetic fields. So once begun, it appeared that electric and magnetic field changes could continue to trigger each other.

Maxwell analyzed what would happen if electric charges were *oscillating* (moving constantly back and forth) and found that the resulting pattern of electric and magnetic fields would spread out and travel rapidly through space. Something similar happens when your finger or a nervous frog moves up and down in a pool of water. The disturbance moves outward and creates a pattern we call a *wave* in the water (Figure 4.2). You might first think that there must be very few situations in nature where electric charges would oscillate, but this is not at all the case. As we shall see, because they are warm, all atoms and molecules (which consist of charged particles) oscillate back and forth all the time. The resulting electromagnetic disturbances are among the most common phenomena in the universe.

Maxwell was able to calculate the speed at which an electromagnetic disturbance moves through space; he

FIGURE 4.2
Making Waves   An oscillation in a pool of water creates an expanding disturbance called a wave.   *(©1993, Comstock, Inc.)*

found that it is equal to the speed of light, which had been measured experimentally. On that basis, he speculated that light was one form of a family of possible electric and magnetic disturbances called **electromagnetic radiation,** a conclusion that was again confirmed in laboratory experiments. When light (reflected from the pages of an astronomy textbook, for example) enters a human eye, its changing electric and magnetic fields stimulate nerve endings, which then transmit the information contained in these changing fields to the brain.

The word *radiation* will be used frequently in this book, so it is important to understand what it means. In everyday language, radiation is often used to describe certain kinds of energetic subatomic particles released by radioactive materials in our environment. (An example is the kind of radiation used to treat cancer.) But this is not what we mean when we speak of radiation in an astronomy text. Radiation, as used in this book, is a general term for light, x rays, and other forms of electromagnetic waves. This radiation provides almost our only link with the universe beyond our own solar system.

4.1.2

The Wave-Like Characteristics of Light

The changing electric and magnetic fields in radiation are similar (as hinted earlier) to the waves that can be set up on a quiet pool of water. In both cases the disturbance travels rapidly outward from the point of origin and can use its energy to disturb other things farther away. (For example, in the case of water, the expanding ripples moving away from our twitching frog could disturb the peace of a grasshopper sleeping on a leaf in the same pool.) In the case of electro-

magnetic waves, the radiation generated by a transmitting antenna full of charged particles at your local radio station can, sometime later, disturb a group of electrons in your home radio antenna and bring you the news and weather while you are getting ready for class or work in the morning.

But the waves generated by charged particles differ from water waves in some profound ways. Water waves require water as a *medium* to travel in. Sound waves, to give another example, are pressure disturbances that require air to travel through. But electromagnetic waves do not require any medium at all: The fields generate each other and so can move through a vacuum (such as outer space). This was such a disturbing idea to 19th-century scientists that they actually made up a medium to fill all of space—one for which there was not a single shred of evidence—just so light waves could have something to travel through: They called it the *aether.* Today we know that there is no aether and that electromagnetic waves have no trouble at all moving through empty space (as all the starlight visible on a clear night must surely be doing).

The other difference is that *all* electromagnetic waves move at the same speed (the speed of light), which turns out to be the fastest possible speed in the universe. No matter where electromagnetic waves are generated and no matter what other properties they have, when they are moving (and not interacting with matter), they move at the speed of light. Yet you know from everyday experience that waves like light are not all the same. For example, we perceive that light waves differ from one another in a property we call color. Let's see how we can denote the differences among the whole broad family of electromagnetic waves.

The nice thing about a wave is that it is a repeating phenomenon. Whether it is the up-and-down motion of a water wave or the changing electric and magnetic fields in a wave of light, the pattern of disturbance repeats in a cyclical way. Thus, any wave motion can be characterized by drawing a series of crests and troughs (Figure 4.3). Moving up a full crest and down a full trough completes one cycle. The horizontal length covered by one cycle—the distance between one crest and the next, for example—is called the **wavelength.** Ocean waves provide an analogy: The wavelength is the distance separating successive wave crests.

Various forms of electromagnetic radiation differ from one another in their wavelengths. Those with the longest waves, ranging up to many kilometers in length, are called *radio waves*. Forms of electromagnetic energy of successively shorter wavelengths are called *infrared, light, ultraviolet, x rays,* and *gamma rays.* All these forms (described in greater detail in Section 4.2) are the same basic kind of energy and could be thought of as different kinds of light.

For visible light, our eyes (and minds) perceive different wavelengths as different colors: Red, for example, is the longest visible wave, and violet is the shortest. The main colors of visible light from longest to shortest wavelength can be remembered using the mnemonic ROY G. BIV—for red, orange, yellow, green, blue, indigo, and violet.

We can also characterize different waves by their **frequency,** the number of wave cycles that pass per second. If you count ten crests moving by each second, for example, then the frequency is 10 cycles per second (cps). In honor of Heinrich Hertz, the physicist who—inspired by Maxwell's work—discovered radio waves, a cps is also called a *Hertz (Hz).* Take a look at the dial of your radio, for example, and you will see the channel assigned to each radio station on the dial characterized by its frequency, usually in units of KHz (kiloHertz, or thousands of Hertz) or MHz (megaHertz, or millions of Hertz).

Wavelength and frequency are related to each other because all electromagnetic waves travel at the same speed. To see how this works, imagine a parade in which everyone is forced by prevailing traffic conditions to move at exactly the same speed. You stand on a corner and watch the waves of marchers come by. First you see row after row of very skinny fashion models. Because they are not very wide, a good number of the models can fit by you each minute; we can say they have a high frequency. Next, however, come several rows of circus elephants. The elephants are quite large, and marching at the same speed as the models, so far fewer of them can march past you per minute: They must be content with a lower frequency. Mathematically, we can express this relationship as

$$c = \lambda f$$

where the Greek letter for "l"—lambda or λ—is used to denote wavelength and c is the scientific symbol for the speed of light. In words, the formula can be expressed as follows: For any wave motion, the speed at which a wave moves equals the frequency times the wavelength. Waves with longer wavelengths have lower frequencies.

Light as a Photon   4.1.3

The electromagnetic wave model of light (as formulated by James Clerk Maxwell) was one of the great triumphs of 19th-century science. When Heinrich Hertz actually made invisible electromagnetic waves on one side of a room and detected them on the other in 1887, it ushered in a new era that led to the modern age of telecommunications. However, by the turn of the century, new, more sophisticated experiments had revealed that light behaves in certain ways that simply cannot be explained by the wave model. Reluctantly, physicists had to accept that sometimes light (and all other electromagnetic radiation) behaves more like a "particle"—or at least a self-contained packet of energy—than a wave. We call such packets of electromagnetic energy **photons.**

The fact that light behaves like a wave in certain experiments and like a particle in others was a very surprising and unlikely idea; at first, it was as confusing to physicists as it seems to students. After all, our common sense says that waves and particles are opposite concepts. A wave is a repeating disturbance that by its very nature is not just in one place, but spreads out. A particle, on the other hand, is something that can only be in one place at any given time. Strange as it sounds, though, countless experiments now confirm that electromagnetic radiation can sometimes behave like a wave and at other times like a particle.

Then again, perhaps we shouldn't be surprised that something that always travels at the "speed limit" of the universe and doesn't need a medium to travel through might not obey our everyday "commonsense" ideas. The confusion that this wave–particle duality of light caused in physics was eventually resolved by the introduction of a more complicated theory of waves and particles, now called *quantum mechanics.* (This is one of the most interesting fields of modern science, but mostly beyond the scope of our book. If you get interested in it, see some of the references at the end of this chapter.)

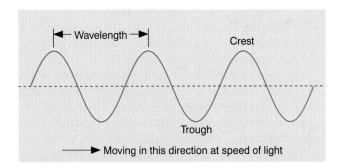

FIGURE 4.3

Characterizing Waves Electromagnetic radiation has wave-like characteristics. The wavelength (λ) is the distance between crests, the frequency (*f*) is the number of cycles per second, and the speed (*c*) is the distance the wave covers over time.

In any case, you should now be prepared when you see scientists (or your authors) sometimes discuss electromagnetic radiation as if it consisted of waves and at other times refer to it as a stream of photons. A photon carries a specific amount of energy (it is a packet of energy after all). That energy is proportional to the frequency of the same radiation when considered as waves. High-frequency waves are the same as photons with high energy; low-frequency waves are the same as photons with low energy. Among the colors of visible light, violet-light photons have the highest energy, red the lowest.

Propagation of Light   4.1.4

Let's think for a moment about how light from a lightbulb moves through space. The expansion of light waves is extremely democratic: They travel away from the bulb not just toward *your* eyes, but in all directions. They must therefore cover an ever-widening space. Yet the total amount of light available can't change once the light has left the bulb. This means that, as the same expanding shell of light covers a larger and larger area, there must be less and less of it in any given place. Light (and all other electromagnetic radiation) gets weaker and weaker as it gets farther and farther from its source.

The increase in the area that the same amount of radiation must cover is proportional to the square of the distance that the radiation has traveled (Figure 4.4). If we stand twice as far from the source, our eyes will intercept two squared, or four times less light. If we stand ten times farther from the source, we get ten squared, or 100 times less light. You can see how this weakening means trouble for sources of light at astronomical distances! A typical star emits about the same total energy as does the Sun, but even the nearest star is about 270,000 times farther away, and so it appears about 73 billion times fainter. No wonder that the stars, which close-up would look more or less like the Sun, look like mere pinpoints of light from far away.

This idea, that the **apparent brightness** of a source (how bright it looks to us) gets weaker with distance in the way we have described, is known as the **inverse-square law** of light propagation. In this respect, the propagation of radiation is similar to the effects of gravity. Remember that the force of gravity between two attracting masses is also inversely proportional to the square of their separation.

THE ELECTROMAGNETIC SPECTRUM

Objects in the universe send us an enormous range of electromagnetic waves. To make discussing them manageable, scientists have divided the full **electromagnetic spectrum** (or range) of such waves into some arbitrary categories. These are shown in Figure 4.5, with some information about the waves in each part or band of the spectrum.

4.2.1

Types of Electromagnetic Radiation

Electromagnetic radiation with the shortest wavelengths, not longer than 0.01 nanometer (nm), is called **gamma rays.** (One nm is 10^{-9} m; see Appendix 5.) The name comes from the third letter of the Greek alphabet; gamma rays were the third kind of radiation discovered coming from radioactive atoms when physicists were first investigating their behavior. Because gamma rays carry a lot of energy, they can be dangerous for living tissues. We shall see that gamma radiation is generated deep in the interior of stars by reactions among atomic nuclei. They are also generated by some of the most violent phenomena in the universe, such as the deaths of stars and the merging of stellar corpses. Gamma rays coming from space interact with atoms high in the Earth's atmosphere and are absorbed before they reach the ground; thus, they can only be studied using instruments in space.

Electromagnetic radiation with wavelengths between 0.01 nm and 20 nm is referred to as **x rays.** These are now familiar to all of us from visits to the doctor and dentist; being more energetic than light, x rays are able to penetrate

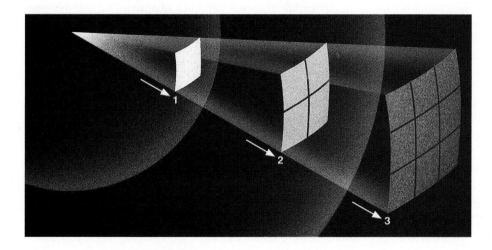

FIGURE 4.4
The Inverse-Square Law for Light
As light energy radiates away from its source, it spreads out in such a way that the energy passing through a unit area decreases as the square of the distance from its source.

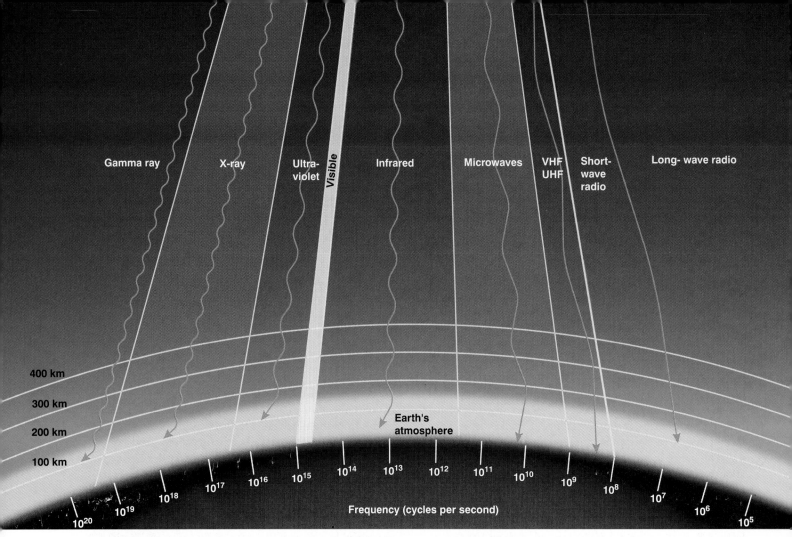

Gamma ray   X-ray   Ultra-violet   Visible   Infrared   Microwaves   VHF UHF   Short-wave radio   Long- wave radio

400 km

300 km

200 km

100 km

Earth's atmosphere

10^{20} 10^{19} 10^{18} 10^{17} 10^{16} 10^{15} 10^{14} 10^{13} 10^{12} 10^{11} 10^{10} 10^{9} 10^{8} 10^{7} 10^{6} 10^{5}

Frequency (cycles per second)

FIGURE 4.5

Radiation and the Earth's Atmosphere   The bands of the electromagnetic spectrum and how well the Earth's atmosphere transmits them. Note that high-frequency waves from space do not make it to the surface and must therefore be observed from space. Some infrared and microwaves are absorbed by water and thus are best observed from high altitudes. Low-frequency radio waves are blocked by the Earth's ionosphere.

soft tissues and so allow us to make images of the shadows of the bones within them. While x rays can penetrate a short length of human flesh, they are stopped by the large numbers of atoms in the Earth's atmosphere with which they interact. Thus, x-ray astronomy (like gamma-ray astronomy) could not develop until humanity had ways of sending instruments above our atmosphere (Figure 4.6).

Radiation intermediate between x rays and visible light is **ultraviolet** (meaning higher energy than violet). Outside the world of science, ultraviolet light is sometimes called "black light" because our eyes cannot see it. Ultraviolet radiation is mostly blocked by a section of the Earth's atmosphere called the *ozone* layer, but a small fraction of the ultraviolet rays from our Sun do penetrate to cause sunburn or, in extreme cases, skin cancer in human beings. Ultraviolet astronomy is also best done from space.

Electromagnetic radiation with wavelengths between roughly 400 and 700 nm is called *visible light,* because

these are the waves that human vision is sensitive to. As we shall see, this is also the band of the electromagnetic spectrum where the Sun gives off the greatest amount of radiation. These two observations are not coincidental: Human eyes evolved to see the kinds of waves that the Sun produced most effectively. The more light there was to see by, the more efficiently we could evade predators and make babies before we got stepped on or eaten (which, in a crude way, is what life is all about from the point of view of your genetic material). Visible light penetrates the Earth's atmosphere effectively, except when it is temporarily blocked by passing clouds.

In 1672, in the first paper that he submitted to the Royal Society, Newton described an experiment in which he permitted sunlight to pass through a small hole and then through a prism. Newton found that sunlight, which gives the impression of being white, is actually made up of a mixture of all the colors of the rainbow (Figure 4.7). Although

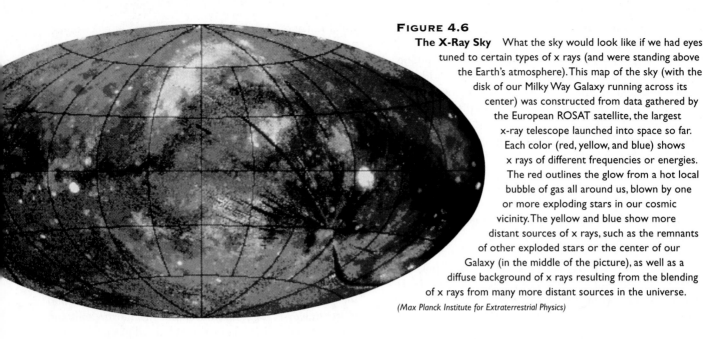

FIGURE 4.6

The X-Ray Sky What the sky would look like if we had eyes tuned to certain types of x rays (and were standing above the Earth's atmosphere). This map of the sky (with the disk of our Milky Way Galaxy running across its center) was constructed from data gathered by the European ROSAT satellite, the largest x-ray telescope launched into space so far. Each color (red, yellow, and blue) shows x rays of different frequencies or energies. The red outlines the glow from a hot local bubble of gas all around us, blown by one or more exploding stars in our cosmic vicinity. The yellow and blue show more distant sources of x rays, such as the remnants of other exploded stars or the center of our Galaxy (in the middle of the picture), as well as a diffuse background of x rays resulting from the blending of x rays from many more distant sources in the universe. (*Max Planck Institute for Extraterrestrial Physics*)

all the colors of light may impinge on objects in our world, not all of them are reflected. When white light from the Sun or from a light fixture hits a pair of blue jeans, the dye in the jeans absorbs all the colors except blue. Some of the blue colors are reflected back into our eyes, and our brain interprets that particular set of wavelengths as blue color.

Between visible light and radio waves are the wavelengths of **infrared** or heat radiation. Astronomer William Herschel first discovered infared in 1800 while trying to measure the temperatures of different colors of sunlight spread out into a spectrum. He noticed that when he accidentally pushed his thermometer beyond the reddest color, it still registered heating due to some invisible energy coming from the Sun. This was our first hint about the existence of the other (invisible) bands of the electromagnetic spectrum, although it would take many decades for our full understanding to develop.

A "heat lamp" radiates mostly infrared radiation, and the nerve endings in our skin are sensitive to this band of the electromagnetic spectrum. Infrared waves are absorbed by water and carbon dioxide molecules in the Earth's atmosphere, and for this reason, infrared astronomy is best done from high mountaintops, high-flying airplanes, and spacecraft.

All the electromagnetic waves longer than infrared are called **radio** waves, but this is so broad a category that we generally divide it into several subsections. Among the most familiar of these are **microwaves,** used in shortwave communication and microwave ovens; radar waves, used at airports and by the military; FM and television waves, which carry the news and entertainment without which it is now impossible to imagine modern culture; and AM radio waves, which were the first to be developed for broadcasting. These different categories range in wavelength from a

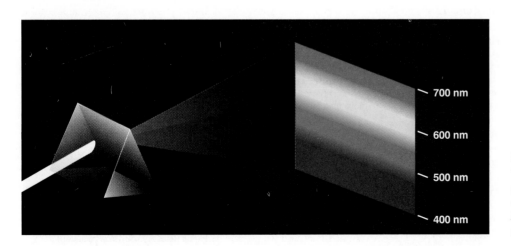

~ 700 nm

~ 600 nm

~ 500 nm

~ 400 nm

FIGURE 4.7

The Action of a Prism When we pass a beam of white sunlight through a prism, we see a rainbow-colored band of light that we call a continuous spectrum.

few millimeters to hundreds of meters. Other radio radiation can have wavelengths as long as several kilometers.

With such a wide range of wavelengths, you will not be surprised to learn that not all radio waves interact with the Earth's atmosphere the same way. Microwaves are absorbed by water vapor (which is what makes them effective in heating foods with high water contents in microwave ovens). FM and TV waves are not absorbed and can travel easily through our atmosphere. AM radio waves are absorbed or reflected by a layer in the Earth's atmosphere called the *ionosphere*. Given some of the truly dumb things people are saying these days on AM radio, maybe it's just as well that this type of radiation can neither get in or out of our planet's shielding atmosphere.

We hope this brief survey has left you with one strong impression. Although visible light is what most people associate with astronomy, the light that our eyes can see is only a tiny fraction of the broad range of waves that the universe sends us. Today we understand that judging some astronomical phenomenon by using only the light we can see is like hiding under the table at a big dinner party and judging all the guests by nothing but their shoes. There's a lot more to each person than meets our eye under the table. It is very important for those who study astronomy today to avoid being *visible-light chauvinists*—those who respect only the information seen by their eyes while ignoring the information gathered by instruments sensitive to other bands of the electromagnetic spectrum.

Table 4.1 summarizes the bands of the electromagnetic spectrum and indicates the temperatures and types of typical astronomical objects that emit each kind of electromagnetic radiation. Many of the objects mentioned in the table will not be familiar to you this early in the course, but you can return to this table as you learn more about the types of objects astronomers study.

Radiation and Temperature

Some astronomical objects emit mostly infrared radiation, others mostly visible light, and still others mostly ultra-violet radiation. What determines the type of electromagnetic radiation emitted by the Sun, stars, and other astronomical objects? The answer often turns out to be their *temperature*.

At the microscopic level, everything in nature is in motion. A *solid* is composed of molecules and atoms that are in continuous vibration: They move back and forth in place, but their motion is much too small for our eyes to make out. A *gas* consists of molecules that are flying about freely at high speed, continually bumping into one another and bombarding the surrounding matter. The energy of these random molecular and atomic motions is called *heat*. The hotter the solid or gas, the more rapid the motion of the molecules or atoms; the temperature of something is thus just a measure of the average energy of the particles that make it up.

This motion at the microscopic level is responsible for much of the electromagnetic radiation on Earth and in the universe. As atoms and molecules move about and collide, or vibrate in place, their charged particles generate electromagnetic radiation (as we saw in Section 4.1). The characteristics of this radiation are determined by the temperature of the atoms and molecules that give rise to them. In a hot solid or gas, for example, the individual particles vibrate in place or move rapidly from collision to collision, so the emitted waves are, on average, more energetic. In very cool material, the particles have low-energy atomic and molecular motions, and thus generate lower energy waves.

Radiation Laws

To understand in more quantitative detail the relationship between temperature and electromagnetic radiation, it is useful to consider an idealized object that (unlike your sweater or your astronomy instructor's head) does not reflect or scatter any radiation, but absorbs all the electromagnetic energy that falls on it. Such an object is called a **blackbody.** The energy that is absorbed heats the blackbody, causing the atoms and molecules in it to begin vibrating or moving

TABLE 4.1
Types of Electromagnetic Radiation

Type of Radiation	Wavelength Range (nm)	Radiated by Objects at This Temperature	Typical Sources
Gamma rays	Less than 0.01	More than 10^8 K	Produced in nuclear reactions. Requires very-high-energy processes.
X rays	0.01–20	10^6–10^8 K	Gas in clusters of galaxies; supernova remnants; solar corona.
Ultraviolet	20–400	10^4–10^6 K	Supernova remnants; very hot stars.
Visible	400–700	10^3–10^4 K	Stars.
Infrared	10^3–10^6	10–10^3 K	Cool clouds of dust and gas; planets; satellites.
Radio	More than 10^6	Less than 10 K	No astronomical objects this cold; radio emission produced by electrons moving in magnetic fields (synchrotron radiation).

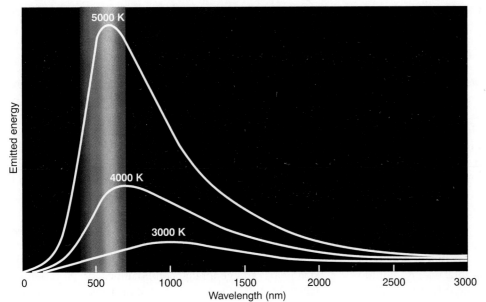

around at increasing speeds. Thus, the blackbody will radiate electromagnetic waves. By absorbing radiation, the blackbody heats up until it is emitting energy at the same rate that energy is being absorbed. We want to discuss such an idealized object, because, as you will see, stars are not bad approximations of blackbodies.

The radiation from a blackbody has several characteristics, as illustrated in Figure 4.8. This figure shows the number of waves (or total energy) given off at each wavelength by blackbodies of different temperatures.

First of all, notice that the white curves show that a blackbody emits some energy at *all* wavelengths within a given range. We say that it gives off a *continuous spectrum* of waves. This is because in any solid or gas, some molecules or atoms vibrate or move between collisions slower than average, and some move faster than average. So when we look at the electromagnetic waves emitted, we find a broad range or spectrum of energies and wavelengths. More waves are emitted at the *average* vibration or motion rate (the tall part of each curve), but if we have a large number of atoms or molecules, some waves will be detected at each wavelength.

Second, note that a blackbody at a higher temperature emits *more* energy at all wavelengths than does a cooler one. In a hot gas (the taller curves in Figure 4.8), for example, the atoms have more collisions and give off more waves at each possible energy or wavelength. In the real world of stars, this means that hotter stars give off more energy at every wavelength than do cooler stars.

Third, the graph shows us that the higher the temperature, the shorter the wavelength at which the maximum energy is emitted. As you might expect, hot objects give off their average waves at shorter wavelengths (higher energies) than do cool objects. You may have observed examples of this rule in everyday life. When a burner on an electric stove is turned on low, it emits only heat, which is

infrared radiation, but does not yet glow with visible light. If the burner is set to a higher temperature, it starts to glow a dull red. At a still higher setting, it glows a brighter orange-red (shorter wavelength). At even higher temperatures, which cannot be reached with ordinary stoves, metal can appear brilliant yellow or even blue-white.

We can use these ideas to come up with a rough sort of "thermometer" for measuring the temperatures of stars. The Sun and stars emit energy that approximates that from a blackbody, so it is possible to estimate temperatures by noting the spectrum of waves they give off. Because many stars give off most of their waves in visible light, the color of light that dominates a star's appearance is a rough indicator of its temperature. If one star looks red and another looks blue, which one has the higher temperature? Because blue is the shorter wavelength color, it is the sign of a hotter star. (Note that the temperatures we associate with different colors in science are not the same as the ones artists use. In art, red is often called a "hot" color and blue a "cool" color, but in nature, it's the other way around.)

We can develop a more precise star thermometer by measuring how much energy a star gives off at each wavelength and by constructing diagrams like Figure 4.8. The graph of the energy emitted by a blackbody at each wavelength is called its *blackbody curve*. The location of the peak (or maximum) in the blackbody curve of each star can tell us its temperature. The temperature at the surface of the Sun, which is where the radiation that we see is emitted, turns out to be 5800 K. (Throughout this text we use the Kelvin or absolute temperature scale. On this scale, water freezes at 273 K and boils at 373 K. All molecular motion ceases at 0 K. The various temperature scales are described in Appendix 5.) There are stars cooler than the Sun and stars hotter than the Sun; there are even some stars so hot that most of their energy is given off at ultraviolet wavelengths.

The wavelength at which a blackbody emits its maximum energy can be calculated according to the equation

$$\lambda_{\max} = 3 \times 10^6/T$$

where the wavelength is in nanometers and the temperature is in Kelvins. This relationship is called **Wien's law.** For the Sun, the wavelength at which the maximum energy is emitted is 520 nm, which is near the middle of that portion of the electromagnetic spectrum called visible light. Characteristic temperatures of other astronomical objects, and the wavelengths at which they emit most of their energy, are listed in Table 4.1.

We can also describe our observation that hotter objects radiate more energy at all wavelengths in a mathematical form. If we sum up the contributions from all parts of the electromagnetic spectrum, we obtain the total energy emitted by a blackbody. That total energy, emitted per second per square meter by a blackbody at a temperature T, is proportional to the fourth power of its absolute temperature. This relationship is known as the **Stefan–Boltzmann law,** which can be written in the form of an equation as

$$E = \sigma T^4$$

where E stands for the energy and σ (the Greek letter sigma) is a constant number.

Notice how impressive this result is. Increasing the temperature of a star would have a tremendous effect on its ability to generate electromagnetic energy. If the Sun, for example, were twice as hot as it is today—that is, if it had a temperature of 11,600 K—it would radiate 2^4, or 16, times more energy than it does now. Tripling the temperature would raise the energy output 81 times!

4.3 SPECTROSCOPY IN ASTRONOMY

As we hinted at the beginning of the chapter, electromagnetic radiation carries a tremendous amount of information about the nature of stars and other astronomical objects. To extract this information, however, astronomers must be able to study the amounts of energy we receive at different wavelengths of visible light (and other radiation)

in fine detail. Let's examine how we can do this and what we can learn.

Optical Properties of Light

Visible light and other forms of electromagnetic energy exhibit certain behaviors that are important to the design of telescopes and other instruments. For example, light is *reflected* from a surface. If the surface is smooth and shiny, as in a mirror, the direction of the reflected light beam can be accurately calculated from a knowledge of the shape of the reflecting surface. Light is also bent, or *refracted,* when it passes from one kind of transparent medium into another, say from the air into a glass lens.

The reflection and refraction of light are the basic properties that make possible all optical instruments—from eyeglasses to giant astronomical telescopes. Such instruments are generally combinations of glass lenses, which bend light according to the principles of refraction, and curved mirrors, which depend on the properties of reflection. Small optical devices, such as eyeglasses or binoculars, generally use lenses, while large telescopes depend almost entirely on mirrors for their main optical elements. In Chapter 5 we will discuss a number of astronomical instruments and their uses. For now, we turn to another behavior of light, one that is essential for the decoding of light.

When light passes from one transparent medium to another, an interesting effect occurs in addition to simple refraction. Because the bending of the beam depends on the wavelength of the light as well as the properties of the medium, different wavelengths or colors of light are bent by different amounts and separated. This phenomenon is called **dispersion.**

Figure 4.7 shows how light can be separated into different colors with a *prism*—a piece of glass in the shape of a triangle. Upon entering one face of the prism, light is refracted once, the violet light more than the red, and upon leaving the opposite face, the light is bent again and further dispersed. If the light leaving the prism is focused on a screen, the different wavelengths or colors that compose white light are lined up side by side (Figure 4.9) as in a rainbow—which is formed by the dispersion of light through raindrops (see "Making Connections: The Rainbow"). Because this array of colors is a spectrum of light, the instrument used to disperse the light and form the spectrum is called a **spectrometer.**

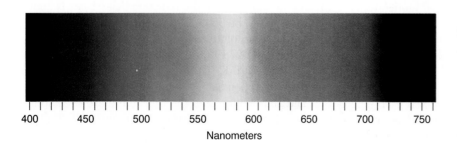

FIGURE 4.9
A Continuous Spectrum When white light passes through a prism, its dispersion forms a continuous spectrum of all the colors. Although it is hard to see in this printed version, in a well-dispersed spectrum, many subtle gradations in color are visible as your eye scans from one end (the violet) to the other (the red).

The Rainbow

Rainbows are an excellent illustration of re-fraction of sunlight. You have a good chance of seeing a rainbow anytime you are between the Sun and a rain shower, and this situation is illustrated in Figure 4A. The raindrops can act like little prisms and break white light into the spectrum of colors. Suppose

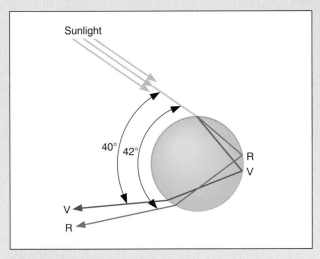

that a ray of sunlight encounters a raindrop and passes into it. The light changes direction—is refracted (Figure 4B)—when it passes from air to water; the blue light is refracted more than the red. Some of the light is then re-flected at the backside of the drop and re-emerges from the front, where it is again re-fracted. As a result, the white light is spread out into a rainbow of colors.

Note that in Figure 4B, violet light lies above the red light after it emerges from the raindrop. When you look at a rainbow, however, it is the red light that is higher in the sky. Why? Look again at Figure 4A. If the observer looks at a raindrop that is high in the sky, the violet light passes over her head and the red light enters her eye. Similarly, if the observer looks at a raindrop that is low in the sky, the violet light reaches her eye and the drop appears violet, whereas the red light from that same drop strikes the ground and is not seen. Colors of intermediate wavelengths are refracted to the eye by drops that are intermediate in altitude be-tween the drops that appear violet and the ones that appear red. Thus, a single rainbow always has red on the outside and violet on the inside.

For an even simpler example of refraction, put a pencil at a slanted angle in a glass of water. What do you see? Can you offer an explanation?

FIGURE 4A

This diagram shows how light from the Sun, which is located behind the observer, can be refracted by raindrops to produce a rainbow.

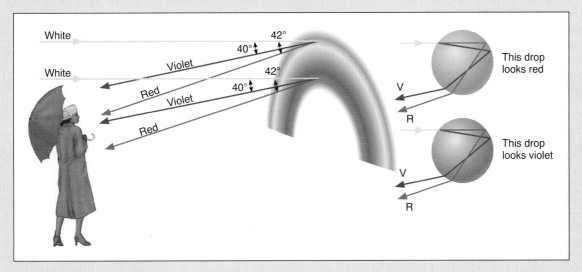

FIGURE 4B

A diagram showing the path of light passing through a raindrop. Refraction separates white light into its component colors.

The Value of Stellar Spectra

If the spectrum of the white light from the Sun and stars were simply a continuous rainbow of colors, astronomers would have little interest in the detailed study of a star's spectrum once they had learned its approximate temperature, as described earlier. When Newton, who first described the laws of refraction and dispersion in optics, observed the solar spectrum, all he could see was a continuous band of colors.

However, in 1802 William Wollaston built an improved spectrometer that included a lens to focus the Sun's spectrum on a screen. With this device Wollaston saw that the colors were not spread out uniformly, but that instead some ranges of color were missing, appearing as dark bands in the solar spectrum. He mistakenly attributed these lines to natural boundaries between the colors. In 1815 the German physicist Joseph Fraunhofer, upon a more careful examination of the solar spectrum, found about 600 such dark lines (Figure 4.10), which pretty quickly ruled out the boundary hypothesis.

These physicists called the missing colors *dark lines* because the rainbow-colored (continuous) spectrum as they viewed it was interrupted at various wavelengths, and in their spectrometers such interruptions looked like dark lines. Later, researchers found that similar dark lines could be produced in the spectra (*spectra* is the plural of *spectrum*) of artificial light sources by passing their light through various apparently transparent substances (usually containers with just a bit of thin gas in them).

These gases turned out not to be transparent at *all* colors: They were quite opaque at a few sharply defined wavelengths. Something in each gas had to be absorbing just a few colors of light and no others. All gases did this, but each different element absorbed a different set of colors. If the gas in a container consisted of two elements, light passing through it was missing the colors (showing dark lines) that go with both of the elements. So it seems that certain lines in the spectrum "go with" certain elements.

What would happen if there were no continuous spectrum for our gases to remove light from? What if, instead, we heated the same gases until they were hot enough to glow with their *own light*? (Bear in mind that the gas containers used in these experiments held gases in pretty low concentrations, so collisions among the molecules or atoms were rare.) When the gases were heated, a spectrometer revealed no continuous spectrum, but several separate *bright lines.* That is, these hot gases *emitted* light only at certain specific wavelengths or colors. Interestingly, the colors at which they emitted when they were heated were the very same colors at which they had absorbed when a continuous source of light was behind them.

When the gas was pure hydrogen, it would absorb or emit one pattern of colors; when it was sodium, it would absorb or emit a different pattern. A mixture of hydrogen and sodium emitted both sets of spectral lines. Thus, scientists began to see that different substances showed distinctive *spectral signatures* by which their presence could be detected (Figure 4.11). Just as your signature allows the bank to identify you, the unique pattern of colors for each type of atom can help us identify which element or elements are in the gas.

Types of Spectra

In such experiments, then, we can distinguish among the three different types of spectra. A **continuous spectrum** (formed when a dense collection of gas or a solid gives off radiation) is an array of all wavelengths or colors of the rainbow. A continuous spectrum can serve as a backdrop from which the atoms of a much less dense gas can absorb light. A dark line, or **absorption line spectrum,** consists of a series or pattern of dark lines—missing colors—superimposed upon the continuous spectrum of a source. A bright line, or **emission line spectrum,** appears as a pattern or series of bright lines; it consists of light in which only certain discrete wavelengths are present.

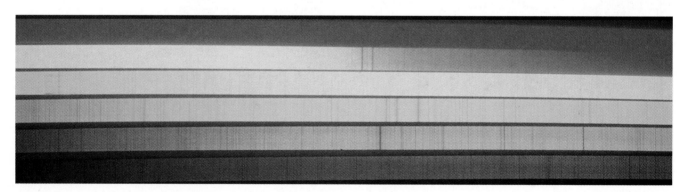

FIGURE 4.10

The Visible Spectrum of the Sun Our star's spectrum is crossed by dark lines produced by atoms in the solar atmosphere that absorb light at certain wavelengths. *(National Solar Observatory/National Optical Astronomy Observatories)*

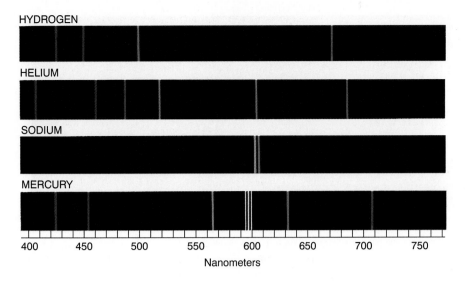

HYDROGEN

HELIUM

SODIUM

MERCURY

400   450   500   550   600   650   700   750

Nanometers

FIGURE 4.11
Line Spectra From Different Elements
Each type of hot glowing gas (each element)
produces its own unique pattern of lines, so
the composition of a gas can be identified by
its spectrum.

(Figure 4.18—a little further ahead in this chapter—shows you a picture of what each of these looks like.)

In the gaseous form, each particular chemical element or compound produces its own characteristic pattern of spectral lines—its spectral signature. No two patterns are alike. In other words, each particular gas can absorb or emit only certain wavelengths of light peculiar to that gas. The temperature and other conditions determine whether the lines are bright or dark, but the wavelengths are the same in either case, and it is the precise pattern of wavelengths that makes the signature of each element unique. Liquids and solids can also generate spectral lines or bands, but they are broader and less well defined, and hence more difficult to interpret.

The dark lines in the solar spectrum thus give evidence of certain chemical elements between us and the Sun, absorbing those wavelengths of sunlight. Because the space between us and the Sun is pretty empty, astronomers quickly realized that the atoms doing the absorbing must be in a thin atmosphere of cooler gas around the Sun. This outer atmosphere is not all that different from the rest of the Sun, just thinner and cooler. Thus, we can use what we learn about its composition as a pretty good indicator of what the whole Sun is made of. Similarly, we can use the presence of absorption and emission lines to analyze the composition of other stars and clouds of gas in space.

Such analysis of spectra is the key to modern astrophysics. Only in this way can we "sample" the stars, which are too far away for us to have much hope of visiting. Even if we could visit them, they are so hot that it would be very difficult to find graduate students willing to go take a sample. But there is no need to ask for such volunteers: Encoded in the electromagnetic radiation from celestial objects is clear information about the chemical makeup of these objects. Only by understanding what the stars were made of could we begin to make theories about what made them shine and how they evolved.

In 1860, the German physicist Gustav Kirchhoff became the first person to use spectroscopy to identify an element in the Sun when he found the spectral signature of sodium gas. In the years that followed, astronomers found many other chemical elements in the Sun and stars. In fact, the element helium was found first in the Sun from its spectrum, and only later identified on Earth. (The word *helium* comes from *helios,* the Greek name for the Sun.) Only in the 20th century, with the development of a model for the atom, did scientists learn why such dark and bright spectral lines exist. We therefore turn next to a closer examination of the atoms that make up all matter.

4.4 THE STRUCTURE OF THE ATOM

The idea that matter is composed of tiny particles called atoms is at least 25 centuries old. It took until the 20th century, however, for scientists to invent instruments that permitted them to probe inside an atom and find that it is not, as had been thought, hard and indivisible. Instead, the atom is a complex structure composed of still smaller particles.

4.4.1

Probing the Atom

The first of these smaller particles was discovered by British physicist J. J. Thomson in 1897. Named the *electron,* this particle is negatively charged. (It is the flow of these particles that produces currents of electricity, whether in lightning bolts or in the wires leading to your hair dryer.) Because an atom in its normal state is electrically neutral, each electron in an atom must be balanced by the same amount of positive charge.

The next problem was to determine where in the atom the positive and negative charges are located. In 1911, British physicist Ernest Rutherford devised an experiment that provided part of the answer to this question. He bombarded an extremely thin piece of gold foil, only about 400 atoms thick, with a beam of alpha particles emitted from a

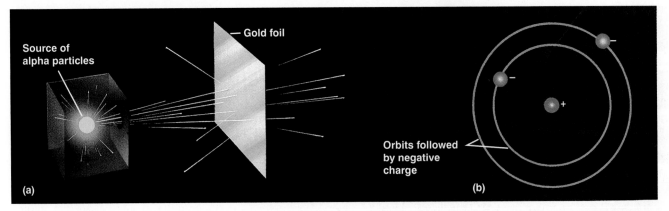

FIGURE 4.12

Rutherford's Experiment (a) When Rutherford allowed alpha particles from a radioactive source to strike a target of gold foil, he found that, although most of them went straight through, some of them rebounded back in the direction from which they came. (b) From this experiment, he concluded that the atom must be constructed like a miniature solar system, with the positive charge concentrated in the nucleus and the negative charge orbiting in the large volume around the nucleus.

radioactive material (Figure 4.12). We now know that alpha particles are helium atoms that have lost all of their electrons and thus are positively charged. Most of the alpha particles passed through the gold foil just as if it and the atoms composing it were nearly empty space. About 1 in 8000 of the alpha particles, however, completely reversed direction and bounced backward from the foil. Rutherford wrote, "It was quite the most incredible event that has ever happened to me in my life. It was almost as incredible as if you fired a 15-inch shell at a piece of tissue paper and it came back and hit you."

The only way to account for the alpha particles that reversed direction when they hit the gold foil was to assume that nearly all of the mass as well as all of the positive charge in each individual gold atom was concentrated in a tiny **nucleus.** When an alpha particle strikes a nucleus, it reverses direction, much as a cue ball reverses direction when it hits another billiard ball. Rutherford's model placed the other type of charge—the negative electrons— in orbit around this nucleus.

Rutherford's model required that the electrons be in motion. Positive and negative charges attract each other, so stationary electrons would fall into the nucleus. Also, because both the electrons and the nucleus are extremely small, most of the atom is empty, which is why nearly all of Rutherford's alpha particles were able to pass right through the gold foil without colliding with anything. Rutherford's model was a very successful explanation of the experiments he conducted, although eventually scientists would discover that the nucleus has structure as well.

4, 4, 2

The Atomic Nucleus

The simplest possible atom (and the most common one in the Sun and stars) is hydrogen: The nucleus of ordinary hydrogen contains a single positively charged particle called a *proton*. Moving around this proton is a single *electron*. The mass of an electron is nearly 2000 times smaller than the mass of a proton, but the electron carries an amount of charge exactly equal to that of the proton but opposite in sign (Figure 4.13). Opposite charges attract each other, so it is electromagnetic force that holds the proton and electron together, just as gravity is the force that keeps the planets in orbit around the Sun.

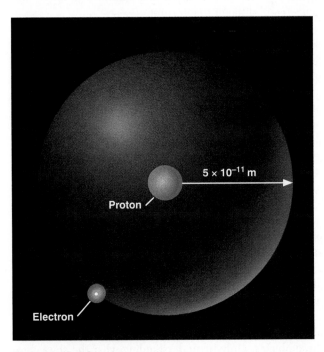

FIGURE 4.13

The Hydrogen Atom This is a schematic diagram of a hydrogen atom in its lowest energy state, also called the ground state. The proton and electron have equal but opposite charges, which exert an electromagnetic force that binds the hydrogen atom together.

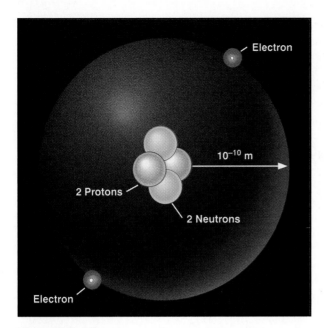

FIGURE 4.14
The Helium Atom Here we see a schematic diagram of a helium atom in its lowest energy state. Two protons are present in the nucleus of all helium atoms. In the most common variety of helium, the nucleus also contains two neutrons, which have nearly the same mass as the proton but carry no charge. Two electrons can be seen orbiting the nucleus.

There are, of course, other types of atoms. Helium, for example, is the second most abundant element in the Sun. Helium has two protons in its nucleus, instead of the single proton that characterizes hydrogen. In addition, the helium nucleus contains two *neutrons*, particles with a mass comparable to that of the proton but with no electric charge. Moving around this nucleus are two electrons, so the total net charge of the helium atom is also zero (Figure 4.14).

From this description of hydrogen and helium, perhaps you have guessed the pattern for building up all the elements (different types of atoms) that we find in the universe. The type of element is determined by the *number of protons* in the nucleus of the atom. Any atom with 6 protons is the element carbon, with 8 protons is oxygen, with 26 is iron, and with 92 is uranium. On Earth, a typical atom has the same number of electrons as protons, and these electrons follow complex orbital patterns around the nucleus. (Deep inside stars, however, it is so hot that the electrons get loose from the nucleus and live separate, productive lives.)

Although the number of neutrons in the nucleus is usually approximately equal to the number of protons, the number of neutrons is not necessarily the same for all atoms of a given element. For example, most hydrogen atoms contain no neutrons at all. There are, however, hydrogen atoms that contain one proton and one neutron, and others that contain one proton and two neutrons. The

various types of hydrogen nuclei with different numbers of neutrons are called **isotopes** of hydrogen (Figure 4.15), and other elements have isotopes as well. You can think of isotopes as siblings, closely related but with different characteristics and behaviors.

The Bohr Atom   4.4.3

There is one serious problem with Rutherford's model for atoms. Maxwell's theory of electromagnetic radiation says that when electrons change either their speed or direction of motion, they must emit energy. Orbiting electrons constantly change their direction of motion, so they should emit a constant stream of energy. According to Maxwell's theory, all electrons should spiral into the nucleus of the atom as they lose energy, and this collapse should happen very quickly—in about 10^{-16} seconds.

It was the Danish physicist Niels Bohr (1885–1962) who solved the mystery of how electrons remain in orbit. He was trying to develop a model of the atom that would also explain certain regularities observed in the spectrum of hydrogen (see Section 4.5). He suggested that the spectrum of hydrogen can be understood if we assume that only orbits of certain sizes are possible for the electron. Bohr further assumed that as long as the electron moves only in one of these allowed orbits, it radiates no energy. Its energy would change only if it moved from one orbit to another.

This suggestion, in the words of science historian Abraham Pais, was "one of the most audacious hypotheses ever introduced in physics." If something equivalent were at work in the everyday world, you might find that, as you went for a walk after astronomy class, nature permitted you to walk 2 steps per minute, 5 steps per minute, and 12 steps per minute, but no speeds in between. No matter how you tried to move your legs, only *certain* walking speeds would be permitted. To make things more bizarre, it would take no effort to walk at one of the allowed speeds, but it would

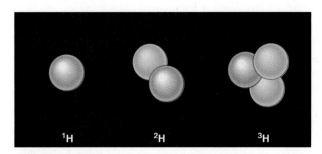

FIGURE 4.15
Isotopes of Hydrogen A single proton in the nucleus defines the atom to be hydrogen, but there may be zero, one, or two neutrons. The most common isotope is the one with only a single proton. A hydrogen nucleus with one neutron is called deuterium; one with two neutrons is called tritium. The symbol for hydrogen is H; the superscript to the left tracks the total number of particles (protons plus neutrons) in the nucleus.

be difficult to change from one speed to another. Luckily, no such rules apply at the level of human behavior, but at the microscopic level of the atom, experiment after experiment has confirmed the validity of Bohr's strange idea. Bohr's suggestions became one of the foundations of the new (and much more sophisticated) model of the subatomic world called quantum mechanics.

In Bohr's model, if the electron moves from one orbit to another closer to the atomic nucleus, it must give up some energy in the form of electromagnetic radiation. On the other hand, if the electron goes from an inner orbit to one farther from the nucleus, it requires some additional energy. One way to obtain the necessary energy is to absorb electromagnetic radiation that may be streaming past the atom from an outside source.

A key feature of Bohr's model is that each of the permitted electron orbits around a given atom has a certain energy value. To move from one orbit to another (which will have *its* own specific energy value) requires a change in the electron's energy—a change determined by the difference between the two energy values. If the electron goes to a lower level, the energy difference will be given off; if the electron goes to a higher level, the energy difference must be obtained from somewhere else. Each jump (or transition) to a different level has a fixed and definite energy change associated with it.

A crude analogy for this situation might be life in a tower of luxury apartments where the rent is determined by the quality of the view. Such a building has certain definite numbered levels or floors on which apartments are located. No one can live on floor 5.37 or 22.5. Second, the rent gets higher as you go up. If you want to exchange an apartment on the 20th floor for one on the 2nd floor, you will not owe as much rent, and you will get a refund from the landlord. But if you want to move from the 3rd floor to the 25th floor, you'd better be prepared to find some extra resources, because your rent will increase. In the atom, too, the "cheapest" place for an electron to live is the lowest possible level, and energy is required to move to a higher level.

Here we have one of the situations where it is easier to think of electromagnetic radiation as **photons** than as waves (see page 84). As electrons move from one level to another, they give off or absorb little packets of energy. When an electron moves to a higher level, it absorbs a photon of just the right energy (provided one is available). When it moves to a lower level, it emits a photon with the exact amount of energy it no longer needs in its "lower-cost living situation."

The photon and wave perspectives must be equivalent: Light is light, no matter how we look at it. Thus each photon carries a certain amount of energy that is proportional to the frequency of the wave it represents. The constant of proportionality, h, called Planck's constant, is named for Max Planck, the German physicist who was one of the originators of the quantum theory. If metric units are used (that is, if energy is measured in joules and frequency

in cycles or waves per second), Planck's constant has the value $h = 6.626 \times 10^{-34}$ joule-s. Higher-energy photons correspond to higher-frequency waves (which have a shorter wavelength); lower-energy photons would be waves of lower frequency.

To take a specific example, consider a calcium atom inside the Sun's atmosphere, in which an electron jumps from a lower level to a higher level. To do this, it needs about 5×10^{-19} joules of energy, which it can conveniently obtain by absorbing a passing photon of that energy coming from deeper inside the Sun. This photon is equivalent to a wave of light whose frequency is about 7.5×10^{14} Hz and whose wavelength is about 3.9×10^{-7} m (393 nm), in the deep violet part of the visible light spectrum. Although it may seem strange at first to switch this way from picturing light as a photon (or energy packet) to picturing it as a wave, such switching has become second nature to astronomers and can be a handy tool for doing calculations about spectra.

4.5 FORMATION OF SPECTRAL LINES

The Hydrogen Spectrum 4.5.1

Now we can use Bohr's model of the atom to understand how spectral lines are formed. Suppose a beam of white light (which consists of photons of all wavelengths) shines through a gas of atomic hydrogen. It turns out that a photon of wavelength 656 nm has just the right energy to raise an electron in a hydrogen atom from the second to the third orbit. Thus, as all the photons of different energies (waves of different wavelengths or colors) stream by the hydrogen atoms, photons with *this* particular wavelength can be absorbed by those atoms whose electrons are orbiting on the second level. When they are absorbed, the electrons on the second level will now be on the third level, and a number of the photons of this wavelength and energy will be missing from the general stream of white light.

Other photons will have the right energies to raise electrons from the second to the fourth orbit, or from the first to the fifth orbit, and so on. Only photons with exactly these correct energies can be absorbed. All of the other photons (with different energies and colors) will stream past the atoms untouched. Thus, the hydrogen atoms absorb light only at certain wavelengths and produce dark lines at those wavelengths in the spectrum we see.

Now suppose we have a container of hydrogen gas through which a whole series of photons is passing, allowing many electrons to move up to higher levels. Next we turn off the light source. These electrons then "fall" back down from larger to smaller orbits and emit photons of light—but, again, only light of those energies or wavelengths that correspond to the energy difference between permissible orbits. The orbital changes of hydrogen electrons giving rise to spectral lines are shown in Figure 4.16.

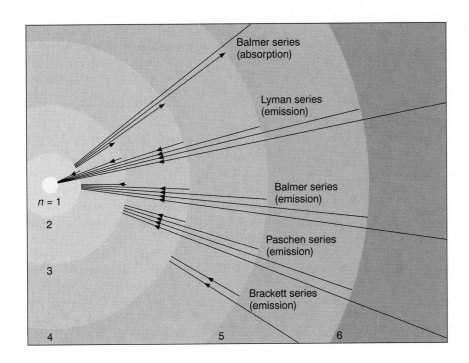

FIGURE 4.16

The Bohr Model for Hydrogen Here we follow the emission or absorption of photons by a hydrogen atom according to the Bohr model. Several different series of spectral lines are shown, corresponding to transitions of electrons from or to certain allowed orbits. Each series of lines that terminates on a specific inner orbit is named for the physicist who studied it. At the top, for example, you see the Balmer series; arrows show electrons jumping from the second orbit ($n = 2$) to the third, fourth, fifth, and sixth orbits. Each time a "poor" electron from a lower level wants to rise a higher position in life, it must absorb energy to do so. It can absorb the energy it needs from passing waves (or photons) of light. The next set of arrows (Lyman series) shows electrons falling down to orbit one from different (higher) levels. Each time an electron goes downward toward the nucleus, it can afford to give off (emit) some energy it no longer needs.

(Note that in a real hydrogen atom the electron orbits are not as evenly spaced as those shown in this diagram.)

Similar pictures can be drawn for atoms other than hydrogen. However, because these other atoms ordinarily have more than one electron each, the orbits of their electrons are much more complicated, and the spectra are more complex as well. For our purposes, the key difference is this: *Each type of atom has its own unique pattern of electron orbits, and no two sets of orbits are exactly alike.* This means that each type of atom will show its own unique set of spectral lines, produced by electrons moving between its unique set of orbits.

If we can learn the lines that go with each element by studying the way atoms absorb and emit light in laboratories here on Earth, we can use this knowledge to identify the elements in celestial bodies. In this way, astronomers today know the chemical makeup of not just any star, but even galaxies of stars so distant that their light started on its way to us long before the Earth had even formed.

Energy Levels and Excitation

4.5.2.

Bohr's model of the hydrogen atom was a great step forward in our understanding of the atom. However, we know today that atoms cannot be represented by quite so simple a picture. Even the concept of sharply defined electron orbits is not really correct. One of the most interesting (and puzzling) results of quantum mechanics is that subatomic particles such as electrons sometimes behave like waves (just as waves of electromagnetic radiation sometimes behave like particles). Because an electron has wave-like characteristics, its position is much harder to pinpoint than you might expect from the Bohr model. We can only estimate the *probability* that it will follow a particular orbit. (If

these ideas are starting to sound a little strange, we assure you that they also sounded strange to physicists when first being proposed. Remember that nature, especially at the subatomic level, is under no obligation to fit "common-sense" human perceptions, which developed in the world of much larger-scale phenomena.)

Luckily, it turns out that the most likely orbits for electrons fall in a fairly narrow range when compared to the size of an atom. This means we can still retain the concept that only certain discrete energies are allowable for an atom. These energies, called **energy levels,** can be thought of as representing certain average distances of the electron's possible orbits from the atomic nucleus.

Ordinarily, an atom is in the state of lowest possible energy, its ground state. In the Bohr model, the **ground state** corresponds to the electron's being in the innermost orbit. An atom can absorb energy, which raises it to a higher energy level (corresponding, in the Bohr picture, to an electron's movement to a larger orbit). The atom is then said to be in an **excited state.** Generally, an atom remains excited for only a very brief time. After a short interval, typically a hundred-millionth of a second or so, it drops back down to its ground state, with the simultaneous emission of light. The atom may return to its lowest state in one jump, or it may make the transition in steps of two or more jumps, stopping at intermediate levels on the way down. With each jump it emits a photon of the wavelength that corresponds to the energy difference between the levels at the beginning and end of that jump.

An energy-level diagram for a hydrogen atom and several possible atomic transitions are shown in Figure 4.17; it shows the same series of emission lines as the diagram of the Bohr model, shown in Figure 4.16. When we measure the energies involved in jumping between levels, we find

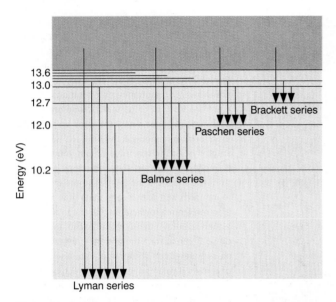

FIGURE 4.17
Energy-Level Diagram for Hydrogen As you get to higher and higher energy levels, they get more and more crowded together, approaching a limit. The shaded region represents energies at which the atom is ionized (the electron is no longer attached to the atom). Each series of arrows represents electrons falling from higher levels to lower ones, releasing photons or waves of energy in the process.

that the transitions to or from the ground state, called the Lyman series of lines, result in the emission or absorption of ultraviolet photons. But the transitions to or from the first excited state (labeled n = 2 in Figure 4.16), called the Balmer series, produce emission or absorption in visible light. In fact it was to explain this Balmer series that Bohr first suggested his model of the atom.

We mentioned that atoms that have absorbed specific photons from a passing beam of white light and have thus become excited generally de-excite themselves and emit that light again in a very short time. You might therefore wonder why *dark* spectral lines are ever produced. In other words, why doesn't this re-emitted light quickly "fill in" the darker absorption lines?

Imagine a beam of white light coming toward you through some cooler gas. Some of the re-emitted light *is* actually returned to the beam of white light you see, but this fills in the absorption lines only to a slight extent. The reason is that the atoms in the gas re-emit light in all directions and only a small fraction of the re-emitted light is in the direction (toward you) of the original beam. In a star, much of the re-emitted light actually goes in directions leading back into the star, which does observers outside the star no good whatsoever.

Figure 4.18 summarizes the different kinds of spectra we have discussed. A lightbulb or the inner (denser) part of a star produces a continuous spectrum (a). When that continuous spectrum is viewed through a thinner cloud of gas,

an absorption line spectrum can be seen superimposed on the continuous spectrum (b). If we look only at a cloud of excited gas atoms (with no continuous source seen behind it), we see that the excited atoms give off an emission line spectrum (c). In both b and c the atoms of thin gas only absorb or give off certain colors, which can be "read off" to tell us what elements are present in the gas.

Atoms in a hot gas are moving at high speeds and continually colliding with one another and with any loose electrons. They can be excited and de-excited by these collisions, as well as by absorbing and emitting light. The speed of atoms in a gas depends on the temperature. When the temperature is higher, so are the speed and energy of the collisions. The hotter the gas, therefore, the more likely that electrons will occupy the outermost orbits, which correspond to the highest energy levels. This means that the level where electrons *start* their upward jumps in a gas can serve as an indicator of how hot that gas is. In this way, the absorption lines in a spectrum can also give astronomers information about the temperature of the regions where the lines originate.

Ionization

We have described how certain discrete amounts of energy can be absorbed by an atom, raising it to an excited state and moving one of its electrons farther from its nucleus. If enough energy is absorbed, the electron can be completely removed from the atom. The atom is then said to be **ionized.** The *minimum amount* of energy required to remove one electron from an atom in its ground state is called its *ionization energy,* or ionization potential.

Still greater amounts of energy must be absorbed by the now-ionized atom (called an **ion**) to remove an additional electron deeper in the structure of the atom. Successively greater energies are needed to remove the third, fourth, fifth, and so on, electrons from the atom. If enough energy is available (in the form of very short-wavelength photons or from a collision with a very fast-moving electron or another atom), an atom can become completely ionized, losing all of its electrons. A hydrogen atom, having only one electron to lose, can be ionized only once; a helium atom can be ionized twice, and an oxygen atom up to eight times. When we examine regions of the cosmos where there is a great deal of radiation, such as the neighborhoods where hot young stars have recently formed, we see a lot of ionization going on.

An atom that has become ionized has lost a negative charge—which was carried away by the electron—and thus is left with a net positive charge. It therefore exerts a strong attraction for any free electron. Eventually, one or more electrons will be captured and the atom will become neutral (or ionized to one less degree) again. During the electron capture process, the atom emits one or more photons, depending on whether the electron is captured at once to the lowest energy level of the atom or stops at one

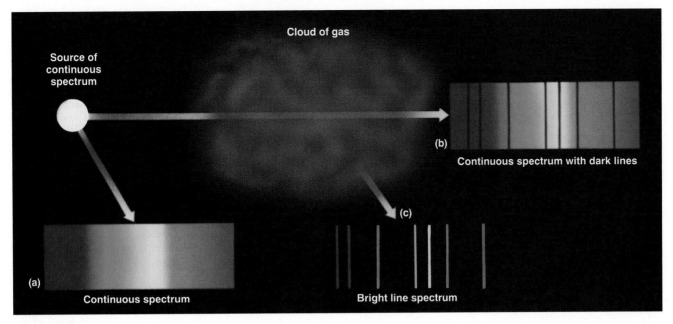

FIGURE 4.18

Three Kinds of Spectra   When we see a lightbulb or other source of continuous radiation (a), all the colors are present. When the continuous spectrum is seen through a thinner gas cloud, the cloud's atoms produce absorption lines in the continuous spectrum (b). When the excited cloud is seen without the continuous source behind it, its atoms produce emission lines (c). We can learn which types of atoms are in the cloud from the pattern of the absorption or emission lines.

or more intermediate levels on its way to the lowest available level.

Just as the excitation of an atom can result from a collision with another atom, ion, or electron (collisions with electrons are usually most important), so can ionization. The rate at which such collisional ionizations occur depends on the speeds of the atoms and hence on the temperature of the gas.

The rate at which ions and electrons recombine also depends on their relative speeds—that is, on the temperature. In addition, it depends on the density of the gas: the higher the density, the greater the chance for recapture, because the different kinds of particles are crowded more closely together. From a knowledge of the temperature and density of a gas, it is possible to calculate the fraction of atoms that have been ionized once, twice, and so on. In the Sun, for example, we find that most of the hydrogen and helium atoms in its atmosphere are neutral, whereas most of the calcium atoms, as well as many other heavier atoms, are ionized once.

The energy levels of an ionized atom are entirely different from those of the same atom when it is neutral. Each time an electron is removed from the atom, the energy levels of the ion, and thus the wavelengths of the spectral lines it can produce, change. This helps astronomers tell the different ions of a given element apart. Ionized hydrogen, having no electron, can produce no absorption lines.

4.6 THE DOPPLER EFFECT

Although the last two sections contain many new concepts, we hope that you have seen one major idea emerge. Astronomers can learn about the elements in stars and galaxies by decoding the information in the spectral lines. There is a complicating factor in learning how to decode the message of starlight, however. If a star is moving toward or away from us, its lines will be in a slightly different place in the spectrum from where they would be in a star at rest. And most objects in the universe do have some motion relative to the Sun.

Motion Affects Waves   4.6.1

In 1842 Christian Doppler pointed out that if a light source is approaching or receding from the observer, the light waves will be, respectively, crowded more closely together or spread out. (This is actually true for all kinds of waves, including sound: Doppler first measured the effect of motion on waves by hiring a group of musicians to play on an open railroad car as it was moving along the track.) The general principle, now known as the **Doppler effect,** is illustrated in Figure 4.19.

In part (a) of the figure, the light source (S) is stationary with respect to the observer. The source gives off a series of

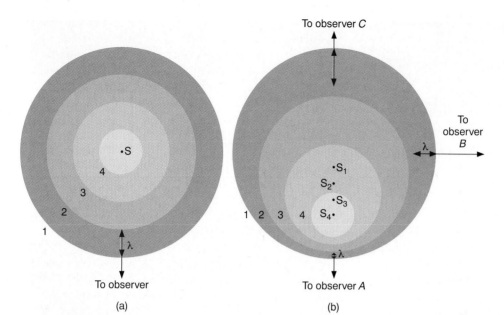

FIGURE 4.19
The Doppler Effect (a) A source S makes waves whose numbered crests (1, 2, 3, 4) wash over a stationery observer. (b) The source S now moves toward observer A and away from observer C. Wave crest 1 was omitted when the source was at position S_1, crest 2 at position S_2, and so forth. Observer A sees the waves compressed by this motion and sees a blueshift (if the waves are light). Observer C sees the waves stretched out by the motion and sees a redshift. Observer B, whose line of sight is perpendicular to the source's motion, sees no change in the waves (and feels left out).

waves, whose crests we have labeled 1, 2, 3, and 4. The light waves spread out evenly in all directions, like the ripples from a splash in a pond. The crests are separated by a distance, λ, where λ is the wavelength. The observer, who happens to be located in the direction of the bottom of the page, sees the light waves coming nice and evenly, one wavelength apart. Observers located anywhere else would see the same thing.

On the other hand, if the source is moving with respect to the observer, as in part (b), the situation is more complicated. Between the time that one crest is emitted and the next is ready to come out, the source has moved a bit—in our case toward the bottom of the page and toward our original observer. By moving, the source has decreased the distance between crests—squeezed the crests together a bit, you might say.

In part (b) we show the source in four positions, S_1, S_2, S_3, and S_4, each corresponding to the emission of one wave crest. To our original observer, now labeled observer A, the waves seem to follow one another more closely, at a decreased wavelength and thus increased frequency. (Remember, because all light waves travel at the speed of light, motion cannot affect the speed. This means that as the wavelength decreases, the frequency must increase. If the waves are shorter, more will be able to fit by during each second.)

The situation is not the same for other observers. Let's look at the situation from the point of view of observer C, located opposite observer A in the figure. For her, the source is moving away from her location. As a result, the waves are not squeezed together, but instead are spread out by the motion of the source. The crests arrive with an increased wavelength and decreased frequency. To observer B, in a direction at right angles to the motion of the source, no effect is observed. The wavelength and frequency remain the same as they were in part (a) of the figure.

We can see from this illustration that the Doppler effect is produced only by a motion toward or away from the observer, a motion called **radial velocity.** Observers between A and B and between B and C would observe, respectively, some shortening or lengthening of the light waves for that part of the motion of the source that is along their line of sight.

You have probably heard the Doppler effect with sound waves in everyday life. When a train whistle or police siren approaches you and then moves away, you will notice a decrease in the pitch (or frequency) of the sound waves. They have changed from slightly more frequent than at rest when coming toward you, to slightly less frequent than at rest when moving away from you. (A nice example of this change in the sound of a train whistle can be heard at the end of the Beach Boys song "Caroline, No" on their album *Pet Sounds*.)

Color Shifts   4.6.2

When the source of waves moves toward you, the wavelengths decrease a bit. If the waves involved are visible light, then the colors of the light change slightly. As wavelengths decrease, they shift a bit toward the blue end of the spectrum: astronomers call this a *blueshift*. (Since the end of the spectrum is really violet, the term should probably be *violetshift*, but blue is a more common color.) When the source moves away from you, and wavelengths get longer, we call the change in colors a *redshift*. Because the Doppler effect was first used with visible light in astronomy, the terms *blueshift* and *redshift* became well established. Today astronomers use these words to describe changes in radio wavelengths or x-ray wavelengths as comfortably as they use them to describe changes in visible light.

The greater the motion toward us or away from us, the greater the Doppler shift. If the relative motion is entirely along the line of sight, the formula for the Doppler shift of light is

$$\Delta\lambda/\lambda = v/c$$

where λ is the wavelength emitted by the source, $\Delta\lambda$ is the difference between λ and the wavelength measured by the observer, c is the speed of light, and v is the relative velocity (speed) of the observer and the source in the line of sight. The variable v is counted as positive if the velocity is one of recession, and negative if it is one of approach. Solving this equation for the velocity, we find

$$v = c\,\Delta\lambda/\lambda$$

If a star approaches or recedes from us, the wavelengths of light in its continuous spectrum appear shortened or lengthened, respectively, as do those of the dark lines. However, unless its speed is tens of thousands of kilometers per second, the star does not appear noticeably bluer or redder than normal. The Doppler shift is thus not easily detected in a continuous spectrum and cannot be measured accurately in such a spectrum. On the other hand, the wavelengths of the absorption lines can be measured accurately, and their Doppler shift is relatively simple to detect.

This may sound like a horrible note on which to end the chapter. If all the stars are moving and motion changes the wavelength of each spectral line, won't this be a disaster for astronomers trying to figure out what elements are present in the stars? After all, it is the precise wavelength (or color) that tells astronomers which lines belong to which element. And we first measure these wavelengths in containers of gas in our laboratories, which are not moving. If every line in a star's spectrum is now shifted by its motion to a different wavelength (color), how can we be sure which lines and which elements we are looking at in a star whose speed we do not know?

This situation sounds worse than it really is. Astronomers rarely judge the presence of an element in an astronomical object by a single line. It is the *pattern* of lines unique to hydrogen or calcium that enables us to determine that those elements are part of the star or galaxy we are observing. The Doppler effect does not change the pattern of lines from a given element—it only shifts the whole pattern slightly toward redder or bluer wavelengths. The shifted pattern is still quite easy to recognize. Best of all, when we do recognize a familiar element's pattern, we get a bonus: The amount the pattern is shifted now tells us the speed of the object in our line of sight.

We hope you can see why the training of new astronomers includes so much work on learning to decode light (and other electromagnetic radiation). We have seen that a skillful "decoder" can learn the temperature of a star, its elemental composition, and even its speed in a direction toward us or away from us. That's really an impressive bit of decoding for a species that is likely to be looking at the stars from afar for some time to come.

SURFING THE WEB

SkyView [skyview.gsfc.nasa.gov/skyview.html]
This multiwavelength "virtual observatory" by Thomas McGlynn allows you to dial up any part of the sky and ask to see what it looks like in various bands of the electromagnetic spectrum. You can also find specific objects, such as the brightest stars. Don't expect pictures as detailed as the ones in this book, but it's fun to see what the sky would look like if you had radio or x-ray eyes.

Tutorial on Temperature
[www.unidata.ucar.edu/staff/blynds/tmp.html]
An introduction to the idea and measurement of temperature, written by astronomer B. D. Lynds, with useful historical information and applications in the realms of astronomy.

Blackbody Radiation Demonstration
[www-astro.phast.umass.edu/courseware/vrml/bb/]
Astronomer Karen Strom has designed this site, which lets you see blackbody curves for different temperatures and then goes on to explore some of the radiation laws we mention briefly in this chapter. Can get technical in places.

SUMMARY

4.1 James Clerk Maxwell showed that whenever charged particles change their motion, as they do in every atom and molecule, they give off waves of energy. Light is one form of this **electromagnetic radiation.** The **wavelength** of light determines the color of visible radiation. Wavelength (λ) is related to **frequency** (f) and the speed of light (c) by the equation $c = \lambda f$. Electromagnetic radiation sometimes behaves like waves, but at other times it behaves as if it were a little packet of energy, called a **photon**. The **apparent brightness** of a source of electromagnetic energy decreases with increasing distance from that source in proportion to the square of the distance, a relationship known as the **inverse-square law.**

4.2 The **electromagnetic spectrum** consists of **gamma rays, x rays,** and **ultraviolet radiation** (all forms of electromagnetic radiation with wavelengths shorter than that of visible light), **visible light,** and **infrared, microwave,** and longer-wave **radio** radiation (the last three with wavelengths longer than that of visible light). Many of these wavelengths cannot penetrate the layers of the Earth's atmosphere and must be observed from space. The emission of electromagnetic radiation is intimately connected to the temperature of the source. The higher the temperature of a **blackbody** (an idealized emitter of electromagnetic radiation), the shorter is the wavelength at which the maximum amount of radiation is emitted. The mathematical equation describing this relationship ($\lambda_{max} = 3 \times 10^6/T$) is known as **Wien's law.** The total energy emitted per square meter increases with increasing temperature. The relationship between emitted energy and temperature ($E = \sigma T^4$) is known as the **Stefan–Boltzmann law.**

4.3 A **spectrometer** is a device that forms a spectrum, often utilizing the optical phenomenon of **dispersion**. The light from an astronomical source can consist of a **continuous spectrum,** a bright line or **emission line spectrum,** or a dark line or **absorption line spectrum.** Because each element leaves its spectral "signature" in the pattern of lines we observe, spectral analyses reveal the composition of the Sun and stars.

4.4 Atoms consist of a **nucleus** containing one or more positively charged protons. All atoms except hydrogen also contain one or more neutrons in the nucleus. Negatively charged electrons orbit the nucleus. The number of protons defines the element (hydrogen, helium, and so on) of the atom. Nuclei with the same number of protons but different numbers of neutrons are different **isotopes** of the same element. According to the Bohr model of the atom, when an electron moves from one orbit to another closer to the atomic nucleus, a photon is emitted and a spectral emission line is formed. Absorption lines are formed when an electron moves to an orbit farther from the nucleus. Since each atom has its own characteristic set of orbits, each is associated with a unique pattern of spectral lines.

4.5 An atom in its lowest **energy level** is said to be in the **ground state.** If an electron is in an orbit other than the least energetic one possible, the atom is said to be **excited.** If an atom has lost one or more electrons, it is called an **ion** and is said to be **ionized.**

4.6 If an atom is moving toward us when an electron changes orbits and produces a spectral line, we see that line shifted slightly toward the blue of its normal wavelength in a spectrum. If the atom is moving away, we see the line shifted toward the red. This shift is known as the **Doppler effect** and can be used to measure the **radial velocities** of distant objects by the formula $v = c(\Delta\lambda/\lambda)$.

INTER-ACTIVITY

A Have your group make a list of all the electromagnetic wave technology that you use during the course of a typical day.

B How many applications of the Doppler effect can your group think of in everyday life? For example, why would the highway patrol find it useful?

C Have members of your group go home and "read" the face of your radio set at home and then compare notes. What do all the words and symbols mean? What frequencies can your radio tune in on? What is the frequency of your favorite radio station? What is its wavelength?

D Suppose astronomers wanted to send a message to an alien civilization that is living on a planet with an atmosphere very similar to that of Earth's. This message must travel through space, make it through the other planet's atmosphere, and be noticeable by the residents of that planet. Have your group discuss what band of the electromagnetic spectrum might be best for this message and why. (Some people, including an earlier Congress, have warned scientists not to send such messages and reveal the presence of our civilization to a possible hostile cosmos. Do you agree with this concern?)

REVIEW QUESTIONS

1. What distinguishes one type of electromagnetic radiation from another? What are the main categories (or bands) of the electromagnetic spectrum?

2. What is a wave? Use the terms *wavelength* and *frequency* in your definition.

3. What is a blackbody? Is this textbook a blackbody? Why or why not? How about your little brother's head? How does the energy emitted by a blackbody depend on its temperature?

4. Where in an atom would you expect to find electrons? Protons? Neutrons?

5. Explain how emission lines and absorption lines are formed. In what sorts of cosmic objects would you expect to see each?

6. Explain how the Doppler effect works for sound waves, and give some familiar examples.

7. What kind of motion for a star does not produce a Doppler effect? Explain why.

8. Describe how Bohr's model used the work of Rutherford and Maxwell. Why was Bohr's model considered a radical notion?

9. Make a list of some of the many practical consequences of Maxwell's theory of electromagnetic waves (television would be one example).

10. Suppose the Sun radiates like a blackbody. Explain how you would calculate the total amount of energy radiated into space by the Sun each second. What information about the Sun would you need to make this calculation?

11. What type of electromagnetic radiation is best suited to observing a star with
    a. a temperature of 5800 K?
    b. a gas heated to a temperature of 1 million K?
    c. a person on a dark night?

12. Why is it dangerous to be exposed to x rays but not (or at least much less) dangerous to be exposed to radio waves?

13. Go outside on a clear night and look carefully at the brightest stars. Some should look red and others blue. The primary factor determining the color of a star is its temperature. Which is hotter, a blue star or a red one? Explain.

14. Water faucets are often labeled with a red dot for hot water and a blue dot for cold. Given Wien's law, does this labeling make sense?

15. The planet Jupiter appears yellowish, and Mars is red. Does this mean that Mars is cooler than Jupiter? Explain your answer.

16. Suppose you are standing at the exact center of a park surrounded by a circular road. An ambulance drives completely around this road, with siren blaring. How does the pitch of the siren change as it circles around you?

17. How could you measure the Earth's orbital speed by photographing the spectrum of a star at various times throughout the year? (*Hint:* Suppose the star lies in the plane of the Earth's orbit.)

18. "Tidal waves," or tsunamis, are waves caused by earthquakes that travel rapidly through the ocean. If tsunamis travel at the speed of 600 km/h and approach a shore at the rate of one wave crest every 15 min, what would be the distance between those wave crests at sea?

19. How many times brighter or fainter would a star appear if it were moved to
    a. twice its present distance?
    b. ten times its present distance?
    c. half its present distance?

20. Two stars with identical diameters are the same distance away. One has a temperature of 5800 K; the other has a temperature of 2900 K. Which is brighter? How much brighter is it?

21. If the emitted infrared radiation from Pluto has a wavelength of maximum intensity at 50,000 nm (50 μm), what is the temperature of Pluto (assuming it behaves like a blackbody)?

22. What is the temperature of a star whose maximum light is emitted at a wavelength of 290 nm?

23. Suppose that a spectral line of some element, normally at 500 nm, is observed in the spectrum of a star to be at 500.1 nm. How fast is the star moving toward or away from the Earth?

Augensen, H. and Woodbury, J. "The Electromagnetic Spectrum" in *Astronomy*, June 1982, p. 6.

Bova, B. *The Beauty of Light.* 1988, Wiley. A readable introduction to all aspects of the production and decoding of light by a science writer.

Connes, P. "How Light Is Analyzed" in *Scientific American,* Sept. 1968.

Darling, D. "Spectral Visions: The Long Wavelengths" in *Astronomy*, Aug. 1984, p. 16; "The Short Wavelengths" in *Astronomy*, Sept. 1984, p. 14.

Gingerich, O. "Unlocking the Chemical Secrets of the Cosmos" in *Sky & Telescope*, July 1981, p. 13.

Hearnshaw, J. *The Analysis of Starlight.* 1986, Cambridge U. Press. A history of spectroscopy.

Newman, J. "James Clerk Maxwell" in *Scientific American,* June 1955.

Sobel, M. *Light.* 1987, U. of Chicago Press. An excellent nontechnical introduction to all aspects of light.

Stencil, R. et al. "Astronomical Spectroscopy" in *Astronomy*, June 1978, p. 6.

Some Introductions to Quantum Mechanics

Gribbin, J. *In Search of Schroedinger's Cat.* 1984, Bantam. Clear, very basic introduction to the fundamental ideas of quantum mechanics, by a British physicist/science writer.

Gribbin, J. *Schroedinger's Kittens and the Search for Reality.* 1995, Little, Brown. Takes off where the above book ends and reports on recent developments.

Pagels, H. *The Cosmic Code: Quantum Physics as the Language of Nature.* 1982, Bantam. An eloquent, very basic introduction by a leading physicist/educator.

This somewhat fanciful artist's conception shows the European Infrared Space Observatory, which operated in Earth orbit from 1995–1998, surveying the sky in wavelengths between 2.4×10^{-6} m and 2.4×10^{-4} m. The 0.6-m aperture telescope on board was cooled to temperatures near absolute zero with liquid helium and was able to make around 26,000 separate observations before the coolant was exhausted. *(ESA painting, courtesy of Peter Bond, Royal Astronomical Society)*

5 Astronomical Instruments

THINKING AHEAD

When you look at the night sky far from city lights on a camping trip, there seem to be an overwhelming number of stars up there. In reality, only about 6000 are visible to the unaided eye. The light from most stars is so weak by the time it reaches Earth that it cannot register on human vision. How can we learn about the vast majority of objects in the universe that our unaided eyes simply cannot see?

"**B**eyond the [faintest stars the eye can see] you will behold through the telescope a host of other stars, which escape the unassisted sight, so numerous as to be almost beyond belief..."

Galileo Galilei in *Siderius Nuncius*, 1610, reporting on his first observations of the night sky with a telescope.

As we saw in the last chapter, we are almost entirely dependent on electromagnetic radiation from space to learn about the universe. Fortunately, objects in the cosmos send us radiation at all wavelengths, from gamma rays to radio waves. Unfortunately for astronomical observations, however, the atmosphere blocks much of the electromagnetic spectrum from reaching the surface.

There are two regions in the spectrum where the Earth's atmosphere is transparent and through which we can observe (see Figure 4.5). One of these is called the *optical* (or visible light) window, and it includes wavelengths from about 300 nm (or 0.3 µm) up to about 30 µm. The other is the *radio* window, which includes wavelengths that range from about 1 mm to about 20 m.

Until recently, astronomers could work only with the radiation in these windows, and all telescopes were designed to receive one or the other of these bands of the spectrum. Today, however, we can also study radiation in parts of the spectrum where the atmosphere is opaque; to observe in these regions, astronomers must put telescopes in orbit.

(a) (b)

FIGURE 5.1
Two Pretelescopic Observatories (a) The Jantar Mantar, built in 1724, by Maharaja Jai
Singh in Delhi, India. (b) Seventeenth-century bronze instruments from the old Chinese
imperial observatory, Beijing. *(David Morrison)*

5.1 TELESCOPES

Many ancient cultures constructed special sites for observing the sky. (Figure 5.1). At these ancient *observatories* they could measure the positions of celestial objects, mostly to keep track of time and date. Many of these ancient observatories had religious and ritual functions as well. Telescopes, which we associate with observatories today, are a relatively late addition, becoming important only in the past 300 years. But their introduction revolutionized our understanding of the universe.

The Astronomical Telescope

Galileo first used telescopes to observe the sky in 1610. As we saw in Chapter 1, his telescopes were simple tubes that could be held in a person's hand, but even so, they got Galileo into a lot of trouble with Church authorities. Astronomical telescopes have come a long way since Galileo's time. Now they tend to be huge devices, constructed at costs of tens of millions of dollars.

The key to understanding telescopes is to realize that celestial objects, such as planets, stars, and galaxies, send much more light to Earth than any human eye (with its tiny opening) can catch. There is plenty of starlight to go around: if you have ever watched the stars with a group of friends, you know that each of you can see each of the stars. If there were a thousand more people watching, each of them would also catch a bit of each star's light. As far as you are concerned, the light not shining into your eye is wasted.

It would be great if some of this "wasted" light could also be captured and brought to your eye. This is precisely what a telescope does.

The most important functions of a telescope are to *collect* the faint light from an astronomical source and to *focus* all this light into an image. Most objects of interest to astronomers are extremely faint; the more light we can collect, the better we can study such objects. You might think of a telescope as a light-gathering "bucket": Just as a bigger bucket allows you to collect more rainwater when it starts to pour, so a bigger telescope can collect more of the light that "rains down" from the night sky. (Bear in mind that, although we use the word *light* in this paragraph, there are types of telescopes that can collect not just visible light, but each of the various forms of electromagnetic radiation. We want no light-chauvinists among our readers!)

In telescopes that collect radiation in the optical window, the light is gathered by a lens or mirror (which we will discuss in a moment). In other types of telescopes, the collecting devices may not look exactly like the lenses or mirrors you know, but they serve the same function. In all types of telescopes, the light-gathering ability is determined by the diameter, or **aperture,** of the device acting as the light-gathering bucket.

To study astronomical objects, it is useful to form an image, which can then be detected, recorded, measured, reproduced, and analyzed in a host of ways. Before the 20th century, astronomers simply viewed images with their eyes, but this was very inefficient and did not provide a very good long-term record. Today, astronomers rarely look through the larger astronomical telescopes. The image is recorded electronically and stored in computers, and this permanent record becomes the object of later detailed study.

5.1.2

Formation of an Image by a Lens or a Mirror

Whether or not you wear glasses, you see the world through lenses; they are key elements of your eyes. A lens is a transparent piece of material that bends parallel rays of light passing through it, bringing them to the same **focus.** Figure 5.2 shows how a simple lens forms an image. If the curvatures of its surfaces are just right, two parallel rays of light (say from a star) are bent, or refracted, in such a way that they converge toward a point, called the focus of the lens. At the focus, an image of the light source appears. The distance of the focus, or image, behind the lens is called the **focal length** of the lens.

As you look at Figure 5.2, you may ask why two rays of light from the same star would be parallel to each other. After all, if you draw a picture of a star, shining in all directions, the rays of light coming from the star don't look parallel at all. Remember that the stars (and other astronomical objects) are all outrageously far away. By the time the rays of light headed our way actually arrive at Earth, they are—for all practical purposes—parallel to each other. Put another way, any rays that were *not* more or less parallel to the ones pointed at Earth are now heading in some very different direction in the universe.

To view the image formed by the primary lens in a telescope, we can install another lens called an *eyepiece.* This lens can magnify the image. Stars are points of light, and magnifying them makes little difference, but the image of a planet or galaxy, which has structure, can often benefit from magnification, just as a tiny image in a magazine can sometimes look clearer through a magnifying glass. Using different eyepieces, we can change the magnification of the image. These days, the eyepiece of a telescope is usually replaced by a camera or electronic light detector (discussed shortly).

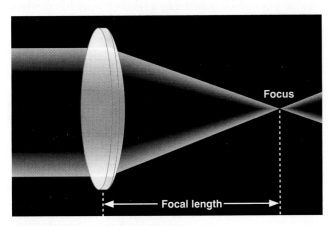

FIGURE 5.2

Formation of an Image by a Simple Lens Parallel rays from a star are bent by the convex lens to a single focus.

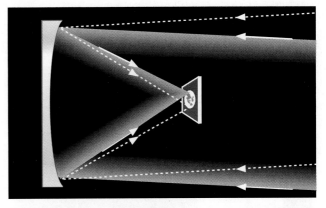

FIGURE 5.3

Formation of an Image by a Mirror Here we see rays from two sides of an object (solid lines and dashed lines). Each set of parallel rays is reflected by the curved surface of the concave mirror, bringing them to a focus at points in front of the mirror.

Rays of light can also be focused to form an image with a concave mirror—one curved like the inner surface of a sphere (Figure 5.3) and coated with metal to make it highly reflecting. If the mirror has the correct concave shape, all parallel rays are reflected back through the same point, the focus of the mirror. Thus, images are produced by a mirror exactly as they are by a lens.

Many people, when thinking of a telescope, picture a long tube with a large glass lens at one end. This design is called a **refracting telescope.** Galileo's telescopes were refractors, as are binoculars or opera glasses today (Figure 5.4.), but refractors (and their lenses) are not very good for most astronomical applications. In large sizes their lenses are difficult to support (because light must pass through them, they can be held at the edge only, not from behind). Also, making a large piece of glass that does not have flaws in it (such as bubbles) is costly and difficult. Most astronomical telescopes (both amateur and professional) use a mirror rather than a lens as their primary optical part; these are called **reflecting telescopes.** Since reflecting telescopes are standard in modern astronomy, we will refer to mirrors in most of the discussion that follows.

The reflecting telescope was conceived by Englishman James Gregory in 1663, and the first successful model was built by Newton in 1668. The concave mirror is placed at the bottom of a tube or open framework. The mirror reflects the light back up the tube to form an image near the front end at a location called the *prime focus.* The image can be observed at the prime focus or various systems of auxiliary mirrors can intercept the light and bring it to a focus at more convenient locations. Two such arrangements are shown in Figure 5.5. In the Newtonian mode, a small secondary mirror is angled to reflect the light off to the side of the telescope tube. In the Cassegrain mode, the secondary mirror reflects the light back down the tube, through a small hole in the primary mirror (which doesn't hurt that mirror's ability to reflect in any significant way), and to a focus below the telescope tube.

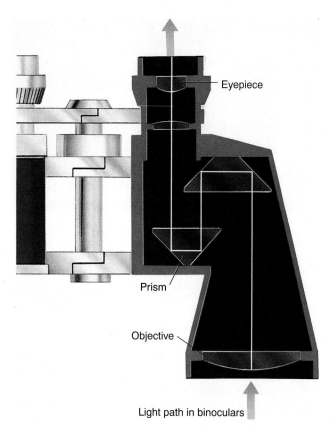

FIGURE 5.4
Binoculars Binoculars are common examples of refracting telescopes. The main light-collecting element in each side is the initial (objective) lens; after following a path through several prisms (used to shorten the length of the instrument), the light is viewed through a magnifying eyepiece. This makes the image both brighter and larger, helping you see what is happening on stage or on the field—even if you can't afford really good tickets.

Properties of Images

The *brightness* of an image is a measure of the amount of energy that is concentrated into a unit area—say, a square millimeter—of the image. The brightness of the image depends on the amount of light focused into it by the primary mirror. Doubling the aperture (size) of the mirror increases its area by a factor of 2×2 or 4, resulting in images that are four times brighter. For example, the two Keck Telescopes (located on a volcanic peak called Mauna Kea in Hawaii) each have an aperture of 10 meters, which is twice the aperture of the older 5-meter Hale telescope at Mt. Palomar (in Southern California) and thus can gather four times as much light from any object. Astronomers who need to see objects that are extremely distant (and thus very faint), such as the most distant galaxies in the universe, will try to get observing time on the largest telescopes they have access to.

Resolution refers to the fineness of detail present in the image. As you can imagine, astronomers are always eager to make out more detail in the images they study,

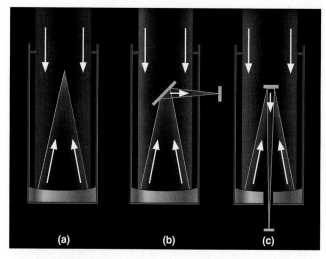

FIGURE 5.5
Focus Arrangements for Reflecting Telescopes (a) Prime focus: Light is detected where it comes to a focus after reflecting from the primary mirror. (b) Newtonian focus: Light is reflected by a small secondary mirror off to one side, where it can be detected. (c) Cassegrain focus: Light is reflected by a small secondary mirror back down through a hole in the primary mirror and detected at a station below.

whether they are following the weather on Jupiter or trying to peer into the violent heart of a cannibal galaxy. The larger the telescope aperture, the sharper the image will be. However, additional limits on resolution are imposed by fluctuations in the Earth's atmosphere, which always introduce a certain amount of blurring in telescopes on the ground. It is important to place observatories at locations where this atmospheric distortion is minimal. (You have probably noticed these fluctuations as the "twinkling" of stars seen from Earth. In space, the light of the stars is steady.)

The resolution of an astronomical image is measured by the angular size of a point source, such as a star. This size is expressed in seconds of arc, or **arcsec,** where 1 arcsec is 1/3600 degree. To give you a sense of how small this unit is, one arcsec is how big a quarter would look when seen from a distance of 5 kilometers. During most of the 20th century, a resolution of about 1 arcsec was considered the standard in astronomy, achieved at major observatory sites only when the atmosphere is unusually steady.

Over the past decade, however, astronomers have been raising their standards. At the best observing locations, such as Mauna Kea in Hawaii, the air is very steady, often yielding resolutions of 0.3 arcsec. New techniques now allow astronomers to further reduce the blurring of images by making tiny but quick changes in the shape of a flexible telescope mirror to compensate for changes in the Earth's atmosphere. With these *adaptive optics*, ground-based telescopes can sometimes match the resolution of the Hubble Space Telescope, which is 0.1 arcsec.

FIGURE 5.6
The 5-Meter Reflector The Hale telescope on Palomar Mountain has a complex mounting structure that enables the telescope (in the open "tube" pointing upwards in this photo) to swing easily into any position. The people in the bottom foreground can give you a sense of scale. *(California Institute of Technology)*

The Complete Telescope

Now that we have seen how an image is produced, we can understand the operation of a telescope. Since its main purpose is to collect light from faint sources, astronomers usually want a telescope with as large an aperture as possible. Thus, they must obtain a large disk of glass, which is laboriously ground and polished (and coated with a reflective layer) to produce a concave mirror of high optical quality. The shape of the mirror must be just right so that all parallel rays of light, no matter where they hit the mirror's surface, come to a single focus.

Once you have a good primary mirror, it must be mounted so that the entire telescope can be pointed toward any object in the sky. When your mirror is 5 meters wide, it has an enormous weight; the mirror alone in the telescope on Mount Palomar weighs 14.5 tons. Thus, you need a complex support structure in order to move it quickly and smoothly from orientation to orientation (Figure 5.6). In addition, since the Earth is rotating, the telescope must have a motorized drive system that moves it backward at exactly the same rate the Earth is moving forward, so it can continue to point at the object being observed. Sophisticated instruments must be built to analyze and record the light collected at the telescope's focus, and all this machinery must be housed in a dome to protect it from the elements.

ASTRONOMY BASICS

How Astronomers Really Use Telescopes

In the popular view, an astronomer spends most nights in a cold observatory peering through a telescope, but this picture, which may have had some validity in previous centuries, is not very accurate today. Most astronomers do not live at observatories, but near the universities or laboratories where they work. A typical astronomer might spend only a week or so each year observing at the telescope and the rest of the time measuring or analyzing the data acquired during that week. Many astronomers use radio telescopes or space experiments, which work just as well during the day. Still others work at purely theoretical problems (often using high-speed computers) and never observe at a telescope of any kind.

Even when astronomers are observing with a large telescope, they seldom peer through it. Electronic detectors record the data permanently for detailed analysis after the observations are completed. At some observatories, it is now possible to conduct observations remotely, with the astronomer sitting at an office computer terminal, which can be thousands of miles away from the telescope.

Time on the major telescopes is at a premium, and an observatory director will typically receive many more requests for telescope time than can be accommodated during the year. Astronomers must therefore write a convincing proposal explaining how they would like to use the telescope and why their observations will be important to the progress of astronomy. A committee of astronomers is then asked to judge and rank the proposals, and time is assigned to those with the most merit. Even if your proposal is among the high-rated ones, you may have to wait many months for your turn. If the skies are cloudy on the nights you have been assigned, it may be more than a year before you get another chance.

Some older astronomers still remember long, cold nights spent alone in an observatory dome, with only music from a tape recorder or all-night radio station for company. The sight of the stars shining brilliantly hour after hour through the open slit in the observatory dome was unforgettable. So was the relief as the first pale light of dawn announced the end of a 12-hour observing session. Astronomy is much easier today, with teams of observers working together, perhaps without ever leaving a computer workstation in a warm room. But some of the romance has gone from the field too.

 OPTICAL DETECTORS AND INSTRUMENTS

As we have seen, the main role of a telescope is to collect as much of the radiation as possible from a given object. But once the radiation has been captured, it must be *detected* and measured. The first detector used for astronomical observations was the human eye, but it suffers from being connected to an imperfect recording and retrieving device, the human brain. Photography and modern electronic detectors have eliminated the foibles and quirks of human memory by making a permanent record of the information from the cosmos.

The eye also suffers from having a very short *integration time*—it takes only a fraction of a second to add light energy together before sending the image to the brain. One important advantage of modern detectors is that the light from astronomical objects can be collected by the detector over longer periods of time; this technique is called "taking a long exposure." Exposures of several hours are required to detect very faint objects in the cosmos.

Astronomers use large telescopes in three basic ways. The first is *imaging*—photographing or otherwise recording the appearance of a small portion of the sky. The second is making an accurate measurement of the *brightness* and *color* of objects. The last is *spectroscopy*—the measurement of the spectra of astronomical sources. All three of these require similar detectors to record and measure the properties of light.

Photographic and Electronic Detectors

Throughout most of the 20th century, photographic film or *plates* served as the prime astronomical detectors, whether for direct imaging or for photographing spectra. In a plate (Figure 5.7), a light-sensitive chemical coating is applied to a piece of glass that, when developed, provides a lasting record of the image. At observatories around the world, vast collections of photographs preserve what the sky has looked like during the past 100 years. Photography represents a huge improvement over the human eye, but it still has serious limitations. Photographic films are inefficient: Only about 1 percent of the light actually falling on the film contributes to the image; the rest is wasted.

FIGURE 5.7

A Photographic Plate   Astronomer Eleanor Helin inspects a 14- × 14-in. glass plate taken at the Palomar Observatory's 48-in. Schmidt telescope. She is searching for trails left by asteroids.

(Kritan Lattu, World Space Foundation)

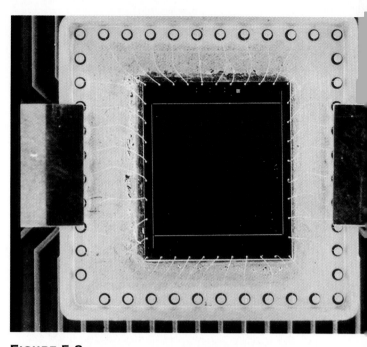

FIGURE 5.8

A CCD Chip   This close-up shows the array of light-sensitive picture elements (called pixels) that can be read off by a computer and used to construct a digital image. These light detectors, chilled to very low temperatures, are far more sensitive to light than the chemicals used in photography.   *(National Optical Astronomy Observatories)*

Astronomers now use much more efficient electronic detectors to record astronomical images, most often **charge-coupled devices (CCDs)**, which are similar to the detectors used in video camcorders (Figure 5.8). In a CCD, photons of radiation generate a stream of charged particles (electrons) that are stored on the detector and counted at the end of the exposure. Because CCDs record as much as 60 to 70 percent of all the photons that strike them, we can detect much fainter objects. They also provide more accurate measurements of the brightness of astronomical objects than photography, and their output can go directly into a computer for analysis.

Infrared Observations

Observing the universe in the infrared band of the spectrum presents some additional challenges. The infrared extends from wavelengths near 1 micrometer, which is about the longwave sensitivity limit of both CCDs and photography, out to 100 mm or longer. Recall from Chapter 4 that infrared radiation is heat radiation. The main challenge to infrared astronomers is to distinguish the tiny amount of heat that reaches the Earth from stars and galaxies from the much greater heat radiated by the telescope itself and our planet's atmosphere.

Typical temperatures on the Earth's surface are near 300 K, and the atmosphere through which observations are made is only a little cooler. According to Wien's law (see Section 4.2), the telescope, the observatory, and even the sky are radiating infrared energy with a peak wavelength of about 10 micrometers. To infrared eyes, everything on Earth is brightly aglow. The challenge is to detect faint cosmic sources against this sea of light. The infrared astronomer must always contend with the situation that a visible-light observer would face if working in broad daylight with a telescope and optics lined with bright fluorescent lights.

To solve this problem, astronomers must protect the infrared detector from nearby radiation, just as you would shield photographic film from bright daylight. Since anything warm radiates infrared energy, the detector must be isolated in very cold surroundings; often it is held near absolute zero (1 to 3 K) by immersing it in liquid helium. The second step is to reduce the radiation emitted by the telescope structure and optics, and to block this heat from reaching the infrared detector.

Like the infrared, each band of the electromagnetic spectrum presents its own challenges to the astronomer. What kind of mirror can you use to reflect such penetrating radiation as x rays and gamma rays, for example? And, as we saw, many kinds of waves from space can be detected only from observatories in orbit, which are costly to launch and sometimes impossible to repair. We will return to these topics later in the chapter.

Spectroscopy

We have discussed detectors as if they were always used to record an image of a portion of the sky, but they can also be used to record a spectrum. Spectroscopy is one of the astronomer's most powerful tools, and more than half the time spent on most large telescopes is used for spectroscopy.

The many different wavelengths present in light can be separated by passing it through a prism to form a spectrum. A *spectrometer* is an instrument designed to record such a spectrum. The design of a simple spectrometer is illustrated in Figure 5.9. Light from the source (actually, the image of a source produced by the telescope) enters the instrument through a small hole or narrow slit and is collimated (made into a beam of parallel rays) by a lens. The light then passes through a prism, producing a spectrum: Different wavelengths leave the prism in different directions because of dispersion. A second lens placed behind the prism focuses the many different images of the slit or entrance hole on the CCD or other detecting device. This collection of images (spread out by color) is the spectrum that astronomers can then analyze at leisure.

In practice, astronomers today are more likely to use a different device, called a *grating*, to disperse the spectrum. A grating is a piece of transparent material with thousands of grooves in its surface. The grooves cause the light waves

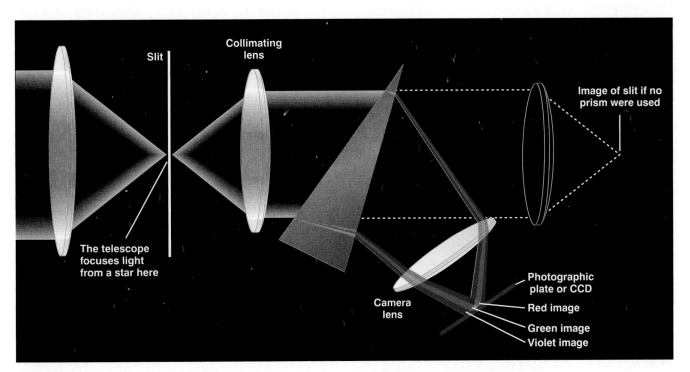

FIGURE 5.9
A Prism Spectrometer The light from the telescope is focused on a slit, so the light of one object at a time enters the spectrometer. A prism (or grating) disperses the light into a spectrum, which is then photographed or recorded electronically.

FIGURE 5.10

Kitt Peak The Kitt Peak National Observatory near Tucson, Arizona, is part of the National Optical Astronomy Observatories. The mountaintop houses a series of telescopes and instruments supported by the National Science Foundation and made available to astronomers from around the country who make strong research proposals. *(NOAO)*

to interfere with each other (slightly differently for each wavelength) with the result that the light also spreads out into a spectrum.

5.3 OPTICAL AND INFRARED OBSERVATORIES

A modern observatory is usually a collection of astronomical telescopes (Figure 5.10). Since the end of the 19th century, astronomers have realized that the best observatory sites are on mountains far from the lights and pollution of cities. Although a number of urban observatories remain, especially in the large cities of Europe, they have become administrative centers or museums. The real action takes place far away, often on desert mountains or isolated peaks in the Atlantic and Pacific Oceans, where we find the living quarters, computers, electronic and machine shops, and of course the telescopes themselves. A large observatory today requires a supporting staff of 20 to 100 people in addition to the astronomers.

Picking the Best Observing Sites

The performance of an optical telescope is determined not only by the size of its mirror but also by its location. The Earth's atmosphere, so vital to life, is the biggest headache to the observational astronomer. In at least four ways the air imposes limitations on the usefulness of telescopes:

1. The most obvious limitation is weather—clouds, wind, rain, and the like. At the best sites the weather is clear as much as 75 percent of the time.
2. Even on a clear night, the atmosphere filters out a certain amount of starlight, especially in the infrared, where the absorption is due primarily to water vapor. Astronomers therefore prefer dry sites, generally found at the highest altitudes.
3. The sky should also be dark. Near cities the air scatters the glare from lights, producing an illumination that hides the faintest stars and limits the distances that can be probed by telescopes. (Astronomers call this effect *light pollution.*) Observatories are best located at least a hundred miles from the nearest large city.
4. Finally, the air is often unsteady; light passing through this turbulent air is disturbed, resulting in blurred star images. Astronomers call these effects "bad **seeing.**" When seeing is bad, images of celestial objects are distorted by the constant twisting and bending of light rays by turbulent air.

The best observatory sites are therefore high, dark, and dry. The world's largest telescopes are found in such remote mountain locations as the Andes mountains of Chile, the desert peaks of Arizona, the Canary Islands in the Atlantic Ocean, and Mauna Kea in Hawaii, a dormant volcano 13,700 ft (4200 m) high (Figure 5.11).

FIGURE 5.11
A High and Dry Site Cerro Paranal, a mountain summit 2.7 km above sea level in Chile's Atacama desert, is the site of the European Southern Observatory's Very Large Telescope. This photograph, taken in late 1990, shows the beginnings of the construction and vividly illustrates that astronomers prefer high and dry sites for their instruments. *(ESO)*

Major New Telescopes

Most of the great telescopes constructed during the first half of the 20th century, including the 100-in. (2.5-m) and 200-in. (5-m) telescopes (see the *Voyagers in Astronomy* box on page 116), were built with private funds and were available only to a small group of astronomers. Following the success of the 200-in. telescope, astronomers from universities in the East and Midwest began a campaign for a national observatory to provide comparable facilities for the rest of the astronomical community. The National Science Foundation eventually agreed, and Kitt Peak National Observatory was built in Arizona (Figure 5.10), with a branch established in the Andes mountains of Chile to provide access to the southern skies. A consortium of European nations founded the European Southern Observatory, also in Chile. The largest telescopes at these new observatories were about 4 m in aperture. Still larger (6 m) was a telescope built by what was then called the Soviet Union, but this instrument was much less productive due to a poorer site and long history of technical problems.

The kinds of research that astronomers are working on today require even bigger light buckets, and construction of ground-based telescopes has proceeding at an unprecedented pace throughout the 1990s. (See Table 5.1, which also includes Web sites for each telescope, in case you want to visit or learn more about them.) New technologies are being used to construct instruments with roughly twice the aperture of the 200-in. telescope, and in some cases to multiply their power further by combining the light from more than one telescope. These instruments have thinner, lighter mirrors and more compact mechanical designs.

The twin 10-m Keck telescopes on Mauna Kea (Figure 5.12) are the first of these new-technology instruments.

FIGURE 5.12
The Keck Telescopes The Keck Observatory is located on Hawaii's Mauna Kea, a tall and (we sure hope) extinct volcano. The domes house twin 10-m telescopes, which (during the 1990s) were the world's largest reflectors—until the components of the Very Large Telescope in Chile were connected. Note how clear the sky is in this picture. *(Wm. Keck Obs.)*

TABLE 5.1
Large Optical Telescopes Being Built or in Operation

Aperture (m)	Telescope Name	Location	Status	Web Address
16.4	Very Large Telescope (four 8.2-m telescopes)	Cerro Paranal, Chile*	First telescope completed 1998	www.eso.org/vlt/
11.8	Large Binocular Telescope (two 8.4-m telescopes)	Mount Graham, Arizona	First light 2002–2003	medusa.as.arizona.edu/btwww/tech/lbtbook.html
10.0	Keck I	Mauna Kea, Hawaii	Completed 1993	astro.caltech.edu/mirror/keck/index.html
10.0	Keck II	Mauna Kea, Hawaii	Completed 1996	astro.caltech.edu/mirror/keck/index.html
9.9	Hobby–Eberly (HET)	Mount Locke, Texas	Completed 1997	www.astro.psu.edu/het/overview.html
8.3	Subaru (Pleiades)	Mauna Kea, Hawaii	First light 1998	www.naoj.org/
8.0	Gemini (North)	Mauna Kea, Hawaii†	First light 1999	www.gemini.edu
8.0	Gemini (South)	Cerro Pachon, Chile†	First light 2000	www.gemini.edu
6.5	Multi-Mirror (MMT)	Mount Hopkins, Arizona	First light 1998	sculptor.as.arizona.edu.edu/foltz/www/
6.5	Magellan	Las Campanas, Chile	First light 1997	www.ociw.edu/~johns/magellan.html
6.0	Large Alt-Azimuth	Mount Pastukhov, Russia	Completed 1976	—
5.0	Hale	Mount Palomar, California	Completed 1948	astro.caltech.edu/observatories/palomar/public/index.html
4.2	William Herschel	Canary Islands, Spain	Completed 1987	www.ast.cam.ac.uk/ING/PR/pr.html
4.2	SOAR	Cerro Pachon, Chile	First light 2002	www.noao.edu/
4.0	Blanco Telescope (NOAO)	Cerro Tololo, Chile†	Completed 1974	www.ctio.noao.edu/ctio.html
3.9	Anglo-Australian (AAT)	Siding Spring, Australia	Completed 1975	www.aao.gov.au/index.html
3.8	NOAO Mayall	Kitt Peak, Arizona†	Completed 1973	www.noao.edu/noao.html
3.8	United Kingdom Infrared (UKIRT)	Mauna Kea, Hawaii	Completed 1979	www.jach.hawaii.edu/UKIRT/home.html
3.6	Canada–France–Hawaii (CFHT)	Mauna Kea, Hawaii	Completed 1979	www.cfht.hawaai.edu/
3.6	ESO	Cerro La Silla, Chile*	Completed 1976	www.ls.eso.org/
3.6	ESO New Technology	Cerro La Silla, Chile*	Completed 1989	www.ls.eso.org/
3.5	Max Planck Institut	Calar Alto, Spain	Completed 1983	www.mpia-hd.mpg.de/CAHA/
3.5	WIYN	Kitt Peak, Arizona†	Completed 1993	www.noao.edu/wiyn/wiyn.html
3.5	Astrophysical Research Corp.	Apache Point, New Mexico	Completed 1993	www.apo.nmsu.edu/
3.0	Shane (Lick Observatory)	Mount Hamilton, California	Completed 1959	www.ucolick.org/
3.0	NASA Infrared (IRTF)	Mauna Kea, Hawaii	Completed 1979	irtf.ifa.hawaii.edu

* Part of the European Southern Observatory (ESO).

† Part of the U.S. National Optical Astronomy Observatories (NOAO).

Instead of a single primary mirror 10 m in diameter, each Keck telescope achieves its large aperture by combining the light from 36 separate hexagonal mirrors, each 1.8 m in width (Figure 5.13). Computer-controlled actuators constantly adjust these 36 mirrors so that the overall reflecting surface maintains just the right shape to collect and focus the light with high accuracy.

Even larger is the European Very Large Telescope (VLT), which consists of four 8-m telescopes operated together (and boasting the light-gathering power of a 16.4-m mirror). The VLT, located on an isolated peak in Chile's Atacama desert called Cerro Paranal, is expected to be fully operational in about 2001 (Figure 5.14). Several additional large telescopes of 8.2-m aperture are under construction by various governments or consortia of universities around the world, including two by the U.S. National Observatory, one on Mauna Kea, and one in Chile. Each of these new instruments is designed to operate in the infrared as well as the visible part of the spectrum. These telescopes will also be equipped with adaptive optics to improve their resolution, as discussed in Section 5.1.

FIGURE 5.13
Thirty-Six Eyes Are Better Than One A close-up of the 10-m Keck telescope's mirror as it was being assembled. The finished mirror is composed of 36 hexagonal sections. In this view, 18 sections have been installed. To get a sense of the mirror's size, note the person working at the center of the mirror assembly. *(California Association for Research in Astronomy)*

FIGURE 5.14
Very Large Telescope Mirror A consortium of European nations is building the world's largest telescope in the mountains of Chile. Four individual telescopes with mirrors 8.2 m in diameter will work together, creating a total light-collecting area equivalent to that of a 16-m telescope. Here we see the mirror for the first telescope in 1995 during testing in France; tests revealed that if the mirror were scaled up to be the size of the Paris metropolitan area (165 km in diameter), the deviation from its required shape would be only about 1 millimeter. *(Photo by Hans-Hermann Heyer, ESO)*

 ## 5.4 RADIO TELESCOPES

In 1931, engineer Karl G. Jansky of the Bell Telephone Laboratories was experimenting with antennas for long-range radio communication when he encountered some mysterious static—radio radiation coming from an unknown source (Figure 5.15). He discovered that this radiation came in strongest about 4 min earlier on each successive day and correctly concluded that since the Earth's sidereal rotation period is 4 min shorter than a solar day (see Section 3.3), the radiation must be originating from some region fixed on the celestial sphere. Subsequent investigation showed that the source of the radiation was part of the Milky Way; Jansky had discovered the first source of cosmic radio waves.

In 1936, Grote Reber, an amateur astronomer and radio ham, built from galvanized iron and wood the first antenna specifically designed to receive these cosmic radio waves. Over the years, Reber built several such antennas and used them to carry out pioneering surveys of the sky for celestial radio sources; he remained active in

FIGURE 5.15
The First Radio Telescope
The rotating radio antenna used by Jansky in his serendipitous discovery of radio radiation from the Milky Way. On June 8, 1998, a memorial sculpture was placed at the site of this 1932 telescope in Holmdel, New Jersey. *(Bell Laboratories)*

George Ellery Hale: Master Telescope Builder

All of the major research telescopes in the world today were built within this century. The giant among early telescope builders was George Ellery Hale (1868–1938). Not once but four times he initiated projects that led to construction of what was at the time the world's largest telescope. And he was a master at finding and winning over wealthy benefactors to underwrite the construction of these new instruments.

Hale's training and early research were in solar physics. In 1892, at age 24, he was named Associate Professor of Astral Physics and director of the astronomical observatory at the University of Chicago. At that time, the largest telescope in the world was the 36-in. refractor at the Lick Observatory near San Jose, California. Taking advantage of an existing glass blank for a 40-in. telescope, Hale set out to raise the money for a larger telescope than the one at Lick. One prospective donor was Charles T. Yerkes, who, among other things, ran the trolley system in Chicago.

Hale wrote to Yerkes, encouraging him to support construction of the giant telescope by saying that "the donor

George Ellery Hale *(Caltech)*

Four times he initiated projects that led to construction of the world's largest telescope.

could have no more enduring monument. It is certain that Mr. Lick's name would not have been nearly so widely known today were it not for the famous observatory established as a result of his munificence." Yerkes agreed, and the new telescope was completed in May 1897; it remains the largest refractor in the world.

Even before the completion of the Yerkes refractor, Hale was not

only dreaming of building a still-larger telescope, he was taking concrete steps to achieve that goal. In the 1890s there was a major controversy about the relative quality of refracting and reflecting telescopes. Hale realized that 40 in. was close to the maximum feasible aperture for refracting telescopes. If telescopes with significantly larger apertures were to be built, they would have to be reflecting telescopes.

Using funds borrowed from his own family, Hale set out to construct a 60-in. reflector. For a site, he left the Midwest for the much better conditions on Mount Wilson, at that time a wilderness peak above the small city of Los Angeles. In 1904, at the age of 36, Hale received funds from the Carnegie Foundation to establish the Mount Wilson Observatory. The 60-in. mirror was placed in its mount in December 1908.

Two years earlier, in 1906, Hale had already approached John D. Hooker, who had made his fortune in hardware and steel pipe, with a proposal to build a 100-in. telescope. The technological risks were substantial. The 60-in. telescope was not yet complete, and the utility of large reflectors for astronomy had yet to be demon-

strated. George Ellery Hale's brother called him "the greatest gambler in the world." Once again, Hale was successful in obtaining funds, and the 100-in. telescope was completed in November 1917. (It was with this telescope that Edwin Hubble was able to establish that the spiral nebulae were separate islands of stars—or galaxies—quite removed from our own Milky Way; see Chapter 25.)

Hale was not through dreaming. In 1926, he wrote an article in *Harper's Magazine* about the scientific value of a still larger telescope. This article came to the attention of the Rockefeller Foundation, which granted $6 million (a very large sum in those days) for the construction of a 200-in. telescope. Hale died in 1938, but the 200-in. (5-m) telescope on Palomar Mountain was dedicated ten years later and is now named in Hale's honor.

The Yerkes 40-inch (1-m) refracting telescope. *(Yerkes Observatory)*

radio astronomy for more than 30 years. During the first decade, he worked practically alone, because professional astronomers had not yet recognized the vast potential of radio astronomy.

5. 4. 1

Detection of Radio Energy from Space

It is important to understand that radio waves cannot be "heard": They are not the sound waves you hear coming out of the radio receiver in your home or car. Like light, radio waves are a form of electromagnetic radiation, but unlike light, we cannot detect them with our senses—we must rely on electronic equipment to pick them up. In commercial radio broadcasting, we encode sound information (music or the voice of a newscaster) into the radio waves, which must be decoded at the other end and turned back into sound by speakers or headphones. There are two familiar ways to *modulate* (encode information into) radio waves: by changing the amplitude of the waves, called *amplitude modulation* or AM radio, and by changing the frequency of the waves, called *frequency modulation* or FM.

The radio waves we receive from space do not, of course, have rock music or other program information encoded in them (although astronomers would love to tune in on a radio broadcast from some distant civilization; see Chapter 15). If cosmic radio signals were translated into sound, they would sound like the static you hear when dialing between stations. Nevertheless, there is information in the radio waves we receive—information that can tell us about the chemistry and physical conditions in the sources of the waves.

Just as vibrating charged particles can produce electromagnetic waves (see Chapter 4), electromagnetic waves can make charged particles move up and down. Radio waves can produce a current in conductors of electricity such as metals. An antenna is such a conductor: It intercepts radio waves, which induce a feeble current in it. The current is then amplified in a radio receiver until it is strong enough to measure or record. Like your television or radio set, receivers can be tuned to select a single frequency (channel). In astronomy today, however, it is more common to use sophisticated data-processing techniques that allow thousands of separate frequency bands to be detected simultaneously. Thus, the astronomical radio receiver operates much like a spectrometer on an optical telescope, giving information about how much radiation we receive at each wavelength or frequency. After computer processing, the radio signals are recorded on magnetic disks for further analysis.

Radio Telescopes 5. 4. 2

Radio waves are reflected by conducting surfaces just as light is reflected from a shiny metallic surface, and according to the same laws of optics. A radio-reflecting telescope consists of a concave metal reflector (called a *dish*), analogous to a telescope mirror. The radio waves collected by the dish are reflected to a focus, where they can then be directed to a receiver and analyzed. Because humans are such visual creatures, radio astronomers often construct a pictorial representation of the radio sources they observe. Figure 5.16 shows such a radio image of a distant galaxy, where radio telescopes reveal vast jets and regions of radio emission that are completely invisible in photographs taken with light.

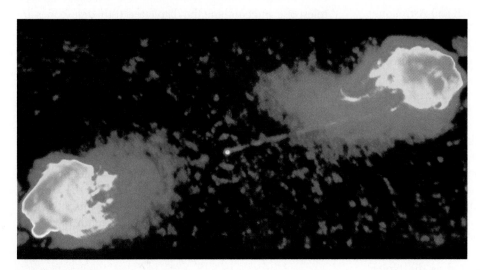

FIGURE 5.16
A Radio Image This image has been constructed of radio observations of a galaxy called Cygnus A. Colors have been added to help the eye sort out regions of different radio intensity. Red regions are the most intense, blue the least. The visible galaxy would be a small dot in the center of the image. The radio image reveals jets of expelled material (over 160,000 LY long) on either side of the galaxy. *(NRAO/AUI)*

FIGURE 5.17
Large Radio Dish The 100-m radio telescope near Bonn, Germany. *(Max Planck Institut für Radioastronomie)*

Radio astronomy is a young field compared with optical astronomy, but it has experienced tremendous growth in recent decades. Several nations lacking high-quality sites for optical telescopes have instead chosen to concentrate their astronomical efforts in the radio part of the spectrum; these include the Netherlands, Britain, Germany, and Japan. The world's largest radio reflectors that can be pointed to any direction have apertures of 100 m. One of these is at the Max Planck Institute for Radio Astronomy, located near Bonn, Germany (Figure 5.17); the other is at the U.S. National Radio Astronomy Observatory in West Virginia. Table 5.2 lists some of the major radio telescopes of the world.

Radio Interferometry

As we discussed earlier, a telescope's ability to show us fine detail (its resolution) depends on its aperture, but it also depends on the wavelength of the radiation the telescope is gathering. The longer the waves, the harder it is to resolve fine detail in the images or maps we make. Because radio waves have such long wavelengths, they present tremendous challenges for astronomers who need good resolution. In fact, even the largest radio dishes on Earth—operating alone—cannot make out as much detail as the typical small optical telescope used in a college astronomy lab. To overcome this difficulty, radio astronomers have learned to sharpen their images by linking two or more radio telescopes together electronically.

Telescopes linked together in this way are called an **interferometer**. This may seem like a strange term, because

the telescopes in an interferometer work cooperatively; they don't interfere with each other. *Interference,* however, is a technical term for the way waves that arrive in our instruments at slightly different times interact with each other, and this interaction allows us to coax more detail out of our observations. The resolution of an interferometer depends on the separation of the telescopes, not on their individual apertures. Two telescopes separated by 1 km provide the same resolution as would a single dish 1 km across (although they are not, of course, able to collect as much radiation as a radio-wave bucket 1 km across).

To get even better resolution, astronomers combine a large number of radio dishes into an **interferometer array.** In effect, such an array works as a large number of two-dish interferometers, all observing the same part of the sky together. Computer processing of the results permits the reconstruction of a high-resolution radio image. The most extensive such instrument is the National Radio Astronomy Observatory's Very Large Array (VLA) near Socorro, New Mexico (see the opening figure in Chapter 4). It consists of 27 movable radio telescopes, each having an aperture of 25 m, spread over a total span of about 36 km. By electronically combining the signals from all of its individual telescopes, this array permits the radio astronomer to make pictures of the sky at radio wavelengths comparable to those obtained with an optical telescope, with a resolution of about 1 arcsec.

Initially, interferometer arrays were limited in size by the requirement that all of the dishes be accurately wired together. The maximum dimensions were thus only a few tens of kilometers. However, larger interferometer separations can be achieved if the telescopes do not require a physical connection. Astronomers have learned to record the radiation coming from space very precisely at each telescope and combine the data later. If the telescopes are as far apart as California and Australia, or as West Virginia and Crimea in Ukraine, the resulting resolution far surpasses that of optical telescopes.

The United States operates the Very Long Baseline Array (VLBA), made up of ten individual telescopes stretching from the Virgin Islands to Hawaii (Figure 5.18). The VLBA, completed in 1993, can form astronomical images with a resolution of 0.0001 arcsec, permitting features as small as 10 AU to be distinguished at the center of our Galaxy. When the VLBA is used in conjunction with a Japanese radio telescope in space, the size of the array extends beyond the dimensions of the planet itself (Figure 5.19). Table 5.2 lists the largest arrays currently in use.

Radar Astronomy

Radar is the technique of transmitting radio waves to an object in our solar system and then detecting the radio radiation that the object reflects back. The time required for the round trip can be measured electronically with great precision. Because we know the speed at which radio

TABLE 5.2
Major Radio Observatories of the World

Observatory	Location	Description	Web Site
Individual Radio Dishes			
Arecibo Telescope (National Astron. & Ionospheric Center)	Arecibo, Puerto Rico	305-m fixed dish	www.naic.edu
Greenbank Telescope (National Radio Astron. Observ.)	Green Bank, West Virginia	100- × 110-m steerable dish	www.gb.nrao.edu/GBT/GBT.html
Effelsberg Telescope (Max Planck Institute für Radioastronomie)	Bonn, Germany	100-m steerable dish	www.mpifr-bonn.mpg.de/effberg.html
Lovell Telescope (Jodrell Bank Radio Observat.)	Manchester, England	76-m steerable dish	www.jb.man.ac.uk/
Goldstone Tracking Station (NASA/JPL)	Barstow, California	70-m steerable dish	gts.gdscc.nasa.gov/
Australia Tracking Station (NASA/JPL)	Tidbinbilla, Australia	70-m steerable dish	tid.cdscc.nasa.gov/
Parkes Radio Observatory	Parkes, Australia	64-m steerable dish	www.parkes.atnf.csiro.au/
Arrays of Radio Dishes			
Australia Telescope	Several sites in Australia	8-element array (seven 22-m dishes plus Parkes 64-m)	www.atnf.csiro.au/
MERLIN	Cambridge, England and other British sites	Network of 7 dishes (the largest of which is 32 m)	www.jb.man.ac.uk/merlin/
Westerbork Radio Observatory	Westerbork, the Netherlands	12-element array of 25-m dishes (1.6-km baseline)	www.nfra.nl/wsrt
Very Large Array (NRAO)	Socorro, New Mexico	27-element array of 25-m dishes (36-km baseline)	www.nrao.edu/doc/vla/html/VLAhome.shtml
Very Long Baseline Array (NRAO)	Ten U.S. sites, Hawaii to Virgin Islands	10-element array of 25-m dishes (9000-km baseline)	www.nrao.edu/doc/vlba/html/VLBA.html
Very-Long-Baseline-Interferom. Space Observ. Program (VSOP)	Connect a satellite to network on Earth	Japanese HALCA 8-m dish in orbit and ≈40 dishes on Earth	sgra.jpl.nasa.gov/
Millimeter-Wave Telescopes			
IRAM	Granada, Spain	30-m steerable mm-wave dish	iram.fr/
James Clerk Maxwell Telescope	Mauna Kea, Hawaii	15-m steerable mm-wave dish	www.jach.hawaii.edu/JCMT/pages/intro.html
Nobeyama Cosmic Radio Observatory	Minamimaki-Mura, Japan	6-element array of 10-m mm-wave dishes	www.nro.nao.ac.jp/~nma/index-e.html
Hat Creek Radio Observatory (University of California)	Cassel, California	6-element array of 5-m mm-wave dishes	bima.astro.umd.edu/bima

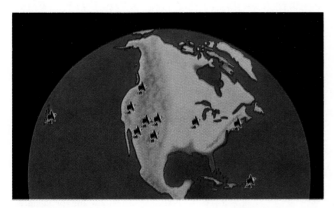

FIGURE 5.18
The Very Long Baseline Array A map showing the distribution of the ten antennas that constitute an array or radio telescopes stretching across the United States. *(National Radio Astronomy Observatory)*

waves travel (the speed of light), we can determine the distance to the object or a particular feature on its surface (such as a mountain).

Radar observations of the Moon and planets have yielded our best knowledge of the distances to these worlds and played an important role in navigating spacecraft throughout the solar system. In addition, as will be discussed in later chapters, radar observations have determined the rotation periods of Venus and Mercury, probed tiny Earth-approaching asteroids, and allowed us to investigate the surfaces of Mercury, Venus, Mars, and the large satellites of Jupiter.

Any radio dish can be used as a radar telescope if it is equipped with a powerful transmitter as well as a receiver. The most spectacular facility in the world for radar astronomy is the 1000-ft (305-m) telescope at Arecibo in Puerto

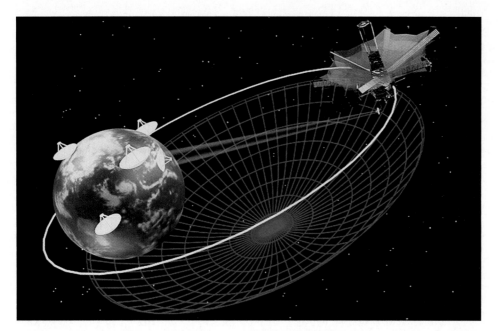

FIGURE 5.19
Very Long Baseline Interferometry Space Observatory This artist's conception shows the Japanese Highly Advanced Laboratory for Communications and Astronomy (HALCA) satellite, which orbits the Earth every 6.3 hours and has an 8-m radio reflector on board. Connected to an array of radio dishes on Earth, it provides radio astronomers with a resolution equivalent to having a telescope larger than our entire planet. With the resolution of the combined telescopes, you would be able to make out a grain of rice in Tokyo from Los Angeles. *(Institute of Space and Astronautical Science, Japan)*

Rico (Figure 5.20). The Arecibo telescope is too large to be pointed directly toward different parts of the sky. Instead, it is constructed in a huge natural "bowl" (more than a mere dish!) formed by several hills and is lined with reflecting metal panels. The Arecibo bowl has a volume roughly equal to the world's annual beer consumption. A limited ability to track astronomical sources is achieved by moving the receiver system, which is suspended on cables 100 m above the surface of the bowl.

5.5 OBSERVATIONS OUTSIDE THE EARTH'S ATMOSPHERE

The Earth's atmosphere blocks most radiation at wavelengths shorter than visible light, so we can make ultraviolet, x-ray, and gamma-ray observations only from space. Getting above the distorting effects of the atmosphere is also an advantage at visible and infrared wavelengths. The stars don't twinkle in space, so the amount of detail you can observe is limited only by the size of your instrument. On the other hand, it is expensive to place telescopes into space and repairs can present a major challenge. This is why astronomers continue to build telescopes for use on the ground as well as for launching into space.

Airborne and Space Infrared Telescopes

Water vapor, the main source of atmospheric interference for making infrared observations, is concentrated in the lower part of the Earth's atmosphere. For this reason a gain of even a few hundred meters in elevation can make an important

FIGURE 5.20
The Largest Radio and Radar Dish The Arecibo Observatory, with its 1000-ft radio dish filling a valley in Puerto Rico, is part of the National Astronomy and Ionosphere Center, operated by Cornell University under a cooperative agreement with the National Science Foundation. *(National Astronomy and Ionosphere Center)*

MAKING CONNECTIONS

Choosing Your Own Telescope

If the astronomy course you are taking whets your appetite for exploring the sky further, you may be thinking about buying your own telescope. The good news is that many excellent amateur telescopes are now on the market, and prices are much more reasonable than they were 20 years ago. The bad news is that a good telescope, like a good camera or video recorder, is still not inexpensive, and some research is required to find the model that is best for you. The best sources of information about personal telescopes are the two popular-level magazines published for amateur astronomers, *Astronomy* and *Sky & Telescope*. Both carry regular articles of advice, reviews, and ads from reputable telescope dealers around the country. (See also some Web sites listed at the end of the chapter.)

In some ways, choosing a telescope is like choosing a car: Personal preference can play a major role in the decision. A certain shape or brand name may be important to some car buyers, but no big deal to others. In the same way, some of the factors determining which telescope is right for you depend on your preferences:

- Will you be setting up the telescope in one place and leaving it there, or do you want an instrument that is portable and can come with you on camping trips? Does it have to be carried some distance by a person, or can it get to its destination in a car?
- Do you want to observe the sky with your eyes only, or do you want to take photographs? (Long-exposure photography, for example, requires a good clock-drive to turn your telescope to compensate for the Earth's rotation.)
- What sorts of objects will you be observing? Are you interested primarily in comets, planets, star clusters, or galaxies? Or do you want to observe all kinds of celestial sights?

You may not know the answers to some of these questions yet. For this reason, our number-one recommendation is that you "test-drive" some telescopes first. Most communities have amateur astronomy clubs that sponsor star parties open to the public. The members of these clubs often know a lot about telescopes and can share their ideas with you. Your instructor may know where the nearest amateur astronomy club meets; many of the clubs in the country are also listed on the Web sites for *Sky & Telescope* (www.skypub.com) and *Astronomy* magazine (www.astronomy.com).

Furthermore, you may already have an instrument like a telescope at home (or have access to one through a relative or friend). Many amateur astronomers recommend starting your survey of the sky with a good pair of binoculars. These are easily carried around and can show you many objects not visible (or clear) to the unaided eye.

When you are ready to purchase a telescope, you might find the following thoughts useful:

- The key characteristic of a telescope is the aperture of the main mirror or lens; when someone says they have a 6-in. or 8-in. telescope, they mean the diameter of the collecting surface. The larger the aperture, the more light you can gather and the fainter the objects you can see or photograph.
- Telescopes of a given aperture that use lenses (refractors) are typically more expensive than those using mirrors (reflectors) because both sides of a lens must be polished to great accuracy. And, because the light passes through it, the lens must be made of high-quality glass throughout. In contrast, only the front surface of a mirror must be accurately polished.
- Magnification is not one of the criteria on which to base your choice of a telescope. As we discussed in the main text, the magnification of the image is done by a smaller eyepiece, so the magnification can be changed by changing eyepieces. However, a telescope will magnify not only the astronomical object you are viewing, but also the turbulence of the Earth's atmosphere. If the magnification is too high, your image will shimmer and shake and be difficult to view. A good telescope will come with a variety of eyepieces that stay within the range of useful magnifications.
- The mount of a telescope is one of its most critical elements. Because a telescope shows a tiny field of view, which is magnified significantly, even the smallest vibration or jarring of the telescope can move the object you are viewing around, or out of your field of view. A sturdy and stable mount is essential for serious viewing or photography (although it clearly affects how portable your telescope can be).
- A telescope requires some practice to set up and use effectively. Don't expect everything to go perfectly on your first try. Take some time to read the instructions. If a local amateur astronomy club is nearby, use it as a resource.

difference in the quality of an observatory site. Given the limitations of high mountains, most of which attract clouds and violent storms, it was natural for astronomers to investigate the possibility of observing infrared waves from airplanes and ultimately from space.

Infrared observations from airplanes have been made since the 1960s, starting with a 15-cm telescope on board a Learjet. From 1974 through 1995, NASA operated a 0.9-m airborne telescope flying regularly out of the Ames Research Center south of San Francisco. Observing from an altitude of 12 km, the telescope was above 99 percent of the atmospheric water vapor. NASA is now constructing (in partnership with the German government) a much larger 2.5-m telescope, called the Stratospheric Observatory for Infrared Astronomy (SOFIA), to fly in a modified Boeing 747.

Getting even higher and making observations from space itself has important advantages for infrared astronomy. First is the elimination of all interference from the atmosphere. Equally important is the opportunity to cool the entire optical system of the telescope in order to nearly eliminate infrared radiation from this source as well. If we tried to cool a telescope within the atmosphere, it would quickly become coated with condensing water vapor and other gases, making it useless. Only in the vacuum of space can optical elements be cooled to hundreds of degrees below freezing and still remain operational.

The first orbiting infrared observatory, launched in 1983, was the Infrared Astronomical Satellite (IRAS), built as a joint project among the United States, the Netherlands, and Britain (Figure 5.21). IRAS was equipped with a 0.6-m telescope cooled to a temperature of less than 10 K. For the first time the infrared sky could be seen as if it were night, rather than through a bright foreground of atmospheric and telescope emissions. IRAS carried out a rapid but comprehensive survey of the entire infrared sky over a ten-month period, cataloging over 250,000 sources of infrared radiation. Since then, two other infrared telescopes have operated in Earth orbit, similar in aperture to the first IRAS system, but with much better sensitivity and resolution due to improvements in infrared detectors. Both NASA and the Japanese Space Agency plan advanced infrared telescopes for the beginning of the next century (see Table 5.3).

Hubble Space Telescope

In April 1990, a great leap forward (or should we say upward) in astronomy was made with the launch of the Hubble Space Telescope (HST) (Figure 5.22). This telescope has an aperture of 2.4 m, the largest put into space so far. (Its aperture was limited by the size of the payload bay in the Space Shuttle, which was its launch vehicle.) It was named for Edwin Hubble, the astronomer who discovered the expansion of the universe in the 1920s.

HST is operated jointly by the NASA Goddard Space Flight Center and the Space Telescope Science Institute in Baltimore. It is the first orbiting observatory designed to be serviced by Shuttle astronauts, and their visits in 1993, 1997, and 1999 have allowed improvements and replacements of its initial instruments.

HST's mirror was ground and polished to a remarkable degree of accuracy. If we were to scale up its mirror to the size of the entire continental United States, there would be no hill or valley larger than about 6 cm in its smooth surface. Unfortunately, after it was launched, scientists discovered that the primary mirror had a slight error in its shape equal to roughly 1/50 the width of a human hair. Small as that sounds, it was enough to ensure that much of the light entering the telescope did not come to a clear focus and that all the images were blurry. In a misplaced effort to save money, a complete test of the optical system had not been carried out before launch, so the error was not discovered until HST was in orbit.

The solution was to do something very similar to what we do for astronomy students with blurry vision: put corrective optics in front of their eyes. In December 1993, in one of the most exciting and difficult space missions ever flown, astronauts captured the orbiting telescope and

FIGURE 5.21

IRAS The Infrared Astronomical Satellite, a joint project of NASA, the Netherlands, and the United Kingdom. Here the spacecraft is being tested before its 1983 launch. *(NASA/JPL)*

TABLE 5.3
Some Observatories in Space

Observatory	Dates of Operation	Bands of the Spectrum	Notes	Web Site
Einstein (HEAO-2)	1978–1981	x rays	first to take x-ray images	heasarc.gsfc.nasa.gov/docs/heasarc/missions/heao2.html
International UV Explorer (IUE)	1978–1996	UV	UV spectroscopy	www.vilspa.esa.es/iue/iue.html
Infrared Astron. Satellite (IRAS)	1983–1984	IR	mapped 250,000 sources	www.ipac.caltech.edu/ipac/iras/iras.html
Hipparcos	1989–1993	optical	measured over 100,000 precise positions	astro.estec.esa.nl/SA-general/Projects/Hipparcos/hipparcos.html
COBE	1989–	IR, mm	observed the 3-degree background radiation	www.gsfc.nasa.gov/astro/cobe/ed_resources.html
Compton Gamma-Ray Observ.	1990–	gamma	gamma-ray sources and spectra	cossc.gsfc.nasa.gov/cossc/cgro.html
Hubble Space Tel. (HST)	1990–	optical UV, IR	2.4-m mirror; images and spectra	oposite.stsci.edu/
Rosat	1990–1998	x rays	x-ray images and spectra	wave.xray.mpe.mpg.de/rosat
Infrared Space Obs. (ISO)	1995–1998	IR	infrared images and spectra	isowww.estec.esa.nl/
Rossi X-Ray Timing Explorer	1995–	x rays	variability of x-ray sources	heasarc.gsfc.nasa.gov/docs/xte/XTE.html
BeppoSAX	1996–	x rays	observes over wide spectral range	www.sdc.asi.it
HALCA	1997	radio	8-m radio dish, makes array with dishes on the ground	www.vsop.isas.ac.jp/
FUSE	1998–	UV	far-UV spectroscopy	fuse.pha.jhu.edu/
Chandra X-Ray Astronomy Facility	2000–	x rays	x-ray images and spectra	xrtpub.harvard.edu/
Stratospheric Observ. for IR Astron. (SOFIA)	2002–	IR	airborne 2.5-m IR telescope	sofia.arc.nasa.gov/

brought it back into the Shuttle payload bay. There they installed a package containing compensating optics as well as a new improved camera before releasing HST back into orbit. Figure 5.23 illustrates the improvement in the clarity of the images that were returned to Earth. We have sprinkled the new images and discoveries from HST throughout this book and invite you to enjoy them for their beauty as well as for their information content.

Now that it's been refurbished, HST is the premier astronomical instrument in the world, used by hundreds of astronomers from around the globe. It is also the most expensive, with an annual operating cost of about a quarter of

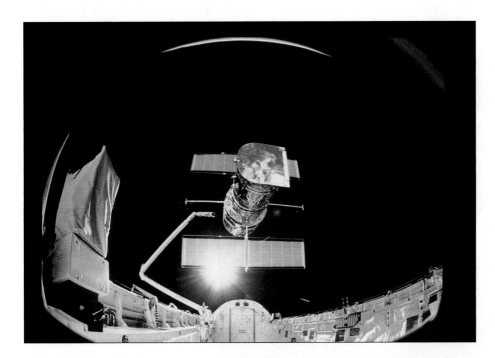

FIGURE 5.22
Launching the Hubble The Hubble Space Telescope being launched in April 1990 from the Space Shuttle Discovery. The robot arm, or Remote Manipulator System, was operated by astronomer Steven Hawley on this historic occasion. The image was taken with a special 70-mm IMAX camera system. You can see the light of the Earth reflected from the closed cover of the telescope. The orange panels are solar arrays, which capture the Sun's light to provide energy for the instruments. *(IMAX/OMNIMAX image © 1990 Smithsonian Institution/Lockheed Corporation)*

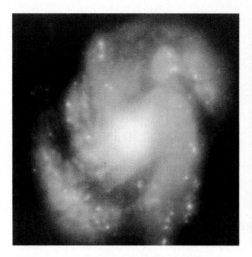

a billion dollars—ten times as much as any ground-based telescope. Its great success has prompted NASA to plan for a larger successor, called the Next Generation Space Telescope (NGST).

High-Energy Observatories

Ultraviolet, x-ray, and gamma-ray observations can only be made from space. Such observations first became possible in 1946, with V2 rockets captured from the Germans. The U.S. Naval Research Laboratory put instruments on these rockets for a series of pioneering flights, used initially to detect ultraviolet radiation from the Sun. Since then, many other rockets have been launched to make x-ray and ultraviolet observations of the Sun, and later of other celestial objects as well.

Beginning in 1960s, a steady stream of high-energy observatories has been launched into orbit to reveal and explore the universe at short wavelengths. The most successful and productive of these were the Einstein x-ray telescope (launched by NASA in 1978), the Roentgen *Satellite* or ROSAT (launched by the Germans in 1990), and the Compton Gamma-Ray Observatory (launched by NASA 1991) (see Table 5.3).

In the x-ray band, for example, Rosat was able to survey and catalog more than 100,000 sources of cosmic x rays and reveal to us details in their structure that previous instruments had been unable to show. The newest and most advanced high-energy observatory is the NASA Chandra X-Ray Astrophysics Facility, launched in early 1999. It is designed to produce x-ray images with unprecedented resolution and sensitivity.

Gamma rays are still more energetic than x rays, and astronomical sources of gamma radiation are thus likely to be very exotic members of the astronomical zoo. Among these are the puzzling objects that give off brief but intense bursts of cosmic gamma rays, first discovered in 1967 by the Vela satellites. (These were launched by the U.S. Department of

FIGURE 5.24
The Compton Gamma-Ray Observatory The deployment of the observatory in April 1991 from the Space Shuttle Atlantis. With a mass of more than 16 tons, it was one of the largest scientific payloads ever launched into space. When the satellite's antenna would not open, two of the astronauts took an unscheduled space walk to free it. Here the observatory is shown high above Africa just before it was released into orbit. *(NASA)*

Defense to carry out worldwide surveillance for possible explosions of nuclear bombs, which emit gamma rays.) The Compton Gamma-Ray Observatory (Figure 5.24) has now cataloged thousands of the high-energy bursts or flashes from all over the sky.

The worldwide reassessment of government priorities in the 1990s has had an impact on astronomy. Both NASA and its European counterpart, the European Space Agency (ESA), have scaled back their commitments to space science, and the Russian space program, once the most ambitious of any nation, has been devastated by budgetary problems. However, the spectacular success of HST and the other space observatories should assure that astronomy will continue to have a place in space. The emphasis is expected to be on smaller, cheaper satellites, but the number of good ideas for astronomy space missions continues to grow.

SURFING THE WEB

Web sites for the major telescopes and observatories on the ground and in space are given in Tables 5.1 through 5.3.

Early Radio Astronomy [www.nrao.edu/intro/ham.connection.html]
A brief summary of the history of radio astronomy, with emphasis on the connection with amateur (or "ham") radio. The rest of the site, maintained by the National Radio Astronomy Observatory, also has some useful background information, along with more technical pages for astronomers who use the observatory's facilities around the country (students can look for the button called Radio Astronomy Fundamentals).

Whispers from the Cosmos [www.ncsa.uiuc.edu/Cyberia/Bima/BimaHome.html]
This on-line "exhibit" about radio astronomy in general, and the Berkeley–Illinois–Maryland Association array of radio telescopes in particular, contains some useful background information about how radio observations can help us learn about a variety of cosmic objects and processes.

Orbiting Astronomical Observatories [www.seds.org/~spider/oaos/oaos.html]
A good annotated list, with fine links, of the telescopes in Earth orbit.

History of High-Energy Astrophysics [heasarc.gsfc.nasa.gov/docs/heasarc/headates/heahistory.html]
The staff of the Laboratory for High-Energy Astrophysics at NASA's Goddard Space Flight Center has assembled a good chronology and set of web links detailing all major experiments and space missions relevant to the study of gamma rays, x rays, or ultraviolet waves from space. Find an instrument you are looking for (including some that are not yet launched), and you can go directly to its home page.

Sites with Information on Buying Your Own Telescope:

Astronomy magazine: [www.kalmbach.com/astro/Hobby/Beginner/firstscope/firstscope.html]

Sky & Telescope magazine: [www.skypub.com/tips/telescopes/telescopes.html]

Frequently Asked Questions about Telescope Buying (by amateur astronomer Dennis Bishop): [www-personal.umich.edu/~dnash/saafaq/faq.html]

SUMMARY

5.1 The astronomical telescope collects light and forms an image, using a convex lens in a **refractor** or a concave mirror in a **reflector** to bring it to a **focus**. The distance from the lens or mirror to the focus is called the **focal length.** The diameter, or **aperture,** of the telescope determines the brightness and **resolution** of the image. Resolution is usually expressed in units of arcseconds.

5.2 Optical (visible light) detectors include the eye, photographic film, and the **charge-coupled device (CCD)** detector. Detectors sensitive to infrared radiation must be cooled to very low temperatures. A spectrometer disperses the light into a spectrum to be recorded for detailed analysis.

5.3 Telescopes are housed in domes and controlled by computers. Observatory sites must be carefully chosen for clear weather, dark skies, low water vapor, and excellent atmospheric **seeing** (low atmospheric turbulence). A new generation of large instruments has been constructed in the 1990s, including the 10-m Keck telescopes at Mauna Kea and the four 8-m instruments that constitute the European Very Large Telescope in Chile.

5.4 In the 1930s, radio astronomy was pioneered by Jansky and Reber. A radio telescope is basically a radio antenna (often a large curved dish) connected to a receiver. Significantly enhanced resolution can be obtained with **interferometers**, including **interferometer arrays** like the 27-element VLA. Expanding to very long baseline interferometers, radio astronomers can achieve resolutions as good as 0.0001 arcsec. **Radar** astronomy involves transmitting as well as receiving. The largest radar telescope is the 305-m bowl at Arecibo.

5.5 Infrared observations are made with telescopes aboard aircraft and in space, as well as from ground-based facilities on dry mountain peaks. Ultraviolet, x-ray, and gamma-ray observations must be made from above the atmosphere. Many orbiting observatories have been flown to observe in these bands of the spectrum in the last few decades. The largest aperture instrument in space is the Hubble Space Telescope (HST). The Compton Gamma-Ray Observatory (GRO) is the most sophisticated instrument for observing high-energy gamma rays ever built. Many additional astronomical satellites are planned, although most are smaller (and less expensive) than HST.

INTER-ACTIVITY

A Most large telescopes get many more proposals for observing projects than there is night observing time available in the course of a year. Suppose your group is the telescope time allocation committee reporting to an observatory director. What criteria would you use in deciding how to give out time on the telescope? What steps could you take to make sure all your colleagues thought the process was fair and people would still talk to you at future astronomy meetings?

B Your group is a committee of nervous astronomers, about to make a proposal to the government of a smaller European country to chip in to build the world's largest telescope in the high dry desert of the Chilean Andes mountains. You expect the government ministers to be very skeptical about supporting this project. What arguments would you make to convince them to participate?

C The same government ministers we met in activity B ask you to draw up a list of the pros and cons of having the world's largest telescope in the mountains of Chile (instead of a mountain in Europe). What would your group list as a pro and as a con?

D Make a list of all the ways that an observing session at a large optical telescope and a large radio telescope might differ. (*One hint:* Bear in mind that because the Sun is not especially bright at many radio wavelengths, observations with radio telescopes can often be done during the day.)

E Another "environmental threat" to astronomy (besides light pollution) comes from the encroachment of terrestrial communications into the "channels"—wavelengths and frequencies—previously reserved for radio astronomy. For example, the demand for cellular phones means that there could be more and more radio channels, used for this purpose. Thus the faint signals from cosmic radio sources will be drowned in a sea of earthly conversation (translated and sent as radio waves). Assume your group is a Congressional committee being lobbied by both radio astronomers who want to save some clear channels for doing astronomy and the companies that stand to make a lot of money from expanding cellular phone use. What arguments would sway you to each side? [For a real-world example of where this issue is being debated, see *Science* magazine, Nov. 28, 1997, p. 1569.]

REVIEW QUESTIONS

1. Name the two spectral windows through which electromagnetic radiation reaches the surface of the Earth, and describe the largest aperture telescope currently in use for each window.

2. List the six bands into which we commonly divide the electromagnetic spectrum, and list the largest-aperture telescope currently in use in each band.

3. When astronomers discuss the apertures of their telescopes, they say bigger is better. Explain why.

4. What are the properties of an image, and what factors determine each?

5. Compare the eye, photographic film, and CCDs as detectors for light. What are the advantages and disadvantages of each?

6. Radio and radar observations are often made with the same antenna, but otherwise they are very different techniques. Compare and contrast radio and radar astronomy in terms of the equipment needed, the methods used, and the kind of results obtained.

7. Why do astronomers place telescopes in Earth orbit? What are the advantages for different spectral regions?

8. What was the problem with the Hubble Space Telescope and how was it solved?

9. Describe the techniques radio astronomers use to obtain a resolution comparable to what astronomers working with visible light can achieve.

THOUGHT QUESTIONS

10. What happens to the image produced by a lens if the lens is "stopped down" with an iris diaphragm—a device that covers its periphery?

11. What would be the properties of an *ideal* astronomical detector? How closely do the actual properties of a CCD approach this ideal?

12. Fifty years ago, the astronomers on the staff of Mount Wilson and Palomar Observatories each received about 60 nights per year for their observing programs. Today an astronomer feels fortunate to get 10 nights per year on a large telescope. Can you suggest some reasons for this change?

13. The largest observatory complex in the world is on Mauna Kea in Hawaii, at an altitude of 4.2 km. This is by no means the tallest mountain on Earth. What are some factors astronomers consider when selecting an observatory site? Don't forget practical ones. Should astronomers, for example, consider building an observatory on Mount McKinley (Denali) or Mount Everest?

14. Another site recently developed for astronomy is the Antarctic plateau. Discuss its advantages and disadvantages.

15. Suppose you are looking for sites for an optical observatory, an infrared observatory, and an x-ray observatory. What are the main criteria of excellence for each? What sites are actually considered the best today?

16. Radio astronomy involves wavelengths much longer than those of visible light, and many orbiting observatories have probed the universe for radiation of very short wavelengths. What sorts of objects and physical conditions would you expect to be associated with radiation emissions of very long and very short wavelengths?

17. The dean of a university located near the ocean proposes building an infrared telescope right on campus and operating it in a nice heated dome so that astronomers will be comfortable on cold winter nights. Criticize this proposal, giving your reasoning.

PROBLEMS

18. The resolution of a radio interferometer is proportional to the maximum spacing of the antennas. The VLA, with a maximum antenna separation of 36 km, achieves a resolution of 1 arcsec at a wavelength of 6 cm.

    a. What is the resolution of a very long baseline interferometer at this wavelength if the antennas are separated by 3600 km?

    b. What is the resolution of the VLBA, with antennas on opposite sides of the Earth?

19. A typical large telescope today requires about $4 million per year to operate. What is the cost per night to use such a telescope? Per hour? Per minute? Can you think of any other activities that have such a high associated cost, in dollars per hour?

20. The HST cost about $1.7 billion for construction and $300 million for its Shuttle launch, and it costs $250 million per year to operate. If the telescope lasts a total of ten years, what is the cost per year? Per day? If the telescope can be used just 30 percent of the time for actual observations, what is the cost per hour and per minute for the astronomer's observing time on this instrument?

SUGGESTIONS FOR FURTHER READING

Chaisson, E. *The Hubble Wars.* 1994, HarperCollins. Controversial story of the building of the HST.

Dyer, A. "What's the Best Telescope for You" in *Sky & Telescope,* Dec. 1997, p. 28.

Florence, R. *The Perfect Machine: Building the Palomar Telescope.* 1994, Harper Perennial. Highly readable account of how the 200-in. telescope was built.

Fugate, R. and Wild, W. "Untwinkling the Stars" in *Sky & Telescope,* May 1994, p. 24; June 1994, p. 20. Discusses adaptive optics.

Gehrels, N. et al. "The Compton Gamma-Ray Observatory" in *Scientific American,* Dec. 1993, p. 68.

Graham, R. "Astronomy's Archangel" in *Astronomy,* Nov. 1998, p. 56. On master mirror-maker Roger Angel of the University of Arizona.

Hawley, S. "Hubble Revisited" in *Sky & Telescope,* Feb. 1997, p. 42. An astronomer–astronaut describes the second servicing mission for HST.

Janesick, J. and Blouke, M. "Sky on a Chip: The Fabulous CCD" in *Sky & Telescope,* Sep. 1987, p. 238.

Junor, B. et al. "Seeing the Details of the Stars with Next Generation Telescopes" in *Mercury,* Sept./Oct. 1998, p. 26. On adaptive optics.

Krisciunas, K. *Astronomical Centers of the World.* 1988, Cambridge U. Press. History of and guide to major observatories.

Krisciunas, K. "Science with the Keck Telescope" in *Sky & Telescope,* Sept. 1994, p. 20. What it's like to use the telescope, and what work is being done with it.

Pilachowski, C. and Trueblood, M. "Telescopes of the 21st Century" in *Mercury,* Sept./Oct. 1998, p. 10. On the telescopes now being built or planned on the ground.

Preston, R. *First Light.* 1987, Atlantic Monthly Press. Superbly written popular book on the work being done at the 200-in. telescope on Mount Palomar.

Shore, L. "VLA: The Telescope That Never Sleeps" in *Astronomy,* Aug. 1987, p. 15.

Sinnott, R. and Nyren, K. "The World's Largest Telescopes" in *Sky & Telescope,* July 1993, p. 27. Several pages of data tables.

Tarenghi, M. "Eyewitness View: First Sight for a Glass Giant" in *Sky & Telescope,* Nov. 1998, p. 47. On the first observations with the Very Large Telescope.

Tucker, W. and K. *Cosmic Inquirers.* 1986, Harvard U. Press. An excellent introduction to the VLA, Einstein x-ray observatory, and other major instruments.

Wakefield, J. "Keck Trekking" in *Astronomy,* Sept. 1998, p. 52. Introduction to the Keck telescopes.

A Rock Called Yogi on Mars In this image taken during the Mars Pathfinder mission in 1997, we see a meter-sized rock on the surface of Mars, nicknamed Yogi. The "super-resolution" image was obtained by adding together seven different photographs and processing the result on a computer. This allowed astronomers to bring out details (especially in the shadows) that could not be seen on any individual image. Yogi was located about 5 meters northwest of the Pathfinder Lander and was one of the first rocks whose composition was investigated by the little six-wheeled robot rover that Pathfinder carried to Mars. The reddish color of the martian soil is caused by the presence of iron oxides (the technical name for rust.) *(NASA/JPL)*

6 Other Worlds: An Introduction to the Solar System

THINKING AHEAD

Surrounding the Sun is a family of worlds, large and small: nine planets, dozens of moons, and smaller pieces too numerous to count. It is a complex system, with a wide range of conditions on the various worlds. Yet the system has some remarkable regularities to it. Most objects circle the Sun in the same direction and in the same plane, for example. What caused the orbits of all these worlds to share a common motion?

"In all of the history of mankind, there will be only one generation that will be the first to explore the Solar System, one generation for which, in childhood, the planets are distant and indistinct discs moving through the night sky, and for which, in old age, the planets are places, diverse new worlds in the course of exploration."

Carl Sagan in *The Cosmic Connection* (1973, Doubleday)

During the past 30 years we have learned more about the solar system than anyone imagined before the space age. In addition to gathering information with powerful new telescopes, we have sent spacecraft directly to many members of the planetary system. (Planetary astronomy is the only branch of our science in which we can—at least vicariously—travel to the objects we want to study.) With evocative names like Voyager, Pioneer, and Pathfinder, our robot explorers have flown past or around every planet except Pluto, returning images and data that have dazzled both astronomers and the public. In the process, we have also investigated dozens of fascinating moons, four ring systems, four asteroids, and two comets.

Probes have penetrated the atmospheres of Venus, Mars, and Jupiter; others have landed on the surfaces of Venus, Mars, and the Moon (Figure 6.1). Humans have even set foot on the Moon and returned samples of its surface soil for analysis in the laboratory.

The planet chapters in today's textbooks are nothing like the ones earlier generations of students were reading. We can now envision the members of the solar system as other worlds like our own, each with its own chemical and geological history, and each with unique sights that interplanetary tourists may someday

FIGURE 6.1
Viking on Mars This 1976 photograph shows the Viking 2 lander on the surface of Mars. The reddish color is the result of iron oxides in the soil. *(NASA)*

even visit. Some have called these last few decades the golden age of planetary exploration, in analogy with another golden age of exploration—the period starting in the 15th century, when great sailing ships plied the Earth's oceans, and humanity began to get familiar with our own planet's surface.

In this chapter we discuss our planetary system, introducing the idea of *comparative planetology*—studying how the planets work by comparing them with each other. We want to get to know the planets not only for their own sakes, but also to see what they can tell us about the origins and evolution of the entire solar system. In the next seven chapters, we will discuss how scientists have learned so much about the planets and describe the better-known members of the solar system in some detail. We will return to the theme of origins and comparative planetology in Chapter 13.

6.1 OVERVIEW OF OUR PLANETARY SYSTEM

The solar system[1] consists of the Sun and many smaller objects: the planets, their satellites and rings, and such "debris" as asteroids, comets, and dust. Decades of observation and spacecraft exploration have revealed that most of these

---

[1] The generic term for a group of planets and other bodies circling a star is *planetary system.* Ours is called the solar system, because our Sun is sometimes called Sol. Strictly speaking, then, there is only one solar system; planets circling other stars are planetary systems.

objects formed together with the Sun about 4.5 billion years ago. They represent aggregations of material that condensed from an enormous cloud of gas and dust. The central part of this cloud became the Sun, and a small fraction of the material in the outer parts eventually formed the other objects.

An Inventory

The Sun, a rather ordinary star, is by far the most massive member of the solar system, as shown in Table 6.1. It is an enormous ball of incandescent gas, 1.4 million km in diameter, with an interior temperature of millions of degrees. The Sun will be discussed later in this book, as our first (and best-studied) example of a star.

Table 6.1 also shows that most of the material of the planets is actually concentrated in the largest one, Jupiter, which is more massive than all the rest of the planets combined. Astronomers were able to measure the masses of the planets centuries ago, using Newton's laws of gravity, by measuring the planets' gravitational effects on each other or on moons that orbit about them. Today we make even more precise measurements of their mass by tracking their gravitational effects on the motion of spacecraft that pass near them.

Besides the Earth, five other planets (Mercury, Venus, Mars, Jupiter, and Saturn) were known to the ancients, and three (Uranus, Neptune, and Pluto) were discovered since the invention of the telescope. The nine planets all revolve in the same direction around the equator of the Sun (Figure 6.2). They orbit in approximately the same plane, like cars traveling on concentric tracks on a giant, flat racing course. As we have seen in earlier chapters, each planet stays in its own traffic lane, following a nearly circular orbit about the Sun and obeying the "traffic laws" discovered by Galileo, Kepler, and Newton.

Each of the planets also rotates (spins) about an axis running through it, and in most cases the direction of rotation is the same as that of revolution about the Sun. The exception is Venus, which rotates very slowly backward (that is, in a retrograde direction). Uranus and Pluto also have

TABLE 6.1
Mass of Members of the Solar System

Object	Percentage of Total Mass
Sun	99.80
Jupiter	0.10
Comets	0.05
All other planets	0.04
Satellites and rings	0.00005
Asteroids	0.000002
Cosmic dust	0.0000001

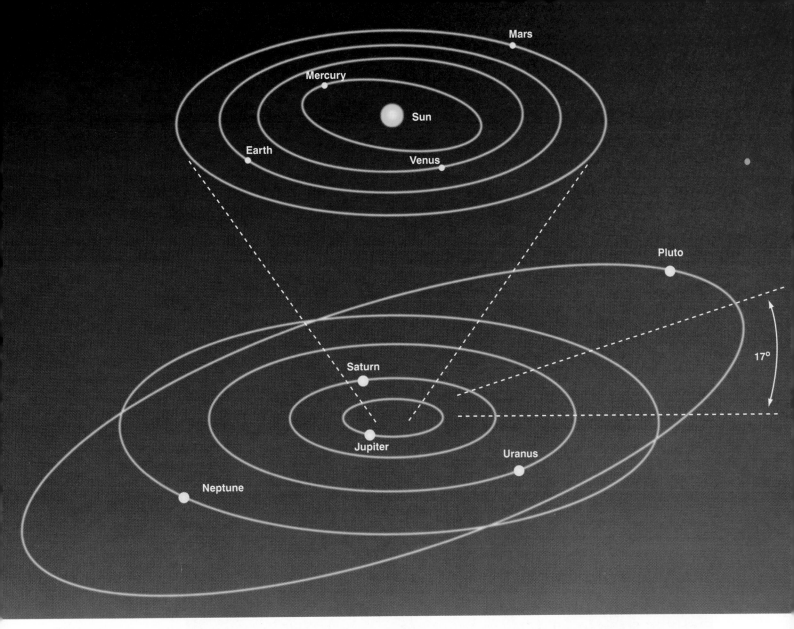

FIGURE 6.2
The Orbits of the Planets All the planets except Pluto orbit the Sun in roughly the same plane.

strange rotations, each spinning about an axis tipped nearly on its side.

The four planets closest to the Sun (Mercury through Mars) are called the inner or **terrestrial planets;** often the Moon is also discussed as a part of this group, bringing the total of terrestrial bodies to five. (We generally call the Earth's satellite the *Moon*, with a capital M, and the other satellites *moons*, with lowercase ms.) The terrestrial planets are relatively small worlds, composed primarily of rock and metal. All of them have solid surfaces that bear the records of their geological history in the form of craters, mountains, and volcanoes (Figure 6.3).

The next four planets (Jupiter through Neptune) are much larger and are composed primarily of lighter ices, liquids, and gases. We call these four the *jovian* (after "Jove," another name for Jupiter in mythology) or **giant planets,** a name they richly deserve (Figure 6.4). Over 1400 Earths could fit inside Jupiter, for example. These planets do not have solid surfaces on which future explorers might land. They are more like vast, ball-shaped oceans with much smaller, dense cores at their centers.

At the outer edge of the system lies little Pluto, which is neither a terrestrial nor a jovian planet; it is similar to the satellites of the outer planets and may have a common origin with them. (Pluto was the last planet to have its mass measured—something that astronomers could not do until Pluto's satellite was discovered in 1978). Table 6.2 summarizes some of the main facts about these nine planets.

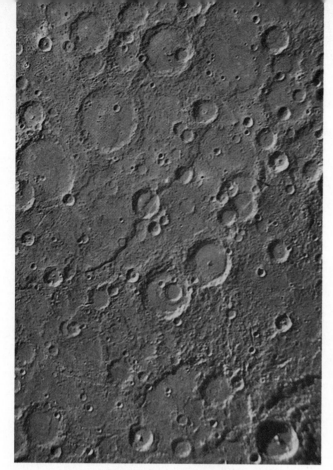

FIGURE 6.3
The Surface of Mercury   The pockmarked face of a terrestrial world, more typical of the inner planets than the watery surface of the Earth. This black and white image, taken with the Mariner 10 spacecraft, shows a region more than 400 km wide.   *(NASA)*

Smaller Members of the Solar System

Most of the planets are accompanied by one or more satellites or moons; only Mercury and Venus move through space alone. There are 64 known satellites (see Appendix 8), and undoubtedly many other small ones remain undiscovered. The largest of the known satellites are as big as small planets and just as interesting; these include, in addition to our Moon, the four largest moons of Jupiter (called the *Galilean satellites* after their discoverer) and the largest moons of Saturn and Neptune (confusingly named Titan and Triton).

Each of the jovian planets also has a system of rings, made up of countless small bodies ranging in size from mountains down to mere grains of dust, all in orbit about the equator of the planet. The bright rings of Saturn are the widest and by far the easiest to see; they are among the most beautiful sights in the solar system (Figure 6.5). But all of the four ring systems are interesting to scientists because of their complicated forms, influenced by the pull of the satellites that also orbit these giant planets.

The solar system has many other, less-conspicuous members. The **asteroids** are rocky and metallic bodies that orbit the Sun like miniature planets, mostly in the space between Mars and Jupiter (Figure 6.6). Most asteroids are remnants of the initial population of the solar system that existed before the planets themselves formed. Some of the smallest satellites of the planets, such as the moons of Mars, are very likely captured asteroids.

Another class of small bodies is composed in large part of ice, made of frozen gases such as water, carbon dioxide, carbon monoxide, and methyl alcohol; these are called

EARTH

FIGURE 6.4
The Four Giant Planets   A montage showing the four giant planets (as photographed by the Voyager cameras) and the Earth to scale.   *(NASA)*

TABLE 6.2
The Planets

Name	Distance from Sun (AU)*	Revolution Period (years)	Diameter (km)	Mass (10^{23} kg)	Density (g/cm^3)
Mercury	0.39	0.24	4,878	3.3	5.4
Venus	0.72	0.62	12,102	48.7	5.3
Earth	1.00	1.00	12,756	59.8	5.5
Mars	1.52	1.88	6,787	6.4	3.9
Jupiter	5.20	11.86	142,984	18,991	1.3
Saturn	9.54	29.46	120,536	5,686	0.7
Uranus	19.18	84.07	51,118	866	1.2
Neptune	30.06	164.82	49,660	1,030	1.6
Pluto	39.44	248.60	2,200	0.01	2.1

* An AU (or astronomical unit) is the distance from the Earth to the Sun.

comets. Comets also are remnants from the formation of the solar system, but they were formed and continue (with rare exceptions) to orbit the Sun in distant, cooler regions—stored in a sort of cosmic deep freeze.

Finally, there are countless grains of rock and dust, which we call simply cosmic dust, scattered throughout the solar system. When these particles collide with the Earth's atmosphere (as millions do each day), they burn up, producing brief flashes of light that we see as shooting stars or

meteors. Occasionally, some larger chunk of material from space survives the passage through the atmosphere to become a **meteorite,** a piece of the solar system that has landed on Earth. (You can see meteorites on display in many natural history museums and can sometimes purchase them from gem and mineral dealers.)

A Scale Model of the Solar System

Astronomy often deals with dimensions and distances that far exceed our ordinary experience. What does 1.4 billion kilometers—the distance from the Sun to Saturn—really mean to anyone? Sometimes it helps to visualize such large systems in terms of a scale model.

In our imagination, let us build a scale model of the solar system, adopting a scale factor of 1 billion (10^9)—that is, reducing the actual solar system by dividing every dimension by a factor of 10^9. The Earth then has a diameter of 1.3 cm, about the size of a grape. The Moon is a pea

FIGURE 6.5
Saturn and Its Rings   Voyager I image of a crescent Saturn and its complex system of rings, taken from a distance of about 1.5 million km. Notice that you can see the edge of the planet through the rings at left.   (NASA/JPL)

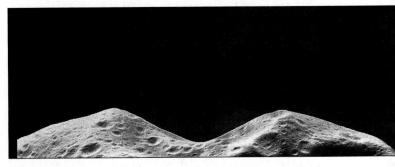

FIGURE 6.6
Asteroid Ida   This image showing part of Ida (taken with the Galileo spacecraft in 1993) is the most detailed close-up of an asteroid we presently have. The spacecraft was 2480 km from the asteroid, about 46 s after its closest approach, when the image was taken. Ida is approximately 50 km in its longest dimension.   (NASA/JPL)

MAKING CONNECTIONS

Names in the Solar System

We humans just don't feel comfortable until something has a name. Types of butterflies, new elements, and mountains of Venus all need names for us to feel we are acquainted with them. How do we give names to objects and features in the solar system?

Planets and satellites are named after gods and heroes in Greek and Roman mythology (with a few exceptions among the moons of Uranus that are drawn from English literature). When William Herschel, a German immigrant to England, first discovered the planet we now call Uranus, he wanted to name it Georgium Sidus (George's star) after King George III of his adopted country. This caused such an outcry among astronomers in other countries, however, that the classical tradition was resumed and has been maintained ever since. Luckily, there were a lot of minor gods in the ancient pantheon, so there are plenty of names left for the many small moons we are discovering around the giant planets. (Appendix 8 lists all of the known satellites; how many names do you recognize?)

Comets are named after their discoverers (giving an extra incentive to comet hunters); asteroids are named by their discoverers after just about anyone or anything they want. Recently, asteroid names have been used to recognize people who have made significant contributions to astronomy, including two of the authors of this book.

That was pretty much all the naming that was needed while our study of the solar system was confined to Earth.

But now our robot probes have surveyed and photographed many worlds in great detail, and on each world there are a host of features that suddenly need names. To make sure that naming things in space remains multinational, rational, and somewhat dignified, astronomers have given the responsibility of approving names to a special committee (including your author David Morrison) of the International Astronomical Union, the body that includes scientists from every country that does astronomy.

This committee has developed a set of rules for naming features on other worlds. For example, many newly discovered features on Venus are being named for women who have made significant contributions to human knowledge and welfare. Volcanic features on Jupiter's moon Io, which is in a constant state of volcanic activity, are being named after gods of fire and thunder from the mythologies of many cultures. Craters on Mercury now commemorate famous novelists, playwrights, artists, and composers. On Saturn's satellite Tethys, all the features have been named after characters in Homer's great epic poem, *The Odyssey*. As we explore further, it may well turn out that there are more places in the solar system needing names than Earth history can provide. Perhaps by then, explorers and settlers on these worlds will be ready to develop their own names for the places that (if only for a while) they will call home.

---

orbiting this grape at a distance of 40 cm, or a bit over a foot away. The Earth–Moon system fits into an attaché case.

In this model the Sun is nearly 1.5 m in diameter, about the average height of an adult, and our Earth is at a distance of 150 m—about one city block—from the Sun. Jupiter is five blocks away from the Sun, and its diameter is 15 cm, about the size of a very large grapefruit. Saturn is 10 blocks from the Sun, Uranus 20 blocks, and Neptune and Pluto each 30 blocks (Pluto's eccentric orbit means that its distance will vary a bit from this average value). Most of the satellites of the outer solar system are the sizes of various kinds of seeds orbiting the grapefruit, oranges, and lemons that represent the outer planets.

In our scale model, a human is reduced to the dimensions of a single atom, and cars and spacecraft to the size of molecules. Sending the Voyager spacecraft to Neptune involves navigating a single molecule from the Earth-grape toward a lemon 5 km away with an accuracy equivalent to the width of a thread in a spider's web.

If that model represents the solar system, where would the nearest stars be? If we keep the same scale, the closest stars would be tens of thousands of kilometers away. If you built our scale model in the city where you live, you would have to place the representations of these stars on the other side of the Earth or beyond.

By the way, model solar systems like the one we just discussed are being built in cities throughout the world. In Sweden, for example, Stockholm's huge Globe Arena has become a model for the Sun, and Pluto is represented by a 12-cm sculpture in the small town of Delsbo 300 km away. Plans are underway to build a model solar system on the Mall between the White House and Congress in Washington, D.C (perhaps proving they are worlds apart?).

 6.2 ## COMPOSITION AND STRUCTURE OF PLANETS

The fact that there are two distinct kinds of planets leads us to believe that they formed under different conditions. Certainly their compositions are dominated by different elements. Let us look at each type in more detail.

The Giant Planets

The two largest planets, Jupiter and Saturn, have nearly the same chemical makeup as the Sun itself; they are composed primarily of the two elements hydrogen and helium, with 75 percent of the mass being hydrogen and 25 percent helium. On Earth, both hydrogen and helium are gases, so Jupiter and Saturn are sometimes called gas planets. But this name is misleading. Jupiter and Saturn are so large that the gas is compressed in their interior until the hydrogen becomes a liquid. Because the bulk of both planets consists of compressed, liquefied hydrogen, we should really call them liquid planets (Figure 6.7).

Under the force of gravity, the heavier elements sink toward the inner parts of a liquid or gaseous planet. Both Jupiter and Saturn thus have cores composed of heavier rock, metal, and ice, but we cannot see these regions directly. In fact, when we look down from above, all we see is the atmosphere with its swirling clouds. We must infer the existence of the denser core inside these planets from studies of each planet's gravity.

Uranus and Neptune are much smaller than Jupiter and Saturn, but each also has a core of metal, rock, and ice. In one way, however, Uranus and Neptune are very similar to Jupiter and Saturn: All four of these giant planets have cores of about the same mass—roughly ten times the entire mass of the Earth. We suspect that these cores formed early on and subsequently attracted the surrounding atmospheres of hydrogen and helium. Uranus and Neptune were less efficient at attracting this gas, so they have much less hydrogen and helium in proportion to their cores.

Chemically, each of the giant planets is dominated by hydrogen and its many compounds. Nearly all of the oxygen present is chemically combined with hydrogen to form H_2O (water). Chemists call such a hydrogen-dominated composition *reduced*. Throughout the outer solar system, we find abundant water (mostly in the form of ice) and reducing chemistry.

The Terrestrial Planets

The terrestrial planets are quite different from the giants. In addition to being much smaller, they are all composed primarily of rocks and metals, which, in turn, are made of elements that are less common in the universe as a whole. The most abundant rocks, called silicates, are made of silicon and oxygen, while the most common metal is iron. We can tell from their densities (see Table 6.2) that Mercury has the highest proportion of metals (which are denser), and the Moon has the lowest. (To refresh your memory about the definition of density, see Section 2.2.)

Earth, Venus, and Mars all have roughly similar bulk compositions: About one third of their mass consists of iron–nickel or iron–sulfur combinations; two thirds is made of silicates. Because they contain very little hydrogen, these planets are largely composed of oxygen compounds (such as the silicate minerals of their crusts). Their chemistry is said to be *oxidized*.

When we look at the internal structure of each of the terrestrial planets, we find that the densest metals are in a central core, with the lighter silicates near the surface. If these planets were liquid like the giant planets, we could understand this effect as the result of the sinking of heavier materials due to the pull of gravity. Thus we believe that, although the terrestrial planets are solid today, they must once have been hot enough to melt.

FIGURE 6.7
Saturn in False Color   This image of Saturn and its ring system were taken by the Hubble Space Telescope on Jan. 4, 1998, celebrating the instrument's 8th anniversary. The colors are not real because the image was taken in infrared light and then computer processed to make a picture our eyes can see. Like the other giant planets, Saturn is made mostly of liquid and gas. Two huge white storm systems are visible near Saturn's equator. The moon Dione is visible at the lower left, and the moon Tethys is seen at the upper right, having just crossed the face of Saturn.   *(E. Kakorchka, U. of Arizona; and NASA)*

FIGURE 6.8
Ganymede A view of Jupiter's satellite Ganymede from the Galileo spacecraft, taken in June 1996. The brownish-gray color of the surface indicates a dusty mixture of rocky material and ice. The bright spots are places where recent impacts have uncovered fresh ice from underneath. *(JPL/NASA)*

Differentiation is the name given to the process by which a planet organizes its interior into layers of different compositions and densities. Once the planet becomes molten, the heavier metals sink to form a core while the lightest minerals float to the surface to form a crust. Later, when the planet cools, this layered structure is preserved. In order for a rocky planet to differentiate, it must be heated to the melting point of rocks, typically above 1300 K. (Temperature was defined in Chapter 4; see Appendix 5 for a discussion of the different temperature scales.)

Moons, Asteroids, and Comets

6.23

Chemically and structurally, the Earth's Moon is like the terrestrial planets, but it is an exception among satellites. Most of the moons in the solar system are found far from the Sun, and they have compositions similar to those of the cores of the giant planets around which they orbit. The three largest satellites, Ganymede and Callisto in the jovian system and Titan in the saturnian system, are composed half of frozen water, half of rocks and metals. These moons differentiated easily during formation, because their interior temperatures had to rise only to the melting point of ice, rather than to that of rock, in order for much of their substance to become liquid. Today these satellites have cores of rock and metal, with upper layers and crusts of very cold and thus very hard ice (Figure 6.8).

The asteroids and comets, as well as the smallest moons, were probably never heated to the melting point. They therefore retain their original composition and structure. This is one of their main attractions for astronomers: They represent relatively unmodified material dating back to the formation of the solar system. In a sense, they are ancient "fossils," helping us to learn about a time whose traces have long been erased on larger worlds.

6.24

Temperatures: Going to Extremes

Generally speaking, the farther a planet or satellite is from the Sun, the cooler it is. The planets are heated by the radiant energy of the Sun, which gets weaker as the square of the distance. You know how rapidly the heating effect of a fireplace or an outdoor "radiant heater" diminishes as you walk away from it, and the same effect applies to the Sun. Mercury, the closest planet to the Sun, has a blistering surface temperature of more than 500 K (230°C), as hot as the cleaning cycle on an electric oven, whereas the surface temperature on Pluto is only about 50 K, colder than liquid air.

Mathematically, the temperatures (expressed on the Kelvin scale) decrease approximately in proportion to the square root of the distance from the Sun. Pluto is 30 AU from the Sun, or 100 times the distance of Mercury. Thus, Pluto's temperature is less than that of Mercury by the square root of 100, or a factor of 10—from 500 K to 50 K.

Because of this strong dependence of temperature on distance from the Sun, we find that the Earth is the only planet where surface temperatures generally lie between the freezing and boiling points of water. Our planet is the only one with liquid water on its surface, and therefore (as far as we know) the only one to support life. All the other worlds in the solar system are too hot or too cold, or else lack air and water like our Moon (see Astronomy Basics below).

ASTRONOMY BASICS

There's No Place Like Home

In the classic film *The Wizard of Oz,* Dorothy, our heroine, concludes after her many adventures in "alien" environments that "there's no place like home." The same can be said of the other worlds in our solar system. There are many fascinating places, large and small, that we might like to visit, but on none of them could human beings survive without a great deal of artificial assistance.

A thick atmosphere keeps the surface temperature on our neighbor Venus at a sizzling 700 K (near 900°F), hot enough to melt several metals. And the pressure of its mostly carbon dioxide atmosphere equals that found almost a kilometer underwater on Earth. Mars, on the other hand, has temperatures generally below freezing, with air (also mostly CO_2) so thin, it's like that found at an altitude of 30 km (100,000 ft) in the Earth's atmosphere. And the red planet is so dry that it has not had any rain for billions of years.

In the outer layers of the jovian planets, it is neither warm enough nor solid enough for human habitation. Any bases we build in the systems of the giant planets may well have to be in space or on one of their satellites—none of which is particularly hospitable to a luxury hotel with swimming pool and palm trees. Perhaps we will find warmer havens deep inside the clouds of Jupiter or in the ocean we suspect exists under the frozen ice of its moon Europa.

All of this suggests that we had better take good care of the Earth, as it is the only site where large numbers of our species can survive unaided. Recent human activity may be reducing the habitability of our planet by adding pollutants to the atmosphere or reducing the fraction of the Earth's surface where vegetation can thrive and continue to supply us with oxygen. In a solar system that seems unready to receive us, making the Earth less hospitable to life may be a grave mistake.

FIGURE 6.9
Jupiter with Huge Dust Clouds The Hubble Space Telescope took this sequence of images of Jupiter in summer 1994, when the fragments of Comet Shoemaker–Levy 9 collided with the giant planet. Here we see the site hit by fragment G, from 5 min to five days after impact. Several of the dust clouds generated by the collisions became larger than the Earth. *(STScI/NASA)*

Geological Activity

The crusts of all of the terrestrial planets, as well as of the larger satellites, have been modified over their histories by both internal and external forces. Externally, each has been battered by a slow rain of projectiles from space, leaving surfaces pockmarked by impact craters of all sizes (see Figure 6.3). We have good evidence that this bombardment was far greater in the early history of the solar system, but it certainly continues to this day. (The collision of more than 20 large pieces of Comet Shoemaker–Levy 9 with Jupiter in the summer of 1994 is one recent example of this process. Figure 6.9 shows the aftermath of these collisions, when mushrooming dust clouds larger than the Earth could be seen in Jupiter's atmosphere.)

At the same time, internal forces on the terrestrial planets have buckled and twisted their crusts, built up mountain ranges, erupted as volcanoes, and generally reshaped the surfaces in what we call *geological activity*. (The prefix *geo* means "Earth," so this is a bit of an "Earth-chauvinist" term, but it is so widely used that your authors bow to tradition.) Among the terrestrial planets, the Earth and Venus have experienced the greatest level of geological activity over their histories, although some of the satellites in the outer solar system are also surprisingly active. In contrast, our own Moon is a dead world where geological activity ceased billions of years ago.

Geological activity on a planet is the result of a hot interior. The forces of volcanism and mountain building are driven by heat escaping from the cores of the planets. As we will see, each of the planets was heated at the time of its birth, and this primordial heat initially powered extensive volcanic activity, even on our Moon. But small objects like the Moon soon cooled off. The larger the planet or satellite, the more likely it is to retain its internal heat, and therefore the more we expect to see surface evidence of continuing geological activity. The effect is similar to our own experience with a baked potato: the larger the potato, the more slowly it cools. If we want a potato to cool quickly, we cut it into small pieces.

For the most part, the history of volcanic activity on the terrestrial planets conforms to the predictions of this simple theory. The Moon, smallest of these objects, is a geologically dead world. Although we know much less about Mercury, it seems likely that this planet, too, ceased most volcanic activity about the same time that the Moon did. Mars represents an intermediate case: It has been much more active than the Moon, but less so than the Earth. Earth and Venus, the largest terrestrial planets, still have molten interiors and high levels of geological activity even today, some 4.5 billion years after their birth.

6.3 DATING PLANETARY SURFACES

How do we know the age of the surfaces we see on planets and satellites? If a world *has* a surface (as opposed to being mostly gas and liquid), we have developed some techniques for estimating how long ago that surface solidified. Note that the age of these surfaces is not necessarily the age of the planet as a whole. On geologically active objects (including the Earth), vast outpourings of molten rock or water have erased evidence of earlier epochs and present us only with a relatively young surface for investigation.

Counting the Craters

The simplest way to estimate the age of a surface is by counting the number of impact craters. This technique works because we have evidence that the rate at which impacts have occurred in the solar system has been roughly constant for several billion years. Thus, in the absence of forces to eliminate craters, the number of craters is simply proportional to the length of time the surface has been exposed. Recently, spacecraft have given us close-up views of many worlds too far away for crater counting by telescope, and this technique has been successfully applied to planets, satellites, and even to asteroids (Figure 6.10).

Bear in mind that crater counts can only tell us the time since the surface experienced a major change. Estimating

FIGURE 6.10
Our Cratered Moon This image of the Moon's surface was taken during the Apollo 16 mission and shows craters of many different sizes. *(NASA)*

ages from crater counts is a little like walking along a sidewalk in a snowstorm after the snow has been falling steadily for a day or more. You may notice that in front of one house the snow is deep while next door the sidewalk may be almost clear. Do you conclude that less snow has fallen in front of Ms. Jones' house than Mr. Smith's? Of course not. Instead, you conclude that Jones has recently swept the walk clean and Smith has not. Similarly, the numbers of craters indicate how long it has been since a planetary surface was last "swept clean" by lava flows or by molten materials ejected when a large impact happened nearby.

Astronomers can thus use the number of craters on *different* parts of the same world to provide important clues about how regions on that world evolved. On a given planet or moon, the more heavily cratered terrain will generally be older (i.e., more time will have elapsed there since something swept the region clean).

Radioactive Rocks

Another way to trace the history of a solid world is to determine the ages of individual rocks. After samples were brought back from the Moon by the Apollo astronauts, the techniques that had been developed to date rocks on Earth could also be applied to establish a chronology for the Moon. In addition, a few samples of material from the Moon, Mars, and the large asteroid Vesta have fallen to Earth as meteorites and can also be examined directly (see Chapter 13).

The ages of rocks can be measured using the properties of natural **radioactivity.** Around the beginning of the 20th century, physicists began to understand that some atomic nuclei are not stable, but can spontaneously split apart (decay) into smaller nuclei. The process of radioactive decay involves the emission of particles such as electrons, or of radiation in the form of gamma rays (see Chapter 4).

For any one radioactive nucleus, it is not possible to say when the decay process might happen. Such decay is random in nature, just as the throw of dice is: As gamblers have all too often found out, it is impossible to say just when dice will come up 7 or 11. But for a very large number of dice tosses, we can calculate the odds that 7 or 11 will come up. Similarly, if you have a very large number of radioactive atoms of one type (say, uranium), there is a specific time period, called its **half-life,** during which the chances are fifty-fifty that decay will occur for any of the nuclei.

A particular nucleus may last a shorter or longer time than its half-life, but in a large sample almost exactly half will have decayed after a time equal to one half-life. Half of those remaining will have decayed in two half-lives, leaving only one half of a half—or one quarter—of the original sample (Figure 6.11).

So if you had a gram of pure radioactive nuclei whose half-life was 100 years, then after 100 years you would have ½ gram, after 200 years 1/4 gram, after 300 years only

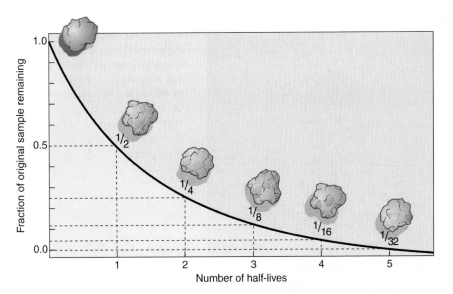

FIGURE 6.11

Radioactive Decay This graph shows (in red) the amount of a radioactive sample that remains after several half-lives have passed. After one half-life, half the sample is left; after two half-lives, one half of the remainder (or 1/4) is left; and after three half-lives, one half of that (or 1/8) is left. Note that in reality, the decay of radioactive materials in a rock sample would not cause any visible change in the appearance of the rock; the splashes of color seen here are shown for educational purposes only.

1/8 gram, and so forth. However, the material does not disappear. Instead the radioactive atoms are replaced with their decay products. Sometimes the radioactive atoms are called *parents*, and the decay products are called *daughter* elements.

In this way, radioactive elements whose half-lives we have determined can provide accurate nuclear clocks. By comparing how much of a radioactive parent element is left in a rock with how much we have of its daughter products, we can learn how long the decay process has been going on and hence how long ago the rock formed. Table 6.3 summarizes the decay reactions used most often to date lunar and terrestrial rocks.

When astronauts first flew to the Moon, one of their most important tasks was to bring back lunar rocks for radioactive age dating. Until then, astronomers and geologists had no real idea of the age of the lunar surface. Counting craters had given relative ages (for example, that the heavily cratered lunar highlands were older than the dark lava plains), but scientists could not agree on the actual age in years. Some thought that the ages were as young

as those of the Earth's surface, implying an active geology on the Moon. Only in 1969, when the first Apollo samples were dated, did we learn that the Moon is an ancient, geologically dead planet. Using such dating techniques, we can determine the ages of both the Earth and the Moon—each was formed about 4.5 billion years ago. (We show this date in the "Cosmic Calendar" at the end of the Prologue.)

By the way, note that the decay of radioactive nuclei generally releases energy in the form of heat. Although the energy from a single nucleus is not very large (in human terms), the enormous numbers of radioactive nuclei in a planet or satellite (especially early on) can be a significant source of internal energy for that world.

 6.4 ORIGIN OF THE SOLAR SYSTEM

Much of astronomy is motivated by a desire to understand the origin of things—to find at least partial answers to the age-old questions of where the universe, the Sun, the Earth, and we ourselves came from. Each planet and satellite is a fascinating place that may stimulate our imaginations as we try to picture what it would be like to visit there. But taken together, the members of the solar system preserve patterns that can tell us about the formation of the entire system. As we begin our exploration of the planets, we want to introduce our modern picture of how the solar system formed. We will return to the question of origins in more detail in Chapter 13.

Looking for Patterns

One way to approach the question of origins is to look for regularities among the planets. We found, for example, that all the planets revolve in the same direction around the Sun, and that they lie in nearly the same flat plane. The

TABLE 6.3
Radioactive Decay Reactions Used to Date Rocks

Parent	Daughter	Half-Life (billion yr)
Samarium-147*	Neodymium-143	106
Rubidium-87	Strontium-87	48.8
Thorium-232	Lead-208	14.0
Uranium-238	Lead-206	4.47
Potassium-40	Argon-40	1.31

* The number after each element is its atomic weight, equal to the number of protons plus neutrons in its nucleus. This specifies the *isotope* of the element; different isotopes of the same element differ in the number of neutrons.

FIGURE 6.12
In the Solar Nebula We see an expert artist's conception of the solar nebula, the flattened cloud of gas and dust from which our planetary system formed. Icy and rocky planetesimals can be seen in the foreground. The bright center is where the Sun is forming. *(Art by planetary astronomer William Hartmann)*

Sun also spins in the same direction about its own axis. Astronomers interpret this pattern as evidence that the Sun and planets formed together from a spinning system of gas and dust that we call the **solar nebula** (Figure 6.12).

The composition of the planets gives another clue. Spectroscopic analysis (see Chapter 4) allows us to determine what elements are present in the Sun and the planets. The Sun has the same hydrogen-dominated composition as Jupiter and Saturn, and therefore appears to have been formed from the same reservoir of material. In comparison, each of the terrestrial planets and satellites is relatively deficient in the light gases and the various ices that form out of the common elements oxygen, carbon, and nitrogen. Instead, on the Earth and its neighbors we see mostly the rarer heavy elements like iron and silicon. This pattern suggests that the processes that led to planet formation in the inner solar system must somehow have rejected much of the lighter materials. These must have escaped, leaving just a residue of heavy stuff.

The reason for this is not hard to guess, bearing in mind the heat of the Sun. The inner planets (which are closer to the Sun) are now made of rock and metal, which can survive heat, but they contain very little ice or gas, which evaporate when the temperatures are high. (To see what we mean, just compare how long a rock and an ice cube each survive when they are placed in the sunlight.) In the outer solar system, where it has always been cooler, the planets and their satellites are composed, in large part, of ice and gas.

The Evidence from Far Away

A second approach to understanding the origin of the solar system is to look outward for evidence that other, similar systems of planets are forming elsewhere. We cannot look back in time to the formation of our own system, but many stars in space are much younger than the Sun. In these systems, the process of planet formation might still be accessible to direct observation. One of the important astronomical discoveries of recent years is the presence of just such planetary nurseries. What we actually observe are other "solar nebulas" or *circumstellar disks*—flattened, spinning clouds of gas and dust surrounding young stars. These disks resemble our own solar system's initial stages of formation billions of years ago. But because the stars are so far away, astronomers have not yet been able to observe the actual formation of planets in these circumstellar disks.

Astronomers have recently begun to develop tools to let them detect planets in orbit around other stars. So far we can find only giant planets, so we know nothing about the presence or absence of terrestrial planets in other planetary systems. But already we know that other planetary systems are not necessarily configured like our own. For example, several of the planets found around distant stars are in highly elliptical orbits, unlike the planets in the solar system, which follow nearly circular orbits. In several cases, giant planets are being found much closer to their stars than Mercury is to the Sun. Future theories for the formation of planetary systems will have to explain not only our own family of planets, but these other intriguing systems as well.

Building Planets

Astronomers can use theoretical calculations to build simulated "model solar nebulas" on a large computer to see how solid bodies might form from the gas and dust in these disks as they cool. These models show that material begins to coalesce by first forming smaller objects,

Carl Sagan: Solar System Advocate

Arguably the best-known astronomer in the world, the late Carl Sagan devoted most of his professional career to studying the planets, and much of his considerable energy to raising public awareness of what we can learn from exploring the solar system. Born in Brooklyn, New York, in 1934, Sagan became interested in astronomy as a youngster, but also credits science fiction stories for sustaining his fascination with what's out there. After earning both his bachelor and PhD degrees from the University of Chicago, he worked as a research associate at the University of California and as an assistant professor at Harvard before becoming professor of astronomy at Cornell University in 1968.

In the early 1960s, when many scientists still thought that Venus might turn out to be a hospitable place, Sagan demonstrated that the atmosphere of our neighborhood planet could act like a giant greenhouse, keeping the heat in and raising the temperature enormously. He showed that the seasonal changes astronomers had seen on Mars were caused not by vegetation, but by wind-blown dust. He was a member of the scientific teams for many of the robotic missions that have explored the solar system and was instrumental in getting NASA to put a message-bearing plaque aboard the Pioneer spacecraft, as well as audio–video records on the Voyager probes (as explained in Chapter 15).

To encourage public interest and public support of planetary exploration, Sagan helped found the Planetary Society, now the largest space-interest organization in the world. He was a tireless and eloquent exponent of the need to study the solar system close-up, and of

Carl Sagan

the value of learning about other worlds in order to take better care of our own.

Sagan did experimental work simulating conditions on the early Earth and demonstrated how some of life's fundamental building blocks might have formed from the primordial "soup" of natural compounds on our planet. He and his students built "Jupiter boxes" to demonstrate how conditions in the

... He was a tireless and eloquent exponent of the need to study the solar system close-up ...

giant planet's atmosphere may in some ways resemble what it was like on Earth before life began. In addition, he and his colleagues made models showing that the consequences of nuclear war for the Earth would be even more devastating than anyone had thought (this is now called the *nuclear winter*

hypothesis) and demonstrating some of the serious consequences of continued pollution of our atmosphere.

Sagan was perhaps best known, however, as a brilliant popularizer of astronomy and the author of many books on science, including the best-selling *Cosmos*, the Pulitzer-prize-winning *Dragons of Eden*, and the recent *Pale Blue Dot*, an evocative tribute to solar system exploration. His book *The Demon Haunted World*, finished just before his death, is perhaps the best antidote to fuzzy thinking about pseudoscience and irrationality in print today. His intriguing science fiction novel, *Contact*, now a successful film, is still recommended by many science instructors as a scenario for making contact with life elsewhere that is much more reasonable than most science fiction.

An appearance on the *Tonight Show* in 1973 (where his lucid explanations caught the attention of then-host Johnny Carson, an amateur astronomer) began a long series of television appearances that culminated in Sagan's 13-part public television series, *Cosmos*. It has been seen by an estimated 500 million people in 60 countries, and has become one of the most-watched series in the history of public broadcasting. Sagan, by the way, was quick to point out that not once in the episodes of *Cosmos* did he say "billions and billions," a phrase now indelibly associated with him (although he did use it in other media appearances over the years). A few astronomers scoffed at a scientist who spent so much time in the public eye, but it is probably fair to say that Sagan's enthusiasm and skill as an explainer won more friends for the science of astronomy than anyone or anything else in this century.

precursors of the planets, which we call **planetesimals** (see Figure 6.12).

Fast computers can now be used to simulate the way millions of planetesimals, each no larger than 100 km in diameter, might have gathered together under their mutual gravity to form the planets we see today. We are beginning to understand that this process was a violent one, with planetesimals crashing into each other and sometimes even disrupting the growing planets themselves. As a consequence of these violent impacts (as well as the heat from radioactive materials in them), all of the planets were heated and differentiated—which explains their present internal structures.

The process of impacts and collisions in the early solar system was complex and often apparently random. The solar nebula model can explain many of the regularities we find in the solar system, but the random collisions of massive collections of planetesimals could be the reason we have some exceptions to the "rules" of solar system behavior. For example, why do the planets Uranus and Pluto spin on their sides? Why does Venus spin slowly and in the opposite direction from all the other planets? Why does the composition of the Moon resemble the Earth in many ways and yet exhibit substantial differences? Although we cannot know for sure, the answers to such questions may well lie in enormous collisions that took place in the solar system long before life on Earth began.

Today, some 4.5 billion years after its origin, the solar system is a much less violent place. As we will see, however some planetesimals have continued to interact and collide, and their fragments move about the solar system as roving "transients" that can "make trouble" for the established members of the Sun's family, such as our own Earth. In Chapter 7 we will examine some of the devastation caused by impacts from space in the past, and some of the dangers that may face us in the future.

SURFING THE WEB

The Nine Planets: A Multimedia Tour of the Solar System [seds.lpl.arizona.edu/nineplanets/nineplanets/nineplanets.html]

Software engineer and amateur astronomer Bill Arnett has put together a site with impressive information about each planet and moon in the solar system. There is also material on space exploration, how objects are named, how to find other planetary images, a glossary, and much more. We like the extensive links to other useful sites at the end of each world's summary.

Views of the Solar System [www.hawastsoc.org/solar/eng/homepage.htm]

Engineer and image processor Calvin Hamilton, formerly with the Los Alamos National Laboratory, has also assembled a good site of information and images about the worlds in our solar system. The site is similar to The Nine Planets, but perhaps a little more focused on graphics. There is useful background information, a data browser, a glossary, and more.

Planetary Sciences Site [nssdc.gsfc.nasa.gov/planetary/planetary_home.html]

The National Space Science Data Center at NASA's Goddard Space Flight Center is a place where NASA data on planetary exploration are stored and disseminated to researchers. David Williams has assembled a somewhat more technical, but still quite accessible site with information and images on each world our spacecraft have explored. The site is strong on data and images, and also includes concise descriptions of the various missions that have explored the planets, including—to the author's credit—those done by other countries.

Gazetteer of Planetary Nomenclature [wwwflag.wr.usgs.gov/USGSFlag/Space/nomen/nomen.html

This is a good resource if you want to find out how objects and features in the solar system get their names. The U.S. Geological Survey branch in Flagstaff, where some of the best maps of solar system objects are made, has brought together the official history and rules from the Working Group of the International Astronomical Union that is charged with approving the names of and on other worlds. Appendices list all the main bodies and when they were discovered and named.

Planetary Photojournal [photojournal.jpl.nasa.gov/]

This site, run by the Jet Propulsion Laboratory and the U.S. Geological Survey Branch of Astrogeology in Flagstaff, is one of the most useful resources on the Web. (We may not be entirely objective, however, since two of the authors of *Voyages* served as advisors for the project to set up the site!) It currently features almost two thousand of the best images from planetary exploration, with detailed captions and excellent indexing. You can dial up images by world, feature name, date, or catalog number, and download images in a number of popular formats.

A Space Library [samadhi.jpl.nasa.gov/]

NASA's Jet Propulsion Laboratory, which is responsible for many of the missions to the planets, has an ambitious program of "space visualization"—using artists and computers to produce scientifically accurate images and animations of the solar system. In this beautiful site, David Seal has collected a large library of solar system art, including scenes from many of the worlds we have explored, solar system maps you can browse through, and art and animations from a number of missions. The most intriguing feature of the site is the "Solar System Simulator," in which you can "dial up" a view of any world from any other. For example, you can view Jupiter as it would be seen from Earth, from Saturn, or one of its own moons. It's really fun to play with!

The Thousand-Yard Model: Earth as a Peppercorn [www.noao.edu/education/peppercorn/pcmain.html]

We give one model for understanding the scale of the solar system in this chapter. Here is another model, by Guy Ottewell of Furman University, in which the Sun is a bowling ball and the Earth a peppercorn. It features careful instructions for doing a "walking tour" of such a scale model and further activities.

6.1 Our solar system consists of the Sun, nine planets, 63 known satellites, and a host of smaller objects. The planets can be divided into two groups: the inner **terrestrial planets** and the outer **giant planets.** Pluto does not fit into either category. The giant planets are composed mostly of liquids and gases. Smaller members of the solar system include **asteroids,** rocky and metallic objects found mostly between Mars and Jupiter; **comets,** made mostly of frozen gases, which generally orbit far from the Sun; and countless smaller grains of cosmic dust. When a **meteor** survives its passage through our atmosphere and falls to Earth, we call it a **meteorite.**

6.2 The giant planets have dense cores roughly ten times the mass of the Earth, surrounded by layers of hydrogen and helium. The terrestrial planets consist mostly of rocks and metals. They were once molten, allowing their structures to **differentiate** (i.e., their denser materials sank to the center). The Earth's Moon resembles the terrestrial planets in composition, but most of the other satellites—which orbit the giant planets—have a large quantity of frozen ices within them. Generally, worlds closer to the Sun have higher temperatures. The surfaces of terrestrial planets have been modified by impacts from space and by varying degrees of geological activity.

6.3 The ages of the surfaces of objects in the solar system can be estimated by counting craters: On a given world, a more heavily cratered region will always be older. We can also use samples of rocks with **radioactive** elements in them to obtain the time since the layer in which the rock formed last solidified. The **half-life** of a radioactive element is the time it takes for half the sample to decay; we determine how many half-lives have passed by how much of a sample remains the radioactive element and how much has become the decay product.

6.4 Regularities among the planets lead astronomers to hypothesize that the Sun and the planets formed together in a giant, spinning cloud of gas and dust called the **solar nebula.** Recent observations show tantalizingly similar circumstellar disks around other stars. We have also found giant planets orbiting a number of nearby stars. Within the solar nebula, material first coalesced into **planetesimals;** many of these gathered together to make the planets and satellites. The remainder can still be seen as comets and asteroids.

A Each group should discuss and make a list of the reasons humanity might have to explore other worlds. Does the group think such missions of exploration are worth the investment? Why?

B Your instructor will assign each group a world. Your task is to think about what it would be like to be there. Feel free to look ahead in the book to the relevant chapters. Discuss where on or around your world would we establish a foothold and what would we need to survive there?

C In Astronomy Basics: *There's No Place Like Home* in this chapter, your authors discuss briefly how human activity is transforming our planet's overall environment. Can you think of other ways that this is happening?

D Some scientists criticized Carl Sagan for "wasting his research time" popularizing astronomy. Do you think scientists should spend their time interpreting their field for the public? Why or why not? Are there ways that scientists who are not as eloquent or charismatic

as Carl Sagan can still contribute to the public understanding of science?

E Your group has been named a special committee by the International Astronomical Union to suggest names of features (like craters, ice floes, trenches, etc.) on a newly explored asteroid. Given the restrictions that any people after whom features are named must no longer be alive, what names or types of names would you suggest? (You are not restricted to names of people, by the way.)

F A member of your group has been kidnapped by a little-known religious cult that worships the planets. They will release him only if your group can tell them which of the planets are currently visible in the sky during the evening and morning. You are forbidden from getting your instructor involved. How else could you find out the information you need? (Be as specific as you can; if your instructor says it's OK, feel free to continue answering this question in a library.)

I. List some reasons that the study of the planets has progressed more in the last few decades than has any other branch of astronomy.

2. Imagine that you are a travel agent in the next century. An eccentric billionaire asks you to arrange a "Guinness Book of Solar System Records" kind of tour. Where would you direct him to find the following (use this chapter and Appendices 7 and 8):

a. the least-dense planet

b. the densest planet

c. the largest moon in the solar system

d. the planet at whose surface you would weigh the most (*Hint:* Weight is directly proportional to surface gravity.)

e. the smallest planet

f. the planet that takes the longest time to rotate

g. the planet that takes the least time to rotate

h. the planet whose diameter is the closest to Earth's

i. the moon that takes the shortest time to orbit its planet

j. the densest moon

k. the moon that orbits the farthest from its planet

l. the most massive moon

m. the most spectacular ring system

3. How do terrestrial and giant planets differ from each other? List as many ways as you can think of.

4. Why are there so many craters on the Moon and so few on the Earth?

5. How do asteroids and comets differ?

6. How and why is the Earth's Moon different from the larger satellites of the giant planets?

7. Where would you look for some "original" planetesimals left over from the formation of our solar system?

8. Describe how we use radioactive elements and their decay products to find the age of a rock sample. Is this necessarily the age of the entire world from which the sample comes? Explain.

9. What was it like in the solar nebula? Why did the Sun form at its center?

THOUGHT QUESTIONS

10. Using some of the astronomical resources in your college library or the World Wide Web, find five names of features on each of three other worlds that are named after people. In a sentence or two, describe who each of these people was.

11. Explain why the planet Venus is differentiated, but asteroid Fraknoi, a very boring and small member of the asteroid belt, is not.

12. Would you expect as many impact craters per unit area on the surface of Venus as on the surface of Mars? Why or why not?

13. Interview a sample of 20 people who are not taking an astronomy class and ask them if they can name a living astronomer. What percentage of those interviewed were able to name one? Typically, the one living astronomer the public knows these days is Stephen Hawking. Why is he better known than most astronomers? How would your results have differed if you had asked the same people to name a movie star or a professional football player?

PROBLEMS

14. Looking at Appendices 7 and 8, find the satellite whose diameter is the largest fraction of the diameter of the planet it orbits.

15. Barnard's Star, the second closest star to us, is about 56 trillion (5.6×10^{12}) km away. Calculate how far it would be using the scale model in Section 6.1.

16. A radioactive nucleus has a half-life of 5×10^8 years. Assuming that a sample of rock (say in an asteroid) solidified right after the solar system formed, approximately what fraction of the radioactive element should be left in the rock today?

SUGGESTIONS FOR FURTHER READING

Barnes-Svarney, P. "The Chronology of Planetary Bombardments" in *Astronomy*, July 1988, p. 21.

Beatty, J. et al. *The New Solar System*, 4th ed. 1999, Sky Publishing. A compendium of articles on planetary astronomy by leading scientists in the field. Occasionally technical.

Greeley, R. and Batson, R. *The NASA Atlas of the Solar System*. 1997, Cambridge U. Press. Over-sized, coffee-table book with the best maps and images of the planets beautifully displayed.

Hartmann, W. "Piecing Together Earth's Early History" in *Astronomy*, June 1989, p. 24.

Jayawardhana, R. "Spying on Planetary Nurseries" in *Astronomy*, Nov. 1998, p. 62.

Kross, J. "What's in a Name?" in *Sky & Telescope*, May 1995, p. 28. Discusses the naming of features in the solar system.

Miller, R. and Hartmann, W. *The Grand Tour: A Traveler's Guide to the Solar System*, 2nd ed. 1993, Workman. Lavishly illustrated beginners' primer.

Morrison, D. *Exploring Planetary Worlds.* 1993, Scientific American Library/W. H. Freeman. Clear, up-to-date, non-technical survey.

Morrison, D. and Owen, T. *The Planetary System,* 2nd ed. 1996, Addison-Wesley. A thorough introduction to our modern knowledge.

O'Dell, R. "Exploring the Orion Nebula" in *Sky & Telescope,* Dec. 1994, p. 20. Discusses planetary nurseries.

O'Meara, S. "The World's Largest Solar System Scale Model" in *Sky & Telescope*, July 1993, p. 99. At the Lakeview Museum in Peoria, IL. (See an update in the March 1998 issue, p. 80, on a wide range of such models.)

Sagan, C. *Pale Blue Dot.* 1994, Random House. Eloquent introduction to planetary exploration and a sequel to Sagan's *Cosmos.*

Sheehan, W. *Worlds in the Sky.* 1992, U. of Arizona Press. History of how we came to understand the planets.

Molten lava fountaining from a volcanic vent in Hawaii—testament to the dynamic nature of the Earth's crust. *(G. Briggs/USGS)*

7 Earth as a Planet

THINKING AHEAD

Airless worlds in our solar system seem peppered with craters large and small. The Earth, on the other hand, has few craters, but a thick atmosphere and much surface activity. We believe that in the past the Earth, too, had many craters that have since been erased by forces in its crust and atmosphere. What can the persistent cratering on so many other worlds tell us about the history of our own planet?

"**B**y coming face to face with alternative fates of worlds more or less like our own, we have begun to better understand the Earth. Every one of these worlds is lovely and instructive. But, so far as we know, they are also, every one of them, desolate and barren. Out there, there are no 'better places.' So far, at least."

Carl Sagan in *Pale Blue Dot*
(1994, Random House)

As our first step in exploring the solar system in more detail, we turn to the most familiar planet, our own Earth. The first humans to see the Earth as a blue sphere floating in the blackness of space were the Apollo 8 astronauts who made the first voyage around the Moon in 1968. For many people, the historic visits to the Moon and the images showing our world as a small, distant globe represent a pivotal moment in human history, when it became difficult for educated human beings to avoid viewing our world with a global perspective.

In previous chapters we discussed the size and orbit of the Earth, and we described terrestrial phenomena, such as seasons and tides, that are related to the motions of the Earth. In this chapter we examine the composition and structure of our planet with its envelope of ocean and atmosphere. We ask how that envelope and the surface of our planet came to be the way they are today. Because we live on it, we know the Earth much better than any other planet, and we can perform experiments on its surface that we cannot yet do elsewhere with any regularity. This direct knowledge helps us use the Earth as a reference point for understanding other worlds where we are mostly limited to remote observations.

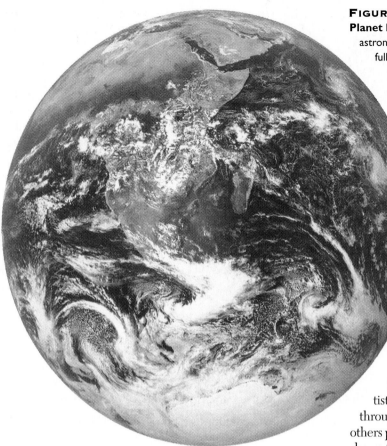

FIGURE 7.1
Planet Earth   The Earth from space, taken by the Apollo 17 astronauts in December 1972. This is one of the rare images of a full Earth taken during the Apollo program; most images show only some of the Earth's disk in sunlight.   *(NASA)*

The Earth is composed largely of metal and silicate rock (see Section 6.2). Most of this material is in a solid state, but some of it is hot enough to be molten. The structure of the material in the Earth's interior has been probed in considerable detail by measuring the transmission of **seismic waves** through the Earth. These are waves produced in the material of our planet by natural earthquakes or by impacts or explosions.

Seismic waves travel through a planet rather like sound waves through a struck bell. Just as the sound frequencies vary depending on what material the bell is made of and how it is constructed, so a planet's response depends on its composition and structure. By monitoring the seismic waves in different locations, scientists can learn about the layers the waves have traveled through. Some of these vibrations travel along the surface; others pass directly through the interior. Seismic studies have shown that our planet's interior consists of several distinct layers with different compositions, illustrated in Figure 7.2.

The top layer is the **crust,** the part of the Earth we know best (Figure 7.3). The crust under the oceans, which cover 55 percent of the surface, is typically about 6 km thick and is composed of volcanic rocks called **basalts.** Produced by the cooling of volcanic lava, basalts are made primarily of the elements silicon, oxygen, iron, aluminum, and magnesium. The continental crust, which covers 45 percent of the surface, is from 20 to 70 km thick and is predominantly composed of a different volcanic class of silicates (rocks made of silicon and oxygen) called **granites.** These crustal rocks typically have densities of about 3 g/cm^3. The crust is the easiest layer for geologists to study, but it makes up only about 0.3 percent of the total mass of the Earth.

7.1   THE GLOBAL PERSPECTIVE

Basic Properties

The Earth is a medium-sized planet, with a diameter of approximately 13,000 km (Figure 7.1). As one of the terrestrial planets, it is composed primarily of a few heavy elements, such as iron, silicon, and oxygen—very different from the composition of the Sun and stars, which are dominated by the light elements hydrogen and helium. The Earth's orbit is nearly circular, and it is close enough to the Sun to support liquid water on its surface. It is the only planet in our solar system that is neither too hot nor too cold, but "just right" for the development of life as we know it. Some of the basic properties of the Earth are summarized in Table 7.1.

Earth's Interior

The interior of a planet—even our own Earth—is difficult to study, and its composition and structure must be determined indirectly. Our only direct experience is with the outermost skin of the Earth's crust, a layer no more than a few kilometers deep. It is important to remember that in many ways we know less about our own planet 5 km beneath our feet than we do about the surfaces of Venus and Mars.

TABLE 7.1
Some Properties of the Earth

Semimajor axis	1.00 AU
Period	1.00 year
Mass	5.98×10^{24} kg
Diameter	12,756 km
Escape velocity	11.2 km/s
Rotation period	23 h 56 m 4 s
Surface area	5.1×10^{8} km^2
Atmospheric pressure	1.00 bar

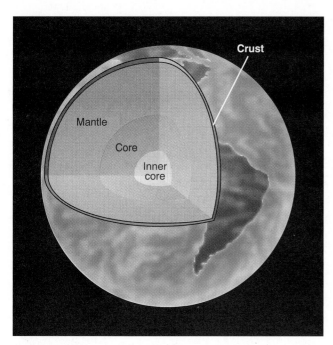

FIGURE 7.2
The Interior Structure of the Earth

The largest part of the solid Earth, called the **mantle,** stretches from the base of the crust down to a depth of 2900 km. The mantle is more or less solid, but at the temperatures and pressures found there, the mantle rock can deform and flow slowly. The density in the mantle increases downward from about 3.5 g/cm^3 to more than 5 g/cm^3 as a result of the compression produced by the weight of overlying material. Samples of upper-mantle material are occasionally ejected from volcanoes, permitting a detailed analysis of its chemistry.

Beginning at a depth of 2900 km, we encounter the dense metallic **core** of the Earth. With a diameter of 7000 km, our core is substantially larger than the entire planet Mercury. The outer core is liquid. The innermost part of the core (about 2400 km in diameter) is extremely dense and probably solid. In addition to iron, the core probably also contains substantial quantities of nickel and sulfur.

The separation of the Earth into layers of different densities is an example of differentiation (a concept explained in Chapter 6). The fact that the Earth is differentiated is evidence that it was once warm enough for the mantle rocks to melt, permitting the heavier metals to sink to the center and form the dense core.

Magnetic Field and Magnetosphere

We can find additional clues about the Earth's interior from its magnetic field. Our planet behaves in some ways as if there were a giant bar magnet inside, approximately aligned with the rotational poles of the Earth. This magnetic field is actually generated by moving material in the Earth's liquid metallic core. As the liquid metal inside the

Earth circulates, it sets up a circulating electric current. And, as discussed in Chapter 4, whenever charged particles are moving, they set up a magnetic field.

The Earth's magnetic field extends into surrounding space. When a charged particle encounters a magnetic field, it begins to spiral around it and becomes trapped in the magnetic zone. Above the Earth's atmosphere, our field is able to trap small quantities of electrons and other atomic particles. This region, called the **magnetosphere,** is defined as the zone within which the Earth's magnetic field dominates over the weak interplanetary field that extends outward from the Sun (Figure 7.4).

Where do the charged particles trapped in our magnetosphere come from? They flow outward from the hot surface of the Sun; astronomers call this flow the *solar wind* (see Chapter 14). It not only provides particles for the Earth's magnetic field to trap; it also elongates our field in the direction pointing away from the Sun. Typically, the Earth's magnetosphere extends about 60,000 km, or ten Earth radii, in the direction of the Sun. But pointing away from the Sun, the magnetic field is shaped like a wind sock and can reach as far as the orbit of the Moon.

The magnetosphere was discovered in 1958 by instruments on the first U.S. Earth satellite, Explorer 1, which recorded the ions (charged particles) trapped in its inner part. The regions of high-energy ions in the magnetosphere

FIGURE 7.3
Earth's Crust This computer-generated globe shows the surface of the Earth's crust with the oceans removed, as determined from satellite images and ocean floor radar mapping. The amounts of vegetation on the continents, as recorded by the instruments aboard the NOAA satellites, are indicated by color, ranging from light brown for the least vegetation, through yellow and orange, to green for the most. (*J. Frawley and NASA*)

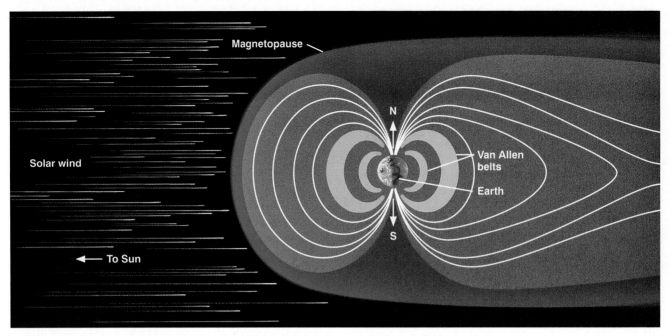

FIGURE 7.4
Earth's Magnetosphere A cross-sectional view of our magnetosphere (or zone of magnetic influence), as revealed by numerous spacecraft missions. Note how the wind of charged particles from the Sun "blows" the magnetic field outward, like a wind sock.

are often called the *Van Allen belts* in recognition of the University of Iowa professor who built the scientific instrumentation for Explorer 1 and correctly interpreted the satellite measurements. Since 1958, hundreds of spacecraft have explored various regions of the magnetosphere. In Chapter 14 we will discuss how the magnetic field and Van Allen belts of the Earth give rise to the brilliant curtains of light called *aurorae,* visible in our atmosphere near the poles.

 THE CRUST OF THE EARTH

Let us now examine our planet's outer layers in more detail. The Earth's crust is a dynamic place. Volcanic eruptions, erosion, and large-scale movements of the continents constantly rework the surface of our planet. Geologically, ours is the most active planet. Many of the geological processes described in this section have taken place on other planets as well, but usually in their distant pasts. Some of the satellites (moons) of the giant planets do have impressive activity (as we will see in Chapter 11). For example, Jupiter's satellite Io has a remarkable number of active volcanoes.

Composition of the Crust

The Earth's crust is largely made up of oceanic basalts and continental granites. These are both examples of **igneous** rock, the term used for any rock that has cooled from a

molten state. All volcanically produced rock is igneous (Figure 7.5).

Two other kinds of rock are familiar to us on the Earth, although it turns out that neither is common on other planets. **Sedimentary** rocks are made of fragments of igneous

FIGURE 7.5
Igneous Rocks The formation of igneous rock as liquid lava cools and freezes. This is a lava flow from a basaltic eruption in Hawaii.
(David Morrison)

rocks or living organisms, deposited by wind or water and cemented together without melting. On Earth, these include the common sandstones, shales, and limestones. **Metamorphic** rocks are produced when high temperature or pressure alters igneous or sedimentary rocks physically or chemically (the word *metamorphic* means "changed in form"). Metamorphic rocks are produced on Earth because geological activity carries surface rocks to considerable depths and then brings them back up to the surface. Without such activity, these changed rocks would not exist.

There is a fourth very important category of rock that can tell us much about the early history of the planetary system. This is **primitive** rock, which has largely escaped chemical modification by heating. Primitive rock represents the original material out of which the planetary system was made. There is no primitive material left on the Earth because the entire planet was strongly heated early in its history. To find primitive rocks, we must look to smaller objects such as comets, asteroids, and small planetary satellites.

A block of marble on Earth is composed of materials that have gone through all four of these states. Beginning as primitive material before the Earth was born, it was heated in the early Earth to form igneous rock, subsequently eroded and redeposited (perhaps many times) to form sedimentary rock, and finally transformed several kilometers below the Earth's surface into the hard white metamorphic stone we see today.

Plate Tectonics

Geology is the study of the Earth's crust, particularly the processes that have shaped its surface throughout history. Heat escaping from the interior provides the energy for the formation of mountains, valleys, volcanoes, and even the continents and ocean basins themselves. But not until the middle of the 20th century did geologists succeed in understanding just how these landforms are created.

Plate tectonics is a theory that explains how slow motions within the mantle of the Earth move large segments of the crust, resulting in a gradual drifting of the continents as well as the formation of mountains and other large-scale geologic features. It is a concept as basic to geology as evolution by natural selection is to biology or gravity is to understanding the orbits of planets.

The Earth's crust and upper mantle (to a depth of about 60 km) are divided into about a dozen major plates that fit together like the pieces of a jigsaw puzzle (Figure 7.6). These plates are also capable of moving slowly with respect to each other. In some places, such as the Atlantic Ocean, the plates are moving apart; in others they are being forced together. The power to move the plates is provided by slow **convection** of the mantle, a process by which heat escapes from the interior through the upward flow of warmer material and the slow sinking of cooler material. (Convection, in which energy is transported from a warm region such as the interior of the Earth to a cooler region such as the upper mantle, is a process we encounter often in astronomy.)

As the plates slowly move, they come into contact with each other and cause dramatic changes to the Earth's crust over time. Four basic kinds of interactions between crustal plates are possible at their boundaries: (1) They can pull apart; (2) one plate can burrow under another; (3) they can slide alongside each other; or (4) they can jam together. Each of these activities is important in determining the geology of the Earth.

Rift and Subduction Zones

Plates pull apart from each other along **rift zones,** such as the Mid-Atlantic Ridge, driven by upwelling currents in the mantle (Figure 7.7a). A few rift zones are found on land, the best known being the central African rift, an area where the African continent is slowly breaking apart. Most

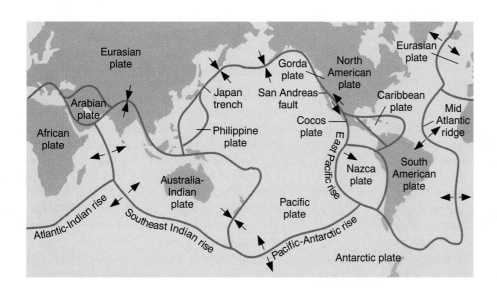

FIGURE 7.6
Earth's Continental Plates   The major plates into which the crust of the Earth is divided. Arrows indicate the motion of the plates.

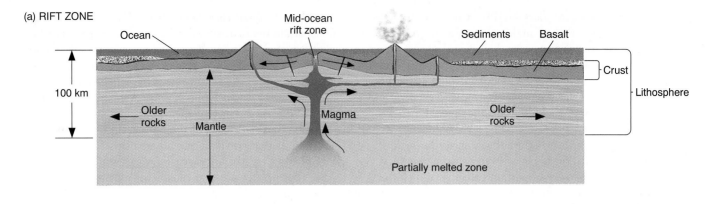

(a) RIFT ZONE

Ocean · Mid-ocean rift zone · Sediments · Basalt · Crust · Lithosphere

100 km · Older rocks · Mantle · Magma · Older rocks · Partially melted zone

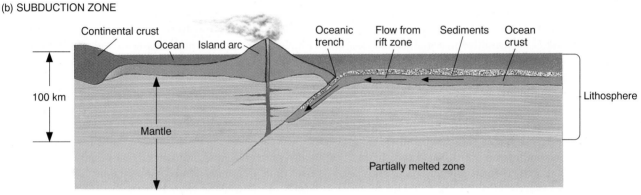

(b) SUBDUCTION ZONE

Continental crust · Ocean · Island arc · Oceanic trench · Flow from rift zone · Sediments · Ocean crust · Lithosphere

100 km · Mantle · Partially melted zone

FIGURE 7.7
Rift Zones (a) and Subduction Zones (b)   The regions (mostly beneath the oceans) where new crust is formed and old crust is destroyed.

rift zones, however, are in the oceans. New material rises from below to fill the space between the receding plates; this is basaltic lava, the kind of igneous rock that forms most of the ocean basins.

From a knowledge of how the sea floor is spreading, we can calculate the average age of the oceanic crust. About 60,000 km of active rifts have been identified, with average separation rates of about 4 cm per year. The new area added to the Earth each year is about 2 km^2, enough to renew the entire oceanic crust in a little more than 100 million years. This is a very short interval in geological time, less than 3 percent of the age of the Earth. The present ocean basins thus turn out to be among the youngest features on our planet.

When two plates come together, one plate is often forced down beneath another in what is called a **subduction zone** (Figure 7.7b). Generally, the thicker continental masses cannot be subducted, but the thin oceanic plates can be rather readily thrust down into the upper mantle. Often a subduction zone is marked by an ocean trench, a fine example being the deep Japan Trench along the coast of Asia. The subducted plate is forced down into regions of high pressure and temperature, eventually melting several hundred kilometers below the surface. Its material is recycled back into a downward-flowing convection current,

ultimately balancing the flow of material that rises along rift zones. The amount of crust destroyed at subduction zones is approximately equal to the amount formed at rift zones.

All along the subduction zone, earthquakes and volcanoes mark the death throes of the plate. Some of the most destructive earthquakes in history have taken place along subduction zones. These include the 1923 Yokohama earthquake and fire, which killed 100,000 people, and the 1976 earthquake that leveled the Chinese city of Tangshan and resulted in nearly half a million deaths.

Fault Zones and Mountain Building

Along much of their lengths, the crustal plates slide parallel to each other. These plate boundaries are marked by cracks or **faults.** Along active fault zones, the motion of one plate with respect to the other is several centimeters per year, about the same as the spreading rates along rifts.

One of the most famous faults is the San Andreas Fault, lying on the boundary between the Pacific Plate and the North American Plate. This fault runs from the Gulf of California to the Pacific Ocean northwest of San Francisco (Figure 7.8). The Pacific Plate, to the west, is moving northward, carrying Los Angeles, San Diego, and parts of the southern California coast with it. In several

FIGURE 7.8
The San Andreas Fault We see part of a very active region in California where one crustal plate is sliding sideways with respect to the other. The fault is marked by the valley running up the center of the photo. *(USGS)*

Alps, for example, are a result of the African Plate's bumping into Europe. As we will see, however, quite different processes produced the mountains on other planets.

Once a mountain range is formed by upthrusting of the crust, its rocks are subject to the erosional force of water and ice. The sharp peaks and serrated edges characteristic of our most beautiful mountains (Figure 7.9) have little to do with the forces that initially make mountains. Instead, they result from the processes that tear them down. Ice is an especially effective sculptor of rock. In a planet without moving ice or running water, mountains remain smooth and dull.

Volcanoes

Volcanoes mark locations where molten rock (called **magma**) rises to the surface. One example is the mid-ocean ridges, long undersea mountain ranges formed by hot material rising from the Earth's mantle at plate boundaries. A second major kind of volcanic activity is associated with subduction zones, and volcanoes sometimes also appear in regions where continental plates are colliding. In each case, the volcanic activity gives us a way to sample some of the material from deeper within our planet.

Another location for volcanic activity on our planet is found above mantle "hot spots," areas far from plate

million years, Los Angeles will be an island off the coast of San Francisco.

Unfortunately for us, the motion along most fault zones does not take place smoothly. The creeping motion of the plates against each other builds up stresses in the crust that are released in sudden, violent slippages, generating earthquakes. Because the average motion of the plates is constant, the longer the interval between earthquakes, the greater the stress and the larger the energy released when the surface finally moves.

For example, the part of the San Andreas Fault near the central California town of Parkfield has slipped every 22 years or so during the past century, moving an average of about 1 m each time. In contrast, the average interval between major earthquakes in the Los Angeles region is about 140 years, and the average motion is about 7 m. The last time the San Andreas slipped in this area was in 1857; tension has been building ever since, and sometime soon it is bound to be released. Sensitive instruments placed within the Los Angeles Basin show that the basin is actually contracting in size as these tremendous pressures mount beneath the surface.

When two continental masses are brought together by the motion of the crustal plates, they are forced against each other under great pressure. The Earth buckles and folds, forcing some rock deep below the surface and raising other folds to heights of many kilometers. This is the way most of the mountain ranges on Earth were formed; the

FIGURE 7.9
Mountains on Earth The Alps, a young region of the Earth's crust where sharp mountain peaks are being sculpted by glaciers. *(David Morrison)*

Alfred Wegener: Catching the Drift of Plate Tectonics

When studying maps or globes of the Earth, many students notice that the coast of North and South America, with only minor adjustments, could fit pretty well against the coast of Europe and Africa. It seems as if these great land masses could once have been together and then were somehow torn apart. The same idea had occurred to a number of scientists (including Francis Bacon as early as 1620). But not until the 20th century could such a proposal be more than speculation. The man who made the case for continental drift in 1912 was a German meteorologist and astronomer named Alfred Wegener.

Born in Berlin in 1880, Wegener was from an early age fascinated by Greenland, the world's largest island, which he dreamed of exploring. He studied at the universities in Heidelberg, Innsbruck, and Berlin, receiving a doctorate in astronomy by re-examining 13th-century astronomical tables. But his interests turned more and more toward the Earth, particularly its weather. He carried out experiments using kites and balloons, becoming so accomplished that he and his brother set a world record in 1906 by flying for 52 hours in a balloon.

Wegener first conceived of continental drift in 1910 while examining a world map in an atlas. But it took two years for him to assemble sufficient data to propose the idea in public: He published the results of his work in book form in 1915. Wegener's evidence went far beyond the congruence in the shapes of the continents. He proposed that the similarities between unusual fossils found only in certain parts of

Alfred Wegener (1880–1930).
(Historical Pictures Service, Chicago)

South America and Africa indicated that these two continents were together at one time. He also showed that resemblances among living animal species on different continents could best be explained by assuming that the continents were once connected in a supercontinent he called *Pangaea* (from Greek elements meaning "all land").

Wegener's suggestion first met with a hostile reaction from scientists. Although he had marshaled an impressive list of arguments for his theory, he was missing a *mechanism:* no one could explain *how* solid continents could drift over thousands of miles. A few

Wegener's suggestion first met with hostile reaction from scientists.

scientists were sufficiently impressed by Wegener's work to continue searching for additional evidence, but many found the notion of moving continents too different from the accepted view of the Earth's surface to take seriously. Developing an understanding of the mechanism (plate tectonics) would take decades of further progress in geology, oceanography, and geophysics.

Wegener was disappointed in the reception to his suggestion, but he continued his research on the weather, the causes of craters on the Moon, and the exploration of Greenland. In 1924 he was appointed to a special meteorology and geophysics professorship created especially for him at the University of Graz. Four years later, on his fourth expedition to his beloved Greenland, he celebrated his 50th birthday with colleagues and then set off toward a different camp on the island. He never made it; he was found a few days later, dead of an apparent heart attack.

Critics of science often point to the resistance to the continental drift hypothesis as an example of the flawed way that scientists regard new ideas. (Many people advancing crackpot theories have claimed that they are being unjustly ridiculed, just as Wegener was.) But we think there is a more positive light in which to view the story of Wegener's suggestion. Scientists in his day maintained a skeptical attitude because they needed more evidence and a clear mechanism that would fit what they understood about nature. Once the evidence and the mechanism were clear, Wegener's hypothesis quickly became the centerpiece of our new view of a dynamic Earth.

boundaries where heat is nevertheless rising from the interior of the Earth. Perhaps the best-known such hot spot is under the island of Hawaii, where it currently supplies the magma to maintain three active volcanoes, two on land and one under the ocean (Figure 7.10). The Hawaiian hot spot has been active for at least 100 million years. As the Earth's plates have moved during that time, the hot spot

has generated a 3500-km-long chain of volcanic islands. The tallest Hawaiian volcanoes are among the largest individual mountains on Earth, more than 100 km in diameter and rising 9 km above the ocean floor. As we saw in Chapter 5, one of these volcanic mountains, the now-dormant Mauna Kea, has become one of the world's great sites for doing astronomy.

FIGURE 7.10
A Volcano Mauna Loa, a large volcano in Hawaii, during eruption. The stars of the Southern Cross are visible to the left of the volcanic plume. *(Dale P. Cruikshank)*

Not all volcanic eruptions produce mountains. If the lava is very fluid and flows rapidly from long cracks, it spreads out to form lava plains. The largest known terrestrial eruptions, such as those that produced the Snake River Basalts in the northwestern United States or the Deccan Plains in India, are of this type. As we will see later, similar lava plains are found on the Moon and the other terrestrial planets.

7.3 THE EARTH'S ATMOSPHERE

We live at the bottom of the ocean of air that envelops our planet. The atmosphere, weighing down upon the Earth's surface under the force of gravity, exerts a pressure at sea level that scientists define as 1 **bar** (a term that comes from barometer, an instrument used to measure atmospheric pressure). A bar of pressure means that each square centimeter of the Earth's surface has a weight of 1.03 kg pressing down on it. Humans have evolved to exist best at this pressure; make the pressure a lot lower or higher and we do not function well at all.

The total mass of our atmosphere is about 5×10^{18} kg; this sounds like a big number, but it is only about a millionth of the total mass of the Earth. The atmosphere represents a smaller fraction of the Earth than the fraction of your mass represented by the hair on your head.

Structure of the Atmosphere

The structure of the atmosphere is illustrated in Figure 7.11. Most of the atmosphere is concentrated near the surface of the Earth, within about the bottom 10 km—where

clouds form and airplanes fly. Within this region, called the **troposphere,** warm air, heated by the surface, rises and is replaced by descending currents of cooler air—another example of convection. This circulation generates clouds and other manifestations of weather. Within the troposphere, temperature drops rapidly with increasing elevation to values near 50°C below freezing at its upper boundary, where the **stratosphere** begins. Most of the stratosphere, which extends to about 50 km above the surface, is cold and free of clouds.

Near the top of the stratosphere is a layer of **ozone (O_3)**, a heavy form of oxygen with three atoms per molecule instead of the usual two. Since ozone is a good absorber of ultraviolet light, it protects the surface from some of the Sun's dangerous ultraviolet radiation, making it possible for life to exist on land. Because ozone is essential to our survival, we react with justifiable concern to evidence that atmospheric ozone is being destroyed. Just such evidence has been accumulating; one especially important agent of this destruction is a series of industrial chemicals called CFCs (chlorofluorocarbons).

Each year a large ozone hole forms above the Antarctic continent, and by now the ozone loss has progressed into the temperate zones. The production of CFCs has

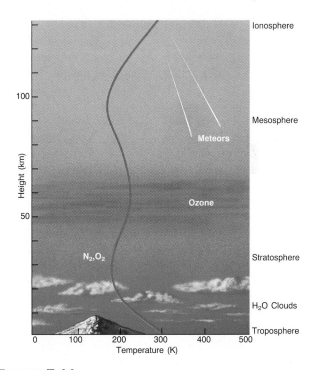

FIGURE 7.11
The Structure of the Earth's Atmosphere Height increases up the left side of the diagram, and the names of the different layers are shown at the right. In the upper ionosphere, ultraviolet radiation from the Sun can strip electrons from their atoms, leaving the atmosphere ionized. (This layer can scatter or reflect AM radio waves, assuring that some of the dumb things said on AM radio never make it out of our planet's atmosphere.) The curving red line shows the temperature; see the scale for temperature at the bottom of the chart.

been banned by international agreement, but these chemicals are destroyed so slowly that we can expect increasing loss of ozone well into the next century.

At heights above 100 km, the atmosphere is so thin that orbiting satellites can pass right through it with very little friction. At these elevations, individual atoms can occasionally escape completely from the gravitational field of the Earth. Thus there is a continuous, slow leaking of atmosphere—especially of lightweight atoms, which move faster than heavy ones. The Earth's atmosphere cannot, for example, hold on to hydrogen or helium, which escape into space.

Atmospheric Composition and Origin

At the Earth's surface, the atmosphere consists of 78 percent nitrogen (N_2), 21 percent oxygen (O_2), and 1 percent argon (A), with traces of water vapor (H_2O), carbon dioxide (CO_2), and other gases. Variable amounts of dust particles and water droplets are also found suspended in the air.

A complete census of the Earth's volatile materials, however, should look at more than the gas now present. *Volatile* materials are those that evaporate at a relatively low temperature. If the Earth were just a little bit warmer, some materials that are now liquid or solid might become part of the atmosphere. Suppose, for example, that our planet were heated to above the boiling point of water (100°C, or 373 K); that's a large change for humans, but a small change compared to the range of possible temperatures in the universe. At 100°C the oceans would boil, and the water vapor would become a part of the atmosphere.

To estimate how much water vapor would be released, we note that there is enough water to cover the entire Earth to a depth of about 3000 m. Since the pressure exerted by 10 m of water is about equal to 1 bar, the average pressure at the ocean floor is about 300 bars. Since water weighs the same whether in liquid or vapor form, if the oceans boiled away the atmospheric pressure of the water would still be 300 bars. Water would therefore greatly dominate the Earth's atmosphere, with nitrogen and oxygen reduced to the status of trace constituents.

On a warmer Earth another source of additional atmosphere would be found in the sedimentary carbonate rocks of the crust. These are minerals that contain abundant carbon dioxide, which if released by heating would generate about 70 bars of CO_2, far more than the current CO_2 pressure of only 0.0005 bar. Thus, the atmosphere of a warm Earth would be dominated by water vapor and carbon dioxide, with a surface pressure close to 400 bars.

Several lines of evidence convince scientists that the composition of the Earth's atmosphere today is not the same as it was originally. We examine this question in more detail in Section 7.4. For now, just note that we do not know the source of the Earth's original atmosphere. Today we see that CO_2, H_2O, sulfur dioxide (SO_2), and other gases are released from deeper within the Earth through the action of volcanoes. Much of this apparently new gas, however, is probably recycled material that has been subducted through plate tectonics.

Three possibilities exist for the original source of the Earth's atmosphere and oceans (since the early Earth was warmer, we can think of today's oceans as liquified early atmosphere, as just discussed): (1) The atmosphere could have been formed with the rest of the Earth, as it accumulated from the debris left over from the formation of the Sun; (2) it could have been released from the interior through volcanic activity, subsequent to the formation of the Earth; or (3) it may have been derived from impacts by comets or other icy materials from the outer parts of the solar system. Current evidence favors the comet hypothesis, but all three mechanisms may have contributed.

Weather and Climate

All planets with atmospheres have *weather*, which is just a name we give to the circulation of the atmosphere. The energy that powers the weather is derived primarily from the sunlight that heats the surface. As the planet rotates, and as the slower seasonal changes cause variations in the amount of sunlight striking different parts of the Earth, the atmosphere and oceans try to redistribute the heat from warmer to cooler areas. Weather on any planet represents the response of its atmosphere to changing inputs of energy from the Sun (see Figure 7.12 for a dramatic example).

Climate is a term used to refer to the effects of the atmosphere through decades and centuries. Changes in climate (as opposed to the random variations in weather from one year to the next) are often difficult to detect over short time periods, but as they accumulate their effect can be devastating. Modern farming is especially sensitive to temperature and rainfall; for example, calculations indicate that a drop of only 2°C throughout the growing season would cut the wheat production by half in Canada and the United States.

The best-documented changes in the Earth's climate are the great ice ages, which have periodically lowered the temperature of the Northern Hemisphere over the past million years or so. The last ice age, which ended about 14,000 years ago, lasted some 20,000 years. At its height, the ice was almost 2 km thick over Boston and stretched as far south as New York City. Today we are in a relatively warm period, interpreted by many scientists as a fairly short-lived interglacial interval between major ice ages.

Scientists think that the ice ages are primarily the result of changes in the tilt of the Earth's rotational axis, produced by the gravitational effects of the other planets. This idea of an astronomical cause of climate changes was first proposed in 1920 by the Serbian scientist Milutin Milankovich. As we will see in Chapter 9, there is also evidence of periodic climate changes on Mars, and modern calculations suggest that these also have their origin in slow changes of that planet's orbit and rotational axis.

FIGURE 7.12
A Storm from Space   A large tropical storm marked by clouds swirling around a low-pressure region, photographed by astronauts orbiting the Earth.   *(NASA)*

 ## LIFE AND CHEMICAL EVOLUTION

As far as we know, Earth seems to be the only planet in the solar system with life. The origin and development of life is an important part of the story of our planet. Life arose early in Earth's history, and living organisms have been interacting with their environment for billions of years. We recognize that life forms have evolved to adapt themselves to the environment on Earth, and we are now beginning to realize that the Earth itself has been changed in important ways by the presence of living matter.

The Origin of Life

The record of the birth of life on Earth has been lost in the restless motions of the crust. By the time the oldest surviving rocks were formed, about 3.8 billion years ago, life already existed. At 3.5 billion years ago, life had achieved the sophistication to build large colonies called stromatolites (Figure 7.13), a form so successful that stromatolites still grow on Earth today. But little crust survives from these ancient times, and abundant fossils have been produced only during the past 600 million years—less than 15 percent of our planet's history.

Any theory of the origin of life must therefore be partly speculative, because there is little direct evidence to go on. All we really know is that the atmosphere of the early Earth, unlike today, contained abundant carbon dioxide but no oxygen gas. In the absence of oxygen, many complex chemical reactions are possible that lead to the production of amino acids, proteins, and other chemical building blocks of life. It now seems relatively certain that life arose from these building blocks very early in Earth's history. Detailed genetic analysis shows how the millions of species on Earth today are related to each other and demonstrates that all terrestrial life is descended from a single common microbial ancestor.

For tens of millions of years after its formation, life (perhaps little more than large molecules, like the viruses of today) probably existed in warm, nutrient-rich seas, living off accumulated organic chemicals. Eventually, however, as this easily accessible food became depleted, life began the long evolutionary road that led to the vast numbers of different organisms on Earth today. As it did so, life began to influence the chemical evolution of the atmosphere.

Alternatively, life may not have formed in warm surface waters at all. We have discovered microbes at deep-sea vents where the temperature is above the boiling point of water, as well as in rocks miles below the surface. Genetic analysis shows that the earliest surviving terrestrial life forms were all adapted for high temperatures, and some biologists think that life might have originated under such circumstances, which seem extreme to us but could have provided the environment needed for life's origin. Yet another possibility is that life began on Mars rather than the Earth and was "seeded" onto our planet by meteorites traveling from Mars to Earth.

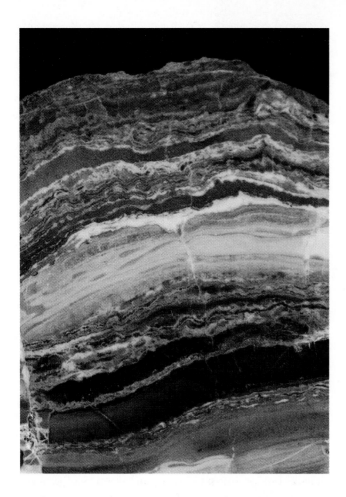

FIGURE 7.13
Cross Sections of Fossil and Contemporary Stromatolites
The layered, dome-like structures are mats of sediment trapped in shallow waters by large numbers of blue-green bacteria that can photosynthesize. Such colonies of microorganisms date back more than 3 billion years. *(NASA/ARC, courtesy of David DesMarais)*

Evolution of the Atmosphere

One of the key steps in the evolution of life on Earth was the development of blue-green algae, a very successful life form that takes in carbon dioxide from the environment and releases oxygen as a waste product. These microorganisms proliferated, giving rise to all the life forms we call plants. Since the energy for making new plant material from chemical building blocks comes from sunlight, we call the process *photosynthesis.*

Studies of the chemistry of ancient rocks show that Earth's atmosphere lacked free oxygen until about two billion years ago, in spite of the presence of plants releasing oxygen by photosynthesis. Apparently, chemical reactions with the crust removed the oxygen gas as quickly as it formed. Slowly, however, the increasing evolutionary sophistication of life led to a growth in plant population and thus increased oxygen production. At the same time, increased geological activity resulted in heavy erosion that buried much of the plant carbon before it could recombine with oxygen to form CO_2.

Free oxygen began accumulating in the atmosphere about two billion years ago, and the increased amount of this gas led to the formation of the Earth's ozone layer, which protects the surface from lethal solar ultraviolet light. Before that, it was unthinkable for life to venture outside the protective oceans, so the land masses of Earth were barren. The presence of oxygen and hence ozone thus allowed the colonization of the land. It also made possible a tremendous proliferation of animals, who lived by taking in and using the organic materials produced by plants as their own energy source.

As animals evolved in an environment increasingly rich in oxygen, they were able to develop techniques for breathing oxygen directly from the atmosphere. We humans take it for granted that plenty of free oxygen is available in the Earth's atmosphere to release energy from the food we take in. Although it may seem funny to think of it this way, we are life forms that have evolved to breathe in the waste product of plants.

On a planetary scale, one of the most important consequences of life has been a decrease in atmospheric carbon dioxide. In the absence of life, Earth would probably have a much more massive atmosphere dominated by CO_2 like that of the planet Venus. But life in combination with high levels of geological activity has effectively stripped us of most of this gas.

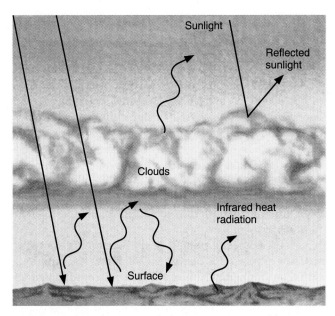

FIGURE 7.14
How the Greenhouse Effect Works   Sunlight that penetrates to the Earth's lower atmosphere and surface is reradiated as infrared or heat radiation, which is trapped by the atmosphere. The result is a higher surface temperature for our planet.

The Greenhouse Effect and Global Warming

We have a special interest in the carbon dioxide content of the atmosphere because of the role this gas plays in retaining heat from the Sun through a process called the **greenhouse effect.** To understand how the greenhouse effect works, consider the fate of the sunlight that strikes the surface of the Earth. It is absorbed, heats the surface layers, and is then re-emitted as infrared or heat radiation (Figure 7.14). However, the CO_2 in our atmosphere, which is transparent to visible light, is largely opaque to infrared energy. As a result, it acts like a blanket, trapping the heat in the atmosphere

and impeding its flow back to space. On average, as much energy reaches the surface from the atmospheric greenhouse effect as from direct sunlight. For this reason, temperatures are only slightly lower at night than in the daytime. Over time this extra heat means that the Earth's surface regions get warmer than they would have without the effect of the CO_2. The more CO_2 there is in our atmosphere, the higher the temperature at which the Earth's surface reaches a new balance.

The greenhouse effect in a planetary atmosphere is similar to the heating of a gardener's greenhouse or the inside of a car left out in the Sun with the windows rolled up. In these examples, the window glass plays the role of carbon dioxide, letting sunlight in but reducing the outward flow of heat radiation. As a result, a greenhouse or car interior winds up much hotter than would be expected from the heating of sunlight alone. On Earth, the current greenhouse effect elevates the surface temperature by about 23°C. Without this greenhouse effect, the average surface temperature would be well below freezing and the Earth would be locked in a global ice age.

That's the good news; the bad news is that the greenhouse effect is increasing. Modern industrial society depends on energy extracted from burning fossil fuels. As these ancient coal and oil deposits are oxidized, additional carbon dioxide is released into the atmosphere. The problem is increased by the widespread destruction of tropical forests, which we depend on to extract CO_2 from the atmosphere and replenish our supply of O_2. So far in this century, the amount of CO_2 in the atmosphere has increased by about 25 percent, and it is continuing to rise at 0.5 percent per year.

By early in the 21st century, the CO_2 level is predicted to reach twice the value it had before the industrial revolution (Figure 7.15). The consequences of such an increase for the Earth's surface and atmosphere (and the creatures who live there) are likely to be complex. Many groups of

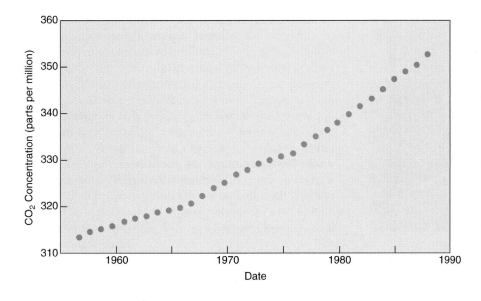

FIGURE 7.15
The Increase of Atmospheric Carbon Dioxide over Time   The amount of CO_2 is expected to double by the middle of the 21st century.   *(Adapted from data obtained at the Mauna Loa Observatory of NOAA)*

scientists are now studying the effects of such *global warming* with elaborate computer models. At the present time we are not sure what all the effects will be, since the Earth's surface is an extremely complicated system and the specific results of increased temperature will surely vary from one location to another.

Already some global warming is apparent. In North America and Europe, summer temperatures throughout the 1980s and 1990s reached record highs. Because they are superimposed on the usual year-to-year fluctuations in weather, these effects are not easy to sort out, but most scientists are convinced that global warming due to an enhanced greenhouse effect is a reality. The increasing greenhouse effect comes at a particularly difficult time: Our planet is already in an unusually warm interglacial period. We are rapidly entering unknown territory where human activities are contributing to the highest temperatures on Earth in more than 50 million years.

7.5 COSMIC INFLUENCES ON THE EVOLUTION OF EARTH

Where Are the Craters on Earth?

In discussing the Earth's geology in Section 7.2, we dealt only with the effects of internal forces, expressed through the processes of plate tectonics and volcanism. In contrast, on the Moon we see primarily craters, produced by the impacts of interplanetary debris (as described in Chapter 6).

FIGURE 7.16
Aftermath of the Tunguska Explosion   This photograph, taken 21 years after the blast, shows a part of the forest that was devastated by the 15-megaton explosion.   *(Novosty)*

Why do we not see more evidence on Earth of the kinds of impact craters that are so prominent on the Moon and other planets?

It is not possible that the Earth escaped being struck by the interplanetary debris that has pockmarked the Moon. From a cosmic perspective, the Moon is almost next door, and it is unlikely that if our closest neighbor was bombarded by projectiles from space over vast periods of time, we could somehow have been immune. Our atmosphere does make small pieces of cosmic debris burn up (which we see as *meteors*—commonly called shooting stars). But the layers of air provide no shield against the large impacts that form craters several kilometers or more in diameter.

In the course of its history, the Earth must therefore have been cratered as heavily as the Moon. The difference is that on the Earth these craters are destroyed by our active geology before they can accumulate. As plate tectonics constantly renews our crust, evidence of past cratering events is destroyed. Only in the past few decades have geologists succeeded in identifying the eroded remnants of many old impact craters. Even more recent is our realization that these impacts have had an important influence on the evolution of life.

Recent Impacts

The collision of interplanetary debris with the Earth is not a hypothetical idea: Evidence of a number of relatively recent impacts can be found on our planet's surface. The best-studied historic collision took place on June 30, 1908, near the Tunguska River in Siberia. In this desolate region, a remarkable explosion took place in the atmosphere about 8 km above the surface. The shock wave flattened more than a thousand square kilometers of forest. Herds of reindeer and other animals were killed, and a man at a trading post 80 km from the blast was thrown from his chair and knocked unconscious (Figure 7.16). The blast wave spread around the world, recorded by instruments designed to measure changes in atmospheric pressure.

Despite this violence, no craters were formed by the Tunguska explosion. Shattered by atmospheric pressure, a stony projectile with a mass of approximately 100,000 tons disintegrated to create a blast equivalent to a 15-megaton nuclear bomb.[1] Had it been smaller or more fragile, the impacting body would have dissipated its energy at high altitude and probably attracted no attention. If it had been larger or made of stronger material (such as metal), it would have penetrated all the way to the surface and formed a crater. Instead, only the shock of the explosion reached the surface, but the devastation it left behind in Siberia bore witness to the power of such impacts. Imagine if the same rocky impactor had exploded over New York

---

[1]A megaton is the explosive energy released by a million tons of TNT—not a million pounds or kilograms, but a million *tons!*

FIGURE 7.17

Meteor Crater in Arizona Here we see a 50,000-year-old impact scar, made by the collision of a 50-meter lump of iron with our planet. While impact craters are common on less-active bodies such as the Moon, this is one of the very few well-preserved craters on the Earth. *(Meteor Crater, Northern Arizona)*

City in 1908; history books might today record it as one of the most deadly events in human history.

The best-known recent crater on Earth was formed 50,000 years ago in Arizona. The projectile in this case was a lump of iron about 50 m in diameter (we will discuss in Chapter 8 how the explosion of a small chunk of material can make a much bigger crater). Now called Meteor Crater and a major tourist attraction on the way to the Grand Canyon, the crater is about a mile across and has all of the features associated with similar-sized lunar impact craters (Figure 7.17). Meteor Crater is one of the few impact features on the Earth that remain relatively intact; other, older craters are so eroded that only a trained eye can distinguish them. Nevertheless, more than 150 have been identified as further evidence that impacts continue to happen to our planet over the millennia (see our list of Web sites if you want to find out more about these other impact scars).

Extinction of the Dinosaurs

The impact that produced Meteor Crater would have been dramatic indeed to any humans who witnessed it, since the energy release was also equivalent to a 15-megaton nuclear bomb. But such explosions are devastating only in their local areas; they have no *global* consequences. Much larger (and rarer) impacts, however, can disturb the ecological balance of the entire planet and thus influence the course of evolution.

The best-documented large impact took place 65 million years ago, at the end of what is now called the Cretaceous era of geological history. This point in the history of life on Earth was marked by a **mass extinction,** in which more than half of the species on our planet died out. There are a dozen or more mass extinctions in the geological record, but this particular event has always intrigued paleontologists because it marks the end of the dinosaur age. For tens of millions of years these great creatures had flourished. Then they suddenly disappeared (along with many other species), and thereafter the mammals began the development and diversification that ultimately led to the readers (and authors) of this book.

The object that collided with the Earth at the end of the Cretaceous era struck a shallow sea in what is now the Yucatan peninsula of Mexico. Its mass must have been more than a trillion tons (Figure 7.18), because scientists

FIGURE 7.18
A 20-km Asteroid Hits the Earth This is the approximate magnitude of the event that led to a mass extinction 65 million years ago. *(Painting by Don Davis)*

have found a worldwide layer of sediment deposited from the dust cloud that enveloped the planet after its impact. First identified in 1979, this sediment layer is rich in the rare metal iridium and other elements that are relatively abundant in asteroids and comets but very rare in the Earth's crust. Even diluted by the terrestrial material excavated by the explosion from the crater it made, this cosmic component is easily identified. In addition, the sediment contains many minerals characteristic of the temperatures and pressures of a gigantic explosion.

The impact released energy equivalent to five billion Hiroshima-sized nuclear bombs, excavating a crater 200 km across and deep enough to penetrate through the Earth's crust. This large crater, named Chicxulub for a small town near its center, has subsequently been buried in sediment, but its outlines can still be identified (Figure 7.19). The explosion that created the Chicxulub crater lifted about *100 trillion tons* of dust into the atmosphere, determined by measuring the thickness of the sediment layer that formed when this dust settled to the surface.

Such a quantity of airborne material would have blocked sunlight completely, plunging the Earth into a period of cold and darkness that lasted several months. Other worldwide effects included large-scale fires (started by the hot debris from the explosion) that destroyed most of the planet's forests and grasslands, and a long period in which rainwater around the globe was acidic. Presumably these environmental effects, rather than the explosion itself,

were responsible for the mass extinction, including the death of the dinosaurs.

A demonstration of such effects took place in July 1994, when pieces of a comet called Shoemaker–Levy 9 impacted the planet Jupiter. Although Jupiter has no solid surface, the comet pieces themselves contained a great deal of dusty material. The comet fragments exploded as they penetrated the giant planet's atmosphere, and vast clouds of dark material were seen to expand over great distances (see Chapter 12). In Figure 7.20 a computer graphic shows what it might have looked like if the largest fragment of the comet (roughly 1 km across) had struck Detroit instead.

Impacts and the Evolution of Life

Several other mass extinctions in the geological record have been tentatively identified with large impacts, although none is so dramatic as the blast that destroyed the dinosaurs. Even without such specific documentation, it is now clear that impacts of this size do occur on all planets, and that their effects can be catastrophic for life. A catastrophe for one group of living things, however, may create opportunities for another group. Following each mass extinction, there is a sudden evolutionary burst as new species develop to fill the ecological niches opened by the event.

Impacts by comets and asteroids represent the only mechanisms we know of that could cause truly global catastrophes and seriously influence the evolution of life all over the planet. According to some estimates, the majority of all species extinctions may be due to such impacts. As paleontologist Stephen Jay Gould of Harvard has noted, such a perspective fundamentally changes our

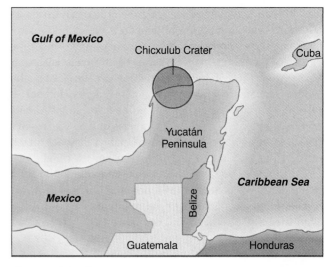

FIGURE 7.19
Site of the Chicxulub Crater Map showing the location of the buried crater from the impact event 65 million years ago on Mexico's Yucatan peninsula. *(NASA)*

FIGURE 7.20

If a Comet Had Hit Earth   Computer simulation by astronomer John Spencer at the Lowell Observatory, showing what the Earth might have looked like 2 hours after impact if the largest fragment of Comet Shoemaker–Levy 9 had hit near Detroit, Michigan. A photograph of the debris cloud from the fragment explosion on Jupiter was superimposed onto a global view of the Earth. If the roughly 1-km-wide comet fragment had actually hit the Earth, the debris would have included a mixture of vaporized Earth solids as well as the material from the comet. Nevertheless, this image makes clear why large impacts can cause planetwide effects.

view of biological evolution. The central issues for the survival of a species must now include more than just its success in competing with other species and adapting to slowly changing environments, as envisioned by Darwinian natural selection. Of at least equal importance is its ability to survive random global ecological catastrophes due to impacts.

Still earlier in its history, the Earth was subject to even larger impacts from the leftover debris of planet formation.

We know the Moon was repeatedly struck by objects more than 100 km in diameter, a thousand times more massive than the object that wiped out most terrestrial life 65 million years ago. The Earth must have experienced similar large impacts during its first 700 million years of existence. Some of them were probably violent enough to strip the planet of most of its atmosphere and to boil away its oceans. Such events would sterilize the planet, utterly destroying any life that had begun. Life may have formed and been wiped out several times before our own ancestors took hold sometime about 4 billion years ago.

The fact that the oldest surviving microbes are thermophiles (adapted to very high temperatures) can also be explained by such large impacts. An impact that was just a bit too small to sterilize the planet would still have destroyed anything that lived in what we consider "normal" environments, and only the creatures that liked high temperatures would survive. Thus the oldest surviving terrestrial life forms are probably the remnants of a sort of evolutionary bottleneck caused by repeated large impacts early in the planet's history.

Today, then, we think of the Earth as a target in a cosmic shooting gallery, subject to random violent events that were unsuspected a few decades ago. We owe our existence to such events. Had the impact 65 million years ago not redirected the course of evolution, mammals might never have become the dominant large animals they are today (and intelligent dinosaur students might be reading this text). But scientists have begun to ask an even more fundamental question: If impacts with comets and asteroids had not occurred throughout our planet's history, could biological evolution in more stable environments have produced the wondrous diversity of life that populates the Earth today?

SURFING THE WEB

🖥 Earth on the Solar System Web Sites

- The Nine Planets Site: [seds.lpl.arizona.edu/nineplanets/nineplanets/earth.html]
- Views of the Solar System Site: [www.hawastsoc.org/solar/eng/earth.htm]
- Planetary Sciences Site: [nssdc.gsfc.nasa.gov/planetary/planets/earthpage.html]

🖥 Tutorial on the Magnetosphere [ssdoo.gsfc.nasa.gov/education/lectures/magnetosphere.html]
Primer on the Earth's zone of magnetic influence by a NASA scientist.

🖥 The Electronic Volcano [www.dartmouth.edu/~volcano]
Everything you wanted to know about volcanoes (but were afraid to ask) is on this site by a Dartmouth librarian and two scientists, or can be reached through the links it provides, including catalogs, maps, current events, and images.

🖥 NASA Observatorium Ozone Guide [observe.ivv.nasa.gov/nasa/exhibits/ozone/Ozone/.html]
A basic introduction to the ozone problem, with annoying typefaces, but useful information and links.

🖥 National Earthquake Information Center [wwwneic.cr.usgs.gov/]
A wealth of information about earthquakes from the U.S. Geological Survey.

🖥 Asteroid and Comet Impact Hazard Page [impact.arc.nasa.gov/]
Voyages co-author David Morrison is behind this informative site about the dangers of asteroids and comets hitting the Earth and what we might do about it. Included are congressional and scientific position papers, a list of objects that pass near the Earth, useful background information, and annotated links to related Web sites around the world.

🖥 Terrestrial Impact Craters Site [www.hawastsoc.org/solar/eng/tercrate.htm]
This site includes a concise introduction to impact craters on our planet, with images, information, and an educator's guide. See also a good site from the Canadian Geological Survey called "Terrestrial Impact Structures" at [gdcinfo.agg.emr.ca/crater/world_craters.html].

SUMMARY

7.1 The Earth is the prototype terrestrial planet. Its interior composition and structure are probed using **seismic waves.** Such studies reveal that we have a metal **core** and silicate **mantle.** The outer layer or **crust** consists primarily of oceanic **basalt** and continental **granite.** A global magnetic field, generated in the core, produces the Earth's **magnetosphere,** which can trap charged atomic particles.

7.2 Terrestrial rocks can be classified as **igneous, sedimentary,** or **metamorphic.** A fourth type, **primitive** rock, is not found on the Earth. Our planet's geology is dominated by **plate tectonics,** in which crustal plates move slowly in response to mantle **convection.** The surface expression of plate tectonics includes continental drift, recycling of the ocean floor, mountain building, **rift zones, subduction**

zones, **faults,** earthquakes, and volcanic eruptions of **magma** from the interior.

7.3 The atmosphere has a surface pressure of 1 **bar** and is composed primarily of N_2 and O_2, plus such important trace gases as H_2O, CO_2, and **ozone** (O_3). Its structure consists of the **troposphere,** the **stratosphere,** and tenuous higher regions. Atmospheric oxygen is the product of photosynthetic life forms on Earth. Atmospheric circulation (weather) is driven by seasonally changing deposition of sunlight. Longer-term climatic variations, such as the ice ages, are probably due to changes in the planet's orbit and axial tilt.

7.4 Life originated on Earth at a time when the atmosphere lacked O_2 and consisted mostly of CO_2. Later, photosynthesis gave rise to free oxygen and ozone. Carbon dioxide in the atmosphere heats the surface through the **greenhouse effect;** today, increasing atmospheric CO_2 is leading to the global warming of our planet.

7.5 Earth, like the Moon and other planets, has been influenced by the impacts of cosmic debris, including such small recent examples as Meteor Crater and the Tunguska explosion. Larger past impacts are responsible for at least some **mass extinctions,** including the large extinction 65 million years ago that ended the Cretaceous period. Such impacts have probably played an important role in the evolution of life.

INTER-ACTIVITY

A If we can predict that lots of ground movement takes place along subduction zones and faults, why do so many people live there? Should we try to do anything to discourage people from living in these areas? What inducement would your group offer people to move? Who would pay for the relocation? (Note that two of the authors of this book live quite close to the San Andreas and Hayward faults. If they wrote this chapter and haven't moved, what are the chances others living in these kinds of areas will move?)

B After your group reads the box on Alfred Wegener, discuss some reasons why his idea did not catch on right away among scientists. From your studies in this course and in other science courses (in college and before), can you cite other scientific ideas that we now accept but that had controversial beginnings? Can you think of any scientific theories that are still controversial today? If your group comes up with some, can you also discuss ways that scientists can decide if each theory on your list is right?

C Suppose we knew that a large chunk of rock or ice (about the same size as the one that hit 65 million years ago) will impact on Earth in about five years. What could or should we do about it? (A recent film, *Deep Impact*, dealt with this theme.) Does your group think that the world as whole should be putting more money into finding and predicting the orbits of cosmic debris near the Earth?

D Carl Sagan pointed out that any defensive weapon that we might come up with to deflect an asteroid *away* from the Earth could be used as an offensive weapon by an unstable government leader in the future to cause an asteroid not heading our way to come *toward* the Earth. The history of human behavior, he noted, showed that most weapons that are built (even with the best of motives) seem to wind up being used. Bearing this in mind, does your group think we should be building weapons to protect the Earth from asteroid or comet impact? Can we afford not to build them? How can we safeguard them?

REVIEW QUESTIONS

1. What are the Earth's core and mantle made of? Explain how we know.

2. Describe the differences among primitive, igneous, sedimentary, and metamorphic rock, and relate these differences to their origins.

3. Explain briefly how the following phenomena happen on Earth, relating your answers to the theory of plate tectonics:
  a. earthquakes
  b. continental drift
  c. mountain building
  d. volcanic eruptions
  e. creation of the Hawaiian island chain

4. Describe three ways in which the presence of life has affected the composition of the Earth's atmosphere.

5. How do impacts by comets and asteroids influence the Earth's geology, its atmosphere, and the evolution of life?

6. Why are there so many impact craters on our neighbor world, the Moon, and so few on the Earth?

7. If you wanted to live where the chances of a destructive earthquake were small, would you pick a location near a fault zone, near a mid-ocean ridge, near a subduction zone, or on a volcanic island such as Hawaii? What are the relative risks of earthquakes at each of these locations?

8. If all life were destroyed on Earth, would new life eventually form to take its place? Explain how conditions would have to change for life to start again on our planet.

9. Why will a decrease in the Earth's ozone be harmful to life?

10. Why are we concerned about the increases in CO_2 and other gases that cause the greenhouse effect in the Earth's atmosphere? What steps could we take in the future to reduce the levels of CO_2 in our atmosphere? What factors stand in the way of taking the steps you suggest (you may include technological, economic, and political factors in your answer).

11. Is there evidence of climate change in your area over the past century? How would you distinguish a true climate change from the random variations in weather that take place from one year to the next?

12. How might the history of the Earth have been different if it were closer to the Sun? Farther from the Sun?

13. What fractions of the volume of the Earth are occupied by the core, the mantle, and the crust?

14. Suppose that the next slippage along the San Andreas Fault in southern California takes place in the year 2001, and that it completely relieves the accumulated strain in this region. How much slippage is required for this to occur?

15. Measurements using Earth satellites have shown that Europe and North America are moving apart by 4 m per century due to plate tectonics. As the continents separate, new ocean floor is created along the Mid-Atlantic Rift. If the rift is 5000 km long, what is the total area of new ocean floor created in the Atlantic each century? Each year?

16. Over the entire Earth there are 60,000 km of active rift zones, with average separation rates of 4 m per century. How much area of new ocean crust is created each year over the entire planet? (This area is approximately equal to the amount of ocean crust that is subducted, since the total area of the oceans remains about the same.)

17. With the information from Problem 16, you can calculate the average age of the ocean floor. First find the total area of ocean floor (equal to about 60 percent of the surface area of the Earth). Then compare this with the area created (or destroyed) each year. The average lifetime is the ratio of these numbers— the total area of ocean crust compared to the amount created (or destroyed) each year.

18. What is the volume of new oceanic basalt added to the Earth's crust each year? Assume that the thickness of the new crust is 5 km, that there are 60,000 km of active rifts, and that the average speed of plate motion is 4 cm per year. What fraction of the Earth's entire volume does this annual addition of new material represent?

19. The sea-level pressure of the atmosphere (1 bar) corresponds to 10^4 kg of mass above each square meter of the Earth's surface. Calculate the total mass of the atmosphere in kilograms and in tons (1 ton equals 1000 kg). Then compute the total mass of ocean, given that the oceans would be 3000 m deep if water covered the globe uniformly. (*Note:* 1 m^3 of water has a mass of 1 ton.) Compare the mass of atmosphere and the mass of the oceans with the total mass of the Earth to determine what percentage of our planet is represented by the atmosphere and oceans.

20. Suppose a major impact that produces a mass extinction takes place on the Earth once every 5 million years. Further suppose that if such an event occurred today, you and most other humans would be killed (this would be true even if the human species as a whole survived). Such impact events are random, and one could take place at any time. Calculate the probability that such an impact will occur within the next 50 years (within your lifetime). This is equal to the probability that you will be killed by this means, rather than dying from an auto accident or heart disease or some other "natural" cause. How do the risks of dying from the impact of an asteroid or comet compare with other risks we are concerned about? (*Hint:* To find the annual risk, go to the library and look up the annual number of deaths from a particular cause in a particular country, and then divide by the population of that country.)

Allegre, C. and Schneider, S. "The Evolution of the Earth" in *Scientific American,* Oct. 1994, p. 66.

Broadhurst, L. "Earth's Atmosphere: Terrestrial or Extraterrestrial" in *Astronomy,* Jan. 1992, p. 38.

Cattermole, P. and Moore, P. *The Story of the Earth.* 1985, Cambridge U. Press.

Chyba, C. "Death from The Sky: Tunguska" in *Astronomy,* Dec. 1993, p. 38. Excellent review article.

Gallant, R. "Journey to Tunguska" in *Sky & Telescope,* June 1994, p. 38. A planetarium director tells of his trip to the site and fills in its history.

Hartmann, W. "Piecing Together Earth's Early History" in *Astronomy,* June 1989, p. 24.

Hartmann, W. and Miller, R. *The History of the Earth.* 1993, Workman. Lavishly illustrated chronicle of our planet's evolution.

Heppenheimer, T. "Journey to the Center of the Earth" in *Discover,* Nov. 1987, p. 86.

Jones, P. and Wigley, T. "Global Warming Trends" in *Scientific American,* Aug. 1990, p. 84. Three hundred years of data analyzed.

Lewis, J. *Rain of Iron and Ice: The Very Real Threat of Comet and Asteroid Bombardment.* 1995, Addison-Wesley. Good overview of impacts.

Morrison, D. and Chapman, C. "Target Earth: It Will Happen" in *Sky & Telescope,* Mar. 1990, p. 261. (See an updated version in *Astronomy,* Oct. 1995, p. 34.)

Murphy, J. and Nance, R. "Mountain Belts and the Supercontinent Cycle" in *Scientific American,* Apr. 1992, p. 84. There has been more than one supercontinent through geologic time.

Orgel, L. "The Origin of Life on Earth" in *Scientific American,* Oct. 1994, p. 77.

York, D. "The Earliest History of the Earth" in *Scientific American,* Jan. 1993, p. 90.

Apollo 11 astronaut Buzz Aldrin on the surface of the Moon. Because there is no atmosphere, ocean, or geological activity on the Moon today, the footprints you see in this image will likely be preserved in the lunar soil for millions of years. *(NASA)*

8 Cratered Worlds: The Moon and Mercury

THINKING AHEAD

The Moon is the only other world human beings have ever visited. What is it like to stand on the surface of our natural satellite? And what can we learn from going there and bringing home pieces of a different world?

"**This is one small step for a man, one giant leap for mankind.**"

What astronaut Neil Armstrong meant to say when he became the first human being to walk on another world. (In his nervousness, he left out the word "a," making the statement a bit less effective.)

We begin our discussion of the planets as worlds with two relatively simple objects: the Moon and Mercury. Unlike the Earth, the Moon is geologically dead, a place that has exhausted its internal energy sources. Because its airless surface preserves the events that happened long ago, the Moon provides us with a window on the earlier epochs of solar system history. The planet Mercury is in many ways similar to the Moon, which is why the two are discussed together. Both are relatively small and are lacking in atmospheres, deficient in geological activity, and dominated by the effects of impact cratering from outside. Still, the processes that have molded their surfaces are not unique to these two worlds; they have acted on many other members of the planetary system as well.

FIGURE 8.1

The Moon from Earth   The Moon as photographed by astronauts from a spacecraft near the Earth. The left side of the image shows part of the hemisphere that faces the Earth; several dark maria are visible. The right side shows part of the hemisphere that faces away from the Earth; it is dominated by highlands. The resolution of such an image is several kilometers, similar to that of high-powered binoculars or a small telescope.   *(NASA)*

but lunar geology hardly existed as a scientific subject. All that changed beginning in the early 1960s.

Initially the Russians took the lead in lunar exploration with Luna 3, which returned the first photos of the lunar far side in 1959, and then with Luna 9, which landed on the surface in 1966, and transmitted pictures and other data to Earth. However, these efforts were overshadowed on July 20, 1969, when the first American astronaut set foot on the Moon.

Table 8.2 summarizes the nine Apollo flights—six that landed and three others that circled the Moon but did not land. The initial landings were on flat plains selected out of safety considerations, but with increasing experience and confidence, NASA targeted the last three missions to more geologically interesting locales. The level of scientific exploration also increased with each mission, as the astronauts spent longer times on the Moon and carried more elaborate equipment. Finally, on the last Apollo Moon landing, NASA included one scientist, geologist Jack Schmitt, among the astronauts (Figure 8.2).

In addition to landing on the lunar surface and studying it at close hand, the Apollo missions accomplished three objectives of major importance for lunar science. First, the astronauts collected nearly 400 kg of samples for detailed laboratory analysis on Earth (Figure 8.3). These samples, still being studied, have probably revealed more about the Moon and its history than all other lunar studies combined. Second, each Apollo landing after the first one

8.1   GENERAL PROPERTIES OF THE MOON

The Moon (Figure 8.1) has only 1/80 the mass of the Earth and about 1/6 of the Earth's surface gravity—too low to retain an atmosphere. If, early in its history, the Moon expelled an atmosphere from its hot interior or collected a temporary envelope of gases from impacting comets, such an atmosphere was lost before it could leave any recognizable evidence of its short existence. Indeed, the Moon is dramatically deficient in a wide range of volatiles, those elements and compounds that evaporate at relatively low temperatures. Some of the Moon's properties are summarized in Table 8.1, along with comparative values for Mercury.

Exploration of the Moon

Most of what we know about the Moon today derives from the U.S. Apollo program, which sent nine spacecraft with people on board to our satellite between 1968 and 1972, landing 12 astronauts on its surface (see the photograph that opens this chapter). Before the era of spacecraft studies, astronomers had mapped the side of the Moon that faces the Earth with telescopic resolution of about 1 km,

TABLE 8.1
Properties of the Moon and Mercury

	Moon	Mercury
Mass (Earth = 1)	0.0123	0.055
Diameter (km)	3476	4878
Density (g/cm³)	3.3	5.4
Surface gravity (Earth = 1)	0.17	0.38
Escape velocity (km/s)	2.4	4.3
Rotation period (days)	27.3	58.65
Surface area (Earth = 1)	0.27	0.38

TABLE 8.2
Apollo Flights to the Moon

Flight	Date	Landing Site	Main Accomplishment
Apollo 8	Dec. 1968	—	First humans to fly around the Moon.
Apollo 10	May 1969	—	First rendezvous in lunar orbit.
Apollo 11	July 1969	Mare Tranquillitatis	First human landing on the Moon; 22 kg of samples returned.
Apollo 12	Nov. 1969	Oceanus Procellarum	First ALSEP; visit to Surveyor 3.
Apollo 13	Apr.1970	—	Landing aborted due to explosion in Command Module.
Apollo 14	Jan. 1971	Mare Nubium	First "rickshaw" on the Moon.
Apollo 15	July 1971	Imbrium/Hadley	First "rover"; visit to Hadley Rille; astronauts traveled 24 km.
Apollo 16	Apr. 1972	Descartes	First landing in highlands; 95 kg of samples returned.
Apollo 17	Dec. 1972	Taurus Mountains	Geologist among the crew; 111 kg of samples returned.

deployed an Apollo Lunar Surface Experiment Package (ALSEP), which continued to operate for years after the astronauts departed. (The ALSEPs were turned off by NASA in 1978 as a cost-cutting measure.) Third, the orbiting Apollo Command Modules carried a wide range of instruments to photograph and analyze the lunar surface from above.

The last human left the Moon in December 1972, just a little more than three years after Neil Armstrong took his "giant leap for mankind." The program of lunar exploration was cut off in midstride due to political and economic pressures. It had cost just about $100 per American, spread over ten years—the equivalent of one large pizza per person per year. Yet for many people, the Moon landings were one of the central events in 20th-century history.

The giant Apollo rockets built to travel to the Moon were left to rust on the lawns of NASA centers in Florida, Texas, and Alabama (Figure 8.4), although recently they have at least been moved indoors to museums. Today no nation on Earth has the capability of returning people to the Moon; NASA estimates that it would require more

FIGURE 8.2
A Scientist on the Moon Geologist (and later U.S. Senator) Jack Schmitt in front of a large boulder in the Litrow Valley at the edge of the lunar highlands. The "rover" vehicle the astronauts used for driving around on the Moon is seen at left. Note how black the sky is on the airless Moon. *(NASA)*

FIGURE 8.3
Handling Moon Rocks Lunar samples collected in the Apollo project are analyzed and stored in NASA facilities at the Johnson Space Center in Houston, Texas. Here a technician is examining a rock sample using gloves in a sealed environment to avoid contaminating the sample. *(NASA)*

FIGURE 8.4
Moon Rocket on Display One of the unused Saturn 5 rockets built to go to the Moon, now a tourist attraction at NASA's Johnson Space Center in Houston. *(NASA)*

even of a lunar base as a way station on the routes to Mars and the rest of the solar system.

Composition and Structure of the Moon

The composition of the Moon is not the same as that of the Earth. With an average density of only 3.3 g/cm^3, the Moon must be made almost entirely of lighter silicate rock. Compared to the Earth, it is depleted in iron and other metals. We also know from the study of lunar samples that water and other volatiles are largely absent. It is as if the Moon were composed of the same basic silicates as the Earth's mantle and crust, with the core metals and the volatiles selectively removed. These differences in composition between Earth and Moon provide important clues concerning the origin of the Moon, a topic we return to later in this chapter.

Probes of the interior carried out with seismometers taken to the Moon as part of the Apollo program confirm the absence of a large metal core. The Moon also lacks a global magnetic field like that of the Earth, a result consistent with the theory that such a field must be generated by motions in a liquid metal core. Not only is the metal lacking, but in addition the Moon is cold and has a solid interior. The level of seismic activity is very low; the total energy released by moonquakes is 100 billion times less than that of earthquakes on our planet.

than a decade to mount such an effort, and the financial and political will to do so are lacking. Having reached our nearest neighbor in space, humanity has for now retreated back to our own planet. It is hard to guess how long it will be before humans again venture out into the solar system.

The scientific legacy of Apollo remains, however, as discussed in the following sections of this chapter. The Apollo results have also been recently supplemented by three small scientific orbiters, one launched by NASA, one by Japan, and the third built by the U.S. military as a by-product of the "star wars" space defense effort (Figure 8.5).

The U.S. Air Force's Clementine satellite took high-resolution photos of the entire Moon (including parts missed by Apollo) and also provided fascinating hints of water ice in deep, permanently shadowed craters near the lunar poles. NASA's Lunar Prospector mission confirmed the ice and estimated its total quantity at roughly 6 billion tons. As liquid, this would only be enough water to fill a lake a few miles across, but compared with the rest of the dry lunar crust, so much water is remarkable. Presumably the water was carried to the Moon by comets and asteroids that hit our satellite. Some small fraction froze into a few extremely cold regions where the Sun never shines. One reason this discovery could be important is that it raises the possibility of human habitations near the lunar poles, or

 8.2 **THE LUNAR SURFACE**

General Appearance

If you look at the Moon through a telescope, you see that it is covered by impact craters of all sizes. However, none of these craters or other topographic features is large enough to be seen without optical aid. The most conspicuous of the Moon's surface features—those that can be seen with the unaided eye and that make up the feature often called "the man in the Moon"—are vast splotches of darker material of volcanic origin.

Centuries ago, the early lunar observers thought that the Moon had continents and oceans, and that it was a possible abode of life. Early names given to the lunar "seas"—Mare Nubium (Sea of Clouds), Mare Tranquillitatis (Tranquil Sea), and so on—are still in use today. In contrast, the "land" areas between the seas do not have names. Thousands of individual craters have been named, however, mostly for great scientists and philosophers. Among the most prominent are those named for Plato, Copernicus, Tycho, and Kepler. Galileo has only a very small crater, however, reflecting his low standing among the Roman Catholic scientists who made some of the first lunar maps.

We know today, however, that the resemblance of lunar features to terrestrial ones is superficial. Even when

Lunar South Pole This mosaic of approximately 1500 images taken with the Clementine spacecraft shows the south pole region of the Moon. Don't let the round border of the image fool you; you are only looking at a small portion of the Moon. For a sense of scale, note that the largest visible impact feature, the Schrodinger Basin at the lower right, is 320 km in diameter. Clementine gave astronomers their first close-up look at the Moon's polar regions. The south pole is of particular interest because ice was found in the permanent shadow zone (center). *(USGS)*

they look somewhat similar, the origins of lunar features such as craters and mountains may be very different from their terrestrial counterparts. The Moon's relative lack of internal activity, together with the absence of air and water, make most of its geological history unlike anything we know on Earth. And much of the lunar surface is older than the rocks of the Earth's crust, as we learned when the first lunar samples were analyzed in the laboratory.

Ages of Lunar Rocks

To trace the detailed history of the Moon or of any planet, we must be able to estimate the ages of individual rocks. Once lunar samples were brought back by the Apollo astronauts, the radioactive dating techniques that had been developed for the Earth (see Section 6.3) were applied to them. The solidification ages of the Apollo samples ranged from about 3.3 to 4.4 billion years, substantially older than most of the rocks on the Earth. As we saw in Chapter 6, it now seems clear that both the Earth and the Moon were formed 4.5 billion years ago.

Geological Features

Most of the surface of the Moon (83 percent) consists of silicate rocks called *anorthosites;* these regions are known as the lunar **highlands.** They are made of relatively low-density rock that solidified on the cooling Moon like slag floating on the top of a smelter. Because they formed so early in lunar history (between 4.1 and 4.4 billion years ago), the highlands are also extremely heavily cratered, bearing the scars of billions of years of impacts by interplanetary debris (Figure 8.6).

The most prominent lunar features are the so-called seas, still called **maria** (Latin for "seas") by scientists. Maria is the plural term, by the way; the singular is *mare* (pronounced "mah-ray"). These dark, round plains (Figure 8.7), which are much less cratered than the highlands, cover just 17 percent of the lunar surface, mostly on the side facing the Earth. Today we know that the maria are not ocean basins at all. They are volcanic plains, laid down in eruptions billions of years ago. They partly fill huge depressions called *impact basins,* which were produced by

FIGURE 8.6
Lunar Highlands   The old, heavily cratered lunar highlands make up 83 percent of the Moon's surface. The width of this image is 250 km.   *(NASA)*

FIGURE 8.8
Mare Orientale   The youngest of the large lunar impact basins is Orientale, formed 3.8 billion years ago. Its outer ring is about 1000 km in diameter, roughly the distance between New York City and Detroit, Michigan. Unlike most of the other basins, Orientale has not been filled in with lava flows, so it retains its striking "bull's-eye" appearance. It is located on the edge of the Moon as seen from Earth; some authors have speculated that if it were facing us, the giant feature might have been seen as a kind of "evil eye" in more superstitious times.   *(NASA)*

FIGURE 8.7
Lunar Maria   About 17 percent of the Moon's surface consists of the maria—flat plains of basaltic lava. This view of Mare Imbrium also shows numerous secondary craters and evidence of material ejected from the large crater Copernicus on the upper horizon. Copernicus is an impact crater almost 100 km in diameter that was formed after the lava in Imbrium had already been deposited.   *(NASA)*

collisions of large chunks of material with the Moon relatively early in its history (Figure 8.8).

The lunar maria are all composed of basalt, very similar in composition to the oceanic crust of the Earth or to the lavas erupted by many terrestrial volcanoes. A series of large eruptions on the Moon between 3.3 and 3.8 billion years ago (dated from laboratory measurements of returned samples) formed smooth flows typically a few meters thick but extending over distances of hundreds of kilometers. Eventually, these lava flows filled in the lowest parts of the basins to form the mare surfaces we see today.

Volcanic activity may have begun very early in the Moon's history, although most evidence of the first half-billion years is lost. What we do know is that the major mare volcanism, which involved the release of highly fluid magma from hundreds of kilometers below the surface, ended about 3.3 billion years ago (Figure 8.9). After that the Moon's interior cooled, and by 3 billion years ago all volcanic activity had ceased. Since then our satellite has been a geologically dead world, changing only slowly as the result of random impacts.

The major lunar mountains are all the result of the Moon's history of impacts. Long, arc-shaped ranges that border the maria are debris ejected from the impacts that formed these giant basins. These mountains have low, rounded profiles that resemble old, eroded mountains on Earth (Figure 8.10). But appearances can be deceiving. The mountains of the Moon have not been eroded, except for the effects of meteoritic impacts. They are rounded because that is the way they formed, and there has been no water or ice to carve them into cliffs and sharp peaks. (Compare Figure 8.10 with Figure 7.9.)

On the Lunar Surface

> The surface is fine and powdery. I can pick it up loosely with my toe. But I can see the footprints of my boots and the treads in the fine sandy particles.
> —Neil Armstrong, Apollo 11 astronaut, immediately after stepping onto the Moon for the first time

The surface of the Moon is buried under a fine-grained soil of tiny, shattered rock fragments. The dark basaltic dust of the lunar maria was kicked up by every astronaut footstep, and this dust eventually worked its way into all of the astronauts' equipment. The upper layers of the surface are porous, consisting of loosely packed dust into which their boots sank several centimeters (Figure 8.11). This lunar dust, like so much else on the Moon, is the product of impacts. Each cratering event, large or small, breaks up the rock of the lunar surface and scatters the fragments. Ultimately, billions of years of impacts reduce much of the surface layer to particles about the size of dust or sand.

In the absence of any air, the lunar surface experiences much greater temperature extremes than the surface of the Earth, even though the Earth is virtually the same distance from the Sun. Near local noon, when the Sun is highest in the sky, the temperature of the dark lunar soil rises to the boiling point of water. During the long lunar night (which lasts two Earth weeks) it drops to about 100 K ($-173°C$). The extreme cooling is a result not only of the absence of air, but also of the porous nature of the dusty soil, which cools more rapidly than solid rock would.

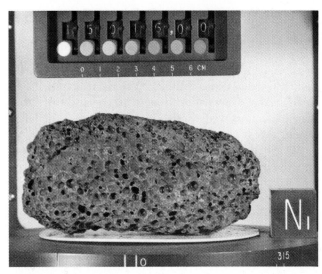

FIGURE 8.9
Rock from a Lunar Mare   Sample of basalt from the mare surface. The gas bubbles are characteristic of rock formed from lava. *(NASA)*

8.3   IMPACT CRATERS

The Moon provides an important benchmark for understanding the history of the planetary system. Most solid worlds show the effects of impacts, often extending back to the era when a great deal of debris from our system's formation process was still present. On the Earth, this long history has been erased by our active geology. On the Moon, in contrast, most of the impact history is preserved. If we can understand what has happened on the Moon, we may be able to extrapolate this knowledge to the other planets and satellites.

FIGURE 8.10
Lunar Mountain   Mt. Hadley on the edge of Mare Imbrium, photographed by the Apollo 15 astronauts. Note the smooth contours of the lunar mountains, which have not been sculpted by water or ice.   *(NASA)*

FIGURE 8.11
Footprint in Moondust   Apollo photo of an astronaut's bootprint in the lunar soil.   *(NASA)*

Volcanic Versus Impact Origin of Craters

Until the middle of the 20th century, scientists did not generally recognize that lunar craters were the result of impacts. Since impact craters are extremely rare on Earth, geologists did not consider them to be the major feature of lunar geology. They reasoned (perhaps unconsciously) that since the craters we have on Earth are volcanic, the lunar craters must have a similar origin.

One of the first geologists to propose that lunar craters were the result of impacts was Grove K. Gilbert, a scientist with the U.S. Geological Survey in the 1890s (Figure 8.12). He pointed out that the large lunar craters, which are mountain-rimmed, circular features with floors generally below the level of the surrounding plains, are larger and have different shapes from known volcanic craters. Terrestrial volcanic craters are smaller, deeper, and almost always occur at the tops of volcanic mountains (Figure 8.13). Gilbert therefore concluded that the lunar craters were not volcanic. The only alternative was an impact origin. His careful reasoning, although not accepted at the time, laid the foundations for the modern science of lunar geology.

Gilbert concluded that the lunar craters were produced by impacts, but he didn't understand why all of them were circular. Stones thrown in to a sand pit, for instance, produce elliptical craters when the projectiles strike at an angle, as many cosmic impactors must. (Try this for yourself, by conducting a scientific experiment at a local playground. Only when you drop a stone into a sandbox from directly overhead will your crater be circular.) The solution lies in the speed with which projectiles approach the Earth or Moon. Attracted by the gravity of the larger body, the incoming chunk strikes with at least escape velocity, which is 11 km/s for the Earth and 2.4 km/s (5400 mi/h) for the Moon. To this escape velocity is added whatever speed the projectile already had with respect to the Earth or Moon, typically 10 km/s or more.

At these speeds the energy of impact produces a violent explosion that excavates a large volume of material in a more or less symmetrical way. Photographs of bomb and shell craters on Earth confirm that explosion craters are always essentially circular. Only following World War I did scientists recognize the similarity between impact craters and explosion craters, but Gilbert did not live to see his impact hypothesis widely accepted.

The Cratering Process

Let's consider how an impact at these high speeds produces a crater. When such a fast projectile strikes a planet, it penetrates two or three times its own diameter before stopping. During these few seconds its energy of motion is transferred into a shock wave, which spreads through the target body, and into heat, which vaporizes most of the projectile and some of the surrounding target. The shock wave fractures the rock of the target, while the hot silicate vapor generates an explosion not too different from that of a

FIGURE 8.12
Grove K. Gilbert, a founding member of the U.S. Geological Survey, was among the first scientists to recognize the importance of impact cratering on the Earth and Moon. *(USGS)*

nuclear bomb detonated at ground level (Figure 8.14). The size of the excavated crater depends primarily on the speed of impact, but generally it is about 10 to 15 times the diameter of the projectile.

An impact explosion of the sort described above leads to a characteristic kind of crater, as illustrated in Figure 8.15. The central cavity is initially bowl-shaped (the word *crater* comes from the Greek word for *bowl*), but the gravitational rebound of the crust partially fills it in, producing a flat floor and sometimes creating a central peak. Around the rim, landslides create a series of terraces.

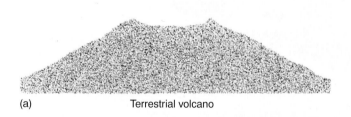

(a)   Terrestrial volcano

(b)   Lunar impact crater

FIGURE 8.13
Volcanic and Impact Craters   Profiles of typical terrestrial volcanic craters (a) and typical lunar impact craters (b) are quite different from each other.

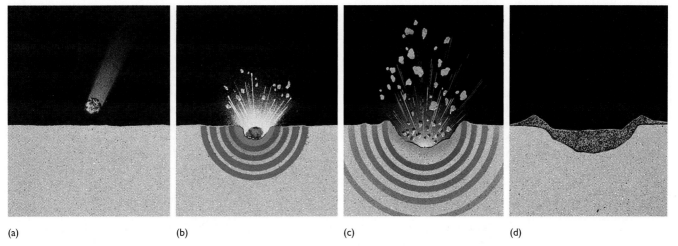

(a) (b) (c) (d)

FIGURE 8.14

Stages in the Formation of an Impact Crater   (a) The impact occurs. (b) The projectile vaporizes and a shock wave spreads through the lunar rock. (c) Ejecta are thrown out of the crater. (d) Most of the ejected material falls back to fill the crater and form an ejecta blanket.

The rim of the crater is turned up by the force of the explosion, so it rises above both the floor and the adjacent terrain. Surrounding the rim is an *ejecta blanket* consisting of material thrown out by the explosion. This debris falls back to create a rough, hilly apron, typically about as wide as the crater diameter. Additional, higher-speed ejecta fall at greater distances from the crater, often digging small *secondary craters* where they strike the surface (see Figure 8.7).

Some of these streams of ejecta can extend from the crater for hundreds or even thousands of kilometers, creating on the Moon the bright *crater rays* that are so prominent in photos taken near full phase (see Seeing for Yourself: *Observing the Moon*). The brightest lunar crater rays are associated with large young craters such as Kepler and Tycho.

Using Crater Counts

As discussed in Chapter 6, if a world has had little erosion or internal activity, like the Moon during the past 3 billion years, it is possible to use the number of impact craters on its surface to estimate the age of that surface. By "age" we mean the time since a major disturbance occurred on that surface (such as the volcanic eruptions that produced the lunar maria).

We can't directly measure the rate at which craters are being formed on the Earth and Moon, since (as we saw in Chapter 7) the average interval between *large* crater-forming impacts is longer than the span of human history. Our best-known example of such a large crater, Meteor Crater in Arizona, is 50,000 years old. However, the cratering rate can be estimated from the number of craters on the lunar maria or calculated from the number of potential "projectiles" (asteroids and comets) present in the solar system today, as will be discussed in Chapter 11. Both lines of reasoning lead to about the same answers.

FIGURE 8.15

Typical Impact Crater   King Crater on the far side of the Moon, a fairly recent lunar crater 75 km in diameter, shows most of the features associated with large impact structures.   *(NASA)*

Observing the Moon

The Moon is one of the most beautiful sights in the sky, and it is the only object close enough to reveal its *topography* (mountains and valleys) to us without a visit from a spacecraft. A fairly small amateur telescope easily shows craters or mountains on the Moon as small as a few kilometers across.

Even as seen through a good pair of binoculars, the appearance of the Moon's surface changes dramatically with its phase (see Chapter 3 for an explanation of the phases). At full phase it shows almost no topographic detail, and you must look closely to see more than a few craters. This is because the sunlight illuminates the surface straight on, and in this flat lighting no shadows are cast. Much more revealing is the view near first or last quarter, when sunlight streams in from the side and topographic features cast sharp shadows. It is almost always more rewarding to study a planetary surface under such oblique lighting, when the maximum information about the surface relief can be obtained.

> At full phase it shows almost no topographic detail, and you must look closely to see more than a few craters.

The flat lighting at full phase does, however, accentuate brightness contrasts on the Moon, such as those between the maria and highlands. Notice on the accompanying photo that several of the large mare craters seem to be surrounded by aprons of white material, and that the light streaks or rays that can stretch for hundreds of kilometers across the surface are clearly visible. These lighter features are ejecta, splashed out from the crater-forming impact.

By the way, there is no danger in looking at the Moon with binoculars or telescopes. The reflected sunlight is never bright enough to hurt you. In fact, the sunlit surface of the Moon has about the same brightness as a sunlit landscape on Earth. So explore away. Several guides to observing the Moon are listed in the recommendations for further reading at the end of this chapter.

One interesting thing about the Moon that you can see without binoculars or telescopes is popularly called "the new Moon in the old Moon's arms." Look at the Moon when it is a thin crescent and you can often make out the faint circle of the entire lunar disk, even though the sunlight shines on only the crescent. The rest of the disk is illuminated not by sunlight but by earthlight—sunlight reflected from the Earth. The light of the full Earth on the Moon is about 50 times brighter than that of the full Moon shining on the Earth.

The Appearance of the Moon at Different Phases. Illumination from the side brings craters and other topographic features into sharp relief. At full phase *(right)* there are no shadows and it is more difficult to see such features. However, the flat lighting at full phase brings out some classes of surface features, such as the bright rays of ejecta that stretch out from a few large young craters. *(Lick Observatory)*

For the Moon these calculations indicate that a crater 1 km in diameter should be produced about every 200,000 years, a 10-km crater every few million years, and one or two 100-km craters every billion years. For the land area of the Earth, the rates of crater formation are about 10 times greater, primarily because of Earth's larger surface area.

If these cratering rates have stayed the same over the millennia, we can figure out how long it must have taken to make all the craters we see crowded onto the lunar maria. Our calculations show that it would have taken several billion years. This result is similar to the age determined for the maria from radioactive dating—3.3 to 3.8 billion years old. The fact that these two calculations agree suggests that our assumption was right: Comets and asteroids in approximately their current numbers have been impacting planetary surfaces for at least 3.8 billion years. Calculations carried out for other planetary bodies (and their satellites) indicate that they also have been subject to about the same number of interplanetary impacts during this time.

Earlier than 3.8 billion years ago, however, we have good reason to believe that the impact rates must have been a great deal higher. This becomes immediately evident when we compare the numbers of craters on the lunar highlands with those on the maria. Typically, there are ten times more craters on the highlands than on a similar area of maria. Yet the radioactive dating of highland samples showed that they are only a little older than the maria, typically 4.2 billion years rather than 3.8 billion years. If the rate of impacts had been constant throughout the Moon's history, the highlands would have had to be at least ten times older. They would thus have had to form 38 billion years ago—long before the universe itself began.

In science, when an assumption leads to an unbelievable conclusion, we must go back and re-examine that assumption—in this case, the constant impact rate. The contradiction is resolved if the impact rate varied over time, with a much heavier bombardment earlier than 3.8 billion years ago (Figure 8.16). This heavy bombardment produced most of the craters we see today in the highlands.

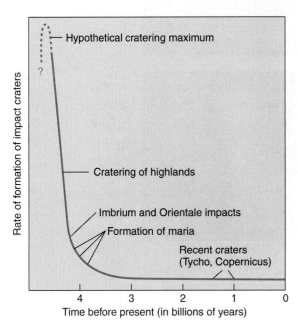

FIGURE 8.16
Cratering Rates over Time How the number of craters being made on the Moon's surface has varied with time over the past 4.3 billion years.

8.4 THE ORIGIN OF THE MOON

Since the Moon is our closest celestial neighbor, the question of where it came from has intrigued philosophers and scientists for centuries. In our time, however, understanding the origin of the Moon has proved to be extremely difficult for planetary scientists. Part of the problem is simply that we know so much about our satellite (quite the opposite of our usual problem in astronomy). The great wealth of data, particularly on the details of the composition of the Moon, presents a challenge to any simple theory of lunar origins. Put in simple terms, our problem is that the Moon is both tantalizingly similar to the Earth, and frustratingly different.

Theories for the Origin of the Moon

Most of the various theories for the Moon's origin follow one of three general ideas:

1. The fission theory—that the Moon was once part of the Earth but separated from it early in their history.
2. The sister theory—that the Moon formed together with (but independently of) the Earth, as we believe many satellites of the outer planets formed.
3. The capture theory—that the Moon formed elsewhere in the solar system and was captured by the Earth.

Unfortunately, there are fundamental problems with each of these ideas. Perhaps the easiest theory to reject is the capture theory. Its primary drawback is that no one knows of any way that the early Earth could have captured a large satellite from elsewhere. One body approaching another cannot go into orbit around it without a substantial loss of energy; this is the reason that spacecraft destined to orbit other planets are equipped with retro-rockets. Further, if such a capture did take place, it would be into a very eccentric orbit rather than the nearly circular orbit the Moon occupies today. Finally, there are too many compositional similarities between the Earth and the Moon, particularly an identical fraction of the major isotopes of oxygen, to justify seeking a completely independent origin.

The fission theory, which states that the Moon separated from the Earth, was suggested in the late 19th century by George Darwin, whom we profiled in Chapter 3. More modern calculations have shown that the sort of spontaneous fission or splitting imagined by Darwin is impossible. Further, it is difficult to understand how a Moon made out of terrestrial material in this way could have developed the many distinctive chemical differences now known to characterize our satellite.

Scientists have therefore been left with the sister theory—that the Moon formed alongside the Earth—or with some modification of the fission theory that finds a more acceptable way for the lunar material to have separated from the Earth. But the more we learn about our satellite, the less these old ideas seem to fit the bill.

The Giant Impact Theory

In an effort to resolve these apparent contradictions, scientists developed a fourth theory for the origin of the Moon, one that involves a giant impact early in the Earth's history. There is increasing evidence that large projectiles—objects of essentially planetary mass—were orbiting in the inner solar system at the time that the terrestrial planets formed. The giant impact theory envisions the Earth being struck obliquely by an object approximately one tenth of the Earth's mass—a "bullet" about the size of Mars. This is very nearly the largest impact the Earth could experience without being shattered.

Computer calculations (Figure 8.17) show that such an impact would disrupt much of the Earth by penetrating into the core, ejecting a vast amount of material into space, and releasing almost enough energy to break the planet apart. The simulations done on large computers indicate that material totaling several percent of the Earth's mass could be ejected in such an impact. Most of this material would be from the stony mantles of the Earth and the impacting body, not from their metal cores. This ejected rock vapor then cooled and formed a ring of material orbiting around the Earth. It was this ring that ultimately condensed into the Moon.

While we do not have any current way of showing that the giant impact hypothesis is the correct model of the Moon's origin, it does offer potential solutions to most of the major problems raised by the chemistry of the Moon. First, since the Moon's raw material is derived from the mantles of the Earth and the projectile, the absence of metals is easily understood. Second, most of the volatile elements would have been lost during the high-temperature

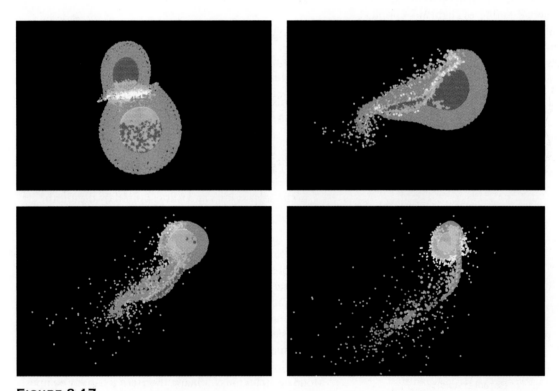

FIGURE 8.17

How an Impact May Have Formed the Moon Four frames from one of a series of computer simulations showing an oblique giant impact on the Earth. Iron is shown as spheres of blue and green. Rock is represented by red, pink, yellow, and white spheres, in order of increasing temperature. In this particular simulation, the impactor has 1/5 of the mass in the system, and the Earth has 4/5. After the impact, a plume of material is ejected into space. It is this plume that, under the right circumstances, can condense to form the Moon. Note that the plume eventually has much more pink (rock) than blue or green (iron), explaining why the Moon resembles the mantle of the Earth. *(Courtesy of A. G. W. Cameron, Center for Astrophysics)*

phase following the impact. Yet by making the Moon primarily of terrestrial mantle material, it is also possible to understand similarities such as identical abundances of various oxygen isotopes.

This idea that large impacts (especially during the early history of the solar system) have played a major role in shaping the worlds we see is not unique to our study of the Moon. As you read through the other chapters about the planets and their satellites, you will see further indications that a number of the present-day characteristics of our system may be due to our violent past.

8.5 MERCURY

The planet Mercury is similar to the Moon in many ways. Like the Moon, it has no atmosphere and its surface is heavily cratered. As described later in this chapter, it also shares with the Moon a violent birth history.

Mercury's Orbit

Mercury is the nearest to the Sun of the nine planets and, in accordance with Kepler's third law, it has the shortest period of revolution about the Sun (88 of our days) and the highest average orbital speed (48 km/s). It is appropriately named for the fleet-footed messenger god of the Romans. Because Mercury remains close to the Sun, it can be difficult to pick out in the sky. As you might imagine, it's best seen when its eccentric orbit takes it as far from the Sun as possible.

The semimajor axis of Mercury's orbit—that is, the planet's average distance from the Sun—is 58 million km, or 0.39 AU. However, because its orbit has the high eccentricity of 0.206, Mercury's actual distance from the Sun varies from 46 million km at perihelion to 70 million km at aphelion. (If some of our terms are new to you, you may want to look at Chapter 2 or the glossary at the end of the book.) Pluto is the only planet with a more eccentric orbit than Mercury. Furthermore, the 7° inclination of Mercury's orbit to the plane of the ecliptic is also greater than that of any other planet except Pluto.

Composition and Structure

Table 8.1 compares some basic properties of Mercury with those of the Moon. Mercury's mass is 1/18 that of the Earth; again Pluto is the only planet with a smaller mass. Mercury is also the second smallest of the planets, having a diameter of only about 4880 km, less than half that of the Earth. Mercury's density is 5.4 g/cm³, much greater than the density of the Moon, indicating that the composition of these two objects differs substantially.

Its composition, in fact, is one of the most interesting things about Mercury; it is unique among the planets. Mercury's high density tells us that it must be composed largely of heavier materials such as metals. The most likely models

FIGURE 8.18
Mercury's Internal Structure   The interior of Mercury is dominated by a metallic core about the same size as our Moon.
(University of Arizona, courtesy Robert Strom)

for Mercury's interior suggest a metallic iron–nickel core amounting to 60 percent of the total mass, with the rest of the planet made up primarily of silicates. The core has a diameter of 3500 km and extends to within 700 km of the surface (Figure 8.18). We could think of Mercury as a metal ball the size of the Moon surrounded by a rocky crust 700 km thick. Unlike the Moon, however, Mercury also has a weak magnetic field. The existence of this field is consistent with a large metal core, and it suggests that at least part of the core must be liquid in order to generate the observed field. (See Section 7.1 for the explanation of how magnetic fields and liquid cores are connected.)

Mercury's Strange Rotation

Visual studies of Mercury's indistinct surface markings were once thought to indicate that the planet kept one face to the Sun (as the Moon does to the Earth). Thus, for many years it was widely believed that Mercury's rotation period was equal to its revolution period of 88 days, making one side perpetually hot while the other was always cold.

Radar observations of Mercury in the mid-1960s, however, showed conclusively that Mercury does not keep one side fixed to the Sun. Recall from Section 5.4 that the frequency of a transmitted radar pulse can be controlled precisely. Any motion of the target then introduces a measurable change in this frequency as a result of the Doppler effect (see Section 4.6). Such target motions can include rotation: If a planet is turning, one side seems to be approaching the Earth while the other is moving away from it. The result is to spread or broaden the precise transmitted frequency

MAKING CONNECTIONS

What a Difference a Day Makes

Mercury rotates three times for each two orbits around the Sun. It is the only planet that exhibits this relationship between its spin and its orbit, and there are some interesting consequences for any observers who might someday be stationed on the surface of Mercury.

Here on Earth, we take it for granted that days are much shorter than years. Therefore, the two ways of defining the local "day"—how long the planet takes to rotate, and how long the Sun takes to return to the same position in the sky (see Chapter 3)—are the same for most practical purposes. But this is not the case on Mercury. While Mercury rotates in 59 Earth days, the time for the Sun to return to the same place in Mercury's sky turns out to be two Mercury years or 176 Earth days. (Note that this result is not intuitively obvious, so don't be upset with yourself if you didn't come up with it.) Thus, if one day at noon a Mercury explorer suggests to her companion that they should meet at noon the next day, this could mean a very long time apart!

To make things even more interesting, recall that Mercury has an eccentric orbit, meaning that its distance from the Sun varies significantly during each mercurian year. By Kepler's law, the planet moves fastest in its orbit when closest to the Sun. Let's examine how this affects the way we would see the Sun in the sky during one 176-Earth-day cycle. We'll look at the situation as if we were standing on the surface of Mercury in the center of a giant basin that astronomers call Caloris (Figure 8.21).

At the location of Caloris, Mercury is most distant from the Sun at sunrise; this means the rising Sun looks smaller

in the sky (although still more than twice the size it appears from the Earth). As the Sun rises higher and higher, it gets bigger and bigger; Mercury is now getting closer to the Sun in its eccentric orbit. At the same time, the apparent motion of the Sun slows down as Mercury's faster motion in orbit begins to catch up with its rotation.

At noon the Sun is now three times larger than it looks from Earth and hangs almost motionless in the sky. As the afternoon wears on, the Sun eventually looks smaller and smaller again, and moves faster and faster in the sky. At sunset, a full Mercury year after sunrise, the Sun is back to its smallest apparent size as it dips out of sight. Then it takes another Mercury year before the sun rises again. (Sunrise and sunset are much more sudden on Mercury, since there is no atmosphere to bend or scatter the rays of sunlight.)

Astronomers call locations like the Caloris basin the "hot longitudes" on Mercury, because the Sun is closest to the planet at noon, just when it is lingering overhead for many Earth days. This makes them the hottest places on the planet.

We bring all this up not because the exact details of this scenario are so important, but to illustrate how many of the things we take for granted on Earth are not the same on other worlds. As we've mentioned before, one of the best things about taking an astronomy class should be ridding you forever of any "Earth chauvinism" you might have. The way things are on our planet is just one of the many ways nature can arrange reality.

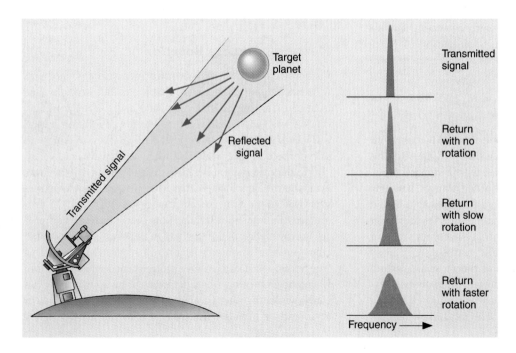

FIGURE 8.19
Doppler Radar Measures Rotation When a radar beam is reflected from a rotating planet, the motion of one side of the planet's disk toward us and the other away from us cause Doppler shifts in the reflected signal. The effect is to cause both a "redshift" and a "blueshift," and thus to widen the spread of frequencies in the radio beam.

into a range of frequencies in the reflected signal (Figure 8.19). The degree of broadening provides an exact measurement of the rotation rate of the planet.

Mercury's sidereal period of rotation (how long it takes to turn with respect to the distant stars) was found to be about 59 days. Shortly after this period was discovered by radar astronomers, the Italian physicist Giuseppe Colombo pointed out that this is two thirds of the planet's period of revolution, and subsequently astronomers found theoretical reasons (connected with *tidal forces,* discussed for Earth in Chapter 3) for expecting that Mercury can rotate stably with a period exactly two thirds that of its revolution—58.65 days.

Mercury is very hot on its daylight side, but because it has no appreciable atmosphere, it gets surprisingly cold during the long nights. The temperature on the surface climbs to 700 K at noontime. After sunset, however, the temperature drops, reaching 100 K just before dawn. The *range* in temperature on Mercury is thus 600 K, more than on any other planet.

The Surface of Mercury

The first close-up look at Mercury came in 1974, when the U.S. spacecraft Mariner 10 passed 9500 km from the surface of the planet and transmitted more than 2000 photographs to Earth, revealing details with a resolution down to 150 m. Mercury strongly resembles the Moon in appearance (Figures 8.20 and 8.21). It is covered with thousands of craters and larger basins up to 1300 km in diameter. Some of the brighter craters are rayed, like Tycho and Copernicus on the Moon, and many have central peaks. There are also scarps (cliffs) over a kilometer high and hundreds of kilometers long, as well as ridges and plains.

Most of the mercurian features have been named in commemoration of artists, writers, composers, and other contributors to the arts and humanities, in contrast with the scientists commemorated on the Moon. Among the most prominent craters are Bach, Shakespeare, Tolstoy, Mozart, and Goethe.

Mercury's larger basins resemble the lunar maria in both size and appearance. They show evidence of flooding from lava, although the flows do not have the distinctive dark color that characterizes the lunar maria. Since Mercury is so difficult to study telescopically and has been visited by only one spacecraft, we actually know very little about the chemistry of its surface or the details of its geological history.

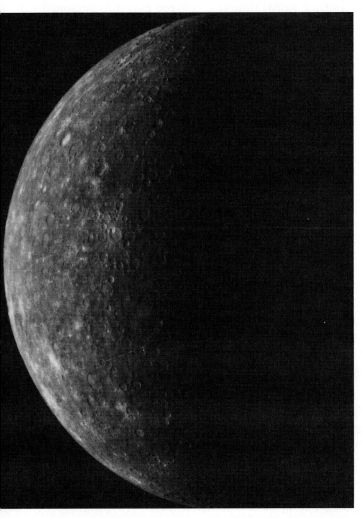

FIGURE 8.20
Mercury Mosaic The planet Mercury as photographed by Mariner 10 in 1974. The heavily cratered surface resembles that of the Moon.
(Mosaic courtesy of the Astrogeology Team, USGS, Flagstaff)

FIGURE 8.21
Caloris Basin Part of the Caloris basin on Mercury, 1300 km in diameter, can be seen at the bottom. This partially flooded impact basin is the largest structural feature on Mercury seen by Mariner 10. Compare photo with the Orientale basin on the Moon (Figure 8.7).
(NASA/JPL)

FIGURE 8.22
Discovery Scarp on Mercury   This long cliff, nearly 1 km high and more than 100 km long, cuts across several craters. We conclude that the compression that made "wrinkles" like this in the planet's surface must have taken place after the craters were formed.
(NASA/JPL)

calculate how internal changes in temperature might have led to this global compression, which has no counterpart on the Moon or the other terrestrial planets.

Although it has become something of a forgotten planet in the two decades since the Mariner 10 flybys, astronomers who study Mercury continue to be surprised. For example, enhanced radar reflectivity from both poles of the planet was discovered in 1992, indicating the presence of ice just beneath the surface. Even though the temperature very near the poles remains low enough for ice to survive, the discovery of these icy polar caps was unexpected, since they seem to imply that the polar regions of Mercury have remained below freezing since the formation of the planet. Perhaps there are cold traps there similar to the deep craters that shelter ice at the poles of our Moon.

The Origin of Mercury

The problem with understanding how Mercury formed is the reverse of that posed by the composition of the Moon. We have seen that, unlike the Moon, Mercury is composed mostly of metal. However, Mercury should have formed with about the same ratio of metal to silicate as that found on the Earth or Venus. How did it lose so much of its rocky material?

The most probable explanation for Mercury's silicate loss may be similar to the explanation for the Moon's lack of a metal core. Mercury is likely to have experienced several giant impacts very early in its youth, and one or more of these may have torn away a fraction of its mantle and crust, leaving a body dominated by its iron core.

Today astronomers recognize that the early solar system was a chaotic place, with the final stages of planet formation characterized by impacts of great violence. Some bodies of planetary mass may have been destroyed, while others could have fragmented and then reformed, perhaps more than once. Both the Moon and Mercury, with their strange compositions, bear testimony to the catastrophes that must have characterized the solar system during its youth.

There is no evidence of plate tectonics on Mercury. Distinctive long scarps, however, can sometimes be seen cutting across craters; this means the scarps must have formed later than the craters (Figure 8.22). These long cliffs appear to have their origin in the slight compression of Mercury's crust. Apparently, at some point in its history this planet shrank, wrinkling the crust, and it must have done so after most of the craters on its surface had already formed. If the standard cratering chronology applies to Mercury, this shrinkage must have taken place during the past 4 billion years. If we understood the interior structure and composition of the planet better, we could probably

---

SURFING THE WEB

A. The Moon

🖥 General Sites about the Moon, with good links:

- The Nine Planets Site: [seds.lpl.arizona.edu/nineplanets/nineplanets/luna.html]
- Views of the Solar System Site: [www.hawastsoc.org/solar/eng/moon.htm]
- Planetary Sciences Site: [nssdc.gsfc.nasa.gov/planetary/planets/moonpage.html]

🖥 Exploring the Moon [cass.jsc.nasa.gov/pub/expmoon/lunar_missions.html]
An extensive site from NASA's Johnson Space Flight Center, with information on U.S. and Russian Moon exploration and good links to many other Web resources.

🖥 Apollo Lunar Surface Journal [www.hq.nasa.gov/office/pao/History/alsj/]
A treasure trove of information, interviews, maps, photos, video and audio clips, and much more on each of the Apollo landing

missions, collected and organized by space historian Eric Jones. Even has edited transcripts of the original conversations between the astronauts and mission control.

🖥 Clementine Mission Sites:

- *Sky & Telescope* Introduction [www.skypub.com/news/special/clemgold.html]
- Official Mission Site [www.nrl.navy.mil/clementine/]
- Lunar Ice [nssdc.gsfc.nasa.gov/planetary/ice/ice_moon.html]

Introductions to the Defense Department mission that discovered the first evidence of ice deep in polar craters on the Moon (and sent back other useful data as well).

🖥 Lunar Prospector Home Page [lunar.arc.nasa.gov/]

This site, from NASA's Ames Research Center, includes not just the results from the recent mission to the Moon (which confirmed the presence of ice—mixed with lunar soil—in deep craters at the poles), but also a history of lunar exploration, science background, and some educational tools.

🖥 Touring the Moon with Binoculars [www.skypub.com/sights/moonplanets/moontour.html]

An introductory guide, with a map, for viewing the Moon with simple equipment.

B. Mercury

🖥 General Sites about Mercury:

- The Nine Planets Site: [seds.lpl.arizona.edu/nineplanets/nineplanets/mercury.html]
- Views of the Solar System Site: [www.hawastsoc.org/solar/eng/mercury.htm]
- Planetary Sciences Site: [nssdc.gsfc.nasa.gov/planetary/planets/mercurypage.html]

🖥 Mercury Unveiled [www.soest.hawaii.edu/PSRdiscoveries/Jan97/MercuryUnveiled.html]

Planetary scientist G. Jeffrey Taylor updates our current understanding of Mercury based on some recently reanalyzed Mariner 10 images, which are shown and well explained.

SUMMARY

8.1 Most of what we know about the Moon derives from the Apollo program, including 400 kg of lunar samples still being intensively studied. The Moon has 1/80 the mass of the Earth and is severely depleted in both metals and volatile materials. It is made almost entirely of silicates like those in the Earth's mantle and crust. Recently, two orbiter spacecraft have found evidence of a small amount of water near the lunar poles, most likely deposited by comet and asteroid impacts.

8.2 Lunar rocks can be dated by their radioactivity. Like the Earth, the Moon was formed about 4.5 billion years ago. The heavily cratered **highlands** are made of rocks more than 4 billion years old. The darker volcanic plains of the **maria** were erupted between 3.3 and 3.8 billion years ago. Generally, the surface is dominated by impacts, including continuing small impacts that produce its fine-grained soil.

8.3 A century ago Gilbert suggested that the lunar craters were of impact origin, but the cratering process was not well understood until more recently. High-speed impacts produce explosions and excavate craters with raised rims, ejecta

blankets, and often central peaks. Cratering rates have been roughly constant for the past 3 billion years but earlier were much greater. Crater counts can be used to derive approximate ages for geological features on the Moon and other planets.

8.4 There are three standard theories for the origin of the Moon: the fission theory; the sister theory; and the capture theory. All have problems, and recently they have been supplanted by the giant impact theory, which ascribes the origin of the Moon to the impact 4.5 billion years ago of a Mars-sized projectile with the Earth.

8.5 Mercury is the nearest planet to the Sun and the fastest moving. Mercury is similar to the Moon in having a heavily cratered surface and no atmosphere, but it differs in having a very large metal core. Early in its evolution, it apparently lost part of its silicate mantle, probably due to one or more giant impacts. Long scarps on its surface testify to a global compression of Mercury's crust during the past 4 billion years.

INTER-ACTIVITY

🔺 In this chapter, we discuss how no nation on Earth now has the capability to send a human being to the Moon, despite the fact that the United States once sent 12 astronauts to land there. How does your group feel about this? Should we continue the exploration of space with human beings? Should we put habitats on the Moon? Should we go to Mars? Does humanity have a "destiny in space"? Whatever your answer to these questions is, make a list of the arguments and facts that would support your position.

B When they first hear about the Giant Impact Hypothesis for the origin of the Moon, many students are intrigued and wonder why we can't cite more evidence for it. Could your group make a list of reasons why we cannot find any traces on the Earth of the great impact that formed the Moon?

C We discuss that the ice (mixed into the soil) that is found on the Moon is most likely ice that was delivered by comets. Have your group make a list of all the reasons why the Moon would not have any ice of its own left from its early days.

D Can your group make a list of all the things that would be different if our Earth had no Moon? Don't restrict your answer to astronomy and geology. (You may want to review Chapter 3.)

E If, one day, humanity decides to establish a colony on the Moon, where should we put it? Make a list of the advantages and disadvantages of locating such a human habitat on the near side, the far side, or at the poles. What site would be best for optical and radio astronomy?

F A member of the class (but luckily, not a member of your group) suggests that he has always dreamed of building a vacation home on the planet Mercury. Can your group make a list of all the reasons why such a house would be hard to build and keep in good repair?

REVIEW QUESTIONS

1. What is the composition of the Moon, and how does it compare to the composition of the Earth? Of Mercury?

2. Outline the main events in the Moon's geological history.

3. Explain how high-speed impacts form circular craters. How can this explanation account for the characteristic features of impact craters?

4. Explain the evidence for a period of heavy bombardment on the Moon about 4 billion years ago. What might have been the source of the impacting debris during this early era?

5. How did our exploration of the Moon differ from that of Mercury (and the other planets)?

6. Summarize the four main theories for the origin of the Moon. Give pros and cons for each.

7. What do current ideas about the origins of the Moon and Mercury have in common?

THOUGHT QUESTIONS

8. One of the primary scientific objectives of the Apollo program was the return of lunar material.
   a. Why was this so important?
   b. What can be learned from samples?
   c. Are they still of value now?

9. Apollo astronaut David Scott dropped a hammer and a feather together on the Moon, and both reached the ground at the same time. What are the two distinct advantages that this experiment on the Moon had over the same experiment as performed by Galileo on the Earth?

10. Galileo thought the lunar maria were seas of water. If you had no better telescope than the one he had, could you prove that they are not composed of water?

11. Why did it take so long for geologists to recognize that the lunar craters had an impact origin rather than a volcanic one?

12. How might a crater made by the impact of a comet with the Moon differ from a crater made by the impact of an asteroid?

13. Why are the lunar mountains smoothly rounded rather than having sharp, pointed peaks (as they were almost always depicted in science fiction illustrations and films before the first lunar landings)?

14. The lunar highlands have about ten times more craters on a given area than do the maria. Does this mean that the highlands are ten times older? Explain your reasoning.

15. Give several reasons that Mercury would be a particularly unpleasant place to build an astronomical observatory.

16. If, in the remote future, we establish a base on Mercury, keeping track of time will be a challenge. Discuss how to define a year on Mercury, and the two ways to define a day. Can you come up with ways that humans raised on Earth might deal with time cycles on Mercury?

17. The Moon has too little iron, Mercury too much. How can both of these anomalies be the result of giant impacts? Explain how the same process can yield such apparently contradictory results.

18. The Moon was once closer to the Earth than it is now. When it was at half its present distance, how long was its period of revolution? (See Chapter 2.)

19. Astronomers believe that the deposit of lava in the giant mare basins did not happen in one flow, but in many different eruptions spanning some time. Indeed, in any one mare we find a variety of rock ages, typically spanning about 100 million years. The individual lava flows as seen in Hadley Rill by the Apollo 15 astronauts were about 4 m thick. Estimate the average time interval between the beginnings of successive lava flows if the total depth of the lava in the mare is 2 km.

20. The Moon requires about one month (0.08 year) to orbit the Earth. Its distance from us is about 400,000 km (0.0027 AU). Use Kepler's third law, as modified by Newton, to calculate the mass of the Earth relative to the Sun.

SUGGESTIONS FOR FURTHER READING

On the Moon

Burns, J. et al. "Observatories on the Moon" in *Scientific American*, March 1990, p. 42.

Chaikin, A. *A Man on the Moon.* 1994, Viking Press. A well-reviewed history of manned lunar exploration.

Coco, M. "Staging a Moon Shot" in *Astronomy*, Aug. 1992, p. 62. Describes photographing the Moon.

Foust, J. "NASA's New Moon" in *Sky & Telescope*, Sept. 1998, p. 48. On results from the Lunar Prospector mission.

Hockey, T. *The Book of the Moon.* 1986, Prentice-Hall. A basic primer on many aspects of the Moon.

Jayawardhana, R. "Deconstructing the Moon" in *Astronomy*, Sept. 1998, p. 40. An update on the giant impact hypothesis for forming the Moon.

Kitt, M. *The Moon: An Observing Guide for Backyard Telescopes.* 1992, Kalmbach. Eighty-page illustrated primer for beginners.

MacRobert, A. "Close-up of an Alien World" in *Sky & Telescope*, July 1984, p. 29. An observing guide.

Morrison, D. and Owen, T. "Our Ancient Neighbor the Moon" in *Mercury*, May/June 1988, p. 66; July/Aug. 1988, p. 98. An overview.

Ryder, G. "Apollo's Gift: The Moon" in *Astronomy*, July 1994, p. 40. Good evolutionary history of the Moon.

Schmitt, H. "Exploring Taurus–Littrow: Apollo 17" in *National Geographic*, Sept. 1973. First-person account by the only scientist to walk on the Moon.

Spudis, P. "The Giant Holes of the Moon" in *Astronomy*, May 1996, p. 50. On the results of the Clementine mission. (See also reports in *Sky & Telescope*, July 1995, p. 32; Feb. 1997, p. 24.)

Taylor, G. "The Scientific Legacy of Apollo" in *Scientific American*, July 1994, p. 40.

On Mercury

Beatty, J. "Mercury's Cool Surprise" in *Sky & Telescope*, Jan. 1992, p. 35.

Chapman, C. "Mercury's Heart of Iron" in *Astronomy*, Nov. 1988, p. 22. Good introduction to our modern view of the planet.

Gingerich, O. "How Astronomers Finally Captured Mercury" in *Sky & Telescope*, Sep. 1983, p. 203. A history of early observations.

Nelson, R. "Mercury: The Forgotten Planet" in *Scientific American*, Nov. 1997, p. 56.

Strom, R. "Mercury: The Forgotten Planet" in *Sky & Telescope*, Sept. 1990, p. 256.

Strom, R. *Mercury: The Elusive Planet.* 1987, Smithsonian Institution Press. Best book on the inner planet.

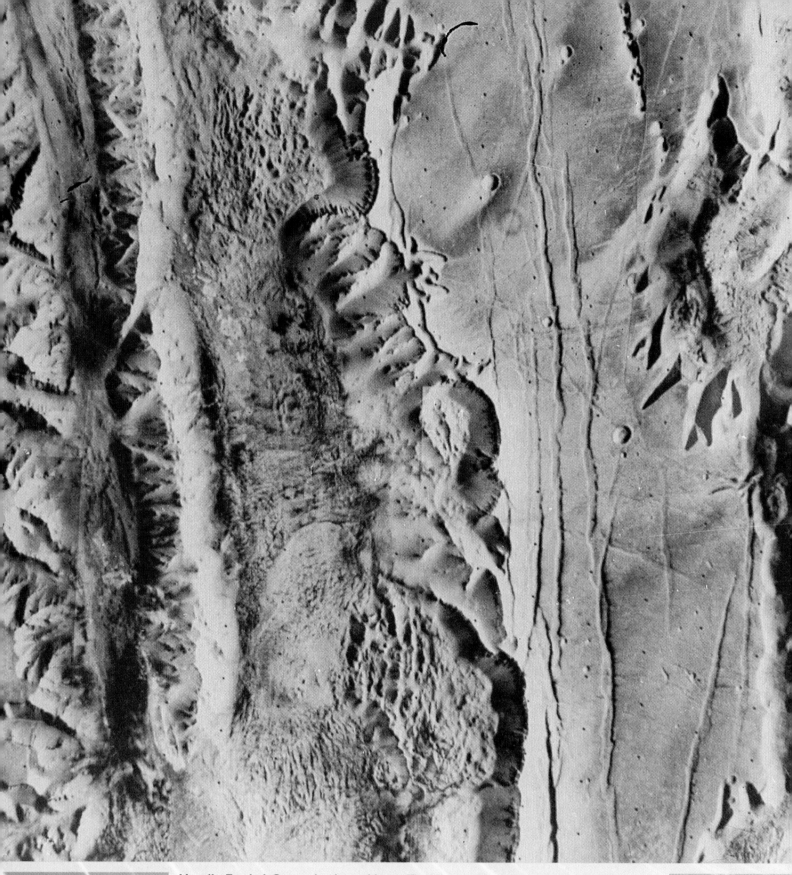

Heavily Eroded Canyonlands on Mars   This Viking spacecraft view, taken from Mars orbit, looks down on a small part of the Valles Marineris canyon complex and shows an area about 60 km across. The 1976 image has been processed by the staff at the U.S. Geological Survey in Flagstaff, Arizona, to bring out some of the marvelous detail in the photograph. *(NASA/USGS, courtesy of Alfred McEwen)*

9 Earth-Like Planets: Venus and Mars

THINKING AHEAD

For decades science fiction writers have conjured up visions of Venus and Mars as Earth-like worlds where humans would someday visit and (perhaps) come face to face with alien life. Now that our robot probes have explored both planets, we know that neither has conditions similar to our own. It seems unlikely that either planet could today sustain life-forms like those that populate the Earth. How did it happen that these three neighboring terrestrial planets have turned out to be so different from each other?

"One deduction from this thin air [on Mars] we must be careful not to make—that because it is thin, it is incapable of supporting intelligent life. That beings constituted physically as we are would find it a most uncomfortable habitat is pretty certain. But lungs are not wedded to logic, as public speeches show, and there is nothing in the world or beyond it to prevent, so far as we know, a being with gills, for example, from being a most superior person."

Percival Lowell, in *Mars* (1895).

The Moon and Mercury are geologically dead. In contrast, the larger terrestrial planets—the Earth, Venus, and Mars— are more active and interesting worlds. We have already discussed the Earth, and we now turn to Venus and Mars. These are the nearest planets and the most accessible to spacecraft. Not surprisingly, the greatest effort in planetary exploration has been devoted to these fascinating worlds. In this chapter we discuss some of the results of more than three decades of scientific exploration of Mars and Venus.

FIGURE 9.1
Venus as Photographed by the Pioneer Venus Orbiter This ultraviolet image shows upper-atmosphere cloud structure that would be invisible at visible wavelengths. Note that there is not a glimpse of the planet's cloud-shrouded surface. *(NASA/ARC)*

In contrast, Mars is both tantalizing and disappointing as seen through a telescope (Figure 9.2). The planet is distinctly red in color, due (as we now know) to the presence of iron oxides in its soil. This may account for its association with war (and bloodletting) in the legends of many early cultures. The best resolution obtainable from telescopes on the ground is about 100 km, or about the same as what we can see on the Moon with the unaided eye. At this resolution, however, no hint of topographic structure can be detected—no mountains, no valleys, not even impact craters. On the other hand, bright polar ice caps can easily be seen, together with dusky surface markings that sometimes change in outline and intensity from season to season. Of all the planets, only Mars has a surface that can be clearly made out from the Earth.

For a few decades around the turn of the 20th century, some astronomers believed they saw evidence of an intelligent civilization on Mars. The controversy began in 1877, when the Italian astronomer Giovanni Schiaparelli announced the he could see long, faint, straight lines on Mars that he called *canale*, or channels. In English-speaking countries, the term was mistakenly translated as *canals*, implying an artificial origin.

Even before Schiaparelli's observations, astronomers had watched the bright polar caps change size with the seasons and had seen variations in the dark surface features as well. With a little imagination, it was not difficult

9.1 THE NEAREST PLANETS: AN OVERVIEW

As you might expect from close neighbors, Mars and Venus are among the brightest objects in the night sky. The average distance of Mars from the Sun is 227 million km, which means it orbits at 1.52 AU, or about half again as far from the Sun as the Earth. However, Mars has a relatively large orbital eccentricity of 0.093, which means its distance varies by 42 million km as it goes around the Sun.

Venus' orbit, on the other hand, is very nearly circular, at a distance of 108 million km (0.72 AU) from the Sun. Like Mercury, Venus sometimes appears as an "evening star" and sometimes as a "morning star." The closest Mars ever gets to Earth is about 56 million km. Venus approaches the Earth more closely than does any other planet: At its nearest it is only 40 million km away from us.

Appearance

Venus looks very bright, and even a small telescope reveals that it goes through phases like the Moon. Galileo discovered that Venus displays a full range of phases, and he used this as an argument to show that Venus must circle the Sun and not the Earth (see Figure 1.15). The planet's actual surface is not visible because it is shrouded by dense clouds that reflect about 70 percent of the sunlight that falls on them, frustrating efforts to study the underlying surface, even with cameras aboard spacecraft in orbit around the planet (Figure 9.1).

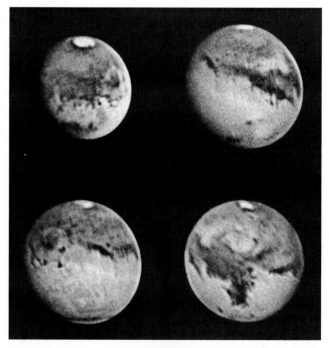

FIGURE 9.2
Mars Seen from the Earth's Surface These are among the best Earth-based photos of Mars, taken in 1988 when the planet was exceptionally close to the Earth. The polar caps and dark surface markings are evident, but not the topographic features. *(Steve Larson, University of Arizona)*

Percival Lowell: Dreaming of an Inhabited Mars

Percival Lowell (1855–1916) was born into the well-known and well-to-do Massachusetts family about whom John Bossidy made the famous toast:

> And this is good old Boston,
> The home of the bean and
>   the cod,
> Where the Lowells talk to
>   the Cabots
> And the Cabots talk only
>   to God.

Percival's brother Lawrence became president of Harvard University, and his sister Amy, who liked to dress in men's clothing and smoke cigars, became a distinguished poet. Percival was already interested in astronomy as a boy: He made observations of Mars at age 13. His undergraduate thesis at Harvard dealt with the origin of the solar system, but he did not pursue this interest immediately. Instead he entered the family business and traveled extensively in Asia. In 1892, however, he decided to dedicate himself to carrying on Schiaparelli's work and solving the mysteries of the martian canals.

In 1894, with the help of astronomers at Harvard but using his own funds, Lowell built an observatory on a high plateau in Flagstaff, Arizona, where he hoped the seeing would be clear enough to show him Mars in unprecedented detail. He and his assistants quickly accumulated a tremendous number of drawings and maps, purporting to show a vast network of martian canals (see Figure 9.2). He elaborated his ideas about the inhabitants of the red planet in several books, including *Mars* (1895) and *Mars and Its Canals* (1906), and in hundreds of articles and speeches. As Lowell put it:

> A mind of no mean order would seem to have presided over the system we see—a mind certainly of considerably more comprehensiveness than that which presides over the various departments of our own public works. Party politics, at all events, have had no part in them; for the system is planet-wide . . . . Certainly what we see hints at the existence of beings who are in advance of, not behind us, in the journey of life.

Lowell's views captured the public imagination and inspired many novels and stories, the most famous of which was H. G. Wells' *War of the Worlds* (1897). In this famous "invasion" novel, the thirsty inhabitants of a dying planet Mars (based entirely on Lowell's ideas) come to conquer the Earth with advanced technology.

Although the Lowell Observatory first became famous for its work on the martian canals, both Lowell and the observatory eventually turned to other projects as well. As we will see in Chapter 11, Lowell became interested in the search for a ninth (and then undiscovered) planet in the solar system. In 1930, Pluto was found at the Lowell Observatory, and it is not a coincidence that the name selected for the new planet starts with Lowell's initials. It was also at the Lowell Observatory that the first measurements were made of the great speed at which galaxies are moving away from us, observations that would ultimately

Percival Lowell in about 1910, observing with his 24-inch telescope at Flagstaff, Arizona. *(Lowell Observatory)*

lead to our modern view of an expanding universe.

Lowell continued to live at his observatory, marrying at age 53 and publishing extensively. He relished the debate his claims about Mars caused far more than the astronomers on the other side, who often complained that Lowell's work was making planetary astronomy a less-respectable field. At the same time, the public fascination with the planets fueled by Lowell's work (and its interpreters) may, several generations later, have helped fan support for the space program and the many missions whose results grace the pages of our text.

---

to picture the canals as crops bordering irrigation ditches that brought water from the melting polar ice to the parched deserts of the red planet. (They assumed the polar caps were composed of water ice, which isn't exactly true, as we will see shortly).

Until his death in 1916, the most effective proponent of intelligent life on Mars was Percival Lowell, a self-made American astronomer and member of the wealthy Lowell family of Boston (see *Voyagers in Astronomy* box). An effective author and speaker, Lowell made what seemed to the public a convincing case for intelligent Martians, who he believed had constructed the huge canals to preserve their existence in the face of a deteriorating climate (Figure 9.3).

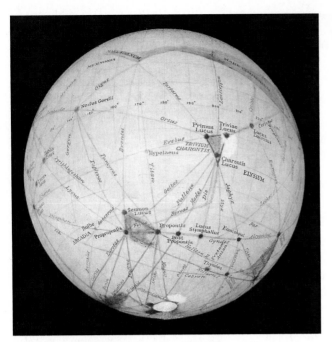

FIGURE 9.3
Lowell's Mars Globe One of the remarkable globes of Mars prepared by Percival Lowell, showing a network of dozens of canals, oases, and triangular water reservoirs that he claimed were visible on the red planet. *(Lowell Observatory)*

The argument for a race of Martians, however, hinged on the reality of the canals, a matter that remained in serious dispute among astronomers. The "canal" markings were always difficult to study, glimpsed only occasionally because atmospheric conditions caused the tiny image of Mars to shimmer in the telescope. Lowell saw canals everywhere (even a few on Venus), but many other observers could not see them at all and remained unconvinced of their existence. When larger telescopes than Lowell's failed to confirm the presence of canals, the skeptics felt vindicated.

Astronomers had given up the idea of martian canals by the 1930s, although it persisted in the public consciousness until the first spacecraft photographs clearly showed that no canals existed. Now it is generally accepted that the straight lines were an optical illusion, the result of the human mind's tendency to see order in random features glimpsed dimly at the limits of the eye's resolution. When we see small, dim dots of surface markings, our minds tend to connect those dots into straight lines.

Rotation of the Planets

Astronomers have determined the rotation period of Mars with great accuracy by watching the motion of permanent surface markings; its sidereal day is 24 h 37 m 23 s, just a little greater than the rotation period of the Earth. This high precision is not obtained by watching Mars for a single rotation, but by noting how many turns it makes over a

long period of time. Good observations of Mars date back more than 200 years, a period during which tens of thousands of martian days have passed. As a result, the rotation period is now known to within a few hundredths of a second.

The rotational axis of Mars has a tilt of about 25°, very similar to the tilt of the Earth's axis. Thus Mars experiences seasons very much like those on Earth. Because of the longer martian year (almost two Earth years), however, each season there lasts about six of our months.

Since no surface detail can be seen on Venus, its rotation period can only be found with radar (as explained for Mercury in the last chapter). The first radar observations of the planet's rotation were made in the early 1960s. Later, topographical surface features were identified on the planet that showed up in the reflected radar signals. The rotation period of Venus, precisely determined from the motion of such "radar features" across its disk, is 243 days. Even more surprising than how *long* Venus takes to rotate is the fact that it spins in a backward or retrograde direction (east to west).

Venus' rotation period, by far the longest in the solar system, is about 19 days longer than its period of revolution about the Sun. The length of a day on Venus—the time the Sun takes to return to the same place in the sky—turns out to be 117 Earth days. Although we do not know the reason for Venus' slow backward rotation, we can guess that it may have suffered one or more extremely powerful collisions during the formation process of the solar system (see Chapter 8).

Basic Properties of Venus and Mars

Before discussing each planet individually, let us compare some of their basic properties with each other and with the Earth (Table 9.1). Venus is in many ways the Earth's twin, with a mass 0.82 times the mass of the Earth and an almost identical density. The level of geological activity is also relatively high, creating one of the most geologically complex surfaces in the solar system. On the other hand, Venus'

TABLE 9.1
Properties of Earth, Venus, and Mars

	Earth	Venus	Mars
Semimajor axis (AU)	1.00	0.72	1.52
Period (year)	1.00	0.61	1.88
Mass (Earth = 1)	1.00	0.82	0.11
Diameter (km)	12,756	12,102	6,790
Density (g/cm³)	5.5	5.3	3.9
Surface gravity (Earth = 1)	1.00	0.91	0.38
Escape velocity (km/s)	11.2	10.4	5.0
Rotation period (hours or days)	23.9 h	2243 d	24.6 h
Surface area (Earth = 1)	1.00	0.90	0.28
Atmospheric pressure (bar)	1.00	90	0.007

atmosphere is not at all like that of Earth. It is much more massive, with a surface pressure nearly 100 times greater than ours. The surface of Venus is also remarkably hot, with a temperature of 730 K (over 850°F), hotter than the self-cleaning cycle of your oven. One of the major challenges presented by Venus is to understand why the atmosphere and surface environment of this twin have diverged so sharply from those of our own planet.

Mars, by contrast, is rather small, with a mass only 0.11 times the mass of the Earth. It is larger than either the Moon or Mercury, however, and unlike them it retains a thin atmosphere. Mars is also large enough to have supported considerable geological activity. But the most fascinating thing about Mars is that long ago it probably had a thick atmosphere and seas of liquid water; there is even a chance that some sort of life flourished there in the distant past.

9.2 THE GEOLOGY OF VENUS

Since Venus has about the same size and composition as the Earth, we might expect its geology to be similar. This is partly true, but Venus does not exhibit the same kind of *plate tectonics* as the Earth, and we will see that its lack of erosion results in a very different surface appearance.

Spacecraft Exploration of Venus

More spacecraft have been sent to Venus than to any other planet. Although the 1962 U.S. Mariner 2 flyby was the first, the Soviet Union launched most of the subsequent missions to Venus. The early Soviet Venera entry probes were crushed by the high pressure of the atmosphere before they could reach the surface, but in 1970 Venera 7 became the first probe to land and broadcast data from the surface of Venus. It operated for 23 minutes before succumbing to the high surface temperature. Additional Venera probes and landers followed, photographing the surface and analyzing the atmosphere and soil.

To understand the geology of Venus, however, we needed to make a global study of its surface, a task made very difficult by the perpetual cloud layers surrounding the planet. The problem resembles the challenge facing air traffic controllers on many an evening at San Francisco airport, when the weather is so foggy that they can't locate the incoming planes visually. The solution is similar in both cases: Use a radar instrument to probe through the obscuring layer.

The first crude global radar map was made by the U.S. Pioneer 12 orbiter in the late 1970s, followed by better maps from the twin Soviet Venera 15 and 16 radar orbiters in the early 1980s. However, most of our information on the geology of Venus is derived from the U.S. Magellan spacecraft, which mapped Venus from orbit between 1991 and 1993. With its powerful *imaging radar*, Magellan was able to study the surface at a resolution of 100 m, much higher than that of previous missions, yielding our first detailed look at the surface of our sister planet (Figure 9.4). (The Magellan spacecraft actually returned more data to Earth than all previous planetary missions combined; each

FIGURE 9.4

Radar Maps of Venus   Two globes of the same face of Venus, constructed from Magellan radar data. Both views show Aphrodite Terra, the largest continental area on Venus. At left, it can be identified by its brightness (the result of rougher terrain). The reddish colors have been added to simulate the red light filtering down to the surface of Venus through the vast cloud layers. At right, the same image (but with contrast slightly improved) was color-coded to resemble the Earth: continental highlands are brown, and the lava basins are blue.   (JPL/USGS)

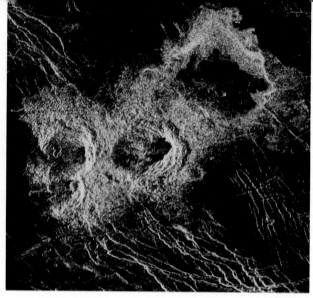

(a)

(b)

FIGURE 9.5

Impact Craters on Venus (a) Large impact craters in the Lavinia region of Venus. Because they are rough, the crater rims and ejecta appear brighter on these radar images than do the smoother surrounding lava plains. The largest of these craters has a diameter of 50 km.
(b) Crater Stein, named after writer Gertrude Stein. The triple impact was caused by the breaking apart of the incoming asteroid during its passage through the thick atmosphere of Venus. We estimate the projectile had an initial diameter of between 1 and 2 km. *(NASA/JPL)*

100 min of data transmission from the spacecraft provided enough information, if translated into characters, to fill two 30-volume encyclopedias.)

Consider for a moment how good a resolution of 100 m really is. It means the radar images from Venus can show anything on the surface larger than a football field. Suddenly a whole host of topographic features on Venus became accessible to our view. As you look at the radar images throughout this chapter, bear in mind that these are constructed from radar reflections, not from visible-light photographs. For example, bright features on these images are an indication of rough terrain, while darker regions are smoother.

Probing Through the Clouds

The radar maps of Venus reveal a planet that looks much the way the Earth might look if our planet's surface were not constantly being changed by erosion and deposition of sediment. Because there is no water or ice on Venus and the surface wind speeds are very low, almost nothing obscures or erases the complex geological features produced by widespread crustal forces and volcanic eruptions. Having finally penetrated below the clouds of Venus, we find its surface to be naked, revealing the history of hundreds of millions of years of geological activity. Venus is in many ways a geologist's dream planet.

About 75 percent of the surface of Venus consists of lowland lava plains. Superficially, these resemble the basaltic ocean basins of the Earth, but they were not produced in quite the same way. There is no evidence of subduction

zones (see Chapter 7) on Venus, indicating that, unlike the Earth, this planet never experienced plate tectonics. Although *convection* (the rising of hot materials) in its mantle generated great stresses in the crust of Venus, they did not initiate plate motion.

The formation of the lava plains of Venus more nearly resembled that of the lunar maria, which were also the result of widespread lava eruptions without the crustal spreading associated with plate tectonics. The venusian lava plains are much younger than the lunar plains, however.

Rising above the lowland lava plains are individual mountains and mountain ranges, as well as two full-scale continents. The largest continent on Venus, called Aphrodite, is about the size of Africa (you can see it stand out in Figure 9.4). Aphrodite stretches along the equator for about one third of the way around the planet. Next in size is the northern highland region Ishtar, which is about the size of Australia. Ishtar contains the highest region on the planet, the Maxwell Mountains, which rise 11 km above the surrounding lowlands. (The Maxwell Mountains are the only feature on Venus named after a man; they commemorate James Clerk Maxwell, whose theory of electromagnetism led to the invention of radar. All other features are named for women, historic or mythological.)

Craters and the Age of the Surface

One of the first questions astronomers addressed with the high-resolution Magellan images was the age of the surface of Venus. Remember that the age of a surface is rarely the age of the world it is on. A young age merely implies an active geology that can resurface the world in short order. As described in the previous chapter, the age can be derived from counting impact craters (Figure 9.5a is an example of what these look like on the Venus radar images). The more densely cratered the surface, the greater its age. The largest crater on Venus (called Mead) is 275 km in diameter, slightly larger than the largest known terrestrial crater (Chicxulub) but much smaller than the lunar impact basins.

You might think that the thick atmosphere of Venus would protect the surface from impacts, burning up the projectiles long before they could reach the surface. But this is only the case for smaller projectiles. The effect of the atmosphere is readily seen in the crater statistics, which show very few craters less than 10 km in diameter, indicating that projectiles smaller than about 1 km (the size that typically produces a 10-km crater) were stopped by the atmosphere.

Those craters with diameters from 10 to 30 km are frequently distorted, apparently because the incoming projectile broke apart and exploded in the atmosphere before it could strike the ground. There are also examples of multiple craters that formed when the projectile broke into several pieces before striking the surface (Figure 9.5b). But if we limit ourselves to impacts that produce craters with diameters of 30 km or greater, crater counts are as useful on Venus for measuring surface age as they are on airless bodies such as the Moon.

There are typically only about 15 percent as many craters in the venusian plains as on the lunar maria, indicating a surface age only about 15 percent as great, or roughly 500 million years old. These results indicate that Venus is indeed a planet with persistent geological activity, intermediate between that of the Earth's ocean basins (which are younger and more active) and that of its continents (which are older and less active).

Almost all of the craters look fresh, with little degradation or filling in by either lava or windblown dust. This is one way we know that the rates of erosion or sediment deposition are very low. We have the impression that relatively little has happened since the venusian plains were resurfaced by large-scale volcanic activity. Apparently Venus experienced some sort of planetwide volcanic convulsion about 500 million years ago.

Volcanoes on Venus

Like the Earth, Venus is a planet with widespread volcanism, and we can see many different types of volcanic features. In the lowland plains, volcanic eruptions are the principal way the surface is renewed, with large flows of highly fluid lava destroying old craters and generating a fresh surface. In addition, there are numerous younger volcanic mountains and other structures associated with surface hot spots—places where convection in the planet's mantle transports the interior heat (and hot magma) to the surface.

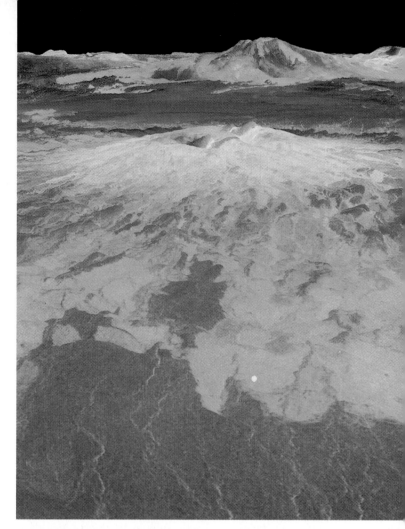

FIGURE 9.6

Venus Volcanoes (Height Exaggerated) This computer-generated view of two large volcanoes on the surface of Venus was assembled from Magellan radar data. In the center foreground is Sapas Mons (named for a Phoenician goddess), which is about 400 km across at its base and 1.5 km high. The image shows a perspective located 4 km above the terrain and 527 km in front of Sapas. Lava flows extending for hundreds of kilometers are visible in the foreground. In the background you can see Maat Mons, 8 km high. The colors have been added from the Russian Venera lander images. The vertical scale has been exaggerated ten times to make the mountains more clearly visible. In real life the mountains would appear to hug the ground much more closely if viewed from this perspective. *(NASA/JPL)*

The largest individual volcano on Venus, called Sif Mons, is about 500 km across and 3 km high—broader but lower than the Hawaiian volcano Mauna Loa. At its top is a volcanic crater or *caldera* about 40 km across, and its slopes show individual lava flows up to 500 km long. Thousands of smaller volcanoes dot the surface, down to the limit of visibility of the Magellan images, which corresponds to cones or domes about the size of a shopping-mall parking lot. Most of these seem similar to terrestrial volcanoes. (Figure 9.6 shows a computer-generated oblique view of two of the

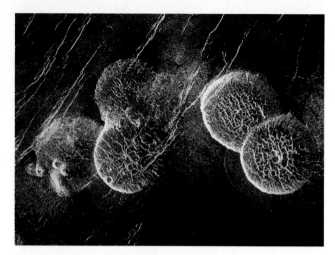

FIGURE 9.7
Pancake-Shaped Volcanoes on Venus These remarkable circular domes, each about 25 km across and about 2 km tall, are the result of eruptions of highly viscous lava. *(NASA/JPL)*

larger volcanoes.) Other volcanoes have unusual shapes, such as the "pancake domes" illustrated in Figure 9.7.

All of this volcanism is the result of eruption of lava onto the surface of the planet. But the hot magma rising from the interior of a planet does not always make it to the surface. On both the Earth and Venus, this magma can collect to produce bulges in the crust. Many of the granite mountain ranges on Earth, such as the Sierra Nevada in California, involve subsurface collections of magma that push up on the material above them. Such subsurface bulges are common on Venus, where they produce large circular or oval features called *coronae* (singular: *corona*) (Figure 9.8).

Tectonic Activity

Convection currents in the mantle of Venus do more than bring magma to the surface. As on the Earth, these convection currents push and stretch the crust. Such forces are called **tectonic,** and the geological features that result from these forces are called *tectonic features.* On the lowland plains, tectonic forces have broken the lava surface to create remarkable patterns of ridges and cracks (Figure 9.9). In a few places the crust has even torn apart to generate rift valleys. The circular features associated with coronae are tectonic ridges and cracks, and most of the mountains of Venus also owe their existence to tectonic forces.

The Ishtar continent, which has the highest elevations on Venus, is perhaps the most dramatic product of these tectonic forces. Ishtar and its tall Maxwell Mountains resemble the Tibetan Plateau and Himalayan Mountains on the Earth. Both are the product of compression of the crust, and both are maintained by the continuing forces of mantle convection.

On Venus' Surface

The successful Venera landers of the 1970s found themselves on an extraordinarily inhospitable planet, with a surface pressure of 90 bars and a temperature hot enough to melt lead and zinc. Despite these unpleasant conditions, the spacecraft were able to photograph their surroundings and collect surface samples for chemical analysis before their instruments gave out. They found that the rock in the landing areas is igneous, primarily basalts. Examples of the Venera photographs are shown in Figure 9.10. Each picture shows a flat, desolate landscape with a variety of rocks, some of which may be ejecta from impacts. Other areas show flat, layered lava flows.

9.3 THE MASSIVE ATMOSPHERE OF VENUS

The atmosphere of Venus causes the high surface temperature and shrouds the surface in a perpetual red twilight. The Sun does not shine directly through the heavy clouds, but the surface is fairly well lit by diffuse light—about the same as the lighting on Earth under a heavy overcast, but

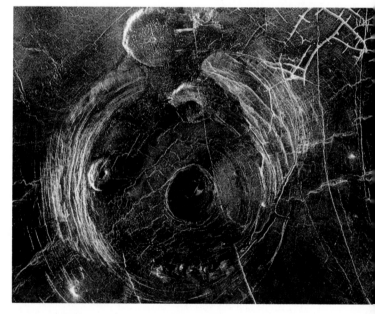

FIGURE 9.8
The "Miss Piggy" Corona Fotla Corona, located in the plains to the south of Aphrodite Terra. This circular feature, about 200 km across, is a region where hot material from deeper inside Venus is pushing up on the surface. Curved fracture patterns clearly show where the material beneath has put stress on the surface. A number of pancake and dome volcanoes are also visible. Fotla was a Celtic fertility goddess. Some students see a resemblance between this corona and Miss Piggy of the Muppets (her right ear is the pancake volcano in the upper center of the image). *(NASA/JPL)*

FIGURE 9.9

Ridges and Cracks This region of the Lakshmi Plains on Venus has been fractured by tectonic forces to produce a cross-hatched grid of cracks and ridges. Be sure to notice the fainter linear features that run perpendicular to the brighter ones! This image shows a region 40 km across. Lakshmi is a Hindu goddess of prosperity. *(NASA/JPL)*

with a strong red tint because the massive atmosphere blocks shorter-wavelength (bluer) colors of light. The weather at the bottom of this deep atmosphere remains perpetually hot and dry, with calm winds. Because of the heavy blanket of clouds and atmosphere, one spot on the surface of Venus is similar to any other as far as weather is concerned.

Composition and Structure

The most abundant gas in the atmosphere of Venus is carbon dioxide (CO_2), which accounts for 96 percent of the air. The second most abundant gas is nitrogen. The predominance of carbon dioxide over nitrogen is not surprising when you recall that the Earth's atmosphere would also be mostly carbon dioxide if this gas were not locked up in marine sediments (see Section 7.3).

Table 9.2 compares the compositions of the atmospheres of Venus, Mars, and the Earth. Expressed in this way, as percentages, the proportions of major gases are very similar for Venus and Mars, but in total quantity their atmospheres are dramatically different. With its surface pressure of 90 bars, the venusian atmosphere is more than 10,000 times more massive than its martian counterpart. Overall, the atmosphere of Venus is very dry; absence of water is one of the important ways that Venus differs from the Earth.

The venusian atmosphere (Figure 9.11) has a huge troposphere that extends up to at least 50 km above the surface. Within the troposphere, the gas is heated from below and circulates slowly, rising near the equator and descending over the poles. Being at the base of the atmosphere of Venus is something like being a kilometer or more below the ocean surface on the Earth. There, also, the mass of water evens out temperature variations and results in a uniform environment.

In the upper troposphere, between 30 and 60 km above the surface, there is a thick cloud layer composed primarily of sulfuric acid droplets. Sulfuric acid (H_2SO_4) is formed from the chemical combination of sulfur dioxide (SO_2) and water (H_2O). In the atmosphere of the Earth, sulfur dioxide is one of the primary gases emitted by volcanoes, but it is quickly diluted and washed out by rainfall. In the dry atmosphere of Venus, this unpleasant substance is apparently stable. Below 30 km the venusian atmosphere is clear of clouds.

FIGURE 9.10

The Surface of Venus Views of the surface of Venus from the Venera 13 spacecraft. The upper images show the scene as recorded by the spacecraft cameras; everything looks orange because the thick atmosphere of Venus absorbs the bluer colors of light. Computer processing has transformed the bottom images to show what the scenes would look like under Earth lighting conditions. The horizon is visible in the upper corner of each image. *(USSR Academy of Science and Carle Pieters, Brown University)*

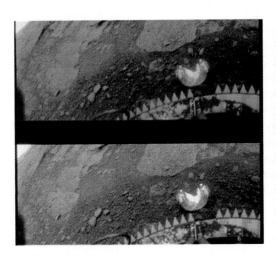
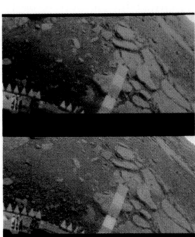

TABLE 9.2
Atmospheric Compositions of Earth, Venus, and Mars (in %)

Gas	Earth	Venus	Mars
Carbon dioxide (CO_2)	0.03	96	95.3
Nitrogen (N_2)	78.1	3.5	2.7
Argon (Ar)	0.93	0.006	1.6
Oxygen (O_2)	21.0	0.003	0.15
Neon (Ne)	0.002	0.001	0.0003

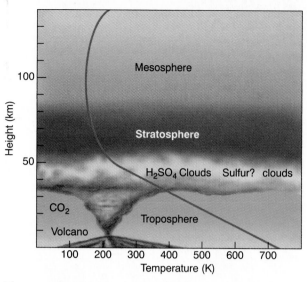

FIGURE 9.11
Venus' Atmosphere   The layers of the massive atmosphere of Venus, based on data from the Pioneer and Venera entry probes. Height is measured along the left axis; the bottom scale shows temperatures, and the red line allows you to read off the temperature at each height. Notice how steeply the temperature rises below the clouds thanks to the planet's huge greenhouse effect.

Surface Temperature

The high surface temperature of Venus was discovered by radio astronomers in the late 1950s and confirmed by the Mariner and Venera probes. How can our neighbor planet be so hot? Although Venus is somewhat closer to the Sun than is the Earth, its surface is hundreds of degrees *hotter* than you would expect from the extra sunlight it receives. Scientists wondered what could be heating the surface of Venus to a temperature above 700 K. The answer turned out to be the *greenhouse effect.*

The greenhouse effect works on Venus just as it does on the Earth (see Section 7.3). But since Venus has so much more CO_2—almost a million times more—the effect is much stronger. Sunlight that diffuses through the atmosphere heats the surface just as on Earth, but the thick CO_2 acts as a blanket, making it very difficult for the infrared radiation to get back into space. As a result, the surface heats up until eventually it is emitting enough to balance the energy it receives from the Sun.

Has Venus always had such a massive atmosphere and high surface temperature, or might it have evolved to such conditions from a climate that was once more nearly Earth-like? The answer to this question is of particular interest to us as we look at the increasing levels of CO_2 in the Earth's atmosphere, as discussed in Section 7.4. Are we in any danger of transforming our own planet into a hellish place like Venus?

Let us try to reconstruct the possible evolution of Venus from an Earth-like beginning to its present state. We imagine that Venus began with moderate temperatures, water oceans, and much of its CO_2 dissolved in the ocean or chemically combined with the surface rocks, as is the case on Earth today. Then we allow for modest additional heating, by a small rise in the energy output of the Sun, for example, or by an increase in atmospheric CO_2. When we make models of how Venus' atmosphere would respond to such effects, it turns out that even a small amount of extra heat can lead to increased evaporation from the oceans and the release of gas from surface rocks.

This in turn means a further increase in the atmospheric CO_2 and H_2O, gases that would amplify the greenhouse effect. That would lead to still more heating and the release of further CO_2 and H_2O. Unless some other processes intervene, the temperature thus continues to rise. Such a situation is called the *runaway greenhouse effect.*

We want to emphasize that the runaway greenhouse effect is not just a larger greenhouse effect; it is an evolutionary process. The atmosphere evolves from having a small greenhouse effect, such as on the Earth, to a situation where greenhouse warming is a major factor, as we see today on Venus. Once the larger greenhouse conditions develop, the planet establishes a new, much hotter equilibrium near its surface. Reversing the situation is difficult, if not impossible.

Note that in our scenario, if the planet's surface has large bodies of water, the runaway greenhouse effect causes their evaporation. This creates an atmosphere of hot water vapor, which itself is a major contributor to the greenhouse effect. Water vapor in the atmosphere is not stable in the presence of solar ultraviolet light, however. It tends to break the molecules of H_2O into their constituent parts—oxygen and hydrogen. The light element hydrogen can escape from the atmospheres of the terrestrial planets, leaving the oxygen behind to combine chemically with surface rock. The loss of water is therefore an *irreversible* process; once the water is gone, it cannot be restored. There is evidence that this is exactly what happened to the water once present on Venus.

While we do not know the point at which a stable greenhouse effect breaks down and turns into a runaway

TABLE 9.3
Some Missions to Explore Mars

Name	Years at Mars	Type	Description
Mariner 9	1971–72	Orbiter	First orbital surveillance
Viking 1 and 2	1976–80	Orbiter and lander	Set up weather station, did soil analysis; color images from surface and orbit
Pathfinder	1997	Lander and rover	Tested rover, analyzed rock composition
Global Surveyor	1997–	Orbiter	High-resolution images
Nozomi	1999–	Orbiter	Explores interaction of atmosphere and magnetosphere with solar wind
Climate Orbiter	1999–	Orbiter	Studies of martian weather
Polar Lander	1999–	Lander	Landing near south polar ice cap
Athena	2003	Lander and rover	Collect samples for return to Earth
Mars Airplane	2003	Airplane	Demonstrate powered flight on Mars
Sample Return	2008	Lander and return capsule	Return samples to Earth (arriving 2008)

greenhouse effect, Venus stands as clear testament to the fact that a planet cannot continue heating indefinitely without a major change in its oceans and atmosphere. It is a conclusion that we and our descendants will surely want to pay close attention to.

9.4 THE GEOLOGY OF MARS

Mars is more interesting to most people than Venus, because it is more hospitable. Although the surface today is dry and cold, there is evidence that Mars once had blue skies and lakes of liquid water. Even today, it is the sort of place we can imagine astronauts visiting and perhaps even setting up permanent bases.

Spacecraft Exploration of Mars

Mars has been intensively investigated by spacecraft, and many additional missions are planned (Table 9.3). The first visitor was the U.S. Mariner 4, which flew by Mars in 1965 and radioed 22 photos to Earth. These pictures showed an apparently bleak planet with abundant impact craters. In those days craters were unexpected; some people who were romantically inclined still hoped to see "canals" or something like them. In any case, newspaper headlines sadly announced that Mars was a "dead planet."

In 1971 Mariner 9 became the first spacecraft to orbit another planet, mapping the entire surface of Mars at a resolution of about 1 km and discovering a great variety of geological features, including volcanoes, huge canyons, intricate layers on the polar caps, and channels that appeared to have been cut by running water. Geologically, Mars didn't look so dead after all. Next came the Viking spacecraft, which were among the most ambitious and successful of all planetary missions.

Two Viking *orbiters* surveyed the planet and served to relay communications for two *landers* on the surface (Figure 9.12). After an exciting and sometimes frustrating

FIGURE 9.12

Mars Globe   This globe is a photographic mosaic assembled by the mapmakers at the U.S. Geological Survey from many Mars images taken with the Viking cameras. Crossing the center of the planet is a 5000-km-long rift valley called Mariner Valley (a name it received when first discovered by the Mariner 9 spacecraft). Three giant volcanoes can be seen at the left edge of the planet. Compare this to Lowell's globe in Figure 9.3. *(JPL/USGS)*

FIGURE 9.13

Panorama from Mars Pathfinder To construct this fantastic 360° view around the Mars Pathfinder lander, mission controllers slowly swiveled the camera in a circle. To identify the rocks and hills in these images, the scientists assigned them names based on some of the favorite cartoon characters they enjoyed while growing up (or, based on the names, not growing up.) The image has been computer processed to sharpen details and colors and correct for distortions as the camera faced in different directions. The scene shows a windswept plain, sculpted long ago when water flowed out of the martian highlands and into the depression where Pathfinder landed. Clearly visible are the yellow ramp that the Sojourner rover used to begin its journey of exploration, the black solar panels above the deflated airbags used to cushion the spacecraft's landing, the tracks the rover made, and the rover itself, probing a rock nicknamed Yogi. *(JPL/NASA)*

search for a safe landing spot, the Viking 1 lander touched down on the surface of Chryse Planitia (the Plains of Gold) on July 20, 1976, exactly seven years after Neil Armstrong's historic first step on the Moon. Two months later, Viking 2 landed with equal success in another plain farther north, called Utopia. The last message from the Viking landers arrived on November 5, 1982, after an enormously successful mission of studying the red planet.

Mars then languished unvisited for two decades after Viking. Two more spacecraft were launched toward Mars, by NASA and the Russian Space Agency, but both failed before reaching the planet. Then in 1997 NASA began a new exploration program, using smaller and less expensive spacecraft. The U.S. plans to launch a lander and an orbiter at every opportunity when Mars and Earth are aligned (approximately once every 26 months), and both Japan and the European Space Agency also plan Mars missions.

The first of the new missions, appropriately called Pathfinder, landed on the martian surface on July 4, 1997 (Figure 9.13). An orbiter called Mars Global Surveyor arrived a few months later and began high-resolution photography of the surface. In 2003 NASA plans to fly an airplane on Mars, and by 2008, NASA hopes to return selected samples of martian rocks and soil to Earth. The ultimate long-term goal is to develop the technology for human flight to Mars, leading perhaps to permanent settlements on another world. Such ambitious piloted missions are beyond the current space budgets of the United States and other space-faring nations, but Mars is well within our reach if we choose to travel there in the 21st century.

Global Properties

Mars has a diameter of 6790 km, just over half the diameter of the Earth, giving it a total surface area very nearly equal to the continental area of our planet. Its density of 3.9 g/cm^3 suggests a composition consisting primarily of silicates but with a small metal core. The planet has no detectable global magnetic field (although it probably had one in the past). This suggests that the red planet has no liquid conducting material in its core today.

Like the Earth, Moon, and Venus, the surface of Mars has continental or highland areas as well as widespread volcanic plains. Approximately half the planet consists of heavily cratered higher-elevation terrain, found primarily in the southern hemisphere. The other half, which is mostly in the north, contains younger, lightly cratered volcanic plains at an average elevation about 4 km lower than the highlands. Remember that we saw a similar pattern on Earth, the Moon, and Venus. A geological division into older highlands and younger lowland plains seems to be characteristic of all the terrestrial planets except Mercury.

One of the main features of martian geology is an impressive uplifted continent the size of North America. This is the 10-km-high Tharsis bulge, a volcanically active region crowned by four great volcanoes that rise another 15 km into the martian sky.

Volcanoes on Mars

The lowland plains of Mars look very much like the lunar maria, and they have about the same number of impact craters. Like the lunar maria, they probably formed between 3 and 4 billion years ago. Apparently Mars experienced extensive volcanic activity at about the same time the Moon did, producing similar basaltic lavas.

The largest volcanic mountains of Mars are found in the Tharsis area (you can see three of them in Figure 9.12), although smaller volcanoes dot much of the surface. The most dramatic volcano on Mars is Olympus Mons (Mount Olympus), with a diameter of more than 500 km and a summit that towers 25 km above the surrounding plains (Figure 9.14). The volume of this immense volcano is nearly 100 times greater than that of Mauna Loa in Hawaii. Placed on the Earth's surface, Olympus would more than cover the entire state of Washington.

Images taken from orbit allow scientists to search for impact craters on the slopes of these volcanoes in order to estimate their age. Many of the volcanoes show fair numbers of such craters, suggesting that they ceased activity a billion years or more ago. However, Olympus Mons has very, very few impact craters. Its present surface cannot be more than about 100 million years old; it may even be much younger. Some of the fresh-looking lava flows might have been formed a hundred years ago, or a thousand, or a million; but geologically speaking they are quite young. This leads geologists to the conclusion that Olympus Mons probably remains intermittently active today—something future Mars land developers may want to keep in mind.

Cracks and Canyons

The Tharsis bulge has many interesting geological features in addition to its huge volcanoes. In this part of the planet the surface itself has bulged upward, forced by great pressures from below, resulting in extensive tectonic cracking of the crust. Among the most spectacular tectonic features are the canyons called the Valles Marineris (or Mariner Valley, named after Mariner 9, which first revealed them to us). They extend for about 5000 km (nearly a quarter of the way around Mars) along the slopes of the Tharsis bulge (Figure 9.15).

The main canyon is about 7 km deep and up to 100 km wide, large enough for the Grand Canyon of the Colorado River to fit comfortably into one of its side canyons. The opening image for this chapter shows a spectacular overview of part of Valles Marineris, and Mars Global Surveyor has obtained high-resolution images of the canyon walls that show layering that might be the result of sediment layers that formed in an ancient sea (Figure 9.16).

The term *canyon* is somewhat misleading here, because the Valles Marineris canyons have no outlets and were not cut by running water. They are basically tectonic cracks, produced by the same crustal tensions that caused the Tharsis uplift. However, water is believed to have played a later role in shaping the canyons, primarily by seeping from deep springs and undercutting the cliffs. This undercutting led to landslides (Figure 9.17), gradually

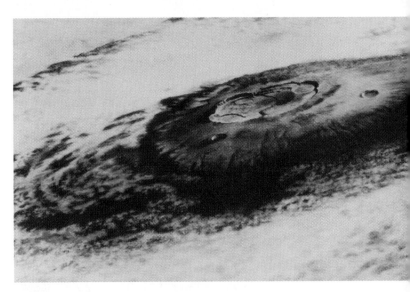

FIGURE 9.14
Olympus Mons The largest volcano on Mars, and probably the largest in the solar system, is Olympus Mons, illustrated in this airbrush painting made from Viking orbiter photographs. Placed on Earth, the base of the Olympus Mons would completely cover the state of Missouri; the caldera, the circular opening at the top, is 65 km across, about the size of Los Angeles. Note the extensive clouds over the lower slopes of the volcano. *(NASA/JPL)*

FIGURE 9.15
Mariner Valley and a U.S. Map The outline of the United States is superimposed to scale on a mosaic of photographs showing the Valles Marineris canyons on Mars. You can see that if it were on Earth, the system would stretch from Los Angeles to Washington, D.C. Two volcanoes are visible in the upper left, which is the direction of the Tharsis bulge. *(Stephen Meszaros and the Astronomical Society of the Pacific)*

widening the original cracks into the great valleys we see today.

While the Tharsis bulge and Valles Marineris are impressive, in general we see fewer tectonic structures on Mars than on Venus. In part, this may reflect a lower general level

of geological activity, as would be expected for a smaller planet. But it is also possible that evidence of widespread faulting has been buried by wind-deposited sediment over much of Mars. Like the Earth, Mars may have hidden part of its geological history under a cloak of soil.

The View on the Surface

Viking 1 and Mars Pathfinder both landed in the Chryse Basin—one of the lowland volcanic plains that may once have held a broad shallow inland sea. The first Viking photos showed desolate but strangely beautiful surroundings, including numerous angular rocks, some more than a meter across, interspersed with dune-like deposits of fine-grained, reddish soil. At the Viking 2 site in Utopia, the surface was somewhat similar but with a substantially greater number of rocks (Figure 9.18). Each Viking lander peered at its surroundings through color stereo cameras, sniffed the atmosphere with a variety of analytical instruments, and poked at nearby rocks and soil with its mechanical arm. As part of its mission of searching for martian life, each lander collected soil samples and brought them on board for analysis, as described in more detail shortly.

The Viking weather stations operated for several years, providing a longer perspective on martian weather. Temperatures varied greatly, due to the absence of moderating oceans and clouds. Typically, the summer maximum at Viking 1 was 240 K (−33°C), dropping to 190 K (−83°C) at the same location just before dawn. The lowest air temperatures, measured farther north by Viking 2, were about 173 K (−100°C). During the winter, Viking 2 also photographed water frost deposits on the ground (Figure 9.19). We make a point of saying "water frost" here, because at some locations on Mars it gets cold enough for

FIGURE 9.18
The View from Viking 2   The surface of Mars as recorded by the Viking 2 cameras. In this image, produced from a black-and-white original, the colors Viking recorded in less-detailed images have been added to this wonderfully detailed scene using computer-assisted techniques. You can see two trenches dug by the Viking collecting scoop.   *(NASA image processing by Mary Dale-Bannister)*

carbon dioxide (dry ice) to freeze out of the atmosphere as well.

Viking found that that the soil consisted of clays and iron oxides, as had long been expected from the red color of the planet. Most of the winds measured at the Viking sites were low to moderate, only a few km per hour. However, Mars is capable of great windstorms that can shroud the entire planet with wind-blown dust. Such high winds can strip the surface of some of its loose, fine covering, leaving the rocks exposed.

Mars Pathfinder was targeted to land in the outwash plain below the mouth of the ancient martian river called Ares Vallis. The Pathfinder cameras revealed a landscape with jumbled angular rocks that looked exactly as if they were emplaced by rapidly flowing water (see Figure 9.13). Presumably, this took place when the Ares Vallis river last flowed more than 3 billion years ago, and the landscape has been little changed since that time. Pathfinder also carried a small rover named Sojourner, about the size of a microwave oven (Figure 9.20). Sojourner, operating on the surface for three months, traveled a total distance of about 100 m and measured the composition of 8 rocks, as well as several soil samples.

Although the Pathfinder found itself in the river mouth as planned, scientists were disappointed to find that all the rocks measured were of volcanic origin and had roughly the same composition. They had hoped that a variety of rocks would be present, including perhaps some *sedimentary* samples—rocks made of fragments of volcanic material deposited by wind or water and cemented

FIGURE 9.20
Sojourner and the Rock Garden   Pathfinder landed in Ares Valles, an ancient flood plain that bears mute witness to Mars' warmer, wetter past. Here we are seeing a cluster of rocks that are tilted in the "downstream" direction of the floods; mission scientists nicknamed this area the "Rock Garden." The Sojourner rover is nestled against a rock they called "Moe" (after one of the Three Stooges).   *(NASA/JPL)*

FIGURE 9.19
Water Frost on Mars   Surface frost photographed at the Viking 2 landing site during late winter.   *(NASA/JPL)*

FIGURE 9.21
Sunset on Mars This beautiful image, taken on martian day 24 of the Pathfinder mission, shows the setting Sun. You can see that the sky nearest the Sun is bluer, and furthest from the Sun is redder. The color of the Sun is not correct on the image, since it was overexposed; it would look white or bluish white. *(Lunar and Planetary Lab, U. of Arizona; JPL/NASA)*

together. But this mission, at least, did not turn up any such examples.

The Pathfinder lander, which NASA soon named the Carl Sagan Memorial Station to honor the well-known astronomer who died while the spacecraft was on the way to Mars (see box in Chapter 6), also had instruments to report back on the martian weather. The temperatures were always below freezing, similar to those measured 20 years earlier by Viking, but each sunny afternoon the surface was heated, and the atmosphere became turbulent as heat waves rose off the surface. This turbulence in turn led to the formation of large dust devils about 200 m across; 20 of these cyclones were measured, an average of one every four days.

The lander camera also carefully measured the brightness of the sky, which showed the effects of a pervasive haze of red dust. Even on a clear day (by martian standards), more sunlight reached the surface by scattering off this haze than by getting through directly, so that shadows were soft and indistinct. Because of the properties of the martian dust, the sky was actually bluest near the Sun and became progressively redder toward the horizon (Figure 9.21).

The Pathfinder measurements continued for 83 martian days, at which point the lander succumbed to the cold temperatures and fell permanently silent. It is possible the rover continued to operate for several more days, but without the lander we could not communicate with it. We can imagine the little rover still moving in the red martian sand and plaintively calling for instructions, cut off forever from its makers on Earth.

Martian Samples

Much of what we know of the Moon, including the circumstances of its origin, comes from studies of lunar samples, but spacecraft have not yet returned martian samples to Earth for laboratory analysis. It is with great interest, therefore, that scientists have concluded that samples of martian material are nevertheless already here on Earth, available for study. About a dozen of these martian rocks exist, all members of a rare class of *meteorites* (Figure 9.22), rocks that have fallen from space (see Chapter 13).

How would rocks have escaped from Mars? Many impacts have occurred on the red planet, as shown by its heavily cratered surface. The larger impacts can eject debris at so great a speed that some fragments actually attain escape velocity from Mars. A long time later (typically a few million years), a very small fraction of these fragments collide with the Earth and survive their passage through our atmosphere, just like other meteorites. (By the way, several rocks from the Moon have also reached our planet as meteorites, although we were only able to demonstrate their lunar origin after the Apollo missions brought back samples directly from the lunar surface.)

Most of the martian meteorites are volcanic basalts; most of them are also relatively young, with ages of about 1.3 billion years. We know from details of their composition that they are not from the Earth or the Moon. Besides, there was no volcanic activity on the Moon to form them as

FIGURE 9.22
Martian Meteorite A fragment of basalt ejected from Mars that eventually arrived on the Earth's surface. *(NASA/JSC)*

recently as 1.3 billion years ago. It would be very difficult for ejecta from impacts on Venus to escape through its thick atmosphere. By process of elimination, the only reasonable origin seems to be Mars, where the Tharsis volcanoes were certainly active at that time.

The martian origin of these meteorites was confirmed by analysis of tiny gas bubbles trapped inside several of them. These bubbles match the atmospheric properties of Mars as measured directly by Viking. Apparently some atmospheric gas was trapped in the rock by the shock of the impact that ejected it from Mars and started it on its way toward Earth.

One of the most exciting results from analysis of these martian samples has been the discovery of both water and organic (carbon-based) compounds in them, which suggest the possibility that Mars may once have had oceans and (according to a few scientists) even life on its surface. As we will see in a moment, there is other evidence for the presence of flowing water on Mars in the remote past.

 ## 9.5 MARTIAN POLAR CAPS AND CLIMATE

The atmosphere of Mars today has an average surface pressure of only 0.007 bar, less than 1 percent that of the Earth. (This is how thin the air is about 30 km above the Earth's surface.) Martian air is composed primarily of carbon dioxide (95 percent), with about 3 percent nitrogen and 2 percent argon. The proportions of different gases are similar to those in the atmosphere of Venus (Table 9.2), but a lot less of each gas is found on Mars.

Clouds on Mars

Several types of clouds can form in the martian atmosphere. First there are dust clouds, raised by winds, which can sometimes grow to produce global-scale dust storms. Second are water-ice clouds similar to those on Earth. These often form around mountains, just as happens on our planet (see Figure 9.14). Finally, the CO_2 of the atmosphere can itself condense at high altitudes to form hazes of dry ice crystals. The CO_2 clouds have no counterpart on Earth, since on our planet temperatures never drop low enough (down to about 150 K) for this gas to condense.

Although the atmosphere contains water vapor and occasional clouds of water ice, liquid water is not stable under present conditions on Mars. Part of the problem is the low temperatures on the planet. But even if the temperature on a sunny summer day rises above the freezing point, liquid water still cannot exist. At a pressure of less than 0.006 bar, only the solid and vapor forms are possible. In effect, the boiling point is as low or lower than the freezing point, and water changes directly from solid to vapor without an intermediate liquid state.

The Polar Caps

Through a telescope the most prominent surface features on Mars are the bright polar caps, which change with the seasons, similar to the seasonal snow cover on Earth. We do not usually think of the winter snow in northern latitudes as a part of our polar caps, but seen from space, the thin snow merges with the Earth's thick, permanent ice caps to create an impression much like that seen on Mars.

The *seasonal caps* on Mars are composed not of ordinary snow, but of frozen CO_2 (dry ice). These deposits condense directly from the atmosphere when the surface temperature drops below about 150 K. The caps develop during the cold martian winters, extending down to about latitude 50° by the start of spring.

Quite distinct from these thin seasonal caps of CO_2 are the *permanent,* or *residual caps* that are always present near the poles (Figure 9.23). As the seasonal cap retreats

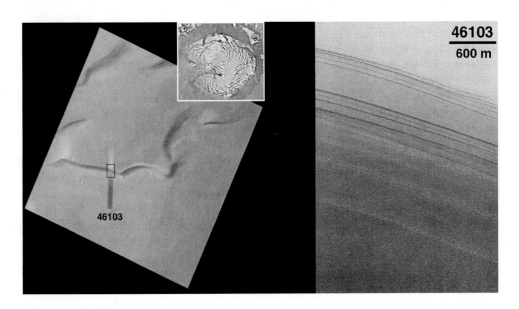

FIGURE 9.23
Layers at the Martian North Pole The small inset in the left image shows a map of the residual north polar cap of Mars, which is about 1000 km across and composed of water ice. The small black box in the middle of the map shows the area covered in the tilted Viking orbiter image at left. The box in that image shows the area of the Global Surveyor high-resolution image at right. On the right image, we see a slope on the edge of the permanent north polar cap, with dozens of layers visible—some thinner than 10 meters.
(NASA/JPL/Malin Space Science Systems)

FIGURE 9.24
Runoff Channels These runoff channels in the old martian highlands are interpreted as the valleys of ancient rain-fed rivers. The width of this image is about 200 km. *(NASA/from Mars Digital Image Map, processing by Brian Fessler, LPI)*

cover the older cratered ground below. Individual layers are typically ten to a few tens of meters in thickness, marked by alternating light and dark bands of sediment. Probably the material in the polar deposits is dust carried by wind from the equatorial regions of Mars.

What do these terraced layers tell us about Mars? Some cyclic process is depositing dust and ice over long periods of time. The time scales represented by the polar layers are tens of thousands of years. Apparently the martian climate experiences periodic changes at intervals similar to those between ice ages on the Earth. Calculations indicate that the causes are probably also similar: The gravitational pull of the other planets produces variations in Mars' orbit and tilt as the great clockwork of the solar system goes through its paces.

Because the polar regions are so interesting scientifically and have much to tell us about the longer-term climate on Mars, a U.S. spacecraft called Mars Polar Lander is targeted to land in the layered terrain near the south pole in December 1999. Its arsenal of equipment includes a robot arm for digging into the soil and cameras to give us close-up views of the landing site.

during spring and early summer, it reveals the brighter, thicker cap beneath. The southern permanent cap has a diameter of 350 km and is composed of frozen CO_2 deposits together with an unknown thickness of water ice. Throughout the southern summer, it remains at the freezing point of CO_2, 150 K, and this cold reservoir is thick enough to survive the summer heat intact.

The northern permanent cap is different. It is much larger, never shrinking below a diameter of 1000 km, and is composed of water ice. Summer temperatures in the north are too high for the frozen CO_2 to be retained. Measurements from Global Surveyor have established the exact elevations in the north polar region of Mars, showing that it is a large basin about the size of our own Arctic Ocean basin. The ice cap itself is about 3 km thick, with a total volume of about 10 million km^3 (similar to that of the Earth's Mediterranean Sea). If Mars ever had extensive liquid water, this north polar basin would have contained a shallow sea. There is some indication of ancient shorelines visible, but better images will be required to verify this suggestion.

In any case, the polar caps represent a huge reservoir of water compared with the very small amounts of water vapor in the atmosphere. The two caps are different because Mars' distance from the Sun varies substantially during the course of its year. This means seasons (and temperatures) on Mars are affected by both the tilt of its axis and its distance from the Sun.

Images taken from orbit also show a distinctive type of terrain surrounding the permanent polar caps, as well as in ice-free areas within the caps themselves (Figure 9.23). At latitudes above 80° in both hemispheres, the surface consists of recent, layered sedimentary deposits that entirely

Channels and Floods

Although no liquid water exists on Mars today, evidence has accumulated that long ago rain fell and rivers flowed on the red planet. Two kinds of geological features appear to be remnants of ancient watercourses. We will examine each of them in turn, but note for the record that both types are far too narrow to have been the "canals" Percival Lowell imagined seeing on Mars.

In the highland equatorial plains there are multitudes of small, sinuous (twisting) channels—typically a few meters deep, some tens of meters wide, and perhaps 10 or 20 km long (Figure 9.24). They are called *runoff channels* because they look like what geologists would expect from troughs that carried the surface runoff of ancient rainstorms. If they were indeed made by rain, these runoff channels seem to be telling us that the planet had a very different climate long ago.

How can we tell when rain might have last fallen on Mars? Crater counts show that this part of the planet is more cratered than the lunar maria but less so than the lunar highlands. Thus the runoff channels are probably older than the lunar maria, presumably at least 3.9 billion years old.

The other water-related features we see are *outflow channels* (Figures 9.25 and 9.26), much larger than the older runoff channels. The largest of these, which drain into the Chryse basin where Pathfinder landed, are 10 km or more in width and hundreds of kilometers long. Many features of these outflow channels have convinced geologists that they were carved by huge volumes of running water, far too great to be produced by ordinary rainfall. Where did such floodwater come from on Mars?

As far as we can tell, the regions where the outflow channels originate contained abundant water frozen in the soil as permafrost. Some local source of heating must have released this water, leading to a period of rapid and catastrophic flooding. Perhaps this heating was associated with the formation of the volcanic plains on Mars, which date back to roughly the same time as the outflow channels.

The older runoff channels, however, require one crucial thing that is absent on the Mars we see today: an atmosphere thick enough to sustain ongoing flows of liquid water.

Climate Change on Mars

The evidence discussed so far suggests that billions of years ago martian temperatures must have been warmer and the atmosphere must have been much more substantial than it is today. What could have changed the climate on Mars so dramatically? We presume that, like the Earth and Venus,

Mars formed with a much thicker atmosphere than it has now, and that this atmosphere maintained a higher surface temperature thanks to the greenhouse effect. But Mars is a smaller planet, and its lower gravity means that gases could escape more easily than from Earth and Venus. As more and more of the atmosphere escaped into space, the temperature on the surface gradually fell.

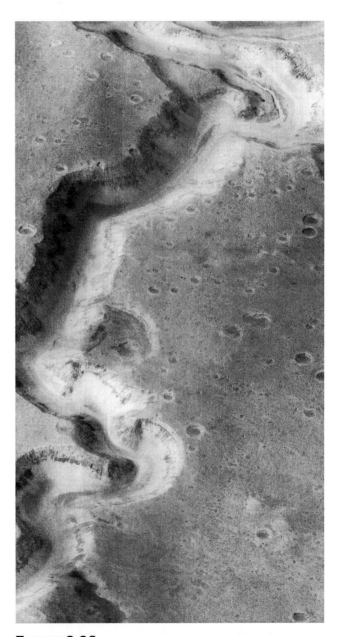

FIGURE 9.26
Possible Evidence of Water on Mars This intriguing channel, called Nanedi Valles, in the Xanthe Terra region near the martian equator, resembles Earth river beds in some (but not all) ways. The tight curves and terraces seen in the channel certainly suggest the sustained flow of a fluid like water. The channel is about 2.5 km across and the entire Global Surveyor image is 10 km wide. We can resolve features as small as 12 m across, a great improvement over the Viking orbiter images. *(NASA/Malin Space Science Systems)*

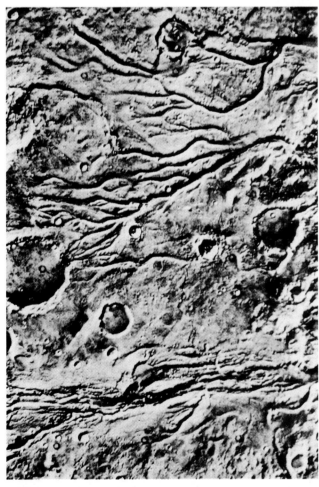

FIGURE 9.25
Outflow Channels Here we see a region of large outflow channels, photographed by Viking. These features appear to have been formed in the distant past from massive floods of water. The width of this image is about 150 km. *(NASA/JPL)*

Astronomy and Pseudoscience: The "Face on Mars"

People like human faces. We humans have developed great skill in recognizing people and interpreting facial expressions. We also have a tendency to see faces in many natural formations, from clouds to the Man in the Moon. One of the curiosities that emerged from the Viking global mapping of Mars was the discovery of a strangely shaped mesa in the Cydonia region that somewhat resembled a human face. Despite later rumors of a cover-up, the "Face on Mars" was, in fact, recognized by Viking scientists and included in one of the early mission press releases. At the low resolution and oblique lighting under which the Viking image was obtained, the mile-wide mesa has something of a Sphinx-like appearance.

Unfortunately, a small band of individuals decided that this formation really was an artificial, carved sculpture of a human face placed on Mars by an ancient civilization that thrived there hundreds of thousands of years ago. A band of "true believers" grew around the face, and tried to deduce the nature of the "sculpture" and who made it. This group linked the "face" to a variety of other pseudoscientific phenomena such as that of the crop circles (patterns in fields of grain, mostly in Britain, now known to be the work of pranksters).

Members of this group accused NASA of covering up evidence of intelligent life on Mars and received a great deal of help in publicizing their perspective from tabloid media. Some of the believers picketed JPL at the time of the failure of the Mars Observer spacecraft, circulating stories that the "failure" of Mars Observer was itself a fake, and that its true (now secret) mission was to photograph the Face.

The high-resolution Mars Observer Camera (MOC) was re-flown on the Mars Global Surveyor mission, which arrived at Mars in 1997. On April 5, 1998, in Orbit 220, the MOC obtained an oblique image of the Face at a resolution of 4.2 m/pixel, a factor of ten improvement in resolution over the Viking image. Immediately released by NASA, the new image showed a low mesa-like hill cut crossways by several roughly linear ridges and depressions, which were misidentified in the 1976 photo as the eyes and mouth of a face. Only with a large dose of imagination can any resemblance to a face be seen in the new image (seen at right), demonstrating how dramatically our interpretation of geology can change with large improvements in resolution.

After 20 years of promoting pseudoscientific interpretations and various conspiracy theories, will the "Face on Mars" believers now accept reality? Probably they will find a way out, perhaps by accusing NASA of faking the new picture. One suggestion they have already made is that the "secret" mission of Mars Observer included a nuclear bomb, which was used to destroy the Face before it could be photographed in greater detail by Global Surveyor.

Space scientists find these suggestions incredible. NASA is spending increasing sums for research on life in the universe, and a major objective of current and upcoming

---

Eventually Mars became so cold that water froze out of the atmosphere, further reducing its ability to retain heat—a sort of *runaway refrigerator effect*, just the opposite of the runaway greenhouse effect that occurred on Venus. The result is the cold, dry Mars we see today. Probably this loss of atmosphere took place within less than a billion years after Mars formed. Based on the absence of runoff channels in the northern plains, it seems that rain has not fallen on Mars for at least 3 billion years.

The Search for Life on Mars

If there was running water on Mars in the past, as the channels imply, perhaps there was life as well; could life, in some form, remain in the martian soil today? Testing this possibility, however unlikely, was one of the primary objectives of the Viking landers. These landers carried, in addition to the instruments already discussed, miniature biological laboratories to test for microorganisms in the martian soil. They looked for evidence of *respiration* by living animals, *absorption of nutrients* offered to organisms that might be present, and an *exchange of gases* between the soil and its surroundings for any reason whatsoever. In various tests, martian soil was scooped up by the spacecraft's long arm and placed into the experimental chambers, where it was isolated and incubated in contact with a variety of gases, radioactive isotopes, and nutrients to see what would happen. A fourth instrument pulverized the soil and analyzed it carefully to determine what organic (carbon-bearing) material it contained.

The Viking experiments were so sensitive that, had one of the spacecraft landed anywhere on Earth (with the possible exception of Antarctica), it would easily have detected life. But, to the disappointment of many scientists and members of the public, no life was detected on Mars. The soil tests for absorption of nutrients and gas exchange did show some activity, but this was most likely caused by chemical reactions that began as water was

Mars missions is to search for evidence of past microbial life on Mars. Conclusive evidence of extraterrestrial life would be one of the great discoveries of science and incidentally might well lead to increased funding for NASA. The idea that NASA or other government agencies would (or could) mount a conspiracy to suppress such welcome evidence is truly bizarre.

Alas, the "Face on Mars" story is only one example of a whole series of conspiracy theories that are kept before the public by dedicated believers, by people out to make a fast buck, and by irresponsible media attention. Others include the "urban legend" that the Air Force has the bodies of extraterrestrials at a secret base, the widely circulated report that UFOs crashed near Roswell, New Mexico (actually it was a balloon carrying an early nuclear test detector), or the notion that alien astronauts helped build the Egyptian pyramids and many other ancient monuments because our ancestors were too stupid to do it alone.

In response to the increase in publicity given to these "fiction science" ideas, a group of scientists, educators, scholars, and magicians (who know a good hoax when they see one) have formed the Committee for the Scientific Investigation of Claims of the Paranormal (CSICOP). For more information about their work delving into the rational explanations for paranormal claims, see their excellent magazine, *The Skeptical Inquirer*, or check out their Web site at www.csicop.org/.

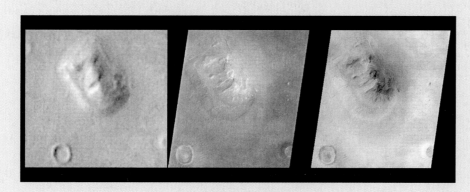

The "Face on Mars" is seen here in low resolution from Viking (left) and with ten times better resolution from Global Surveyor: The middle image is the direct view photographed by Surveyor. At the right, the same image has been processed to simulate the lighting conditions of the Viking image for easier comparison. *(NASA/Malin Space Science Systems)*

added to the soil, and had nothing to do with life. In fact, these experiments showed that martian soil seems much more chemically active than terrestrial soils because of its exposure to solar ultraviolet radiation (since Mars has no ozone layer).

The organic chemistry experiment showed no trace of organic material, which is apparently destroyed by the sterilizing effect of this ultraviolet light. While the possibility of life on the surface has not been eliminated, most experts consider it negligible. Although Mars has the most Earth-like environment of any planet in the solar system, the sad fact is nobody seems to be home today.

However, there is no reason to think that life could not have begun on Mars about 4 billion years ago, at the same time it started on Earth. The two planets had very similar surface conditions then. Thus the attention of scientists has shifted to the search for *fossil* life on Mars. One of the primary questions to be addressed by future spacecraft is whether Mars once supported its own life-forms, and, if so, how this martian life compared with that on our own planet. In 1996, several scientists reported that one of the martian meteorites found in the Antarctic contained evidence of fossil microorganisms; but further examination of this evidence has left most scientists unconvinced. Future sample return missions will focus on returning sedimentary rocks from sites (such as ancient lake beds or hot springs) that once held water and thus perhaps ancient life.

9.6 DIVERGENT PLANETARY EVOLUTION

Venus, Mars, and our own planet Earth form a remarkably diverse triad of worlds. Although all three orbit at about the same distance from the Sun and all apparently started with about the same chemical mix of silicates and metals, their evolutionary paths have diverged. As a result, Venus became hot and dry, Mars became cold and dry, and only the Earth ended up with what we consider a hospitable climate.

We have discussed the runaway greenhouse effect on Venus and the runaway refrigerator effect on Mars. But we do not understand exactly what started these two planets down these evolutionary paths. Was the Earth ever in danger of a similar fate? Or might it still be diverted onto one of these paths, perhaps due to stress on the atmosphere generated by human pollutants? One of the reasons for studying Venus and Mars is to seek insight into these questions.

Some people have even suggested that if we understood the evolution of Mars and Venus better, we could possibly reverse their evolution and restore more Earth-like environments. This process, which is highly speculative, is called *terraforming*. While it seems unlikely that humans could ever make either Mars or Venus into a replica of the Earth, considering such possibilities is a useful part of our more general quest to understand the delicate environmental balance that distinguishes our planet from its two neighbors.

In Chapter 13 we will return to the comparative study of the terrestrial planets and their divergent evolutionary histories.

SURFING THE WEB

A. Venus

🖥 General Venus sites:

- The Nine Planets Site: [seds.lpl.arizona.edu/nineplanets/nineplanets/venus.html]
- Views of the Solar System Site: [www.hawastsoc.org/solar/eng/venus.htm]
- Planetary Sciences Site: [nssdc.gsfc.nasa.gov/planetary/planets/venuspage.html]

🖥 Magellan Mission to Venus [www.jpl.nasa.gov/magellan/]
For four years, the Magellan radar mapper gave us unprecedented amounts of information about the surface of our sister planet. Here you can find the images, the news releases, and mission background, as well as technical information and links.

🖥 Venus Hypermap [www.ess.ucla.edu/hypermap/Vmap/top.html]
You can wander around on a map constructed from Magellan radar images, zooming in on specific regions of interest.

B. Mars

🖥 General Mars sites:

- The Nine Planets Site: [seds.lpl.arizona.edu/nineplanets/nineplanets/mars.html]
- Views of the Solar System Site: [www.hawastsoc.org/solar/eng/mars.htm]
- Planetary Sciences Site: [nssdc.gsfc.nasa.gov/planetary/planets/marspage.html]

🖥 Center for Mars Exploration Home Page [cmex-www.arc.nasa.gov/]
This rich site, developed by a research institute at NASA's Ames Research Center, is a fine place to begin your Web exploration of the red planet, with background information, maps and images, Mars mission summaries, educational resources, and lots of links.

🖥 Mars Pathfinder Mission Home Page [www.jpl.nasa.gov/mpfmir/default.html]
Images, mission overview, updates, and educational activities from the mission. (See also the Virtual Reality Animations and Models from Pathfinder at: mpfwww.arc.nasa.gov/vrml/vrml.html)

🖥 Mars Global Surveyor Home Page [www.jpl.nasa.gov/mgs/]
News, images, mission overview for orbiting spacecraft that is mapping Mars with far greater resolution than we have done before.

🖥 Mars Explorer for Armchair Astronauts [www-pdsimage.wr.usgs.gov/PDS/public/mapmaker]
Browse through a map of Mars made by the cartographers at the U.S. Geological Survey, zoom in on an area of your choice, and let the mapmaker program create a detailed map for you.

🖥 On the Question of the Mars Meteorite [cass.jsc.nasa.gov/lpi/meteorites/mars_meteorite.html]
A site dedicated to following the controversy about whether one of the meteorites from Mars has evidence of fossil life in it. You can follow links from this site to many others.

🖥 Mars Climate Orbiter and Mars Polar Lander Missions [mars.jpl.nasa.gov/msp98/index.html]
All about the NASA missions that arrive at Mars in 1999, including plans, updates, simulations, etc.

SUMMARY

9.1 Venus, the nearest planet, is a great disappointment through the telescope, due to its impenetrable cloud cover. Mars is more tantalizing, with dark markings and polar caps. Early in the 20th century it was widely believed that the "canals" of Mars indicated intelligent life there. Mars has only 11 percent the mass of the Earth, but Venus is nearly our twin in size and mass. Mars rotates in 24 hours and has seasons like the Earth; Venus has a retrograde rotation period of 243 days. Both planets have been extensively explored by spacecraft.

9.2 Venus has been mapped by radar, especially with the Magellan spacecraft; its crust consists of 75 percent lowland lava plains, numerous volcanic features, and many large coronae, which are the expression of subsurface volcanism. The planet has been modified by widespread **tectonics** driven by mantle convection, forming complex patterns of ridges and cracks and building high continental regions such as Ishtar. The surface is extraordinarily inhospitable, with pressure of 90 bars and temperature of 730 K, but several Russian Venera landers investigated it successfully.

9.3 The atmosphere of Venus is 96 percent CO_2. Thick clouds at altitudes of 30 to 60 km are made of sulfuric acid, and a CO_2 greenhouse effect maintains the high surface temperature. Venus presumably reached its current state from more Earth-like initial conditions as a result of a runaway greenhouse effect, which included the loss of large quantities of water.

9.4 Much of what we know about Mars is derived from six spacecraft: two Viking orbiters and the Mars Global Surveyor in orbit, and two Viking landers and the Mars Pathfinder operating on the surface. Mars has heavily cratered highlands in its southern hemisphere but younger, lower volcanic plains over much of its northern half. The Tharsis bulge, as big as North America, includes several huge volcanoes; Olympus Mons is 25 km high and 500 km in diameter. The Valles Marineris canyons are tectonic features widened by erosion. The Viking and Pathfinder landers revealed barren, windswept plains at Chryse and Utopia. Currently there is great interest in martian meteorites, which are samples of the martian crust and may preserve information from the time that Mars was warmer and wetter and could even have supported life.

9.5 The martian atmosphere has a surface pressure of less than 0.01 bar and is 95 percent CO_2. It has dust clouds, water clouds, and carbon dioxide (dry ice) clouds. Liquid water is not possible today, but there may be subsurface permafrost. Seasonal polar caps are made of dry ice, but the northern residual cap is water ice. Evidence of a very different climate in the past is found in water erosion features, both runoff channels and outflow channels, the latter carved by catastrophic floods. The Viking landers searched for martian life in 1976, with negative results, but life might have flourished long ago. The search for fossil life is one of the prime motivations for future sample return missions.

9.6 Earth, Venus, and Mars have diverged in their evolution from what may have been similar beginnings. We need to understand why if we are to protect the environment of the Earth.

INTER-ACTIVITY

A Your group has been asked by high NASA officials to start planning the first human colony on Mars. Begin by making a list of what sorts of things humans would need to bring along to be able to survive for a period of years on the surface of the red planet.

B As a publicity stunt, the mayor of Venus, Texas (there really is such a town), proposes that NASA fund a mission to Venus with humans on board. Clearly, the good mayor neglected to take an astronomy course in college. Have your group assemble a list of as many reasons as possible why it is unlikely that humans will soon land on the surface of Venus.

C Even if humans would have trouble surviving on the surface of Venus, this does not mean we could not learn a lot more about our "veiled sister planet." Have your group brainstorm a series of missions (pretend cost is no object) that would provide us with more detailed information about the Venus atmosphere, surface, and interior.

D Some time late in the 21st century, when travel to Mars has become somewhat routine, a very wealthy couple ask you to plan a honeymoon tour of Mars that includes the most spectacular sights on the red planet. Constitute your group as the Percival Lowell Memorial Tourist Agency, and come up with a list of not-to-be missed tourist stops on Mars. (If you have time to complete this assignment outside of class and have access to the Web, be sure to check out some of the Mars Web sites as part of your research.)

E Astronomers have been puzzled and annoyed about the extensive media publicity that was given the small group of "true believers" who claimed the "Face on Mars" was not a natural formation (see "Making Connections: Astronomy and Pseudoscience"). Have your group make a list of the reasons why many of the media were so enchanted by this story. What do you think astronomers could or should do to get the skeptical, scientific perspective about such issues before the public?

F Assume your group is a "Long-Term Planetary Environment Planning Committee" at the Environmental Protection Agency during a future administration. Make a list of ways in which human activity is affecting the Earth's climate, and discuss what might be done about it.

I. List several ways that Venus, Earth, and Mars are similar, and several ways that they are different.

2. Although neither Venus nor Mars has oceans today, both are considered to have "continents" somewhat like those on the Earth. Describe the continents on each planet. What fraction of the surface area do they take up on each world?

3. Compare the current atmospheres of Earth, Venus, and Mars in terms of
   a. composition
   b. thickness (and pressure at the surface)
   c. greenhouse effect

4. How might Venus' atmosphere have evolved to its present state through a runaway greenhouse effect?

5. The Magellan mission has taught us the most about Venus, while the Viking missions have been the source of the largest amount of information about Mars. Compare the instruments on each mission, and describe how each helped us understand the planet it surveyed.

6. Describe the current atmosphere on Mars. What evidence suggests that it must have been different in the past?

7. Explain the "runaway refrigerator effect" and the role it may have played in the evolution of Mars.

8. What are the advantages of using radar imaging rather than ordinary cameras to study the topography of Venus? What are the relative advantages of these two approaches to mapping the Earth or Mars?

9. Venus and Earth are nearly the same size and distance from the Sun. What are the main differences in the geology of the two planets? What might be some of the reasons for these differences?

10. Why is there so much more carbon dioxide in the atmosphere of Venus than in that of the Earth? Why so much more than on Mars?

11. If the Viking missions were such a rich source of information about Mars, why have we sent the Pathfinder, Global Surveyor, and other more recent missions to Mars? Make a list of questions about Mars that still puzzle astronomers.

12. Compare Mars with Mercury and the Moon in terms of overall properties. What are the main similarities and differences?

13. Contrast the mountains on Mars and Venus with those on Earth and the Moon.

14. We believe that all of the terrestrial planets had similar histories when it comes to impacts from space. Explain how this idea can be used to date the formation of the martian highlands, the martian basins, and the Tharsis volcanoes. How certain are the ages derived for these features?

15. Is it likely that life ever existed on either Venus or Mars? Justify your answer in each case.

16. Suppose that, decades from now, astronauts are sent to Mars and Venus. In each case, describe what kind of protective gear they would have to carry, and what their chances for survival would be if their spacesuits ruptured.

17. We believe that Venus, Earth, and Mars all started with a significant supply of water. Explain where that water is now for each planet.

18. At its nearest approach, Venus comes within about 40 million km of the Earth. How distant is it at its farthest?

19. If you weigh 150 lb on the surface of the Earth, how much would you weigh on Venus? On Mars?

20. Calculate the relative land areas—that is, the amounts of the surface not covered by liquids—of the Earth, Moon, Venus, and Mars. (*Note:* 70 percent of the Earth is covered with water.)

21. Where is the water on Mars? Try to estimate how much might be present in various forms, such as in the polar caps (using the dimensions given in the text) or in subsurface permafrost (assuming various permafrost thicknesses, from 1 to 10 km, and a concentration of ice in the permafrost of 10 percent by volume).

On Venus

Cooper, H. *The Evening Star: Venus Observed.* 1993, Farrar, Straus, and Giroux. Good primer on Venus exploration, with a focus on the Magellan mission.

Grinspoon, D. "Venus Unveiled" in *Astronomy,* May 1997, p. 44.

Grinspoon, D. *Venus Unveiled.* 1997, Addison-Wesley. Well-written introduction.

Kargel, J. "Rivers of Venus" in *Sky & Telescope,* Aug. 1997, p. 32. On lava channels.

Kasting, J. et al. "How Climate Evolved on the Terrestrial Planets" in *Scientific American,* Feb. 1988. Compares Venus with the other inner planets.

Saunders, S. "Venus: A Hellish Place Next Door" in *Astronomy,* Mar. 1990, p. 18.

Robinson, C. "Magellan Reveals Venus" in *Astronomy,* Feb. 1995, p. 32.

Stofan, E. "The New Face of Venus" in *Sky & Telescope,* Aug. 1993, p. 22.

On Mars

Bell, J. "Mars Pathfinder: Better Science?" in *Sky & Telescope,* July 1998, p. 36. Good review of mission results.

Golombeck, M. "The Mars Pathfinder Mission" in *Scientific American,* July 1998, p. 40. Good review by the mission's chief scientist.

Hoyt, W. *Lowell and Mars.* 1976, U. of Arizona Press. Probably the best study of the canal controversy.

Kargel, J. and Strom, R. "The Ice Ages of Mars" in *Astronomy,* Dec. 1992, p. 40.

Kerr, R. "Requiem for Life on Mars? Support for Microbes [in Meteorites] Fades" in *Science,* vol. 282, p. 1398 (Nov. 20, 1998).

McKay, C. "Did Mars Once Have Oceans?" in *Astronomy,* Sept. 1993, p. 27.

Newcott, W. "Return to Mars" in *National Geographic,* Aug. 1998. Report on the Pathfinder mission with great color images, some in 3D.

Raeburn, P. *Uncovering the Secrets of the Red Planet.* 1998, National Geographic Society. Beautiful coffee-table book with the Pathfinder images and results.

Robinson, M. "Surveying Scars of Ancient Martian Floods" in *Astronomy,* Oct. 1989, p. 38.

Wilford, J. *Mars Beckons.* 1990, Random House. A New York Times reporter looks at Mars past, present, and future.

The Jovian Planets This montage shows the four jovian planets to scale. Each image was recorded by one of the Voyager spacecraft that flew by these giant outer worlds. *(NASA/JPL)*

10 The Giant Planets

"What do we learn about the Earth by studying the planets? Humility."

Andrew Ingersoll in discussing the Voyager missions in 1986

Beyond Mars and the asteroid belt we encounter a new region of the solar system: the realm of the giants. Here planets are much larger, the distances between them are vastly increased, and worlds are accompanied by extensive systems of satellites and rings. From many perspectives the outer solar system is where the action is, and the giant planets are the more important members of the Sun's family. When compared to these outer giants, the little cinders of rock and metal that orbit closer to the Sun can seem like insignificant afterthoughts.

FIGURE 10.1

Jupiter This Hubble Space Telescope image of Jupiter, taken on Feb. 13, 1995, from a distance of 961 million km, shows a remarkable amount of detail in the turbulent atmosphere of the giant planet. The Great Red Spot is visible to the lower right; just below and to the left of it are three white oval storms being swept eastward by prevailing winds while the Red Spot moves westward. The Hubble Space Telescope now permits astronomers to make routine "weather reports" from Jupiter. *(R. Beebe, A. Simon, and NASA)*

10.1 EXPLORING THE OUTER PLANETS

Five planets—Jupiter, Saturn, Uranus, Neptune, and Pluto—are found in the outer solar system. The first four of these are the giant or jovian planets, the subject of this chapter. In contrast, Pluto is a small world, by far the smallest in the planetary system. Pluto is physically similar to the satellites of the giant planets, and is therefore discussed with them in Chapter 11.

Composition and Chemistry

As we saw in Chapter 6, most of the mass in the planetary system is in the outer solar system. Jupiter alone exceeds the mass of all the other planets combined (Figure 10.1). The chemistry of the outer solar system is also different, with hydrogen rather than oxygen dominating. When the planets and the Sun were forming, it was cooler in the parts of the solar nebula farther from the Sun. This allowed water ice and other volatile compounds to condense, whereas these materials remained as gas (and eventually dissipated) in the inner solar system.

As a result, the outer planets formed with a significantly greater mass of solid material. This meant that the core bodies for the giant planets had more gravity to capture the gases hydrogen and helium, which were abundant in the solar nebula. Hence, the outer planets were able to grow significantly bigger than those closer to the Sun. Table 10.1 lists the main materials that were available in the outer solar nebula to form planets, based on the elements we observe making up the Sun.

With so much hydrogen available, the chemistry of the outer solar system became *reducing* (see Chapter 6). Most of the oxygen combined with hydrogen to make H_2O and was thus unavailable to form oxidized compounds with other elements. As a result, the compounds detected in the atmospheres of the giant planets are hydrogen-based gases such as methane (CH_4) and ammonia (NH_3), or more complex hydrocarbons (combinations of hydrogen and carbon) such as ethane (C_2H_6) and acetylene (C_2H_2).

Exploration of the Outer Solar System

Six spacecraft, five from the United States and one from Europe, have penetrated beyond the asteroid belt, and a seventh is on the way. The challenges of probing so far away from Earth are considerable. Flight times to the outer planets are measured in years to decades rather than the few months required to reach Venus or Mars. Spacecraft must be highly reliable and capable of a fair degree of independence and autonomy. This is because, even at the speed of light, messages take several hours to pass between Earth and the spacecraft. If a problem develops near Saturn, for example, the spacecraft computer must deal with it directly. To wait hours for the alarm to reach Earth and instructions to be routed back to the spacecraft could spell disaster.

These spacecraft also must carry their own electrical energy sources, since sunlight is too weak to supply energy through solar cells. Heaters are required to keep instruments at proper operating temperatures, and spacecraft must have powerful radio transmitters and large antennas if their precious data are to be transmitted to receivers on Earth a billion kilometers or more away. Table 10.2 sum-

TABLE 10.1
Abundances in the Outer Solar Nebula

Material	Percent (by mass)
Hydrogen (H_2)	75
Helium (He)	24
Water (H_2O)	0.6
Methane (CH_4)	0.4
Ammonia (NH_3)	0.1
Rock and metal	0.3

TABLE 10.2
Missions to the Outer Solar System

Planet	Spacecraft	Encounter Date
Jupiter	Pioneer 10	Dec. 1973
	Pioneer 11	Dec. 1974
	Voyager 1	Mar. 1979
	Voyager 2	Jul. 1979
	Ulysses*	Dec. 1991
	Galileo	Dec. 1995 (orbiter)
	Cassini	Dec. 2002
Saturn	Pioneer 11	Sept. 1979
	Voyager 1	Nov. 1980
	Voyager 2	Aug. 1981
	Cassini	Jul. 2004 (orbiter)
Uranus	Voyager 2	Jan. 1986
Neptune	Voyager 2	Aug. 1989

* The Ulysses spacecraft was designed to study the polar regions of the Sun. To get away from the ecliptic, it needed a gravity boost from flying by Jupiter.

marizes the encounter dates for the spacecraft missions to the outer solar system.

The first spacecraft to the outer solar system were Pioneers 10 and 11, launched in 1972 and 1973 as pathfinders to Jupiter. One of their main objectives was simply to determine whether a spacecraft could actually navigate through the belt of asteroids that lies beyond Mars without colliding with small particles in the belt. Another was to measure the radiation hazards in the enormous *magnetosphere* (or zone of magnetic influence) of Jupiter. Both spacecraft passed through the asteroids without incident, but the energetic particles in Jupiter's magnetic field nearly wiped out their electronics, providing information necessary for the safe design of subsequent missions.

Pioneer 10 flew past Jupiter in 1973, after which it sped outward toward the limits of the solar system. Pioneer 11 undertook a more ambitious program, using the gravity of Jupiter during its 1974 encounter to divert it toward Saturn, which it reached in 1979. The two small craft eventually became the first human-built objects to leave the realm of the planets.

The most productive scientific missions to the outer solar system were Voyagers 1 and 2, launched in 1977 (Figure 10.2). The Voyagers each carried 11 scientific instruments, including cameras and spectrometers, as well as devices to measure the characteristics of planetary magnetospheres. Voyager 1 reached Jupiter in 1979 and used a gravity assist from that planet to take it on to Saturn in 1980 (see Figure 2.12). Voyager 2 arrived at Jupiter four months later and then followed a different path to complete a grand tour of the outer planets, reaching Saturn in 1981, Uranus in 1986, and Neptune in 1989. Much of the information in this chapter and in Chapter 11 is derived from the Voyager missions.

FIGURE 10.2
The Voyager Spacecraft The camera (at bottom) is on an arm that swivels. To the left of the white main antenna for sending data back to Earth, you can see a gold disk; this is the cover for an audiovisual recording of the sights and sounds of Earth, included for the remote possibility that some other civilization might find the capsule someday. The entire spacecraft weighs about a ton on Earth. *(NASA/JPL)*

The productive trajectory of the Voyagers was made possible by the approximate alignment of the four giant planets on the same side of the Sun. About once every 175 years, these planets are in a position that allows a single spacecraft to visit them all by using gravity-assisted flybys to adjust its course for each next encounter. We are fortunate that this opportunity was seized. Because of the alignment, every planet in the outer solar system except Pluto has been visited by spacecraft; without it, our basic reconnaissance of the planetary system would probably have waited until well into the 21st century.

All of these missions were flybys of the giant planets: One quick look and the spacecraft moved on. For a more detailed study, we must send spacecraft that can go into orbit around a planet: This is the mission of Galileo, launched to Jupiter in 1989, and Cassini, sent to Saturn in 1997.

Galileo and Cassini

The Galileo spacecraft arrived at Jupiter in December 1995, beginning its investigations by deploying a small

FIGURE 10.3
Galileo Probe Fall into Jupiter This artist's impression shows the Galileo probe descending into the clouds on its parachute just after the protective heat shield separated. The probe made its measurements of Jupiter's atmosphere on December 7, 1995. *(NASA/ARC)*

survived, descending about 200 km while being carried another 500 km sideways by the strong jovian winds.

The transmitter switched off due to increasing temperature as the probe reached hotter layers of the jovian atmosphere. All of these events were controlled by the computers on the probe and orbiter spacecraft, without any intervention from controllers on the Earth. A few minutes later the polyester parachute melted, and within a few hours the main aluminum and titanium structure of the probe vaporized to become a part of Jupiter itself. About two hours after the receipt of the final probe data, Galileo fired its retro-rockets to slow the spacecraft so it could be captured into orbit around the planet, where its primary objectives were to study Jupiter's large and often puzzling moons (see Chapter 11).

The Cassini mission to Saturn (Figure 10.4), a cooperative venture between NASA and the European Space Agency, is similar to Galileo in its twofold approach. Arriving in 2004, Cassini will also deploy an entry probe before beginning an orbital tour of Saturn and its satellites. But in this case, the probe will not explore Saturn itself, but its intriguing moon Titan, whose surface is veiled by a thick, opaque atmosphere (see Chapter 11).

10.2 THE JOVIAN PLANETS

Let us now examine the four jovian planets in more detail. Our approach is not just to catalog their characteristics, but to compare them with each other, noting their similarities

entry probe into the planet for the first direct studies of its gaseous outer layers. Like a flaming meteor, the 339-kg probe plunged at a shallow angle into the jovian atmosphere, traveling at a speed of 50 km/s (more than 160,000 km/h—fast enough to fly from New York to San Francisco in 100 seconds).

Atmospheric friction slowed the probe within two minutes, reaching a maximum deceleration of nearly 230 *g* and producing temperatures at the front of its heat shield as high as 15,000°C. This high speed was a result of acceleration by the strong gravitational attraction of Jupiter, which presents the most difficult spacecraft entry condition anywhere in the solar system (except for the Sun itself). As the probe speed dropped to 2500 km/h, the remains of the glowing heat shield were jettisoned, and a parachute was deployed to lower the instrumented probe spacecraft more gently into the atmosphere (Figure 10.3).

While the probe sank into the clouds of Jupiter, the Galileo main spacecraft was passing overhead at about half the Earth–Moon distance. The data from the probe instruments were sent to the orbiter over a tiny 23-watt radio transmitter, to be recorded and later relayed to the Earth. The probe made its measurements during the 57 min it

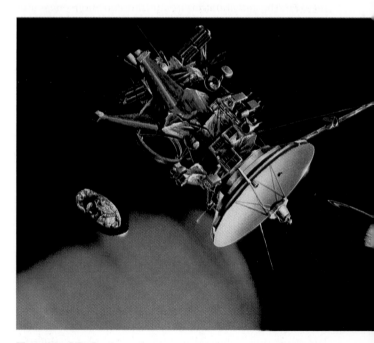

FIGURE 10.4
Cassini Mission Drops Its Probe In this artist's conception, the Huygens probe drops from the Cassini orbiter into the hazy clouds of Saturn's moon Titan. *(NASA/JPL)*

Engineering and Space Science: Teaching an Old Spacecraft New Tricks

By the time Voyager 2 arrived at Neptune in 1989, 12 years after its launch, the spacecraft was beginning to show almost human signs of old age. The arm on which the camera and other instruments were located was "arthritic"—it could no longer move easily in all directions. The communications system was "hard of hearing"—part of its radio receiver was not operating anymore. The "brains" had significant "memory loss"—some of the computer memory on board had failed. And the whole spacecraft was beginning to run out of energy—its radioactive generators had begun showing serious signs of wear.

To make things even more of a challenge, the craft's mission at Neptune was in many ways the most difficult of all four flybys. For example, since sunlight at Neptune is 900 times weaker than at Earth, the on-board camera had to take longer exposures in this light-starved environment. This was a nontrivial requirement, given that the spacecraft was shooting by Neptune at ten times the speed of a rifle bullet, or 42,000 mi/h (at this speed, it could have gone around the Earth's equator in only 36 min).

The solution was to swivel the camera backward at a rate that would exactly compensate for the forward motion of the spacecraft. But since the arm on which the camera was mounted could not move with the ease and grace it once had (and because instructions from Earth took about 4 hours to reach Neptune), engineers had to preprogram the ship's computer to execute an incredibly complex series of maneuvers for each image. The beautiful illustrations you see in this chapter are a testament to their ingenuity.

Another problem that confronted the mission scientists was the sheer distance of the craft from its controllers on Earth. Voyager 2 communicates its results (including a pattern of dark and light dots that make up each image) by means of its on-board radio transmitter. By the time the radio beam arrived at the Earth in 1989, it had traveled (and spread out over) about 4.8 billion km. The power that reached us was approximately 10^{-16} W, or 20 billion times less power than it takes to operate a digital watch. Thirty-eight different antennas on four continents were used by NASA to collect the faint signals from the spacecraft and decode the precious information they contained.

and differences, and attempting to relate their properties to their differing masses and distances from the Sun.

Basic Characteristics

The giant planets are much farther from the Sun than we are. The median distance of Jupiter from the Sun is 778 million km, 5.2 times the Earth's distance (or 5.2 AU); thus it takes just under 12 years to circle the Sun. Saturn is about twice as far away as Jupiter and takes very nearly the standard human generation of 30 years to complete one revolution.

As we saw in Chapter 2, Uranus and Neptune were not discovered until larger telescopes came into regular use. Uranus orbits at 19 AU with a period of 84 years, while Neptune at 30 AU requires 165 years for each circuit of the Sun. Not until 2010 will a full Neptune "year" have passed since its discovery in 1845.

Jupiter and Saturn have many similarities in composition and internal structure, although Jupiter is nearly four times more massive. In contrast, Uranus and Neptune are smaller worlds that differ in composition and structure from their larger siblings. Some of the main properties of these four planets are summarized in Table 10.3.

TABLE 10.3
Basic Properties of the Jovian Planets

Planet	Distance (AU)	Period (years)	Diameter (km)	Mass (Earth = 1)	Density (g/cm³)	Rotation (hours)
Jupiter	5.2	11.9	142,800	318	1.3	9.9
Saturn	9.5	29.5	120,540	95	0.7	10.7
Uranus	19.2	84.1	51,200	14	1.2	17.2
Neptune	30.1	164.8	49,500	17	1.6	16.1

Jupiter, the giant among giants, has enough mass to make 318 Earths. Its diameter is about 11 times the Earth's (and about one tenth of the Sun's). Jupiter's average density is 1.3 g/cm^3, much lower than that of any of the terrestrial planets. (Recall that water has a density of 1 g/cm^3.) Jupiter's material is spread out over a volume so large that more than 1400 Earths could fit within it.

Saturn's mass is 95 times that of the Earth, and its average density is only 0.7 g/cm^3—the lowest of any planet. Since this is less than the density of water, Saturn would be light enough to float if a bathtub large enough to contain it existed (but imagine the ring it would leave)!

Uranus and Neptune each have a mass about 15 times that of the Earth and hence only 5 percent as great as Jupiter. Their densities of 1.2 g/cm^3 and 1.6 g/cm^3, respectively, are much higher than that of Saturn, in spite of their smaller mass and weaker gravity. This tells us that their composition must be fundamentally different, consisting for the most part of heavier materials than hydrogen and helium, the primary constituents of Jupiter and Saturn. We will return to this question of their composition shortly.

Appearance and Rotation

When we look at the giant planets, we see only their atmospheres, composed primarily of hydrogen and helium gas (see the image that opens this chapter). The uppermost cloud deck of Jupiter and Saturn, the part we see when looking down at these planets from above, is composed of ammonia crystals. On Neptune the upper cloud deck is made of methane. On Uranus we see no obvious cloud deck at all, but only a deep and featureless haze.

Seen through a telescope, Jupiter is a colorful and dynamic planet. Distinct details in its cloud patterns allow us to determine the rotation rate of its atmosphere at the cloud level, although such atmospheric rotation may have little to do with the spin of the underlying planet. Much more fundamental is the rotation of the mantle and core; these can be determined by periodic variations in radio waves coming from Jupiter, which are controlled by its magnetic field. Since the magnetic field originates deep inside the planet, it shares the rotation of the interior. This rotation period of 9 h 56 m gives Jupiter the shortest "day"

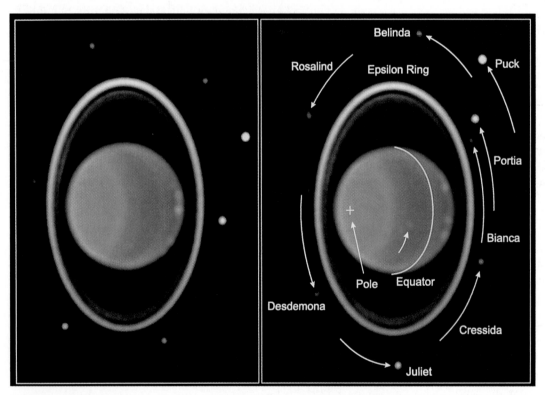

FIGURE 10.5

Infrared Image of Uranus The new infrared camera on the Hubble Space Telescope took this false-color image of the planet Uranus, its ring system, and satellites on July 28, 1997. The south pole of the planet (marked with a + sign on the right image) faces the Sun; its green color shows a strong local haze. The two images were taken 90 min apart, and during that time the five reddish clouds can be seen to rotate around the parallel to the equator. The rings (which are very faint in visible light, but prominent in infrared) and eight satellites can be seen around the equator. This was the "bull's-eye" arrangement that Voyager saw as it approached Uranus in 1986. *(Erich Karkoschka, U. of Arizona and NASA)*

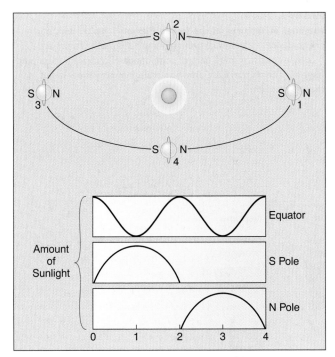

FIGURE 10.6

The Strange Seasons on Uranus (a) The orbit of Uranus as seen from above. At the time Voyager 2 arrived (position 1), the south pole was facing the Sun. As we move counterclockwise in the diagram, we see the planet 21 years later at each step. (b) The amount of sunlight seen at the poles and the equator of Uranus over the course of its 84-year revolution around the Sun.

of any planet. In the same way, we can measure that the underlying rotation period of Saturn is 10 h 40 m. Uranus and Neptune have slightly longer rotation periods of about 17 hours, also determined from the rotation of their magnetic fields.

Remember from Chapters 3 and 9 that the Earth and Mars have seasons because their axes of spin, instead of "standing up straight," are tilted relative to the Sun's orbital plane. This means that, as the Earth revolves around the Sun, sometimes one hemisphere and sometimes the other "leans into" the Sun.

What are the seasons like for the giant planets? The axis of rotation of Jupiter is tilted by only 3°, so there are no seasons to speak of. Saturn, however, does have seasons, since its axis of rotation is inclined at 27° to the perpendicular to its orbit. Neptune has about the same tilt as Saturn (29°); therefore, it experiences similar seasons (only more slowly). The strangest seasons of all are on Uranus, which has an axis of rotation tilted by 98° with respect to the north direction. Practically speaking, we can say that Uranus orbits on its side, and its ring and satellite system follows the same pattern, orbiting about Uranus' equator. This meant that when Voyager 2 approached Uranus in the plane of the planets' orbits, it was aimed like a well-thrown dart into the giant bull's-eye of the Uranus system (Figure 10.5).

We don't know what caused Uranus to be tipped over like this, but one possibility is a collision with a large planetary body when our system was first forming. Whatever the cause, this unusual tilt creates dramatic seasons. When Voyager 2 arrived at Uranus, its south pole was facing directly into the Sun. The southern hemisphere was having a roughly 21-year sunlit summer, while during that same period the northern hemisphere was plunged into darkness. For the next 21-year season (Figure 10.6), the Sun shines on Uranus' equator, and both hemispheres go through cycles of light and dark. Then there are 21 years of an illuminated northern hemisphere and a dark southern hemisphere. If you were to install a floating platform at the south pole of Uranus, for example, it would experience 42 years of light and 42 years of darkness. Any future astronauts crazy enough to set up camp there could spend most of their lives without ever seeing the Sun.

Composition and Structure

Although we cannot see into these planets, astronomers are confident that the interiors of Jupiter and Saturn are composed primarily of hydrogen and helium. Of course, these gases have been measured only in their atmospheres, but calculations first carried out 50 years ago showed that these two light gases are the only possible materials out of which a planet with the observed masses and densities of Jupiter and Saturn could be constructed.

Having said that much, we should add that the precise internal structure of the two planets is not so easy to predict. This is mainly because the planets are so big that the hydrogen and helium in their centers becomes tremendously compressed and behaves in ways that these gases can never behave on Earth. The best theoretical models we have of Jupiter's structure predict a central pressure of over 100 million bars and a central density of about 31 g/cm^3. (By contrast, the Earth's core has a central pressure of 4 million bars and a central density of 17 g/cm^3.)

At the pressures inside the giant planets, familiar materials can take on strange forms. A few thousand kilometers below Jupiter and Saturn's visible clouds, pressures become so large that hydrogen changes from a gaseous to a liquid state. Still deeper, this liquid hydrogen is further compressed and begins to act like a metal, something it never does on Earth. (In a metal, electrons are not firmly attached to their parent nuclei, but can wander around. This is why metals are such good conductors of electricity.) On Jupiter, the greater part of the interior is liquid metallic hydrogen.

Because Saturn is less massive, it has only a small volume of metallic hydrogen, but most of its interior is liquid. Uranus and Neptune are probably too small to reach internal pressures sufficient to liquify hydrogen. We will return to the question of the metallic hydrogen layers when we examine the magnetic fields of the giant planets.

Each of these planets has a core composed of heavier materials, as demonstrated by detailed analyses of their

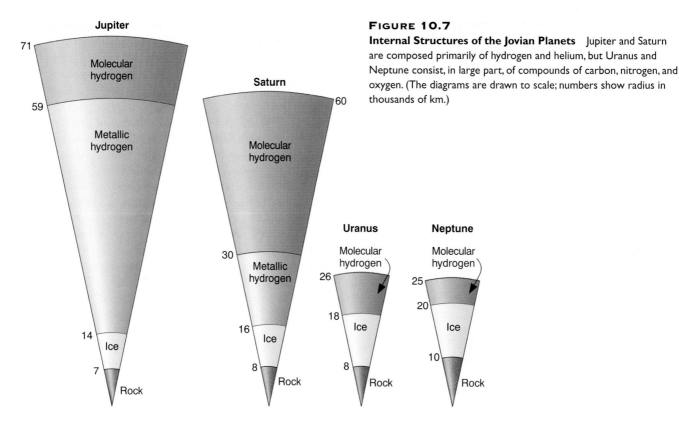

FIGURE 10.7

Internal Structures of the Jovian Planets Jupiter and Saturn are composed primarily of hydrogen and helium, but Uranus and Neptune consist, in large part, of compounds of carbon, nitrogen, and oxygen. (The diagrams are drawn to scale; numbers show radius in thousands of km.)

gravitational fields. Presumably these cores are the original rock-and-ice bodies that formed before the capture of gas from the surrounding nebula (see Chapters 6 and 13). The cores exist at pressures of tens of millions of bars. While scientists speak of the giant planet cores being composed of rock and ice, we can be sure that neither rock nor ice assumes any familiar forms at such pressures and the accompanying high temperatures. What is really meant by *rock* is any material made up primarily of iron, silicon, and oxygen. The term *ice* denotes materials composed primarily of the elements carbon, nitrogen, and oxygen in combination with hydrogen.

Figure 10.7 illustrates the interior structures of the four jovian planets. It appears that all four have similar cores of rock and ice. On Jupiter and Saturn, the cores constitute only a few percent of the total mass, consistent with the initial composition of raw materials shown in Table 10.1. However, most of the mass of Uranus and Neptune resides in these cores, demonstrating that the two planets were unable to attract massive quantities of hydrogen and helium when they were first forming.

Internal Heat Sources

Because of their large sizes, all of the giant planets were strongly heated during their formation by the collapse of surrounding material onto their cores. Jupiter, being the largest, was the hottest. Some of this primordial heat can still remain inside such large planets. In addition, it is possible for giant, largely gaseous planets to generate heat after formation by slowly contracting. (With so large a mass, even a minuscule amount of shrinking can generate significant heat.) The effect of these internal energy sources is to raise the temperatures in the interiors and atmospheres of the planets higher than we would expect from the heating effect of the Sun alone.

Jupiter has the largest internal energy source, amounting to 4×10^{17} W (i.e., it is glowing with the equivalent of 4 million billion 100-W light bulbs). This is about the same as the total solar energy absorbed by Jupiter. The atmosphere of Jupiter is therefore something of a cross between a normal planetary atmosphere (like the Earth's), which obtains most of its energy from the Sun, and the atmosphere of a star, which is entirely heated by an internal energy source. Most of the internal energy of Jupiter is primordial heat, left over from the formation of the planet 4.5 billion years ago.

Saturn has an internal energy source about half as large as that of Jupiter, which means (since its mass is only about one quarter as great) that it is producing twice as much energy per kilogram of material as does Jupiter. Since Saturn is expected to have much less primordial heat, there must be another source at work generating most of this 2×10^{17} W of power. This source is believed to be the separation of helium from hydrogen in Saturn's interior. In the liquid hydrogen mantle, the heavier helium forms drops that sink toward the core, releasing gravitational energy. In effect, Saturn is still differentiating—letting lighter material rise and heavier material fall.

Uranus and Neptune are different. Neptune has a small internal energy source, while Uranus does not emit a measurable amount of internal heat. As a result, these two planets have almost the same temperature, in spite of Neptune's greater distance from the Sun. No one knows why these two planets differ in their internal heat.

Magnetic Fields

Each of the giant planets has a strong magnetic field, generated by electric currents in its rapidly spinning interior. Associated with the magnetic fields are the planets' *magnetospheres,* which are defined as the regions around the planet within which the planet's own magnetic field dominates over the general interplanetary magnetic field. The magnetospheres of these planets are their largest features, extending millions of kilometers into space.

In the late 1950s, astronomers discovered that Jupiter was a source of radio waves that got more intense at longer rather than at shorter wavelengths—just the reverse of what is expected from thermal radiation (radiation caused by the normal vibrations of particles within all matter; see Chapter 4). Such behavior is typical, however, of the radiation emitted when high-speed electrons are accelerated by a magnetic field. We call this **synchrotron radiation** because it was first observed on Earth in particle accelerators, called synchrotrons. This was our first hint that Jupiter must have a strong magnetic field.

Later observations showed that the radio energy originates from a region surrounding Jupiter with a diameter several times that of the planet itself (Figure 10.8). The evidence suggested, therefore, that a vast number of charged atomic particles must be circulating around Jupiter, spiraling around the lines of force of a magnetic field associated with the planet. This is just what we observe happening, but on a smaller scale, in the Van Allen belt around the Earth (see Chapter 7). The magnetic fields of Saturn, Uranus and Neptune, discovered by the spacecraft that first passed close to these planets, work in a similar way.

Inside the magnetosphere, charged particles are trapped into spiraling around in the magnetic field; as a result they can be accelerated to high energies. The charged particles come from the Sun or from the neighborhood of the planet itself. In Jupiter's case, Io, one of its moons, turns out to have a constant series of volcanic eruptions that propel charged particles into space.

The axis of Jupiter's magnetic field (the line that connects the magnetic north pole with the magnetic south pole) is not aligned exactly with the axis of rotation of the planet; rather, it is tipped by about 10°. Uranus and Neptune have even greater magnetic tilts of 60° and 55° respectively. Saturn's field, on the other hand, is perfectly aligned with its rotation axis.

The physical processes around the jovian planets turn out to be milder versions of what astronomers find in many distant objects, from the remnants of dead stars to the puzzling distant powerhouses we call quasars. One reason to study the magnetospheres of the giant planets and the Earth is that they provide nearby accessible analogues of more energetic and challenging cosmic processes.

 ATMOSPHERES OF THE JOVIAN PLANETS

The atmospheres of the jovian planets are the parts we can observe or measure directly. Since these planets have no

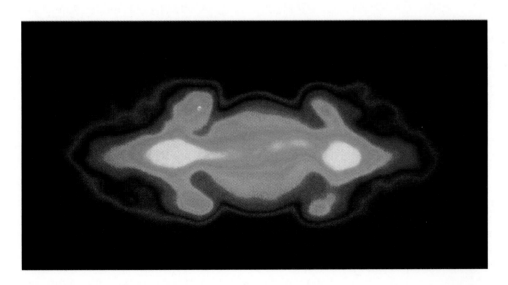

FIGURE 10.8
Jupiter in Radio Waves This false-color image of Jupiter was made with the Very Large Array (of radio telescopes) in New Mexico. We see part of the magnetosphere, brightest in the middle because the largest number of charged particles are in the equatorial zone of Jupiter. The planet itself is slightly smaller than the circular green region in the center. Different colors are used to indicate different intensities of synchrotron radiation. *(Imke dePater/NRAO)*

James Van Allen: Several Planets under His Belt

The career of physicist James Van Allen has spanned the birth and growth of the Space Age; in fact he played a major role in its development. Born in Iowa in 1914, Van Allen received his PhD from the University of Iowa. He then worked for several research institutions and served in the Navy during World War II. In 1951 he was appointed Professor of Physics at the University of Iowa, where he has remained ever since.

After the war Van Allen and his collaborators began using the V2 rockets captured from the Germans, as well as rockets built in the United States, to explore cosmic radiation in the Earth's outer atmosphere. Van Allen designed a technique, nicknamed "the rockoon," in which a small rocket is lifted up by a balloon before being launched, enabling it to reach a higher altitude.

Over dinner one night in 1950, Van Allen and several colleagues came up with the idea of the International Geophysical Year (IGY), an opportunity for scientists around the world to coordinate their investigations of the physics of the Earth, especially research done at high altitudes. Stretched to 18 months, the IGY took place between July 1957 and December 1958. In 1955, the United States and the Soviet Union each committed themselves to launching an Earth-orbiting satellite during IGY, a competition that began what came to be known as the Space Race.

When the Soviet Union won the first lap of the race by launching Sputnik 1 in October 1957, the U.S. government spurred its scientists and engineers to even greater efforts to get something into space and thereby maintain the country's prestige. However, the main U.S. satellite program, called Vanguard, was running into all sorts of difficulties, and each of its early launches crashed or exploded. In the meantime, a second team of rocket engineers and scientists had quietly been working on a military launch vehicle called Jupiter-C. Van Allen spearheaded the design of the instruments aboard the small satellite, called Explorer I, that this rocket would carry. On January 31, 1958, Explorer I became the first U.S. satellite in space.

Unlike Sputnik, Explorer I was equipped to make scientific measurements. Van Allen had placed an instrument aboard Explorer I that could

James Van Allen   *(Photo by D. Hokanson)*

After the war, Van Allen began using V2 rockets captured from the Germans.

detect high-energy charged particles above the atmosphere, and it was this instrument that discovered what we now call the Van Allen belts around the Earth. This was the first scientific discovery of the space program and made Van Allen's name a household word around the world.

Van Allen and his colleagues continued making measurements of the magnetic and particle environment around our planet with more sophisticated spacecraft and instruments. When Pioneers 10 and 11 made the first exploratory surveys of the Jupiter and Saturn environments, they again carried instruments designed by Van Allen. Some scientists therefore refer to the charged-particle zones around those planets as Van Allen belts as well. (Once, when Van Allen was giving a lecture at the University of Arizona, the graduate students in planetary science asked him if he would leave his belt at the school. It is now proudly displayed as the "Van Allen belt.")

Van Allen has continued to be a strong supporter of space science and has been an eloquent senior spokesman for the American scientific community, warning NASA not to put all its efforts into human exploration but to remember how productive and (relatively) inexpensive its robot probes and instruments have always been in helping us learn more about the solar system.

---

solid surfaces, their atmospheres are more representative of their general compositions than is the case with the terrestrial planets. These atmospheres also present us with some of the most dramatic examples of weather patterns in the solar system. As we will see, storms on these planets can grow bigger than the entire planet Earth.

Atmospheric Composition

When sunlight reflects from the atmospheres of the giant planets, the atmospheric gases leave their "fingerprints" in the spectra (as we saw in Chapter 4). Spectroscopic observations of the jovian planets began in the 19th century, but

for a long time astronomers were not able to interpret the spectra they observed. As late as the 1930s, the most prominent features photographed in these spectra remained unidentified. Then, better spectra revealed the presence of methane (CH_4) and ammonia (NH_3) in the atmospheres of Jupiter and Saturn.

At first we thought that methane and ammonia might be the main constituents of these atmospheres, but now we know that hydrogen and helium are actually the dominant gases. The confusion arose because neither hydrogen nor helium possesses easily detected spectral features. It was not until the Voyager spacecraft measured the far-infrared spectra of Jupiter and Saturn that a reliable abundance for the elusive helium could be found. The compositions of the two atmospheres are generally similar, except that on Saturn there is less helium—the result of the precipitation of helium that contributes to Saturn's internal energy source. The most precise measurements of composition were made on Jupiter by the Galileo entry probe in 1995, and as a result, we know the abundances of the elements in the jovian atmosphere even better than those in the Sun.

Clouds and Atmospheric Structure

The clouds of Jupiter are among the most spectacular sights in the solar system, much beloved by makers of science-fiction films. They range in color from white to orange to red to brown, swirling and twisting in a constantly changing kaleidoscope of patterns (Figure 10.9). Saturn shows similar but much more subdued cloud activity: Instead of

vivid colors, its clouds have a nearly uniform butterscotch hue (Figure 10.10).

Different gases freeze at different temperatures. At the temperatures and pressures of the upper atmospheres of Jupiter and Saturn, methane remains a gas, but ammonia can condense, just as water vapor condenses in the Earth's atmosphere, to produce clouds. The primary clouds that we see around these planets, whether from a spacecraft or through a telescope, are composed of frozen ammonia crystals. The ammonia

FIGURE 10.10
Saturn and Its Rings The clouds of Saturn, as photographed by Voyager, are less colorful than those of Jupiter, but the structure and motions of the atmosphere are similar. *(NASA/JPL)*

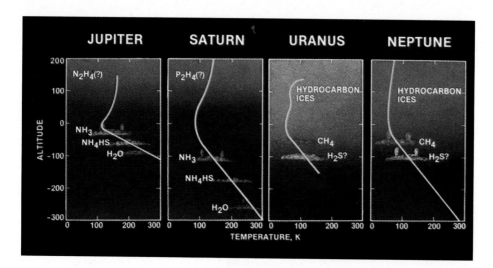

FIGURE 10.11

Atmospheric Structure for the Jovian Planets In each diagram, the yellow line shows how the temperature (see the scale on the bottom) changes with altitude. The location of the main cloud decks on each planet is also shown. *(NASA/JPL)*

cloud deck marks the upper edge of the planets' *tropospheres;* above it lies the cold *stratosphere.* (These layers were defined, for the Earth, in Chapter 7.)

Figure 10.11 is a diagram showing the structure and clouds in the atmospheres of all four jovian planets. On both Jupiter and Saturn, the temperature near the cloud tops is about 140 K (only a little cooler than the polar caps of Mars). On Jupiter this cloud level is at a pressure of about 0.1 bar, but on Saturn it occurs lower in the atmosphere, at about 1 bar. Because the ammonia clouds lie so much deeper on Saturn, they are more difficult to see, and the overall appearance of the planet is much more bland than is Jupiter.

Within the troposphere of these planets, the temperature and pressure both increase with depth. Through breaks in the ammonia clouds, we can see tantalizing glimpses of other cloud layers that can form in these deeper regions of the atmosphere—regions that were sampled directly by the Galileo probe for Jupiter.

As it descended to a pressure of 5 bars, the probe should have passed into a region of frozen water clouds, then below that into clouds of liquid water droplets, perhaps similar to the common clouds of the terrestrial troposphere. At least this is what scientists expected. But the probe saw no water clouds, and it measured a surprisingly low abundance of water vapor in the atmosphere. We now understand that it was just our luck that the probe happened to descend through an unusually dry, cloud-free region of the atmosphere—the most prominent hole in the clouds visible on the entire planet. Andrew Ingersoll of Caltech, who believes that water clouds are common over most of the planet, has called this entry site the "desert" of Jupiter.

The probe continued to make measurements to a pressure of 22 bars, but found no other cloud layers before its instruments stopped working. It also detected lightning storms, but only at great distances, suggesting that the probe itself was in a region of clear weather.

Above the visible ammonia clouds in Jupiter's atmosphere, we find a region that is clear and cold, reaching a minimum temperature near 120 K. At still higher altitudes temperatures rise again, just as they do in the upper atmosphere of the Earth, because here the molecules absorb ultraviolet light from the Sun. This input of energy causes chemical reactions in the air, a process we call **photochemistry.** In Jupiter's upper atmosphere, photochemical reactions create a variety of fairly complex compounds of hydrogen and carbon that form a thin layer of smog far above the visible clouds. We show this smog as a fuzzy orange region in Figure 10.11; however, this thin layer does not block our view of the clouds beneath it.

One puzzle of the main jovian cloud deck that we must mention is the mystery of its colors. The ammonia-condensation clouds identified on the planet should be white, like water clouds on Earth, yet we see beautiful and complex patterns of pastel red, orange, and brown. Some additional chemical or chemicals must be present to create such colors, but we do not know what they are. Various photochemically produced organic compounds have been suggested, as well as sulfur and red phosphorus. But there are no firm identifications, nor any immediate prospects of solving this mystery.

The atmospheric structure on Saturn is similar to that of Jupiter. Temperatures are somewhat colder, however, and the atmosphere is more extended, due to Saturn's lower surface gravity. Thus the layers are stretched out over a longer distance (as you can see in Figure 10.11). Overall though, the same atmospheric regions, condensation clouds, and photochemical reactions that we see on Jupiter should be present on Saturn.

Unlike Jupiter and Saturn, Uranus is almost entirely featureless as seen at wavelengths that range from the ultraviolet to the infrared (see its rather boring image in the opening collage for this chapter). Calculations indicate that the basic atmospheric structure of Uranus should resemble that of Jupiter and Saturn, although its upper clouds (at the

FIGURE 10.12
Neptune's Upper Atmosphere   The planet Neptune is seen here as photographed in 1989 by Voyager 2. The blue color, exaggerated with computer processing, is caused by the scattering of sunlight in the upper atmosphere.   *(NASA/JPL)*

10.13, a remarkable close-up of Neptune's outer layers that could never have been obtained from Earth).

Winds and Weather

Thick atmospheres like the ones on the jovian planets have many regions of high pressure (where there is more air) and low pressure (where there is less). Just as it does on Earth, air flows between these regions, setting up wind patterns that are then distorted by the rotation of the planet. By observing the changing cloud patterns on the jovian planets, we can measure wind speeds and track the circulation of their atmospheres.

The atmospheric motions we see on these planets are fundamentally different from those of the terrestrial planets. For one thing, the giants spin faster and their rapid rotation tends to smear out the circulation of the air into horizontal (east–west) patterns parallel to the equator.

In addition, there is no solid surface below the atmosphere against which the circulation patterns can rub and lose energy (which is how tropical storms on Earth ultimately die out when they come over land). And, as we have seen, on all the giants except Uranus, heat from the inside contributes about as much energy to the atmosphere as sunlight from the outside. This means that deep currents of rising hot air and falling cooler air circulate throughout the atmospheres of the planets in the vertical direction; these are called *convection currents*. All three of these

1-bar pressure level) are composed of methane rather than ammonia. However, the absence of an internal heat source suppresses up-and-down movement and leads to a very stable atmosphere with little visible structure. In addition, the troposphere is hidden from our view by a deep, cold, hazy stratosphere.

Neptune differs dramatically from Uranus in its appearance (Figure 10.12), although the basic atmospheric temperatures are almost identical. The upper clouds are composed of methane, which forms a thin cloud deck near the top of the troposphere at a temperature of 70 K and a pressure of 1.5 bars. Most of the atmosphere above this level is clear and transparent, with less haze than is found on Uranus. The scattering of sunlight by gas molecules lends Neptune a pale blue color similar to that of the Earth's atmosphere. Another cloud layer, perhaps composed of hydrogen sulfide ice particles, exists below the methane clouds at a pressure of 3 bars.

Unlike Uranus, Neptune has an atmosphere in which convection currents emanate from the interior, powered by the planet's internal heat source. These currents carry warm gas above the 1.5-bar cloud level, forming additional clouds at elevations about 75 km higher. These high-altitude clouds can be seen in the Voyager images: They form bright white patterns against the blue planet beneath. They can even cast distinct shadows on the methane cloud tops, permitting their altitudes to be calculated (see Figure

FIGURE 10.13
High Clouds in the Atmosphere of Neptune   These bright narrow cirrus clouds are made of methane ice crystals. From the shadows they cast on the thicker cloud layer below, we can measure that they are about 75 km higher than the main cloud decks.
(NASA/JPL)

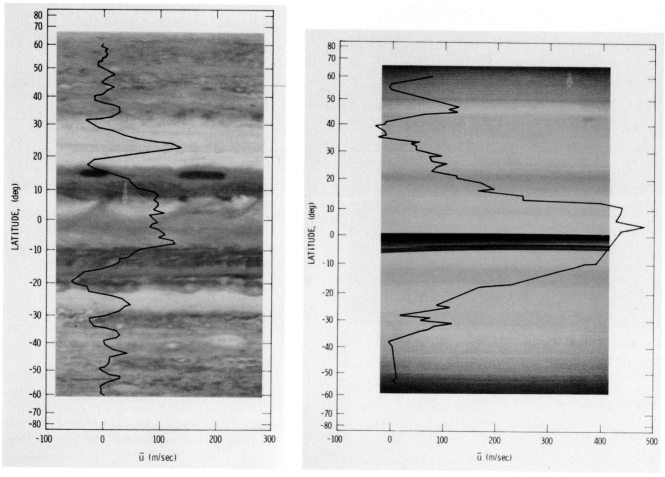

FIGURE 10.14

Winds on Jupiter (left) and Saturn (right) The speed and direction of the east–west winds on Jupiter and Saturn, as measured by Voyager, are shown on these photographic maps. Latitude is shown in the left column of each chart, with the equator being 0°. The dark line shows the wind speed at each latitude (measured relative to the planet's core). Notice that Saturn has much stronger winds at the equator. *(NASA/JPL)*

factors help explain some of the features we see in the Voyager and Hubble Space Telescope images.

The main features of Jupiter's visible clouds (see Figures 10.1 and 10.9), for example, are alternating dark and light bands that stretch around the planet parallel to the equator. These bands are semipermanent features, although they shift in intensity and position from year to year. Consistent with the small tilt of Jupiter's axis, the pattern does not change with the seasons.

More fundamental than these bands are underlying east–west wind patterns in the atmosphere, which do not appear to change at all, even over many decades. These are displayed in Figure 10.14, which shows maps of Jupiter and Saturn made from Voyager images, and indicates how strong the winds are at each latitude. At Jupiter's equator, a jet stream flows eastward with a speed of about 90 m/s (300 km/h), similar to the speed of jet streams in the Earth's upper atmosphere. At higher latitudes there are alternating east- and west-moving streams, with each hemisphere an

almost perfect mirror image of the other. Saturn shows a similar pattern, but with a much stronger equatorial stream flowing at a speed of 1300 km/h.

Generally, the light zones on Jupiter are regions of upwelling air capped by white ammonia cirrus clouds. They apparently represent the tops of upward-moving convection currents. The darker belts are regions where the cooler atmosphere moves downward, completing the convection cycle; they are darker because fewer ammonia clouds mean we can see deeper into the atmosphere, perhaps down to a region of ammonium hydrosulfide (NH_4SH) clouds.

In spite of the strange seasons induced by the 98° tilt of its axis, Uranus' basic circulation is parallel with its equator, as is the case on Jupiter and Saturn. The mass of the atmosphere and its capacity to store heat are so great that the alternating 40-year periods of sunlight and darkness have little effect. In fact, Voyager measurements show that the atmospheric temperatures are a few degrees

FIGURE 10.15
The Great Red Spot   This is the largest storm system on Jupiter, as seen during the Voyager spacecraft flyby, with the Earth superimposed to give a sense of scale. Below and to the right of the Red Spot is one of the white ovals, which are similar but smaller high-pressure features. The colors on the Jupiter image have been exaggerated so astronomers (and astronomy students) can study their differences more effectively. See Figure 10.1 to get a better sense of the colors your eye would actually see near Jupiter.   (S. Meszaros and the Astronomical Society of the Pacific; JPL/NASA)

higher on the dark winter side than on the hemisphere facing the Sun. Here is another indication that the behavior of such giant planet atmospheres is a complex problem: We simply do not understand the seasonal effects on Uranus in detail.

Neptune's weather is characterized by strong east–west winds generally similar to those observed on Jupiter and Saturn. The highest wind speeds near its equator reach 2100 km/h (600 m/s), nearly twice as fast as the peak winds on Saturn. The Neptune equatorial jet stream actually approaches supersonic speeds (speeds higher than the speed of sound in Neptune's air).

Storms

Superimposed on the regular atmospheric circulation patterns we have just described are many local disturbances—weather systems or *storms*, to borrow the term we use on Earth. The most prominent of these are large oval-shaped high-pressure regions on both Jupiter and Neptune.

The largest and most famous "storm" on Jupiter is the Great Red Spot, a reddish oval in the southern hemisphere that was almost 30,000 km long when the Voyagers flew by—big enough to hold more than two Earths side by side (Figure 10.15). First seen 300 years ago, the Red Spot has changed in size but has not disappeared; it is clearly much longer-lived than storms in our own atmosphere. It also differs from terrestrial storms in being a high-pressure region; on our planet such storms are regions of lower pressure. The Red Spot's counterclockwise rotation has a period of six days. Three similar but smaller disturbances (about as

big as the Earth) formed on Jupiter in the 1930s. They look like white ovals and can clearly be seen below and to the left of the Great Red Spot in Figure 10.1. In 1998 the *Galileo* spacecraft watched as two of these ovals collided and merged into one (Figure 10.16).

We don't know what causes the Great Red Spot or the white ovals, but we do understand how they can last so long once they form. On Earth, the lifetime of a large oceanic hurricane or typhoon is typically a few weeks, or even less when it moves over the continents and encounters friction with the land. Jupiter has no solid surface to slow down an atmospheric disturbance, and furthermore the sheer size of the disturbances lends them stability. We can calculate that on a planet with no solid surface, the lifetime of anything as large as the Red Spot should be measured in centuries, while lifetimes for the white ovals should be measured in decades, which is pretty much what we have observed.

Despite Neptune's smaller size and different cloud composition, the 1989 Voyager images showed that it had an atmospheric feature surprisingly similar to the jovian Great Red Spot. Neptune's Great Dark Spot (Figure 10.17) was nearly 10,000 km long. On both planets, the giant storms formed at latitude 20° S, had the same shape, and took up about the same fraction of the planet's diameter. The Great Dark Spot rotated with a period of 17 days, versus about 6 days for the Great Red Spot.

When the repaired Hubble Space Telescope examined Neptune in June 1994, however, astronomers could find no trace of the Great Dark Spot on their images. It is not clear whether the Spot has disappeared or merely faded in

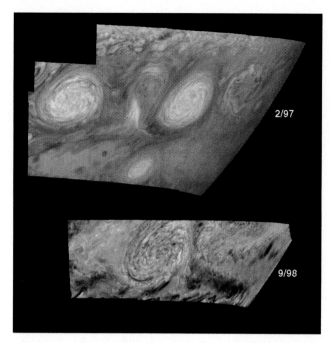

FIGURE 10.16

Merging Storms of Jupiter The two white ovals seen in the upper Galileo image (from February 1997) had merged into a single storm by the time the lower image was taken (September 1998). The colors in the top image are not real, but represent observations in different infrared wavelengths. The blue clouds are high and thin, for example, while the reddish ones are deeper in Jupiter's atmosphere. The winds blow counterclockwise around the white ovals, which is characteristic of a high pressure system. The pear-shaped region between the ovals in the upper picture had clockwise winds, indicating a low-pressure area. These pictures were taken from a range of about 1 million km; the resolution is a few tens of kilometers. *(JPL/NASA)*

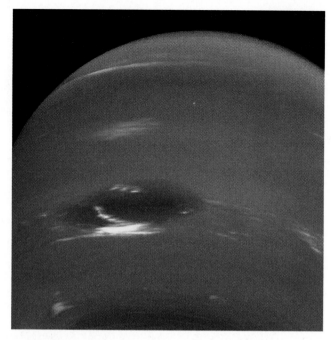

FIGURE 10.17

A Giant Storm on Neptune Here we see the Great Dark Spot of Neptune as photographed by Voyager in 1989. The spot was accompanied by streamers of bright methane cirrus clouds that form around it but at a higher altitude. *(NASA/JPL)*

brightness, as Jupiter's Red Spot also does over time. In November 1994, a new dark spot was seen in Neptune's northern hemisphere, but it had faded somewhat by 1995, indicating that such large storm systems may form and dissipate more quickly on Neptune than on Jupiter.

Large storms on Saturn are much rarer and appear to be connected with the seasons that result from the large tilt of its axis. Approximately once every 30 years observers have seen outbreaks of spots in the equatorial regions of Saturn, with the most recent such outbreak occurring in 1990. These seasonal storms on Saturn appear to be large bubbles or plumes of warmer air that rise into the stratosphere and then gradually dissipate in the strong equatorial jet streams.

Although many of the details of the weather on the jovian planets are not yet understood, it is clear that if you are a fan of dramatic weather, these worlds are the place to look. We study the features in these atmospheres not only for what they have to teach us about conditions in the jovian planets, but also because we hope they can help us understand the weather on Earth just a bit better.

SURFING THE WEB

🖥 General Jupiter Sites

- The Nine Planets Site: [seds.lpl.arizona.edu/nineplanets/nineplanets/jupiter.html]
- Views of the Solar System Site: [www.hawastsoc.org/solar/eng/jupiter.htm]
- Planetary Sciences Site: [nssdc.gsfc.nasa.gov/planetary/planets/jupiterpage.html]

🖥 General Saturn Sites

- The Nine Planets Site: [seds.lpl.arizona.edu/nineplanets/nineplanets/saturn.html]

- Views of the Solar System Site: [www.hawastsoc.org/solar/eng/saturn.htm]
- Planetary Sciences Site: [nssdc.gsfc.nasa.gov/planetary/planets/saturnpage.html]

🖥 General Uranus Sites

- The Nine Planets Site: [seds.lpl.arizona.edu/nineplanets/nineplanets/uranus.html]
- Views of the Solar System Site: [www.hawastsoc.org/solar/eng/uranus.htm]

- Planetary Sciences Site: [nssdc.gsfc.nasa.gov/planetary/planets/uranuspage.html]

General Neptune Sites

- The Nine Planets Site: [seds.lpl.arizona.edu/nineplanets/nineplanets/neptune.html]
- Views of the Solar System Site: [www.hawastsoc.org/solar/eng/neptune.htm]
- Planetary Sciences Site: [nssdc.gsfc.nasa.gov/planetary/planets/neptunepage.html]

Galileo Mission Site [www.jpl.nasa.gov/galileo]

All the new Galileo images and results can be found here, together with educational activities and background information.

Voyager Project Home Page [vraptor.jpl.nasa.gov/voyager/voyager.html]

In 1998, Voyager 1 became the most distant human-made object in space. This site chronicles the planetary flybys of the two Voyagers and gives updates on their continuing mission to explore deep space and search for the heliopause. There is a good science summary for each planet the Voyagers investigated.

Cassini: Voyage to Saturn [www.jpl.nasa.gov/cassini/]

Information about the mission and its engineering and science objectives. There are images of mission equipment and goals and several pages for educators. The Huygens probe has its own Web site at sci.esa.int/huygens/.

SUMMARY

10.1 The outer solar system contains the four jovian (or giant) planets: Jupiter, Saturn, Uranus, and Neptune. Their chemistry is generally reducing, and Jupiter and Saturn have overall compositions similar to that of the Sun. These planets have been explored by the Pioneer, Voyager, and Galileo spacecraft. Voyager 2, perhaps the most successful of all space-science missions, explored Jupiter (1979), Saturn (1981), Uranus (1986), and Neptune (1989)—a grand tour of the jovian planets. Galileo successfully deployed a probe into the atmosphere of Jupiter in 1995 before beginning a 5-year orbital mission.

10.2 Jupiter is 318 times more massive than the Earth. Saturn is about 25 percent, and Uranus and Neptune only 5 percent, as massive as Jupiter. All four have deep atmospheres and opaque clouds, and all rotate quickly (with periods from 10 to 17 hours). Jupiter and Saturn have extensive mantles of liquid hydrogen. Uranus and Neptune are depleted in hydrogen and helium relative to Jupiter and Saturn (and the Sun). Each jovian planet has a core of "ice" and "rock" of about 10 Earth masses. Jupiter, Saturn, and Neptune have major internal heat sources, obtaining as much (or more) energy from their interiors as by radiation from the

Sun. Uranus has no measurable internal heat. Jupiter has the strongest magnetic field and largest magnetosphere of any planet, first discovered by radio astronomers from observations of **synchrotron radiation.**

10.3 The four jovian planets have generally similar atmospheres, composed mostly of hydrogen and helium. The atmospheres of the jovian planets contain small quantities of methane and ammonia gas, both of which also condense to form clouds. Deeper (invisible) cloud layers consist of water and possibly ammonium hydrosulfide (Jupiter and Saturn) and hydrogen sulfide (Neptune). In the upper atmospheres, hydrocarbons and other trace compounds are produced by **photochemistry.** We do not know what colors the clouds of Jupiter. Atmospheric motions on the giant planets are dominated by east–west circulation. Jupiter displays the most active cloud patterns, with Neptune second. Saturn is generally bland, and Uranus is featureless (perhaps due to its lack of an internal heat source). Large storms (oval-shaped high-pressure systems such as the Great Red Spot on Jupiter and the Great Dark Spot on Neptune) can be found in some of the giant planet atmospheres.

INTER-ACTIVITY

A A new member of Congress has asked your group to investigate why the Galileo probe launched into the Jupiter atmosphere in 1995 only survived 57 minutes. Make a list of all the reasons the probe did not last longer, and why it was not made more durable. (Remember that the probe had to hitch a ride to Jupiter!)

B Select one of the jovian planets and get your group to write a script for an evening news weather report for the planet you chose. Be sure you specify roughly how high

in the atmosphere the region lies for which you are giving the report.

C Now that the Galileo probe has "given its life" to tell us more about the outer layers of Jupiter, what does your group think might be the next step to learn more about the jovian planets. Put cost considerations aside for a moment: What kind of mission would you recommend to NASA to learn more about these giant worlds?

D Suppose that an extremely dedicated (and slightly crazy) astronomer volunteers to become a human probe into Jupiter (and somehow manages to survive the trip through Jupiter's magnetosphere alive). As he enters the upper atmosphere of Jupiter, would he fall faster or slower than he would doing the same suicidal jump into the atmosphere of the solid Earth? Groups that have some algebra background could even calculate the force he would feel compared to the force on Earth. (*Bonus question:* If he were in a capsule, falling into Jupiter feet first, and the floor of the capsule had a scale, what would the scale show as his weight?)

REVIEW QUESTIONS

1. Describe the differences in the chemical makeup of the inner and outer parts of the solar system. What is the relationship between what the planets are made of and the temperature where they formed?

2. How did the giant planets grow to be so large?

3. Why is it difficult to drop a probe like Galileo's gently into the atmosphere of Jupiter? How did the Galileo engineers solve this problem?

4. What are the visible clouds on the four jovian planets composed of, and why are they different from each other?

5. Describe the seasons on the planet Uranus.

6. Compare the atmospheric circulation (weather) for the four jovian planets.

7. What are the main atmospheric heat sources of each of the jovian planets?

THOUGHT QUESTIONS

8. Jupiter is denser than water, yet composed for the most part of two light gases, hydrogen and helium. What makes Jupiter so dense?

9. Would you expect to find free oxygen gas in the atmospheres of the giant planets? Why or why not?

10. Why would a tourist brochure (of the future) describing the most dramatic natural sights of the jovian planets have to be revised more often than one for the terrestrial planets?

11. The water clouds believed to be present on Jupiter and Saturn exist at temperatures and pressures similar to those in the clouds of the terrestrial atmosphere. What would it be like to visit such a location on Jupiter or Saturn? In what ways would the environment differ from that in the clouds of Earth?

12. Describe the different processes that lead to substantial internal heat sources for Jupiter and Saturn. Since these two objects generate much of their energy internally, should they be called stars instead of planets? Justify your answer.

13. Use sources in your college library to learn more about the Galileo mission. What technical problems occurred between the mission launch and the arrival of the craft in the jovian system, and how did the mission engineers deal with them? (Good sources of information include *Astronomy* and *Sky & Telescope* magazines.)

PROBLEMS

14. Calculate how many Earths would fit into the volumes of Saturn, Uranus, and Neptune.

15. As the Voyager spacecraft penetrated into the outer solar system, the illumination from the Sun declined. Relative to the situation at Earth, how bright is the sunlight at each of the jovian planets?

16. Jupiter's Great Red Spot rotates in 6 days and has a circumference equivalent to a circle with radius 10,000 km.

Neptune's Great Dark Spot was one fourth as large and rotated in 17 days. For each, calculate the wind speeds at the outer edges of the spots.

17. The ions in the inner parts of the jovian magnetosphere rotate with the same period as Jupiter. Calculate how fast they are moving at the orbit of Io (see Appendix 8). Will these ions strike Io from behind or in front as it moves about Jupiter?

Beatty, J. "Into the Giant" in *Sky & Telescope,* Apr. 1996, p. 20. On the Galileo probe.

Beebe, R. *Jupiter: The Giant Planet.* 1994, Smithsonian Institution Press. Clear, nontechnical summary.

Beebe, R. "Queen of the Giant Storms" in *Sky & Telescope,* Oct. 1990, p. 359. Excellent review of the Red Spot.

Burgess, E. *Far Encounter: The Neptune System.* 1992, Columbia U. Press. Survey of Voyager flyby by a veteran science journalist.

Cowling, T. "Big Blue: The Twin Worlds of Uranus and Neptune" in *Astronomy,* Oct. 1990, p. 42. Nice long review of the two planets.

Gore, R. "Neptune: Voyager's Last Picture Show" in *National Geographic,* Aug. 1990, p. 35.

Littmann, M. *Planets Beyond: Discovering the Outer Solar System,* 2nd ed. 1989, John Wiley. Good introduction to Uranus, Neptune, and Pluto.

Littmann, M. "The Triumphant Grand Tour of Voyager 2" in *Astronomy,* Dec. 1988, p. 34.

Lunine, J. "Neptune at 150" in *Sky & Telescope,* Sept. 1996, p. 38. Nice review.

Miner, E. *Uranus.* 1990, Ellis Horwood/Simon and Schuster. Definitive introduction to the planet.

Rogan, J. "Bound for the Ringed Planet" in *Astronomy,* Nov. 1997, p. 36. On the Cassini mission.

Sagan, C. *Pale Blue Dot.* 1994, Random House. Chapters 6 through 9 explore the outer solar system.

Smith, B. "Voyage of the Century" in *National Geographic,* Aug. 1990, p. 48. Beautiful summary of the Voyager mission to all four outer planets.

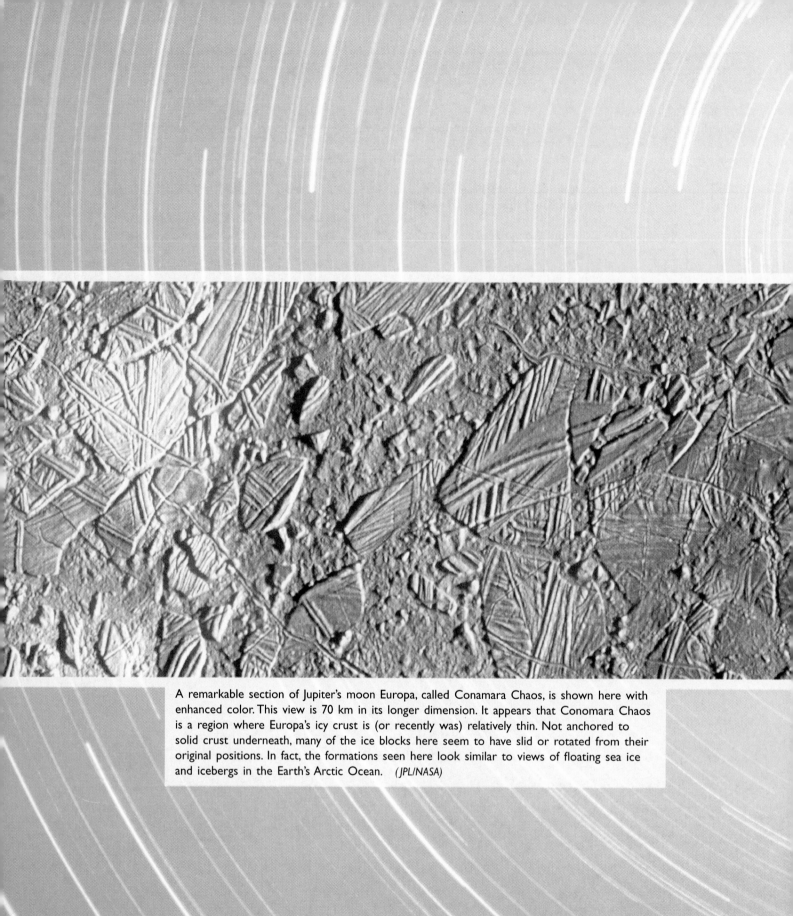

A remarkable section of Jupiter's moon Europa, called Conamara Chaos, is shown here with enhanced color. This view is 70 km in its longer dimension. It appears that Conomara Chaos is a region where Europa's icy crust is (or recently was) relatively thin. Not anchored to solid crust underneath, many of the ice blocks here seem to have slid or rotated from their original positions. In fact, the formations seen here look similar to views of floating sea ice and icebergs in the Earth's Arctic Ocean. *(JPL/NASA)*

11 Rings, Moons, and Pluto

"*O*ur imaginations always
fall short of anticipating the
beauty we find in nature."

Geologist Laurence Soderblom,
discussing the 1989 Voyager
encounter with Neptune's moons

THINKING AHEAD

On an active world, a map is always temporary. The boundaries of towns change, new roads are built, rivers slowly alter their channels. Over geological time, even the Earth's mountains, seas, and continents change shape and form. But what kind of map would be worth making of a world where the surface landforms change substantially in only a decade? Such a world can be found orbiting the planet Jupiter.

All of the giant planets are accompanied by moons or satellites (astronomers use the two terms interchangeably), which orbit about them like planets in a miniature solar system. Over 60 satellites are known in the outer solar system, too many to discuss individually in any detail, but we list their characteristics in Appendix 8. Astronomers suspect that other small satellites will be found in the future. Before the Voyager missions to the outer planets, many of these satellites were mere points of light in our telescopes. Suddenly, in less than a decade, we had close-up images (and much nonvisual information as well); they then became individual worlds for us, each with unique features and peculiarities.

In this chapter we focus our attention on a few of the more interesting satellites and on the tiny planet Pluto, which is smaller than a number of the outer planets' moons. Then we examine the ring systems that gird each of the four giant planets. The processes we see at work in these rings have applications ranging from the formation of planetary systems to the spiral structure of galaxies.

RING AND SATELLITE SYSTEMS

General Properties

As discussed in Chapter 6, the rings and satellites of the outer solar system have compositions different from objects in the inner solar system. We should expect this, since they formed in regions of lower temperature, cool enough that large quantities of water ice were available as building materials. Most of these objects also contain dark, organic compounds mixed with their ice and rock. The presence of this dark, primitive material often means that the objects reflect very little light. Don't be surprised, therefore, to find many objects in the ring and satellite systems that are both icy and black.

FIGURE 11.1

The Galilean Satellites This montage, assembled from individual Galileo and Voyager images, shows a "family portrait" of Jupiter (with its giant red spot) and its four large satellites. From top to bottom, we see Io, Europa, Ganymede, and Callisto. *(NASA/JPL)*

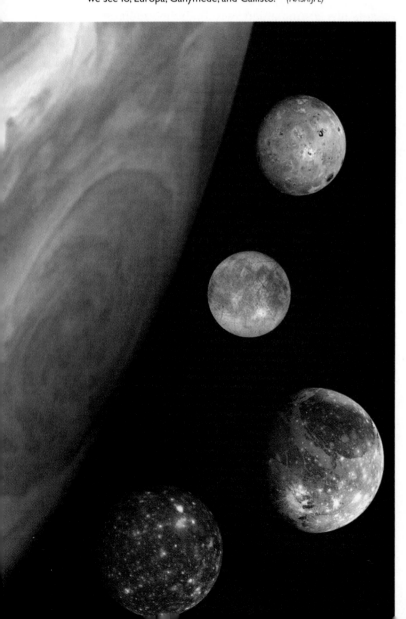

Most of the satellites in the outer solar system are in *direct* or regular orbits; that is, they revolve about their parent planet in a west-to-east direction and in the plane of the planet's equator. Such objects probably formed at about the same time as the planet by processes similar to those that formed the planets in orbit around the Sun (see Chapter 13).

In addition to the regular satellites, irregular satellites orbit in a *retrograde* (east-to-west) direction or else have orbits of high eccentricity (more elliptical than circular) or high inclination (moving in and out of the planet's equatorial plane). These are usually smaller satellites, located relatively far from their planet; they were probably formed far away and subsequently captured by the planet they now orbit.

The Jupiter System

Jupiter has 16 satellites and a faint ring. The satellites include four large moons—Callisto, Ganymede, Europa, and Io (Figure 11.1)—discovered in 1610 by Galileo and therefore often called the *Galilean satellites.* The smaller of these, Europa and Io, are about the size of our Moon, while the larger, Ganymede and Callisto, are bigger than the planet Mercury. We will discuss the Galilean satellites individually in a moment and will try to understand the interesting differences among them.

The other 12 jovian moons are much smaller. The inner 4 all circle the planet inside the orbit of Io; 3 of them were discovered by Voyager. The outer satellites consist of 4 in direct but highly inclined orbits, and 4 farther out in retrograde orbits. These 8 are believed to be captured objects. The two groupings may indicate two parent bodies that were broken up in a collision early in the history of the jovian system.

The Saturn System

Saturn has 19 known satellites in addition to a magnificent set of rings (Figure 11.2). The largest of the satellites, Titan, is almost as big as Ganymede in the jovian system, and it is the only satellite with a substantial atmosphere. We discuss it in detail in the next section. Saturn has 6 other regular satellites with diameters between 400 and 1600 km, and a handful of small moons orbiting in or near the rings. There are also 2 distant irregular satellites, one of which has a retrograde orbit.

The rings of Saturn, one of the most impressive sights in the solar system, are broad and flat, with only a few gaps. They are not solid, but rather a huge collection of icy fragments, all orbiting the equator of Saturn in a traffic pattern that makes rush hour in a big city look simple by comparison. Individual ring particles are composed of water ice and are typically the size of ping-pong, tennis, and basketballs.

FIGURE 11.2
Saturn Family Portrait Another NASA montage of Voyager images shows Saturn and several of its satellites. The satellites are not to scale: Giant Titan is the tiny orange ball toward the back. The moon in the foreground is Enceladus. The color of Saturn is a bit intensified by the image processing. *(NASA/JPL)*

The Uranus System

The ring and satellite system of Uranus is tilted at 98°, just like the planet itself. It consists of 11 rings and 18 satellites. The 5 largest satellites are similar in size to the 6 regular satellites of Saturn, with diameters of 500 to 1600 km, while the 13 smaller satellites and the ring particles are very dark, reflecting only a few percent of the sunlight that strikes them. (The outermost two moons were only found in September 1997, using the Hale 5-m telescope on Mount Palomar with modern electronic detectors.)

Discovered in 1977, the rings of Uranus are narrow ribbons of material with broad gaps in between—very different from the broad rings of Saturn. Astronomers suppose that the ring particles are confined to these narrow paths by the gravitational effects of numerous small satellites, most of which we have not yet glimpsed.

The Neptune System

Neptune has eight satellites: six regular satellites close to the planet, and two irregular satellites farther out. The most interesting of these is Triton, a relatively large moon in a retrograde orbit. Triton has an atmosphere, and active volcanic eruptions were discovered there by Voyager in its 1989 flyby. We compare it in detail with the planet Pluto, which it resembles, later in this chapter. The rings of Neptune are narrow and faint. Like those of Uranus, they are composed of dark materials and are thus not easy to see.

11.2 THE GALILEAN SATELLITES AND TITAN

From 1996 to 1999, the Galileo spacecraft careered through the jovian system on a complex but carefully planned trajectory that provided repeated close encounters with the Galilean satellites. Although scientists had thought they understood the basics of these satellites from the two Voyager flybys of 1979, the new data brought many surprises, and the high-resolution Galileo cameras offered stunning views that far exceeded the Voyager images in their scientific value. This wealth of new information provides us the opportunity, quite literally, to rewrite the textbooks on the Galilean satellites. Beginning in 2004, we can expect a similar bonanza of information about Titan, obtained from the Cassini spacecraft.

We discuss each of these five large satellites in turn, but to compare them with each other and with the other large satellites of the solar system, we first list some basic facts about them (and our own Moon) in Table 11.1.

Callisto, An Ancient, Primitive World

We begin our discussion of the Galilean satellites with the outermost one, Callisto—not because it is remarkable, but because it is not. This makes it a convenient object with which other, more active, worlds can be compared. Its distance from Jupiter is about 2 million km, and it orbits the planet in 17 days. Like our own Moon, Callisto rotates in the same period as it revolves, so it always keeps the same

TABLE 11.1
The Largest Satellites

Name	Diameter (km)	Mass (Moon = 1)	Density (g/cm³)	Reflectivity (%)
Moon	3476	1.0	3.3	12
Callisto	4820	1.5	1.8	20
Ganymede	5270	2.0	1.9	40
Europa	3130	0.7	3.0	70
Io	3640	1.2	3.5	60
Titan	5150	1.9	1.9	20

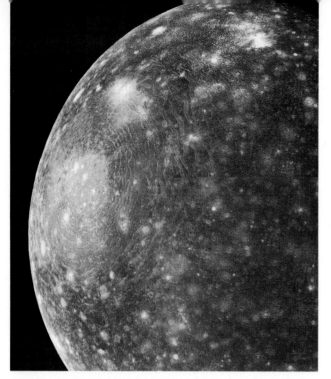

FIGURE 11.3

Callisto Jupiter's outermost large satellite shows a heavily cratered surface of dark ice in this Voyager image. The multiringed feature at center left is a giant impact basin called Valhalla. *(NASA/JPL)*

face toward Jupiter. Callisto's day thus equals its month: 17 days. Its noontime surface temperature is only 130 K (about 140° C below freezing), so water ice is stable on its surface the year around.

Callisto has a diameter of 4820 km, almost the same as the planet Mercury. Its mass is only one third as great, which means its density (the mass per unit volume) must be only one third as great as well. This tells us that Callisto has far less of the rocky metallic materials found in the in-

ner planets and must instead be an icy body through much of its interior. Such a satellite can show us how the geology of an icy object compares with one made primarily of rock.

Unlike the worlds we have studied so far, Callisto has not fully *differentiated* (separated into layers of different density materials). We can tell that it lacks a dense core from the details of its gravitation pull on the Galileo spacecraft during several very close flybys. This fact surprised scientists, who expected that all the big icy moons would be differentiated. It is much easier for an icy body to differentiate than for a rocky one, since the melting temperature of ice is so low. Only a little heating will soften the ice and get the process started, allowing the rock and metal sink to the center and the slushy ice floats to the surface. Yet Callisto seems to have frozen solid before the process of differentiation was complete.

Like the lunar highlands (Figure 11.3), the surface of Callisto is covered with impact craters. The survival of these craters tells us that an icy object can form and retain impact craters in its surface. Callisto is unique among the planet-sized objects of the solar system in the absence of interior forces to drive geological evolution. The satellite was still-born, and it has remained geologically dead for more than 4 billion years.

In thinking about ice so far from the Sun, it is important not to judge its behavior from the much warmer ice we know and love on Earth. At the temperatures of the outer solar system, ice on the surface is nearly as hard as rock, and it behaves similarly. Ice on Callisto does not deform or flow like ice in glaciers on the Earth.

In general form, the icy craters of Callisto look very much like the rocky craters of the Moon and Mercury. Yet in close-up photographs from Galileo, the similarity with lunar craters evaporates (Figure 11.4). Some process has been at work to break down crater walls and bury parts of the surface. Although the results of this erosion look rather

FIGURE 11.4

Worn Craters on Callisto This mosaic of three Galileo images shows an area about 45 km across in the Valhalla region of Callisto. Details as small as 10 m across can be seen. Sunlight is illuminating the surface from the left. We see part of a prominent chain of craters, probably left by a projectile that broke into many pieces (just as Comet Shoemaker–Levy did in 1992). Note that the impact craters are eroded even though Callisto has no atmosphere. *(Arizona State U./JPL/NASA)*

FIGURE 11.5

Ganymede This global view of Ganymede, the largest moon in the solar system, was assembled from images taken by the Galileo spacecraft. The large circular dark region at the upper right is called *Galileo Regio.* Such darker places are older, more heavily cratered regions, while the lighter areas are younger. Really bright spots are sites of geologically recent impacts. *(Galileo Imaging Team; JPL/NASA)*

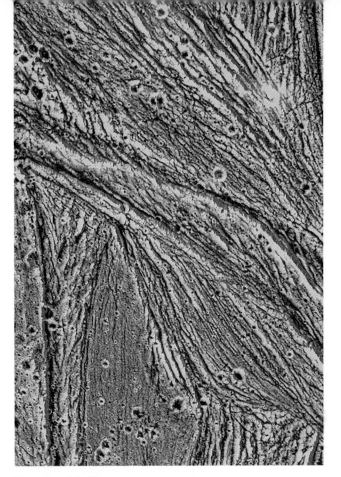

FIGURE 11.6

Close-up of Ganymede Details as small as 75 m across can be seen in this image of Jupiter's largest moon, taken by the Galileo spacecraft cameras. We see an area 55 by 35 km in the Uruk Sulcus region, with a complex assortment of ridges, grooves, craters, and evidence of crustal shifting. The line-like sets of features are parallel sunlit ridges. We can see signs that the icy surface has pulled apart and slid sideways in shaping the landscape. *(JPL/NASA)*

like the effect of wind-blown dust on Earth and Mars, this explanation cannot be correct, since Callisto has no atmosphere. Instead, it is the slow process of evaporation and sublimation of ice that breaks down the craters over billions of years and redistributes the dusty surface material that was mixed with the ice.

Ganymede, the Largest Satellite

Ganymede, the largest satellite in the solar system, is also cratered, but less so than Callisto (Figure 11.5). About one quarter of its surface seems to be as old and heavily cratered as Callisto; the rest formed more recently, as we can tell by the sparse covering of impact craters as well as the relative freshness of the craters. Judging from crater counts, this younger terrain on Ganymede is about a billion years old, younger than the lunar maria or the martian volcanic plains.

The differences between Ganymede and Callisto are more than skin deep. Ganymede is a differentiated world, like the terrestrial planets. Measurements of its gravity field tell us that the rock and metal sank to form a core about the size of our Moon, with a mantle and crust of ice "floating" above it. In addition, the Galileo spacecraft discovered that Ganymede has a magnetic field, the sure signature of a partially molten interior. Ganymede is not a dead world, but rather a place of continuing geological activity powered by an internal heat source. Much of its surface may be as young as that of Venus (half a billion years).

The younger terrain is the result of tectonic and volcanic forces (Figure 11.6). Some features formed when the crust cracked, flooding many of the craters with water from the interior. Extensive mountain ranges were formed from compression of the crust, forming long ridges with parallel valleys spaced 1–2 km apart. In some places older impact craters were split and pulled apart. There are even indications of large-scale crustal movements that are similar to the plate tectonics of the Earth.

Why is Ganymede different from Callisto? Possibly the small difference in size and internal heating between the two led to this divergence in their evolution. But more likely the gravity of Jupiter is to blame for Ganymede's continuing geological activity. Ganymede is close enough to Jupiter that *tidal forces* from the giant planet may have episodically heated its interior and triggered major convulsions on its crust.

As we discussed in Section 3.6, a tidal force results from the unequal gravitational pull on two sides of a body. In a complex kind of modern dance, the large satellites of Jupiter are caught in the varying gravity grip of both the giant planet and each other, leading to a kind of gravitational flexing or kneading in their centers, which can heat them. (A fuller explanation can be found in "Io, a Volcanic Satellite," on page 242.) We will see as we move inward to Europa

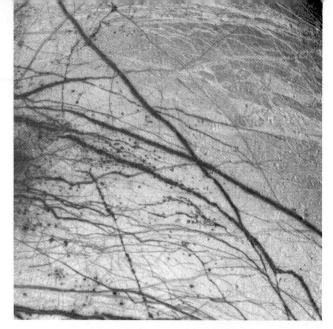

▲ (a) A wide-angle view (about 1260 km across) of the surface in significantly enhanced color. Scientists enhance the color to bring out subtle differences in composition between the different types of terrain. Europa's crust is nearly pure water ice, with very few hills or other topographic features. In fact, overall, Europa is as smooth as a billiard ball. The color is the result of small quantities of contaminants (such as sulfur dust from the volcanoes of Io).

FIGURE 11.7
Europa's Intriguing Surface (a–g) Jupiter's moon Europa is the only place in the solar system beyond Earth that appears to have liquid water—an ocean beneath a thick crust of ice. It is also one of the strangest-looking worlds we have photographed, with a highly active surface geology—all the result of internal heating from tides raised by Jupiter's gravity. *(JPL/NASA)*

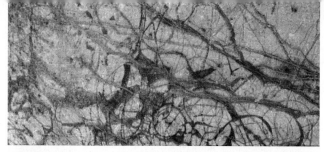

▲ (b) Another wide-angle view, but here we can see that the floating ice crust has been broken up and torn in the past, creating many darker lanes of smooth surface. The area shown is about 360 × 770 km, roughly the size of the state of Nebraska.

▲ (c) We know that the surface of Europa is geologically young because there are so few impact craters. One of the largest craters on Europa is Pwyll, about 26 km in diameter. Even though Pwyll is the result of a recent impact, its crater rim is nearly gone, fading away as the icy crust bends and readjusts over time. The darker material in and outside the crater was probably dredged up from deeper inside Europa by the force of the impact.

and Io that the role of jovian tides becomes more critical for satellites close to the planet.

Europa, the Satellite with an Ocean

Europa and Io, the inner two Galilean satellites, are not icy worlds like most of the satellites of the outer planets. With densities and sizes similar to our Moon, they appear to be predominantly rocky objects. How did they fail to acquire a majority share of the ice that must have been plentiful in the outer solar system at the time of their formation?

The most probable cause is Jupiter itself, which became hot and radiated a great deal of infrared energy during the first few million years after its formation. This heated the disk of material near the planet that would eventually coalesce into the closer moons. Thus any ice near Jupiter evaporated, leaving Europa and Io with compositions more appropriate to bodies in the inner solar system. In addition, the tidal forces from Jupiter have continued to act on these objects, heating their interiors and perhaps leading to further loss of volatile materials such as water and carbon dioxide.

Despite its mainly rocky composition, Europa has an ice-covered surface, as astronomers have long known from examining spectra of sunlight reflected from Europa. In this it resembles the Earth, which also has a layer of water on its surface—but in Europa's case, most of the water may be frozen. There are very few impact craters, indicating that the surface of Europa is in a continual state of geologic self-renewal. Judging from crater counts, the surface must be no more than a few million years old, and perhaps substantially less. In terms of its ability to erase impact craters, Europa is more geologically active than the Earth.

When we look at close-up photos of Europa, we see a strange surface (see the images in Figure 11.7 and the figure that opens this chapter). For the most part the satellite is extremely smooth, but it is criss-crossed with cracks and low ridges that often stretch for thousands of kilometers across the icy plains. Some of these long lines are single, but most are double or multiple, looking rather like the remnants of a colossal freeway system.

It is very difficult to make straight lines on a planetary surface. In Chapter 9, we discussed that when Percival Lowell saw what appeared to him to be straight lines on

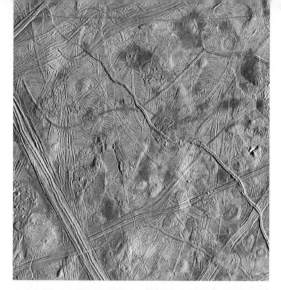

▲**(d)** The surface at higher resolution. The ice is wrinkled and criss-crossed by long ridges. Where these ridges intersect, we can see which are older and which are younger: The younger ones cross over the older ones. To some observers, this system of ridges resembles a giant freeway system on Europa, but the ridges are much wider than our freeways and are a natural result of the flexing of the moon. (The ridge in the left bottom corner, for example, is 8 km wide.)

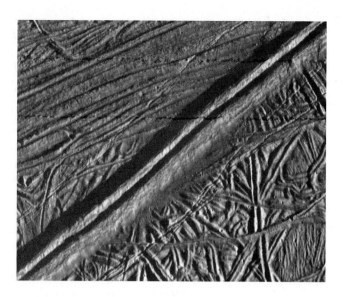

▲**(e)** A very high-resolution image of one young double ridge. The area in this picture is only 17 km across, and the ridge itself is about 3 km wide and 300 m high. It appears to have formed when viscous icy material was forced up through a long straight crack in the crust.

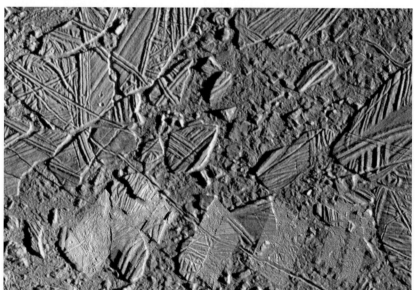

▲**(f)** The image that opens this chapter shows a section of Conamaras Chaos region, where the crust of Europa is especially thin. An even higher resolution view of part of this terrain is seen here, where features as small as 20 m can be distinguished. Many "icebergs," typically about 1 km across, appear to have once floated in a sea of liquid water or slushy ice. The icebergs can be identified by the remnants of ridges and grooves on their surfaces, while the now-frozen "sea ice" between them has a rough, fine-grained structure when seen at this resolution.

▲**(g)** A tiny part of Europa photographed at the highest resolution possible with the Galileo cameras, from an altitude of only 500 km. The region shown is only about a mile wide, and it shows features as small as a house. But what do these features mean? Looking at just a single snapshot of such an alien surface, scientists are not yet able to provide any meaningful geological interpretation. For now, we can simply relish the incredible detail in these images of a puzzling, icy world.

Mars (the martian "canals"), he attributed them to the engineering efforts of intelligent beings. We now know that the lines on Mars were optical illusions, but the lines on Europa are real. If the crust consists of ice floating without much friction on an ocean of liquid water, then such long cracks would be expected. To examine this exciting possibility more closely, NASA redirected the Galileo spacecraft during the third and fourth year of its mission to pass nearer to Europa and send back very high-resolution images of its intriguing surface.

The Galileo images appear to confirm the existence of a global ocean. In many places the surface of Europa looks just as we would expect for a thick layer of ice that was broken up into giant icebergs and ice floes, then refrozen in place. When the ice breaks, water or slush from below may be able to seep up through the cracks and make the ridges and multiple-line features we observe. Many episodes of ice cracking, shifting, rotating, and refreezing are required to explain the complexity we see. The icy crust might vary in thickness from a kilometer or so up to 20 km. In some places the ice appears to be free-floating; in others it seems to be grounded against the rocky ocean floor.

If Europa really has a large ocean of liquid water under its ice, then it is the only place in the solar system, other than Earth, with such an environment. To remain liquid, this ocean must be warmed by heat escaping from the interior of Europa. Hot (or at least warm) springs might be active there, analogous to those we have discovered in the deep oceans of the Earth. However, to have large quantities of liquid water, there must be substantial heating from Europa's interior—more than can be explained by the energy from radioactive rocks (the mechanism that heats the interiors of the terrestrial planets).

Such energy can be generated by the tidal heating mechanism we discussed earlier. Scientists still don't know how much tidal energy is produced inside Europa and whether this heating is steady enough to maintain a global ocean at all times. At some times, the water may be in the form of an icy slush, depending on subtle changes in the orbit of Europa.

What makes the idea of an ocean on Europa with warm springs especially exciting is the discovery in Earth's oceans of ecosystems of primitive life around marine hot springs. Such life derives all its energy from the mineral-laden water and thrives independent of the sunlight or atmosphere at the Earth's surface. Is it possible that similar ecosystems could exist today under the ice of Europa? Some people have suggested that Europa is the most likely place beyond the Earth to find life in the solar system. This possibility ensures that Europa will be a continuing target of planetary missions, even after the end of the Galileo project.

Io, a Volcanic Satellite

Io, the innermost of Jupiter's Galilean satellites, is a close twin of our Moon, with nearly the same size and density. We might therefore expect it to have experienced a similar history. Its appearance, as photographed from space, tells us another story, however (Figure 11.8). Instead of being a dead, cratered world, Io turns out to have the highest level of volcanism in the solar system, greatly exceeding that of the Earth.

Io's active volcanism was discovered by the Voyager spacecraft. Eight volcanoes were actually seen erupting when Voyager 1 passed in March 1979, and six of these were still active four months later when Voyager 2 passed. With the improved instruments carried by the Galileo spacecraft, more than 50 eruptions were found during the calendar year 1997 alone. Many of the eruptions produce graceful plumes that extend to heights of hundreds of kilometers into space (Figure 11.9).

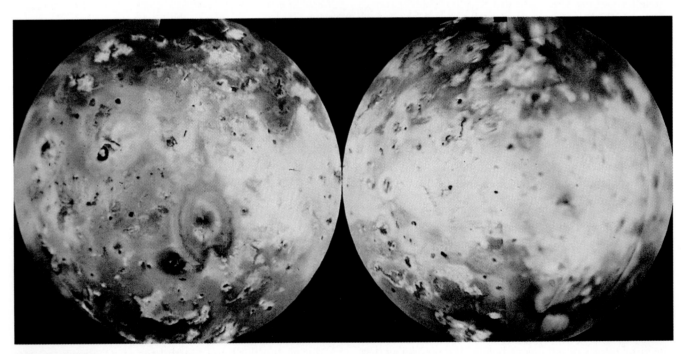

FIGURE 11.8

Io Mosaic   A composite of Voyager images assembled by the U.S. Geologic Survey to show both sides of the volcanically active moon Io. The orange colored deposits are sulfur snow; the white color is sulfur dioxide. (Carl Sagan once quipped that Io looks like it desperately needs a shot of penicillin.)   *(NASA/USGS)*

FIGURE 11.9
Volcanic Plume on Io The large plume rising about 250 km above Io's surface (seen in white above the edge of the moon) erupted from the center of the enormous hoofprint-shaped volcanic feature that takes up most of this image. It is named Pele, after the Hawaiian goddess of fire, and is about the size of the state of Alaska. *(NASA/JPL)*

The Galileo data show that most of the volcanism on Io consists of hot silicate lava, like the volcanoes on Earth. These lavas can exhibit temperatures greater than 1000° C. But sometimes the hot lava on its way up encounters frozen deposits of of sulfur and sulfur dioxide. When these icy deposits are suddenly heated, we get the great eruptive plumes whose dramatic display captured our attention when the Voyager spacecraft first saw them. As the rising plumes cool, the sulfur and sulfur dioxide recondense as solid particles, which fall back to the surface in colorful "snowfalls" that extend as much as a thousand kilometers from the vent. Io also has a variety of volcanic hot spots, one of them a sort of "lava lake" 200 km in diameter.

Galileo also gave us dramatic confirmation of the idea that Io is a world where mapmakers would be in constant demand. Planetary scientists, comparing Galileo images to earlier Voyager maps, have found at least a dozen areas on Io larger than the entire state of Connecticut that have been resurfaced by volcanic activity in the intervening years. Major new surface features were even seen to appear between Galileo orbits, as shown in Figure 11.10.

How can Io maintain this remarkable level of volcanism, in spite of its small size? The answer, as we hinted earlier, lies in gravitational heating of the satellite by Jupiter. Io is about the same distance from Jupiter as our Moon is from the Earth. Yet Jupiter is more than 300 times more massive than Earth, causing forces that pull the satellite into an elongated shape, with a several-kilometer-high bulge extending toward Jupiter.

If Io always kept exactly the same face turned toward Jupiter, this bulge would not generate heat. However, Io's orbit is not exactly circular, due to gravitational perturbations from Europa and Ganymede. In its slightly eccentric orbit, Io twists back and forth with respect to Jupiter, at the same time moving nearer and farther from the planet on each revolution. The twisting and flexing heats Io, much as repeated flexing of a wire coat hanger heats the wire.

After billions of years, constant flexing and heating have taken their toll on Io, driving away water and carbon dioxide and other gases, until now sulfur and sulfur compounds are the most volatile materials remaining. The inside is entirely melted, and the crust itself is constantly recycled by volcanic activity.

The Galileo spacecraft found that the volcanic eruptions are concentrated near the *subjovian* point on the surface, where Jupiter would appear directly overhead in the sky. This provides further support for the hypothesis that the volcanism is the result of tidal heating.

In moving inward toward Jupiter from Callisto to Io, we have encountered more and more evidence of geological activity and tidal heating. Callisto has a surface more than 4 billion years old; Ganymede has been active as recently as the past billion years; Europa has an icy crust at most a few million years old; and parts of the visible surface of Io were deposited in the brief 16-year interval between Voyager and Galileo. Just as the character of planets in our solar system depends in large measure on their distance from the Sun (and on the amount of heat they receive), so it appears that distance from a giant planet like Jupiter can play a large role in the composition and evolution of its satellites (at least partly due to differences in internal heating by Jupiter's unrelenting tides).

Titan, a Satellite with Atmosphere

Titan, discovered in 1655 by the Dutch astronomer Christian Huygens, is the largest satellite of Saturn. It was the

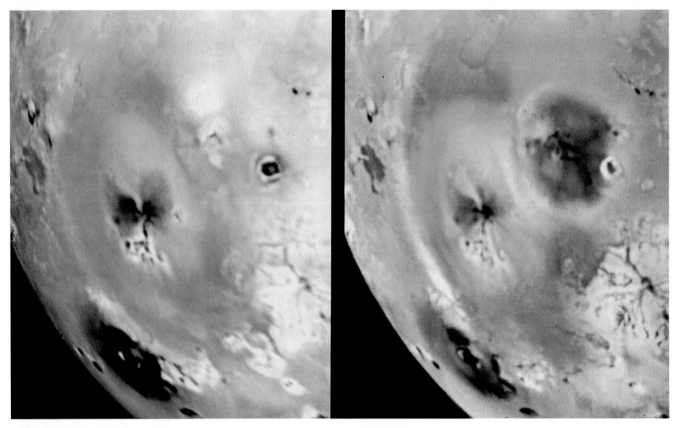

FIGURE 11.10

Changes on Io This pair of images from Galileo shows a dark deposit, about the size of the state of Arizona, that formed between April 4 *(left frame)* and September 19, 1997 *(right frame)*. The 400-km-wide dark region surrounds a volcanic center called Pillan Patera, which produced an eruptive plume some 120 km high between the time these two images were taken. The reddish oval is Pele, which is also visible on Figure 11.9. The colors are somewhat exaggerated by computer processing, and the angle of illumination is different in the two pictures. *(U. of Arizona; JPL/NASA)*

first satellite discovered after Galileo saw the four large moons of Jupiter. Titan has roughly the same diameter, mass, and density as Callisto or Ganymede. Presumably it also has a similar composition—about half ice and half rock, although we don't know whether this satellite has differentiated. What is remarkable and unique about this satellite is the presence of a substantial atmosphere, discovered by Gerard Kuiper in 1944 (Figure 11.11). He noticed that spectra of Titan showed absorption features indicating the presence of substantial amounts of methane gas.

The 1980 Voyager flyby of Titan determined that the atmospheric surface pressure on this satellite is 1.6 bars (1.6 times the air pressure at sea level on Earth), higher than that on any other satellite, and, remarkably, even higher than that of the terrestrial planets Mars and Earth. The atmospheric composition is primarily nitrogen, an important way in which Titan's atmosphere resembles Earth's.

A variety of additional compounds have been detected in Titan's upper atmosphere, including carbon monoxide (CO), hydrocarbons (compounds of hydrogen and carbon) such as methane (CH_4), ethane (C_2H_6), and propane (C_3H_8), and nitrogen compounds such as hydrogen cyanide (HCN), cyanogen (C_2N_2), and cyanoacetylene (HC_3N). Their presence indicates an active chemistry, in which sunlight interacts with atmospheric nitrogen and methane to create a rich mix of organic molecules. This is very much the way we believe organic compounds were formed on the Earth when it still had its original atmosphere. The discovery of HCN on Titan is particularly interesting, since this molecule is the starting point for formation of some of the components of DNA, the fundamental genetic molecule essential to life on Earth.

Titan has multiple layers of clouds (Figure 11.12). The lowest are within the bottom 10 km of the atmosphere; these are condensation clouds, probably composed of methane. Much higher is a dark reddish haze or smog consisting of complex organic chemicals created by the energy of sunlight falling on the molecules of Titan's atmosphere. Formed at an altitude of several hundred kilometers, this aerosol slowly settles downward to the surface, where it may have built up a deep layer of tar-like organic chemicals.

FIGURE 11.11
Titan's Thick Atmosphere To the Voyager cameras, Titan looked like a big fuzzy orange ball (see Figure 10.2). In this enhanced close-up view, we see the main orange haze layer that blocks our view of the surface, but also the blue haze that lies higher in the moon's atmosphere. *(NASA/JPL)*

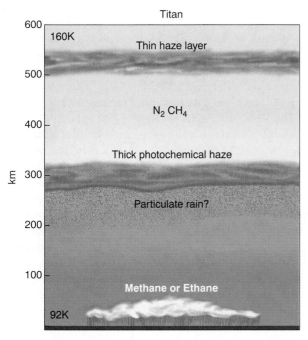

FIGURE 11.12
The Structure of Titan's Atmosphere Heights above the surface are given on the left border. Titan has the most Earth-like atmosphere in the solar system, although it is much colder than our planet. *(NASA/JPL)*

Much to the disappointment of the Voyager scientists, the spacecraft revealed that the reddish haze completely blankets Titan, hiding its surface from our view. Still, the Voyager data allow us to make models of what it might be like to land on this intriguing world. For example, we know Titan's surface temperature is about 90 K, held uniform by the blanketing atmosphere. At such a low temperature, methane plays the same role that water does on Earth: the gas is only a minor constituent of the atmosphere, but it can exist in gas, liquid, or solid form. Methane condenses to form clouds, and there may be seas or lakes of liquid methane and ethane, as well as solid methane ice.

Organic compounds are chemically stable in Titan's cold environment, unlike those on the warmer, oxidizing Earth. Titan's surface, therefore, probably records a chemical history that goes back billions of years. Many scientists believe that this satellite will provide more insights into the early history of Earth's atmosphere, and even into the origin of life, than any other object in the solar system. We hope to know more about the Titan environment when the Huygens probe carried by the Cassini spacecraft parachutes down to the surface in 2004.

11.3 TRITON AND PLUTO

In this section we discuss two apparently similar objects: Neptune's satellite Triton and the planet Pluto. Seeking to understand the ways in which these two worlds are similar to each other is an example of the comparative approach to studying astronomy. Pluto is the only planet not visited by a spacecraft; comparing it with Triton, which was studied by Voyager in 1989, is one way we can learn more about it.

Triton and Its Volcanoes

Triton (don't get it confused with Titan) has a diameter of 2720 km and a density of 2.1 g/cm³, indicating a probable composition of 75 percent rock mixed with 25 percent water ice. Measurements indicate that Triton's surface has the coldest temperature of any of the worlds our robot representatives have visited. Because its reflectivity is so high (about 80 percent), Triton reflects most of the solar energy falling on it, resulting in a surface temperature between 35 and 40 K.

The surface material of Triton is made of frozen water, nitrogen, methane, and carbon monoxide. Although methane and nitrogen exist as gas in most of the solar system, they are frozen at Triton's temperatures. Only a small quantity of nitrogen vapor persists to form an atmosphere. Although the surface pressure of this atmosphere is only 16 millionths of a bar, this is sufficient to support thin haze or cloud layers.

Triton's surface, like that of many other satellites in the outer solar system, reveals a long history of geological evolution (Figure 11.13). While some impact craters are found, so are many regions that have been flooded by the

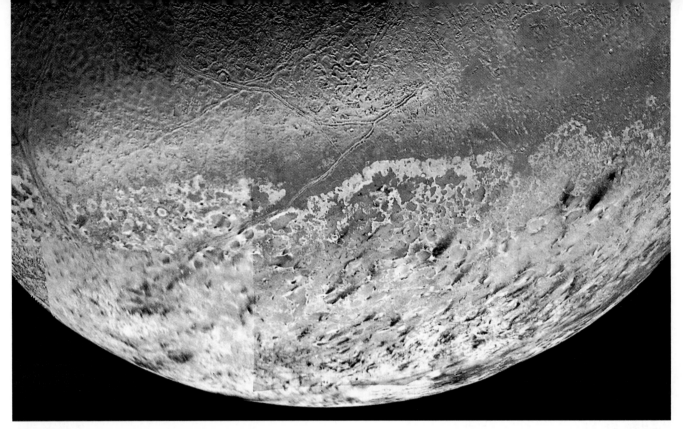

FIGURE 11.13
Neptune's Moon Triton   This mosaic of Voyager images of Triton shows a wide range of surface features. The pinkish area at the bottom is Triton's large southern polar cap. The south pole of Triton currently faces the Sun, and the slight heating effect is driving some of the material northward, where it is colder.   *(NASA/Image processing at USGS)*

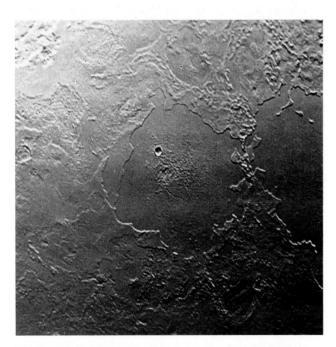

FIGURE 11.14
Lava Lakes   Old flooded "lava lakes" on Triton can be seen in this Voyager close-up. Now frozen solid, these features, 100 to 200 km across, date from an earlier period of volcanic activity that may have poured water onto Triton's surface.   *(NASA/JPL)*

local version of "lava" (perhaps water or water–ammonia mixtures). The evidence for such low-temperature volcanism includes a number of frozen "lava lakes" more than 100 km across (Figure 11.14). There are also mysterious regions of jumbled or mountainous terrain.

The Voyager flyby of Triton took place at a time when the satellite's southern pole was tipped toward the Sun, allowing this part of the surface to enjoy a period of relative warmth. (Remember that warm on Triton is still outrageously colder than anything we experience on Earth.) A polar cap covers much of Triton's southern hemisphere, apparently evaporating along the northern edge. This polar cap may consist of frozen nitrogen that was deposited during the previous winter. Remarkably, the evaporation of this polar cap generates geysers or volcanic plumes of nitrogen gas that make fountains about 10 km high. These plumes differ from the volcanic plumes of Io in their composition, and also in their energy source. They derive their energy from sunlight warming the surface rather than from internal heat.

Discovery of Pluto

Now we turn to Pluto, a small planet that resembles Triton in many ways and is quite different from the giant planets in the outer solar system. Pluto was discovered through a careful, systematic search, unlike Neptune, whose position was calculated from gravitational theory. Nevertheless, the history of the search for Pluto began with indications that Uranus and Neptune had slight departures from their predicted orbits (you may want to review the last section of Chapter 2 for more on what such departures mean). Early in the 20th century several astronomers—most notably Percival Lowell, then at the peak of his fame as an advocate of intelligent life on Mars (see Chapter 9)—became interested in this problem.

At the time Lowell made his calculations, Neptune had moved such a short distance since its discovery that it could not be used effectively to search for perturbations by an unknown ninth planet. Therefore Lowell and his contemporaries based their calculations primarily on the tiny remaining irregularities in the motion of Uranus. Lowell's computations indicated two possible locations for a perturbing planet; the more likely of the two was in the constellation of Gemini. He predicted a mass for the planet intermediate between those of the Earth and Neptune (his calculations gave about 6.6 Earth masses). Other astronomers, however, obtained other solutions from the tiny orbital irregularities, including one model that indicated two exterior planets.

At his Arizona observatory, Lowell searched without success for the unknown planet from 1906 until his death in 1916. Subsequently, his brother donated to the observatory a 33-cm photographic telescope that could record a

12° × 14° area of the sky on a single photograph. The new camera went into operation in 1929.

In February 1930, a young observing assistant named Clyde Tombaugh (see *Voyagers in Astronomy* box), comparing photographs he made on January 23 and 29 of that year (Figure 11.15), found an object whose motion appeared to be about right for a planet far beyond the orbit of Neptune. It was within 6° of the position Lowell had predicted for the unknown planet. The new planet was named for Pluto, the Roman god of the underworld, who dwelt in remote darkness, just like the new planet. The choice of this name, among hundreds suggested, was helped by the fact that the first two letters were Percival Lowell's initials.

Although at the time the discovery of Pluto appeared to be a vindication of gravitational theory similar to the triumph of Adams and Leverrier in predicting the position of Neptune, we now know that Lowell's calculations were wrong. When the mass of Pluto was finally measured, it was found to be much less than that of the Moon. It could not possibly have exerted any measurable pull on either Uranus or Neptune. To the degree that the orbital discrepancies of these two planets are real, they must be due to some other cause.

A survey of the entire sky in 1983 by the Infrared Astronomical Satellite (IRAS), however, has revealed no hidden "Planet X." Nor have we found any measurable pull from an unknown planet on the Voyager and Pioneer spacecraft whose paths are taking them beyond the orbit of

FIGURE 11.15
Pluto's Motion Portions of the two photographs on which Clyde Tombaugh discovered Pluto in 1930. The left one was taken on January 23 and the right one on January 29. Note that Pluto, indicated by an arrow, has moved among the stars during those six nights. Note also that if we hadn't put an arrow next to it, you probably would never have spotted the dot that moved. *(Lowell Observatory)*

Clyde Tombaugh: From Farm to Fame

Clyde Tombaugh discovered Pluto when he was 24 years old, and his position as staff assistant at the Lowell Observatory was his first paying job. Tombaugh had been born on a farm in Illinois, but when he was 16 his family moved to Kansas. There, with his uncle's encouragement, he observed the sky through a telescope the family had ordered from the Sears catalog. He later constructed a larger telescope on his own and devoted his nights (when he wasn't too tired from farm work) to making detailed sketches of the planets.

In 1928, after a hailstorm ruined the crop, Tombaugh decided he needed a job to help support his family. Although he had only a high school education, he thought of becoming a telescope builder. He sent his planet sketches to the Lowell Observatory, seeking advice about whether such a career choice was realistic. By a wonderful twist of fate, his query arrived just when the Lowell astronomers realized that a renewed search for a ninth planet would require a very patient and dedicated observer.

The large photographic plates (pieces of glass with photographic emulsion on them) that Tombaugh was hired to take at night and search during the day contained an average of about 160,000 star images each. How to find Pluto among them? The technique involved taking two photographs about a week apart. During that week, a planet would move a tiny bit, while the stars remained in the same place relative to each other. A new instrument called a "blink comparator" could quickly alternate the two images in an eyepiece. The stars, being in the same position on the two plates, would not appear to change as the two images were "blinked." But a moving object would appear to wiggle back and forth as the plates were alternated.

After examining more than 2 million stars (and many false alarms), Tombaugh found his planet on February 18, 1930. The astronomers at the

Clyde Tombaugh at the Lowell Observatory at the time of his discovery of Pluto in 1930. *(Lowell Observatory)*

> *He later constructed a larger telescope on his own and devoted his nights to making detailed sketches of the planets*

observatory checked his results carefully, and the find was announced on March 13, the 149th anniversary of the discovery of Uranus. Congratulations and requests for interviews poured in from around the world. Visitors descended on the observatory by the scores, wanting to see the place where the first new planet in almost a century had been discovered, as well as the person who had discovered it.

In 1932 Tombaugh took leave from Lowell, where he had continued to search and blink, to get a college degree. Eventually he received a master's degree in astronomy and taught navigation for the Navy during World War II. In 1955, after working to develop a rocket-tracking telescope, he became a professor at New Mexico State University, where he helped found the astronomy department. He died in 1997.

Pluto and into interstellar space. Today it is generally accepted that the supposed perturbations of Uranus and Neptune are not, and never were, real.

Pluto's Motion and Satellite

Pluto's orbit has the highest inclination to the ecliptic (17°) of any planet, and also the largest eccentricity (0.248). Its average distance from the Sun is 40 AU, or 5.9 billion km, but its perihelion distance is under 4.5 billion km, within the orbit of Neptune. Even though the orbits of these two

planets cross, there is no danger of collision because of the high inclination of Pluto's orbit.

Pluto completes its orbital revolution in a period of 248.6 years; since its discovery in 1930, it has traversed less than one quarter of its long path around the Sun. But almost 50 years after that discovery, astronomers found that it is not alone in its distant travels.

In 1978, astronomer James Christy of the U.S. Naval Observatory found that Pluto has a satellite, which was named Charon after the boatman who ferried the dead into the realm of Pluto in mythology. The exact nature of its orbit was not confirmed until 1985, when the system had turned enough that the satellite and the planet began to go behind each other as seen from the Earth. These observations showed that the satellite was in a retrograde orbit and had a diameter of about 1200 km, more than half the size of Pluto itself. This makes Charon the satellite whose size is the largest fraction of its parent planet. Seen from Pluto, Charon would be as large as eight full moons on Earth, side by side in the sky.

Furthermore, Pluto orbits "on its side" like Uranus, and Charon circles in its equatorial plane around its rotation axis. The gravitational interaction between the two worlds has locked them in the tightest of embraces. Not only does Charon take the same time to rotate and revolve, thus keeping the same face toward Pluto, but this period is the same length as Pluto's day. This means that if you were on the Charon-facing side of Pluto, you would see the moon hover in the same place all the time, while from Pluto's opposite hemisphere Charon would never be visible.

The Nature of Pluto

Pluto has not been visited by spacecraft, and it is so faint that the world's largest telescopes are required to study it. The diameter of Pluto is 2190 km, only 60 percent that of the Moon. From the diameter and mass, we find a density of 2.1 g/cm^3, suggesting that Pluto is a mixture of rocky materials and water ice in about the same proportions as in Triton.

Pluto's surface is highly reflective, and its spectrum demonstrates the presence of frozen methane, carbon monoxide, and nitrogen (Figure 11.16). The surface temperature ranges from about 50 K when Pluto is farthest from the Sun to 60 K when it's closest. Even this small difference is enough to cause a partial evaporation of the methane and nitrogen ice to generate an atmosphere when Pluto is closer to the Sun. During recent decades, Pluto has been in its warmest period, near perihelion, and its atmosphere is accordingly near maximum size and density.

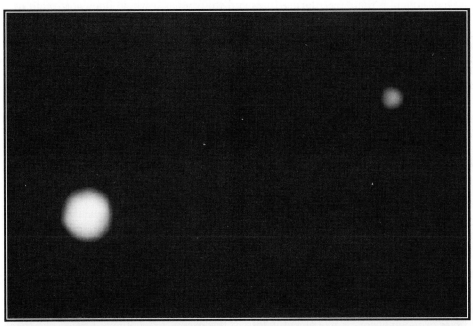

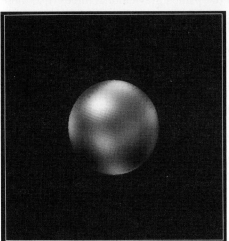

FIGURE 11.16
Pluto and Charon *(Top)* An image of Pluto and its satellite Charon, taken with the Faint Object Camera on the Hubble Space Telescope. The clarity of the image allowed astronomers to measure the diameters of the two worlds to an accuracy of 1%.
(R. Albrecht, ESA/ESO and NASA)
(Bottom) A global map of the two hemispheres of Pluto, constructed using computer processing techniques from images taken during Pluto's full 6.4-day rotation period. The tile pattern is an artifact of the image enhancement techniques. Pluto was over 4.8 billion km from Earth at the time the images were taken (Summer 1994). A number of bright and dark features are visible, including a prominent northern polar cap.
(A. Stern, M. Buie, NASA and ESA)

Observations of a distant star seen through that thin atmosphere suggest that the surface pressure is about a ten-thousandth of the Earth's. Because Pluto is a few degrees warmer than Triton, its atmospheric pressure is about ten times greater.

In February 1999, Pluto crossed the orbit of Neptune and became the ninth planet once more (it had been eighth from the Sun since 1979). Now it will slowly swing further and further from the Sun, and more and more of its atmosphere will slowly freeze out.

The Origin of Pluto

We have a long way to go to understand how Pluto arrived at its unusual place in the outer solar system. In some ways Pluto is like the odd cousin whom everyone hopes will not show up at the next family reunion. Neither its path around the Sun nor its size resembles the jovian planets. The presence of a moon as large as Charon is also a mystery. As we have seen, Pluto is much more similar to a moon such as Triton than to a planet such as Saturn. One thought is that Pluto and Charon are merely the largest examples of a whole family of distant asteroid- or comet-like bodies (which is like discovering that your odd cousin is actually not related to you). As we will discuss in Chapter 12, we have recently discovered some smaller siblings in that family.

It is interesting to note that Triton is the only large satellite in the solar system with a retrograde orbit. The most distant of Neptune's satellites, Nereid, the one beyond Triton, has an orbit that is direct but very large and highly elliptical. Some astronomers speculate that the strange orbits of Pluto, Charon, Nereid, and Triton are the result of violent collisions during the early history of the outer solar system.

11.4 PLANETARY RINGS

All four of the giant planets have rings, with each ring system consisting of billions of small particles or moonlets orbiting close to their planet. Each of these rings displays a complicated structure that seems related to interactions between the ring particles and the larger satellites. However, the four ring systems are very different from each other in mass, structure, and composition (Table 11.2).

Saturn's large ring system is made up of icy particles spread out into several vast, flat rings containing a great deal of fine structure. The Uranus and Neptune ring systems, on the other hand, are nearly the reverse of Saturn's: They consist of dark particles confined to a few narrow rings with broad empty gaps in between. Jupiter's ring and at least one of Saturn's are merely transient dust bands, constantly renewed by erosion of dust grains from small satellites. In this section we focus on the two most massive ring systems, those of Saturn and Uranus.

TABLE 11.2
Properties of Ring Systems

| Planet | Outer Radius | | Mass | Reflectivity |
	(km)	(R_{planet})	(kg)	(%)
Jupiter	128,000	1.8	10^{10}(?)	?
Saturn	140,000	2.3	10^{19}	60
Uranus	51,000	2.2	10^{14}	5
Neptune	63,000	2.5	10^{12}	5

What Causes Rings?

A ring is a collection of vast numbers of particles, each obeying Kepler's laws as it follows its own orbit around the planet. Thus the inner particles revolve faster than those farther out, and the ring as a whole does not rotate as a solid body. In fact, it is better not to think of a ring rotating at all, but rather to consider the revolution of its individual moonlets.

If the ring particles were widely spaced, they would move independently, like separate small satellites. However, in the main rings of Saturn and Uranus the particles are close enough to each other to exert mutual gravitational influence, and occasionally even to rub together or bounce off each other in low-speed collisions. Because of these interactions, we see phenomena such as waves that move across the rings the way water waves move over the surface of the ocean.

There are two basic theories of how such rings come to be. First is the breakup theory, which suggests that the rings are the remains of a shattered satellite. The second theory, which takes the reverse perspective, suggests that the rings are made of particles that were unable to come together to form a satellite in the first place. In either theory, an important role is played by the gravitation of the planet. Close to the planet, gravitational forces (like those that cause tides on the Earth) can tear bodies apart or inhibit loose particles from coming together. The rings of both Saturn and Uranus lie within the region in which large satellites are probably not stable against such forces (Figure 11.17).

In the breakup theory of ring formation, we can imagine a satellite or even a passing comet coming too close and being torn apart. The fragments then remain in orbit as one or more rings. A more likely variant of this idea suggests that a small satellite close to the planet might have broken apart in a collision, with the fragments dispersing into a disk. We do not know which explanation holds for the rings, although many scientists have concluded that at least a few of the rings are relatively young and must therefore be the result of breakup.

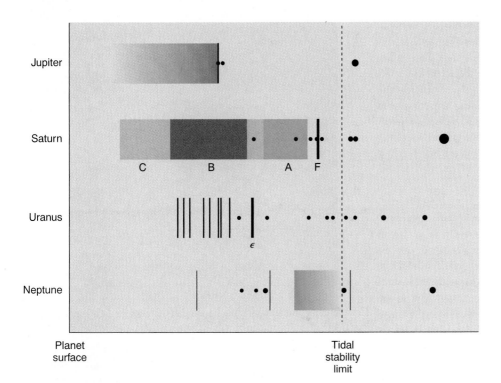

FIGURE 11.17
Four Ring Systems   A diagram showing the location of the ring systems of the four giant planets. The left axis represents the planet's surface. The dotted vertical line is the limit inside which gravitational forces can break up satellites. Notice that each planet's system is drawn *to a different scale* so that this stability limit lines up for all four of them. The black dots are the inner satellites of each planet on the same scale as its rings. Notice that only really small bodies survive inside the stability limit.

Rings of Saturn

The rings of Saturn are one of the most beautiful sights in the solar system (Figure 11.18). From outer to inner, the three brightest rings are labeled with the extremely romantic names of A, B, and C Ring. In Table 11.3 the dimensions of the rings are given in both kilometers and in units of the radius of Saturn, R_s. The B Ring is the brightest and has the most closely packed particles, whereas the A and C Rings are translucent. The total mass of the B Ring, which is probably close to the mass of the entire ring system, is about equal to that of an icy satellite 250 km in diameter. Between the A and B Rings is a wide gap named the Cassini Division after Gian Domenico Cassini, who first glimpsed it in 1675.

The rings of Saturn (which you can see clearly in Figure 10.10) are very broad and very thin. The width of the main rings is 70,000 km, yet their thickness is only about 20 m. If we made a scale model of the rings out of paper the thickness of the pages in this book, we would have to make them 1 km across—about eight city blocks. On this scale, Saturn itself would loom as high as an 80-story building.

The ring particles are composed primarily of water ice, and they range from grains the size of sand up to house-sized boulders. An insider's view of the rings would probably resemble a bright cloud of floating snowflakes and

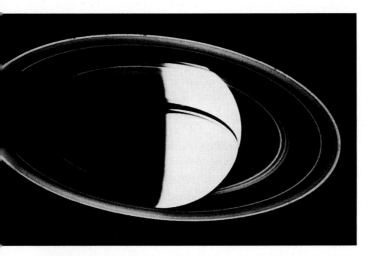

FIGURE 11.18
Saturn's Rings from Underneath   Since you are looking up at sunlight filtering through the rings in this Voyager image, the brightest parts of the ring system are the gaps and the least dense regions. The denser B Ring is not transmitting much light.   *(NASA/JPL)*

TABLE 11.3
Rings of Saturn

Ring name	Outer Edge		Width (km)
	(R_S)	(km)	
F	2.324	140,180	90
A	2.267	136,780	14,600
Cassini Division	2.025	122,170	4,590
B	1.949	117,580	25,580
C	1.525	92,000	17,490

hailstones, with a few snowballs and larger objects, many of these loose aggregates of smaller particles (Figure 11.19).

In addition to the broad A, B, and C Rings, Saturn has a handful of very narrow rings, no more than 100 km wide. The most substantial of these, which lies just outside the A Ring, is called the F Ring; its surprising appearance is discussed later in this section. In general, Saturn's narrow rings resemble the rings of Uranus and Neptune.

Rings of Uranus and Neptune

The rings of Uranus are narrow and black, making them almost invisible from the Earth. The nine main rings were discovered in 1977 from observations made of a star as Uranus passed in front of it. We call the passage of one astronomical object in front of another an *occultation*. During the 1977 occultation, astronomers expected the star's light to disappear as the planet moved across it. But in addition the star dimmed briefly several times before Uranus even reached it, as each narrow ring passed between the star and the telescope. Thus the rings were mapped out in detail even though they could not be seen or photographed directly. When Voyager approached Uranus in 1986, it was able to study the rings at close range; the spacecraft also photographed two new rings (Figure 11.20).

The outermost and most massive of the rings of Uranus is called the Epsilon Ring. It is only about 100 km wide and probably no more than 100 m thick (similar to the F Ring of Saturn). The Epsilon Ring encircles Uranus at a distance of 51,000 km, about twice the radius of Uranus. This ring probably contains as much mass as all of Uranus' other ten rings combined; most of them are narrow ribbons less than 10 km wide—just the reverse of the broad rings of Saturn.

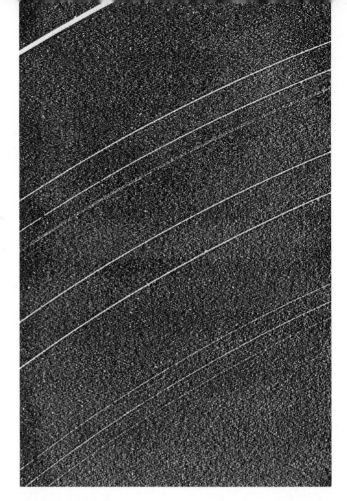

FIGURE 11.20
The Narrow Rings of Uranus The Voyager team had to expose this image for a long time to get a glimpse of these dark rings; you can see the grainy structure of "noise" in the electronics of the camera in the picture background. *(NASA/JPL)*

FIGURE 11.19
Ring Pieces An artist's idealized version of the sizes of typical particles inside the A Ring of Saturn. The ball at the center is roughly the size of an adult's head. In the real rings, the particles are not spherical but are likely to have irregular shapes. *(NASA/JPL)*

The individual particles in the uranian rings are nearly as black as lumps of coal. While astronomers do not understand the composition of this material in detail, it seems to consist in large part of black carbon and hydrocarbon compounds. Organic material of this sort is rather common in the outer solar system. Many of the asteroids and comets (discussed in Chapter 12) also are composed of dark, tar-like materials. In the case of Uranus, its ten small inner satellites have a similar composition, suggesting that one or more satellites might have broken up to make the rings.

The rings of Neptune are generally similar to those of Uranus but even more tenuous (Figure 11.21). There are only four of them, and the particles are not uniformly distributed along their lengths. Because these rings are so difficult to investigate from the Earth, it will probably be a long time before we understand them very well.

Satellite–Ring Interactions

Much of our current fascination with planetary rings is a result of the intricate structures discovered by the Voyager

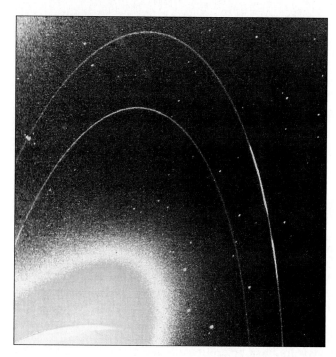

FIGURE 11.21
Neptune's Rings A long exposure of the rings of Neptune, as photographed by Voyager in 1989. Note the two denser regions of the outer ring. *(JPL/NASA)*

spacecraft. We now understand that most of these structures owe their existence to the gravitational effect of satellites. Without satellites the rings would be flat and featureless. Indeed, without satellites there would probably be no rings at all, since left to themselves, thin disks of matter gradually spread and dissipate.

Most of the gaps in Saturn's rings, and also the location of the outer edge of the A Ring, result from gravitational resonances with small inner satellites. A **resonance** takes place when two objects have orbital periods that are exact ratios of each other, such as one-to-two or one-to-three. For example, any particle in the gap at the inner side of the Cassini Division of Saturn's rings would have a period equal to half that of Saturn's satellite Mimas. Such a particle would be nearest Mimas in the same part of its orbit every second revolution. The repeated gravitational tugs of Mimas, acting always in the same direction, would perturb it, forcing it into a new orbit outside the gap and therefore no longer representing a resonance.

One of the most interesting rings of Saturn is the narrow F Ring, which contains several apparent ringlets within its 90-km width. In places the F Ring breaks up into two or three parallel strands that sometimes show bends or kinks. Most of the rings of Uranus and Neptune are also narrow ribbons like the F Ring of Saturn. Clearly, the gravity of some objects must be keeping the particles in these rings from spreading out.

The best theory is that the rings are controlled gravitationally by small satellites orbiting very close to them. This certainly seems to be the case for Saturn's F Ring: Voyager showed that it is bounded by the orbits of two satellites, now called Pandora and Prometheus (Figure 11.22). These two small objects (each about 100 km in diameter) are referred to as *shepherd satellites* since their gravitation serves to "shepherd" the ring particles and keep them confined to a narrow ribbon. A similar situation applies to the Epsilon Ring of Uranus, which is shepherded by the satellites Cordelia and Ophelia. These two shepherds, each about 50 km in diameter, orbit about 2000 km inside and outside the ring.

Theoretical calculations suggest that the other narrow rings in the uranian and neptunian systems should also be controlled by shepherd satellites, but none has been located. The calculated diameter for such shepherds—about 10 km—was just at the limit of detectability for the Voyager cameras, so it is impossible to say if they are present or not. (Given all the narrow rings we see, many scientists still hope to find another, more satisfactory mechanism for keeping them confined.)

Just recently, astronomers have found a relatively narrow ring of dust around a star (called HR 4796A) and have applied the same kind of analysis to suggest that there must be shepherding *planets* on either side of the ring to keep it so narrow. (At the distance of the star, such planets cannot be seen with our present telescopes—the same situation we face for the smaller shepherd moons around the outer planets.) This is a wonderful example of how techniques developed in one branch of astronomy often find application in another.

The main rings of Saturn include several narrow gaps that are the result of embedded satellites, which clear lanes in a manner similar to that of the shepherd satellites. As each small satellite moves through its gap, it produces waves in

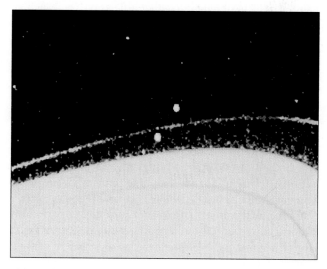

FIGURE 11.22
Saturn's F Ring and Its Shepherds This Voyager image shows the narrow, complex F Ring of Saturn, with its two shepherd satellites Pandora and Prometheus. *(NASA/JPL)*

the surrounding ring material like the wake left by a moving ship. One of these satellites, called Pan, was discovered in 1991 from its wake; calculations based on the waves it produced pinpointed its location, and when the appropriate ten-year-old Voyager picture was examined, there it was!

Studies of planetary rings, and of the gravitational interactions between rings and small satellites, have come a long way in the past few years, but many problems in understanding these complex phenomena remain unsolved. Without additional spacecraft missions to the outer planets, however, it may be difficult to obtain the data necessary to test the new dynamical theories now being developed. The Cassini mission to Saturn—which left Earth in 1998 and is scheduled to begin orbiting the planet in 2004—should provide a wealth of new information on planetary rings.

SURFING THE WEB

Note: Many of the sites about planets and planetary missions listed for Chapters 6 and 10 also include good information about the satellites of the planets. For example, the Nine Planets site has separate sections for many of the main moons in the solar system, and the Galileo mission site has a terrific collection of images of and information about the Galilean moons of Jupiter.

⌨ Europa Orbiter Mission Site
[www.jpl.nasa.gov/ice_fire/euorpao.htm]
Gives the latest plans on a mission to orbit the intriguing moon Europa and try to probe under its ice.

⌨ General Pluto Sites

- The Nine Planets Site: [seds.lpl.arizona.edu/nineplanets/nineplanets/pluto.html]
- Views of the Solar System Site: [www.hawastsoc.org/solar/eng/pluto.htm]
- Planetary Sciences Site: [nssdc.gsfc.nasa.gov/planetary/planets/plutopage.html]

⌨ The Clyde Tombaugh Web Site
[www.klx.com/clyde/index.html]
The discoverer of Pluto died in January 1997 at age 90. Jeff Faust at MIT has brought together articles on Tombaugh and his work, information about Pluto, and links that feature Tombaugh.

⌨ Pluto Home Page
[dosxx.colorado.edu/plutohome.html]
A site from Fran Bagenal at the University of Colorado with good images, diagrams (with captions), brief information, and links about Pluto.

⌨ Pluto-Kuiper Express Mission
[www.jpl.nasa.gov/pluto/pkexprss.htm]
The Jet Propulsion Laboratory is developing plans for possible robotic exploration of the Pluto system in the next century. This site can keep you up to date on plans and technology for such a mission.

⌨ The Planetary Rings Node
[ringside.arc.nasa.gov/]
Mark Showalter at NASA's Ames Research Center maintains this somewhat technical site devoted to the rings of the jovian planets. It includes lots of images, movies, links, and data-sets.

SUMMARY

11.1 The four jovian planets are accompanied by impressive systems of satellites and rings. Jupiter has 16 satellites, Saturn 19, Uranus 18, and Neptune 8, many discovered by Voyager. Most of these satellites are composed, in part, of water ice, with varying amounts of rock and dark primitive material mixed in. Of the four ring systems, Saturn's is the largest; Uranus and Neptune have narrow rings of dark material; and Jupiter has a tenuous ring of dust.

11.2 The largest satellites are Ganymede, Callisto, and Titan, all low-density objects that are composed of more than half water ice. Callisto has an ancient cratered surface while Ganymede shows evidence of extensive tectonic and vol-canic activity, persisting until perhaps 1 billion years ago. Voyager was not able to see Titan's surface because it is unique among jovian-planet moons in having a thick atmosphere (surface pressure of 1.6 bar) with an opaque haze layer. Io and Europa are denser and smaller, each about the size of the Moon. Europa is the only other object in the solar system thought to have a global ocean of liquid water (under an ice crust). Io is the most volcanically active object in the solar system; its eruptions of silicate lava and sulfur compounds are powered by tidal interactions with Jupiter.

11.3 Triton, Neptune's retrograde satellite, and Pluto resemble each other in size, composition, and temperature.

Triton has a very thin atmosphere, and we can see eruptions of nitrogen when surface layers of its extensive polar cap evaporate while facing the Sun. The planet Pluto, discovered by Clyde Tombaugh in 1930, has never been visited by spacecraft. Observations indicate the presence of frozen methane, carbon monoxide, and nitrogen on its surface, and a thin (perhaps temporary) atmosphere. Pluto's moon Charon has a diameter more than half of Pluto's, complicating our theories of how these bodies formed.

11.4 Rings are composed of vast numbers of individual particles (ranging in size), orbiting so close to a planet that its gravitational forces could have broken larger pieces apart or kept pieces from gathering together. Saturn's rings are broad, flat, and nearly continuous, except for a handful of gaps. The particles are mostly water ice, with typical dimensions of a few centimeters. The rings of Uranus are narrow ribbons separated by wide gaps and contain much less mass. They were discovered in 1977 during a stellar occultation. Neptune's rings are similar, but contain even less material. Much of the complex structure of the rings is due to waves and **resonances** induced by satellites within the rings or orbiting outside them. Voyager found that several thin rings were accompanied by shepherd moons that kept the ring material confined.

INTER-ACTIVITY

A Imagine it's the distant future and humans can now travel easily among the planets. Your group is a travel agency, with the task of designing a really challenging tour of the Galilean satellites for a group of sports enthusiasts. What kind of activities are possible on each world? How would rock climbing on Ganymede, for example, differ from rock climbing on Earth? (If you design an activity for Io, you had better bring along very significant radiation shielding. Why?)

B This chapter covered only a few of the dozens of satellites in the outer solar system. Using the Web or your college library, organize your group into a research team and find out more about one of the moons that was not covered in any detail. Your authors' favorites include Uranus' Miranda with its jigsaw puzzle surface, Saturn's Mimas with a "knock-out" crater called Herschel, and Saturn's Enceladus, whose two hemispheres differ significantly; but do be sure to pick a world you are attracted to. Prepare a report to attract tourists to the world you selected.

C In a novel entitled *2010*, science fiction writer Arthur C. Clarke, inspired by the information coming back from the Voyager spacecraft, had fun proposing a life-form under the ice of Europa that was evolving toward intelligence. Suppose future missions do indeed find some sort of life (not necessarily intelligent, but definitely alive) under the ice of Europa—life that evolved completely independently of life on Earth. Have your group discuss what effect such a discovery would have on humanity's view of itself. What should be our attitude toward such a life-form? Do we have an obligation to guard it against contamination by our microbes and viruses? Or, to take an extreme position, should we wipe it out before it becomes competitive with Earth life or contaminates our explorers with microorganisms we are not prepared to deal with? Who should be in charge of making such decisions?

D Work together to make a list of all the reasons you can think of why it is hard to send a mission to Pluto. How would you design a mission that will get there despite your list of reasons?

E Your group has been asked by NASA to come up with one or more missions to learn about Europa. Review what we know about this satellite so far, then design a robotic mission that would answer some of the questions we have. You can assume that budget is not a factor, but your instruments have to be realistic. (Bear in mind that Europa is cold and far from the Sun.)

F Imagine your group is the first landing party on Pluto (let's hope you remembered to bring long underwear!) You land in a place where Charon is visible in the sky and you observe Charon for one Earth week. Describe what Charon will look like during that week. Now you move your camp to the opposite hemisphere of Pluto. What will Charon look like there during the course of a week?

REVIEW QUESTIONS

1. What are the satellites of the outer planets made of, and why is their composition different from that of the Moon?

2. Compare the geology of Callisto, Ganymede, and Titan. Why do we think Titan has an atmosphere while the other two large satellites do not?

3. Explain the energy source that powers the volcanoes of Io.

4. Compare the properties of Titan's atmosphere with those of the Earth's atmosphere.

5. How was Pluto discovered? Why did it take so long to find it?

6. How are Triton and Pluto similar?

7. Describe and compare the rings of Saturn and Uranus.

8. Three possibilities were suggested in the text for the origin of Saturn's rings. List them, and briefly summarize the arguments in favor of each.

9. Why do you think the outer planets have such extensive systems of rings and satellites, while the inner planets do not?

10. Which would have the longer orbital period, a satellite 1 million km from the center of Jupiter or a satellite 1 million km from the center of Earth? Why?

11. Ganymede and Callisto were the first icy objects to be studied from a geological point of view. Summarize the main differences between their geology and that of the rocky terrestrial planets.

12. Compare the properties of the volcanoes on Io with those of terrestrial volcanoes. Give at least two similarities and two differences.

13. Would you expect to find more impact craters on Io or Callisto? Why?

14. Why is it unlikely that humans will be traveling to Io? (*Hint:* Review the information about Jupiter's magnetosphere in Chapter 10.)

15. Where did the nitrogen in Titan's atmosphere come from? Compare its origin with that of the nitrogen in our atmosphere.

16. Do you think there are many impact craters on the surface of Titan? Why or why not?

17. Why do you suppose the rings of Saturn are made of bright particles, whereas the particles in the rings of Uranus and Neptune are black?

18. Suppose you miraculously removed all of Saturn's satellites. What would happen to its rings?

PROBLEMS

19. Saturn's A, B, and C Rings extend about 75,000 to 137,000 km from the center of the planet. Use Kepler's third law to calculate the difference between how long the inner edge and the outer edge of the three-ring system take to revolve about the planet.

20. Use the information in Appendix 8 to calculate what you would weigh on Titan, Io, and Uranus' satellite Miranda.

21. Occultations of stars by the rings of Uranus have yielded resolutions of 10 km in determining ring structure. What angular resolution (in arcsec) would a space telescope have to achieve to obtain equal resolution from Earth orbit? How close to Uranus would a spacecraft have to get to obtain equal resolution with a camera having an angular resolution of 2 arcsec? To solve this problem, you need the "small-angle formula" to relate angular and linear size in the sky. It is usually written as

$$\frac{\text{Angular diameter}}{206{,}265} = \frac{\text{Linear diameter}}{\text{Distance}}$$

where angular diameter is expressed in arcsec.

SUGGESTIONS FOR FURTHER READING

On Satellites

Carroll, M. "Europa: Distant Ocean, Hidden Life?" in *Sky & Telescope*, Dec. 1997, p. 50.

Elliot, J. "The Warming Wisps of Triton" in *Sky & Telescope*, Feb. 1999, p. 42.

Hartmann, W. "View from Io" in *Astronomy*, May 1981, p. 17.

Johnson, T. et al. "The Moons of Uranus" in *Scientific American*, Apr. 1987.

Milstein, M. "Diving into Europa's Ocean" in *Astronomy*, Oct. 1997, p. 38.

Morrison, D. "An Enigma Called Io" in *Sky & Telescope*, Mar. 1985, p. 198.

Rothery, D. *Satellites of the Outer Planets: Worlds in Their Own Right.* 1992, Oxford U. Press.

Talcott, R. "Jumping Jupiter" in *Astronomy*, June 1998, p. 40. On Galileo mission results.

Talcott, R. "The Violent Volcanoes of Io" in *Astronomy*, May 1993, p. 41.

On Pluto

Binzel, R. "Pluto" in *Scientific American*, June 1990.

Burnham, R. "At the Edge of Night: Pluto and Charon" in *Astronomy*, Jan. 1994, p. 41.

Harrington, R. and Harrington, B. "The Discovery of Pluto's Moon" in *Mercury*, Jan./Feb. 1979, p. 1.

Hoyt, W. *Planets X and Pluto.* 1980, U. of Arizona Press. Superb history of the search for Pluto and other outer planets.

Levy, D. *Clyde Tombaugh: Discoverer of Pluto.* 1991, U. of Arizona Press.

Stern, A. and Milton, J. *Pluto and Charon.* 1998. Wiley. Good summary of our modern understanding.

Tombaugh, C. "The Discovery of Pluto" in *Mercury*, May/June 1986, p. 66, and Jul./Aug. 1986, p. 98.

On Rings

Elliot, J. et al. "Discovering the Rings of Uranus" in *Sky & Telescope,* June 1977, p. 412.

Elliot, J. and Kerr, R. *Rings: Discoveries from Galileo to Voyager.* 1985, MIT Press.

Esposito, L. "The Changing Shape of Planetary Rings" in *Astronomy,* Sep. 1987, p. 6.

Sobel, D. "Secrets of the Rings" in *Discover,* Apr. 1994, p. 86. Discusses the outer planet ring systems.

See also the books on the giant planets recommended in Chapter 10 and the books on the planetary system in general listed in Chapter 6.

Comet Hale–Bopp was one of the most attractive and easiest to see comets of the 20th century. It is shown here as it appeared in the sky on March 8, 1997, near the much more distant North America Nebula (the reddish glowing cloud of gas). You can see the comet's long blue ion tail and the shorter white dust tail. *(Tony Hallas, AstroPhoto)*

12 Comets and Asteroids: Debris of the Solar System

THINKING AHEAD

Hundreds of smaller members of the solar system, called asteroids and comets, are known to have crossed the Earth's orbit in the past, and many others will do so in the centuries ahead. What could we do if we knew a few years in advance that one of these bodies (with significant mass) would hit the Earth?

"The day will yet come when posterity will be amazed that we remained ignorant of things that will to them seem so plain.... Men will someday be able to demonstrate in what regions comets have their paths [and] what is their size and constitution. Let us be satisfied with what we have discovered, and leave a little truth for our descendants to find out."

Lucius Annaeus Seneca, in *Natural Questions* (about 63 A.D.)

To understand the early history of life on Earth, scientists need fossils from as long ago as possible. To piece together the early history of the solar system, we need some cosmic fossils—pieces formed when our system was very young. When it comes to answering our questions about origins, the planets themselves are largely mute. Melted, battered by giant impacts, twisted by tectonic forces, they retain little evidence of their births. Reconstructing their early history is almost as difficult as determining the circumstances of human birth by merely looking at an adult. Instead, we must turn to clues provided by the surviving remnants of the creation process—ancient but smaller bodies with which we share our cosmic neighborhood.

We have rather arbitrarily divided these bodies into two categories. **Asteroids** are small, rocky objects containing little volatile (easily evaporated) material. They differ from the terrestrial planets primarily in size and are sometimes even called minor planets. **Comets** are small, icy pieces containing frozen water and other volatile materials (but with solid grains mixed in). Comets are most easily noticed when they get near the Sun and form a visible (but temporary) atmosphere and an extended tail of gas and dust.

FIGURE 12.1

Asteroid Trails Time exposure showing trails left by asteroids (arrows). The telescope moved backward to compensate for the rotation of the Earth; thus the stars remained fixed. But in the time the image was taken, the asteroids moved relative to the stars and left a trail. *(Yerkes Observatory)*

 ASTEROIDS

Discovery and Orbits of the Asteroids

The orbits of most of the asteroids lie between those of Mars and Jupiter, in a region called the **asteroid belt.** The asteroids are too small to be seen without a telescope; hence the first of them was not discovered until the beginning of the 19th century. At that time, astronomers were hunting for an additional planet they thought should exist in the large gap between the orbits of Mars and Jupiter. Some German astronomers even organized themselves into a group they called "the celestial police" to search for this "missing" member of the solar system.

In January 1801, the Sicilian astronomer Giovanni Piazzi thought he had found this missing planet when he discovered the first asteroid, which he named Ceres, orbiting at 2.8 AU from the Sun. However, his discovery was followed the next year by that of another little planet in a similar orbit, and two more were found in 1804 and 1807. Clearly, there was not a single missing planet between Mars and Jupiter, but rather a whole group of objects, each no more than 1000 km across. By 1890 more than 300 had been discovered by sharp-eyed observers. In that year, Max Wolf of Heidelberg introduced astronomical photography to the search for asteroids, greatly accelerating the discovery of additional objects (Figure 12.1). More than 10,000 asteroids now have well-determined orbits.

Asteroids are given both a number (corresponding to the order of discovery) and a name. Originally, the names of asteroids were chosen from goddesses in Greek and Roman mythology. These, however, were soon used up. After exhausting other female names (including those of wives, friends, flowers, cities, colleges, pets, and the like), astronomers turned to the names of colleagues whom they wish to honor for contributions to the field. For example, asteroids 2410 and 4859 are named Morrison and Fraknoi, respectively, for two of the authors of this text.

It would be a formidable task to discover, determine orbits for, and catalog all the asteroids bright enough to be photographed with modern telescopes. Nevertheless, the total number of such objects can be estimated by systematically sampling regions of the sky. These studies indicate that there are approximately a million (10^6) asteroids with diameters greater than 1 km.

The largest asteroid is Ceres, with a diameter just under 1000 km. Two, Pallas and Vesta, have diameters near 500 km, and about 15 more are larger than 250 km (Table 12.1). The number of asteroids increases rapidly with decreasing size; there are about 100 times more objects 10 km across than there are 100 km across. The total mass in all the asteroids, which probably represent just a tiny fraction of the original asteroid population, is less than that of the Moon. The orbits of the four largest asteroids were illustrated in Figure 2.10.

The asteroids all revolve about the Sun in the same west-to-east direction as the planets, and most of their orbits lie near the plane in which the Earth and other planets circle. We define the asteroid belt as the region that contains all asteroids with semimajor axes (see Chapter 2) in the range 2.2 to 3.3 AU. Asteroids at these distances take 3.3 to 6 years to orbit the Sun (Figure 12.2). Although more than 75 percent of the known asteroids are in the main belt, they are not closely spaced. The volume of the belt is actually very large, and the typical spacing between objects (down to 1 km in size) is several million kilometers. (This is fortunate for spacecraft like Galileo and Cassini that need to travel through the belt without a collision.)

In 1917, the Japanese astronomer Kiyotsuga Hirayama found that a number of the asteroids fall into *families*, groups with similar orbital characteristics. He hypothesized that each family may have resulted from an explosion of a larger body or, more likely, from the collision of two bodies. Slight differences in the speed with which the various fragments left the collision scene account for the small spread in orbits now observed for the different asteroids in a given family. Several dozen such families exist, and observations have shown that individual members of the larger ones are physically similar, just as we would expect if they were fragments of a common parent. The existence of these families shows that asteroids must have collided frequently during the history of the solar system.

Composition and Classification

Asteroids are as different as black and white. The majority are very dark, with reflectivities of only 3 to 4 percent, like

TABLE 12.1
The Largest Asteroids

Name	Year of Discovery	Semimajor Axis (AU)	Diameter (km)	Class*
Ceres	1801	2.77	940	C
Pallas	1802	2.77	540	C
Vesta	1807	2.36	510	†
Hygeia	1849	3.14	410	C
Interamnia	1910	3.06	310	C
Davida	1903	3.18	310	C
Cybele	1861	3.43	280	C
Europa	1868	3.10	280	C
Sylvia	1866	3.48	275	C
Juno	1804	2.67	265	S
Psyche	1852	2.92	265	M
Patientia	1899	3.07	260	C
Euphrosyne	1854	3.15	250	C

* C = carbonaceous; S = stony; M = metallic.

† Vesta has a very unusual (once thought unique) basaltic surface.

a lump of coal. However, another large group has typical reflectivities of 15 to 20 percent, a little greater than that of the Moon, and still others have reflectivities as high as 60 percent. To understand more about these differences, astronomers can study the spectra of the light reflected from asteroids for clues about their composition.

The dark asteroids are revealed from spectral studies to be *primitive* bodies (chemically unchanged since the beginning of the solar system) composed of silicates mixed with dark, organic carbon compounds. Two of the largest asteroids, Ceres and Pallas, are primitive, as are almost all of the objects in the outer third of the belt. Most of the primitive asteroids are classed as C asteroids, where C stands for carbonaceous or carbon-rich, but several other classes of primitive objects with different minerals have also been identified.

The second most populous asteroid group is that of the S asteroids, where S stands for a stony composition. Here the dark carbon compounds are missing, resulting in higher reflectivities and clearer spectral signatures of silicate minerals. It appears that most of the S-type asteroids are also primitive.

Asteroids of a third class, much less numerous than those of the first two, are composed primarily of metal and are called M asteroids. Spectroscopically, the identification of metal is difficult, but for at least the largest M asteroid, Psyche, this identification has been confirmed by radar. Since a metal asteroid, like an airplane or ship, is a much better reflector of radar than is a stony object, Psyche seems bright when we aim a radar beam at it.

How did such metal asteroids come to be? We suspect that each came from a parent body large enough for its molten interior to settle out or differentiate. The heavier metals sank to the center. When this parent body shattered in a later collision, the fragments from the core were rich in metals. There is enough metal in even a 1-km M-type asteroid to supply the world with iron and most other industrial metals for the foreseeable future, if only we could bring one safely to Earth. The iron and nickel in one of the world's largest mines, in Sudbury, Canada, originated in just such a collision with a metallic asteroid more than a billion years ago.

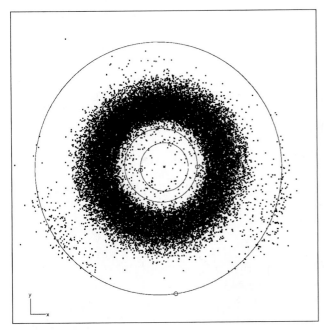

FIGURE 12.2
Asteroid Positions The positions of more than 6000 asteroids in February 1990, seen from above. Also shown are the planets Earth, Mars, and Jupiter. *(Edward Bowell, Lowell Observatory)*

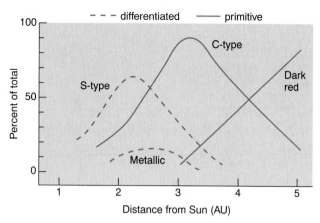

FIGURE 12.3
Where Different Types of Asteroids Are Found Asteroids of different composition are distributed at different distances from the Sun. *(Adapted from work carried out by D. Tholen and J. Gradie, University of Hawaii, and E. Tedesco, JPL)*

In addition to the M asteroids, a few other asteroids show signs of early heating and differentiation. These have basaltic surfaces like the volcanic plains of the Moon and Mars; the large asteroid Vesta (discussed in a moment) is in the latter category. Why only a small percentage of the total number of asteroids have these characteristics, we do not know.

The different classes of asteroids are grouped together at different distances from the Sun (Figure 12.3). Apparently the asteroids are still located near their birthplaces. By tracing how their compositions vary with distance from the Sun, we can reconstruct some of the properties of the solar nebula from which they originally formed.

Vesta: A Volcanic Asteroid

Vesta is one of the most interesting of the asteroids. It orbits the Sun with a semimajor axis of 2.4 AU, and its relatively high reflectivity of almost 30 percent makes it the brightest of the main belt objects, visible to the unaided eye if you know just where to look. But its real claim to fame is the fact that its surface is covered with basalt, indicating that Vesta was once volcanically active in spite of its small size (about 500 km in diameter).

You may be surprised to learn that samples of Vesta's surface are available for direct study in the laboratory (and we thus know a great deal about this asteroid). Scientists have long suspected that *meteorites*—chunks of solar system material that survived their trip to the Earth's surface; see Chapter 13)—come from the asteroids, but there is generally no way to identify the particular source of a given meteorite that strikes the Earth. However, because of Vesta's unusual surface composition, the identification of its fragments seems fairly firm. (As we discussed in Chapter 9, we are also able to determine which meteorites come from Mars because we find pockets of trapped Mars air in them.)

The meteorites believed to come from Vesta are a group of about 30 basaltic meteorites very similar in composition (Figure 12.4). Chemical analysis has shown that they cannot have come from the Earth, Moon, or Mars. On the other hand, their spectra (measured in the laboratory) perfectly match the spectrum of Vesta obtained telescopically. The age of the lava flows from which these meteorites derived has been measured at 4.4 to 4.5 billion years, very soon after the formation of the solar system. This age is consistent with what we might expect for Vesta; whatever process heated such a small object was probably intense and short-lived.

Asteroids Up Close

On the way to its 1995 encounter with Jupiter, the Galileo spacecraft was targeted to fly close to two main-belt asteroids called Gaspra and Ida. Gaspra, the first target, is a member of the Flora family of asteroids and therefore probably a fragment from the collision that formed that family. It is classed as an S-type asteroid, and astronomers knew from variations in its brightness that it had a rotation period of 7 hours. (Its name, incidentally, comes from a resort where 19th-century Russians enjoyed vacationing; the U.S. equivalent would be to call an asteroid Niagara or Mazatlan.)

The Galileo camera revealed that Gaspra is 16 km long and highly irregular, as befits a fragment from a catastrophic collision (Figure 12.5). The detailed images have allowed us to count the craters on Gaspra and to estimate the time its surface has been exposed to collisions. From the rather sparse number of craters, the Galileo scientists have concluded that Gaspra is only about 200 million years old (that is, the collision that formed Gaspra and other

FIGURE 12.4
A Piece of Vesta Photograph of one of the meteorites believed to be a volcanic fragment from the crust of asteroid Vesta. *(NASA)*

FIGURE 12.5

Gaspra Galileo spacecraft image of the small main-belt asteroid, Gaspra, from a distance of 1600 km, with a resolution of about 100 m. The color is highly exaggerated to bring out subtle differences in surface composition. The dimensions of Gaspra are approximately 16 × 11 × 10 km. *(NASA/JPL)*

members of the Flora family took place about 200 million years ago). Calculations suggest that an asteroid the size of Gaspra can expect another catastrophic collision sometime in the next billion years, at which point it will be disrupted to form another generation of still-smaller fragments.

The second target was Ida, a larger S-type asteroid 56 km in length (Figure 12.6). It is much more heavily cratered than Gaspra, suggesting that its surface has been exposed to small impacts for a longer time period, probably more than a billion years. The greatest surprise of the Galileo flyby of Ida, and the most important result scientifically, was the discovery of a satellite—named Dactyl—in orbit about the asteroid.

Although only 1.5 km in diameter, smaller than many college campuses, Dactyl provides scientists with something otherwise beyond their reach—a measurement of the mass and density of Ida using Kepler's laws. The satellite's distance of about 100 km and its orbital period of about 24 hours indicate that Ida has a density approximately 2.5 g/cm^3, which matches the density of primitive rocks. Ida is primitive in composition. By inference, the other S-type asteroids are also primitive, a point that had been in dispute among scientists before the discovery of Dactyl.

Phobos and Deimos, the two satellites of Mars, are probably captured asteroids. They were first studied at close range by the Viking orbiters in 1977, and later by Mars Global Surveyor. Both are rather irregular, somewhat elongated, and heavily cratered (Figures 12.7 and 12.8). Their largest dimensions are about 25 km and 13 km, respectively. Both are dark brownish-gray, and spectral analysis suggests that they are made of primitive materials similar to those that compose most asteroids.

In 1996, NASA launched the Near Earth Asteroid Rendezvous (NEAR) spacecraft, designed to match orbit with an asteroid rather than merely flying past at high speed. Its prime target is Eros, an asteroid about the size of Gaspra and one of the largest of the Earth-approaching asteroids (see Section 12.2). NEAR will map Eros in detail and determine its surface composition and density with high precision.

On the way to Eros, the NEAR spacecraft flew past a much larger C-type asteroid in the main belt, called Mathilde (Figure 12.9). By using its gravitational effect on the spacecraft's orbit, scientists were able to measure the mass of the asteroid and hence its density. To our surprise, the density turned out to be very low, about 1.6 g/cm^3. No rock has this low a density, so scientists interpret the result to mean that Mathilde is a loosely bound collection of smaller rocks with substantial voids between them—a sort of an orbiting gravel bank. Another clue came from the fact that Mathilde has huge craters (as you can see in Figure 12.9); the impacts that formed them could well have shattered a denser body, but a porous collection of pieces might well have survived such hits.

FIGURE 12.6

Ida and Dactyl Asteroid Ida and its moon Dactyl photographed by the Galileo spacecraft in 1993. Irregularly shaped Ida is 56 km in its longest dimension, while Dactyl is about 1.5 km across. *(NASA/JPL)*

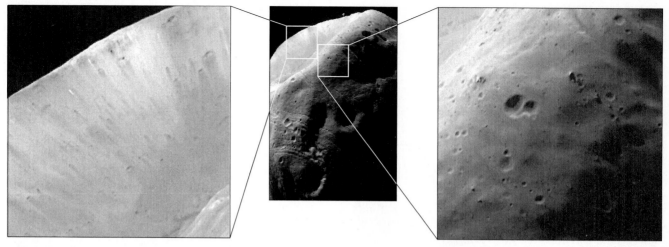

FIGURE 12.7

Phobos Mars Global Surveyor took these images of Phobos, the inner moon of Mars, in August 1998 from a distance of only about 1,000 km. The giant crater you see at the upper left of the main image is called Stickney (the maiden name of Asaph Hall's wife, whose urging helped him persist in finding the satellites.) With a diameter of 10 km, the crater is nearly half the longest dimension of Phobos. The main image shows white boxes at the locations of the two close-ups, each about 1.9 km square. In the left close-up you can see light and dark streaks made by material going downhill into the crater. Going downhill takes on new meaning on Phobos, where gravity is so weak that a 150-lb person would weigh about 2 oz! In the right close-up you can see boulders, the largest of which is about 50 m across. These boulders were presumably ejected by the impact that formed Stickney. *(Malin Space Systems/JPL & NASA)*

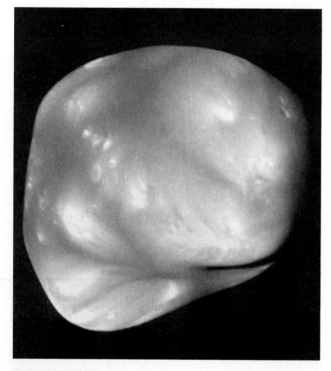

FIGURE 12.8

Deimos Deimos, Mars' outer satellite, is seen here at full phase, resembling a whale from this perspective. It is about 13 km long and, like Phobos, is most likely a captured asteroid. *(NASA/JPL, courtesy of Peter Thomas, Cornell University)*

FIGURE 12.9

Close-up of Mathilde This mosaic consists of four images taken of asteroid Mathilde in June 1997 by the NEAR spacecraft, from a distance of 2400 km. Sunlight is coming from the upper right. This side of the asteroid is about 59 × 47 km across. The deeply shadowed crater at the center is more than 10 km deep! You can see the edge of another large crater at the upper left. *(NEAR MSI/NIS Team and Johns Hopkins University)*

Not all of the asteroids are in the main asteroid belt. In this section, we consider some of the special groups of asteroids that stray outside the belt's boundaries.

The Trojans

The Trojan asteroids are located far beyond the main belt, orbiting the Sun at 5.2 AU, about the same distance as Jupiter. The gravity of the giant planet makes most orbits of asteroids near it unstable, but calculations first carried out in the 18th century show that there are two points in the orbit of Jupiter near which an asteroid can remain almost indefinitely. These two points make equilateral triangles with Jupiter and the Sun (Figure 12.10). Between 1906 and 1908, four such asteroids were found; the number has now increased to several hundred. These asteroids are named for the Homeric heroes from the *Iliad* and are collectively called the Trojans.

Measurements of the reflectivities and spectra of the Trojans show that they are dark, primitive objects like those in the outer part of the asteroid belt. They appear faint because they are so dark and far away, but actually the larger Trojans are quite sizable. Four of them—Hektor, Diomedes, Agamemnon, and Patroclus—have diameters between 150 and 200 km.

In 1990 the first asteroids in Trojan-type orbits were discovered in association with Mars. Like Mars, these asteroids have semimajor axes of 1.5 AU. Other planets, including the Earth, may have their own collection of small Trojan asteroids.

Asteroids in the Outer Solar System

There may be many asteroids with orbits that carry them far beyond Jupiter, but they are difficult to detect and only a few have been discovered. The largest of these mysterious objects is Chiron, with a path that carries it from just inside the orbit of Saturn at its closest approach to the Sun, out to almost the distance of Uranus. The diameter of Chiron is estimated to be about 200 km. In 1992, a still-more-distant object named Pholus was discovered, with an orbit that takes it 33 AU from the Sun, beyond the orbit of Neptune. Pholus has the reddest surface of any object in the solar system, indicating a strange (and still-unknown) surface composition. As more objects are discovered in these distant reaches, astronomers have decided that they too will be given the names of Centaurs from classical mythology; this is because the Centaurs were half human, half horse, and these new objects display some of the properties of both asteroids and comets.

In 1988, as it was reaching its closest approach to the Sun, Chiron was seen to brighten by about a factor of two. This is not the sort of behavior we expect from self-respecting asteroids. Astronomers realized that Chiron (and perhaps the other Centaurs) likely contained abundant volatile materials such as water ice or carbon monoxide ice. As the object gets closer to the Sun, these materials begin to evaporate and produce bright reflective atmospheres of gas and dust. As we will see in the next section, this is precisely the behavior we associate with *comets*. But Chiron and Pholus are much larger than any known comet. What a spectacular show they would put on if either were diverted into the inner solar system! Chiron has been

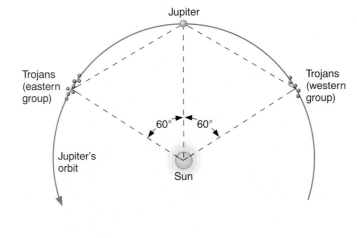

(a)

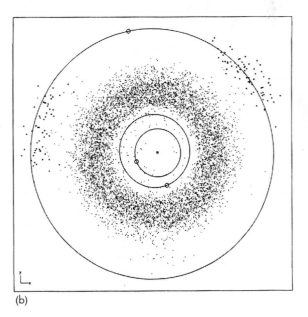

(b)

FIGURE 12.10
Trojan Asteroids (a) Locations of the Trojan points of the orbit of Jupiter. (b) Plot of the actual positions of the 132 known Trojan asteroids in February 1990. *(Part b courtesy of Edward Bowell, Lowell Observatory)*

shown to be in an unstable orbit, so this may yet happen sometime in the distant future.

Earth-Approaching Asteroids

Not just astronomers, but many readers of this text, political leaders, and even military planners have an interest in the asteroids with orbits that come close to or cross the orbit of the Earth. Some of these briefly become the closest celestial objects to us. In 1989 a 200-m object passed within 800,000 km of the Earth, and in 1994 a tiny 10-m object was picked up passing just 105,000 km away. Some of these objects have collided with the Earth in the past, and others will continue to do so, as we saw in Chapter 7. Together with any comets that come close to our planet, they are known collectively as **Near-Earth Objects (NEOs).**

At present fewer than 500 NEOs have been located, although the total population of such objects numbers in the millions (with approximately 2000 larger than 1 km in diameter). Searches for additional members of this group result in the discovery of several new objects each month.

The orbits of Earth-approaching asteroids are unstable. These objects will meet one of two fates: Either they will impact one of the terrestrial planets, or they will be ejected gravitationally from the inner solar system due to a near-encounter with a planet. The probabilities of these two outcomes are about the same. The time scale for impact or ejection is only about 100 million years, very short compared with the age of the solar system. Calculations show that approximately one quarter of the current Earth-approaching asteroids will eventually end up crashing into the Earth itself. The larger of these impacts will generate environmental catastrophes for our planet as described in Section 7.5 (where we also discuss possible responses to such a threat). This is a strong argument for additional study of NEOs; at the very least, we should have a complete catalog of the larger objects and their orbits in order to anticipate any collisions.

If the current population of Earth-approaching asteroids will be removed by impact or ejection in a hundred million years, there must be a continuing source of new objects to replenish our supply. Some of them probably come from the asteroid belt, where collisions between asteroids can eject fragments into Earth-crossing orbits. Others may be dead comets that have exhausted their volatile materials (see next section). So far we have no way of distinguishing between asteroidal fragments and the solid remnants of former comets.

If an Earth-approaching asteroid comes close enough, it can be studied with radar as well as with optical telescopes. Radar can even be used to produce an image of the asteroid. Steven Ostro and his colleagues at the NASA Jet Propulsion Laboratory have succeeded in imaging several such objects with sufficient resolution to establish their shapes and sizes. Their greatest success was with a 5-km-

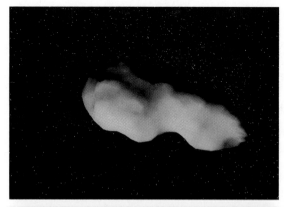
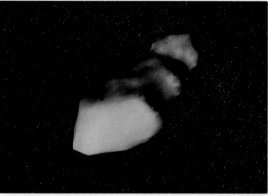
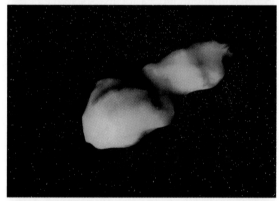
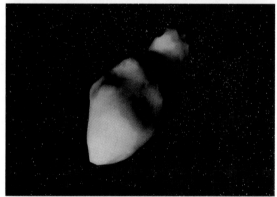

FIGURE 12.11

Radar Images of Toutatis Reconstruction of the size and shape of asteroid Toutatis obtained from radar data collected during its close flyby of Earth in December 1992. Toutatis appears to consist of two irregular, lumpy bodies rotating in contact with each other. Its maximum dimension is about 5 km. *(NASA/JPL, courtesy of Steven Ostro)*

long NEO called Toutatis, which approached to within 3 million km of the Earth in 1992—less than ten times the distance to the Moon. Several radar images of Toutatis are shown in Figure 12.11, obtained with the NASA radar system in California. It appears to be a double object, consisting of two irregular lumps, with diameters of 3 km and 2 km, squashed together.

12.3 THE "LONG-HAIRED" COMETS

Appearance of Comets

Comets have been observed from the earliest times; accounts of spectacular comets are found in the histories of virtually all ancient civilizations. The typical comet is not spectacular, however, but has the appearance of a rather faint, diffuse spot of light somewhat smaller than the Moon and many times less brilliant. Comets are relatively small chunks of icy material that develop an atmosphere as they get closer to the Sun. Later there may be a very faint, nebulous **tail** extending several degrees away from the main body of the comet.

Like the Moon and planets, comets appear to wander among the stars, slowly shifting their positions in the sky from night to night. Unlike the planets, however, most comets appear at unpredictable times, perhaps explaining why they have frequently inspired fear and superstition. They typically remain visible for periods that vary from a few days to a few months.

Today we recognize comets as the best-preserved, most primitive material available in the solar system. Stored in the deep freeze of space until some cosmic accident hurls them our way, these icy objects provide us with unique access to the initial material from which the planets formed 4.5 billion years ago.

Comet Orbits

The study of comets as members of the solar system dates from the time of Isaac Newton, who first suggested that their orbits were extremely elongated ellipses. Newton's colleague Edmund Halley (see *Voyagers in Astronomy* box) developed these ideas, and in 1705 he published calculations of 24 cometary orbits. In particular, he noted that the orbits of the bright comets appearing in the years 1531, 1607, and 1682 were so similar that the three could well be the same comet, returning to perihelion at average intervals of 76 years. If so, he predicted that the object should next return about 1758. When the comet appeared as predicted, it was given the name Comet Halley (rhymes with valley) in honor of the man who first recognized it as a permanent member of our solar system.

Comet Halley has been observed and recorded on every passage near the Sun at intervals from 74 to 79 years

FIGURE 12.12

Comet Halley This composite of three images (one in red, one in green, one in blue) shows Comet Halley as seen with a large telescope in Australia in 1985. During the time the three images were taken in sequence, the comet moved among the stars. The telescope was moved to keep the image of the comet steady, causing the stars to become colored streaks in the background. *(Anglo-Australian Observatory)*

since 239 B.C. The period varies somewhat because of orbital changes produced by the pull of the jovian planets. In 1910 the Earth was brushed by the comet's tail, causing much needless public concern. Comet Halley last appeared in our skies in 1986 (Figure 12.12), when it was met by several spacecraft that gave us a wealth of information about its makeup; it will return in 2061.

Observational records exist for about a thousand comets. Today, new comets are discovered at an average rate of five to ten per year. Most never become conspicuous and are visible only on photographs made with large telescopes. Every few years, however, a comet may appear that is bright enough to be seen easily with the unaided eye; Table 12.2 lists some well-known comets. We are fortunate to have been visited by two bright comets in recent years; probably you had the chance to see at least one of them. First came Comet Hyakutake, with a very long tail; it was visible for about a month in March 1996. A year later Hale–Bopp appeared; it was as bright as the brightest stars

Edmund Halley: Astronomy's Renaissance Man

Edmund Halley (1656–1742). *(Yerkes Observatory, University of Chicago)*

Halley, a brilliant astronomer who made contributions in many fields of science and statistics, was by all accounts a generous, warm, outgoing man. In this he was quite the opposite of his good friend Isaac Newton, whose great work, the *Principia* (see Chapter 2), Halley encouraged, edited, and helped pay to publish. Halley himself published his first scientific paper at age 20, while still in college. As a result, he was given a royal commission to go to Saint Helena (a remote island off the coast of Africa where Napoleon would later be exiled) to make the first telescopic survey of the southern sky. After returning, he received the equivalent of a master's degree and was elected to the prestigious Royal Society in England, all at the age of 22.

In addition to his work on comets, Halley was the first astronomer to recognize that the so-called "fixed" stars move relative to each other, by noting that several bright stars had changed their positions since the ancient Greek catalogs published by Ptolemy. He wrote a paper on the possibility of an infinite universe, proposed that some stars may be variable, and discussed the nature and size of *nebulae* (glowing cloud-like structures visible in telescopes). While in Saint Helena, he observed the planet Mercury going across the face of the Sun and developed the mathematics of how such *transits* could be used to establish the scale of the solar system.

In other fields, Halley published the first table of human life expectancies (the precursor of life-insurance statistics); wrote papers on monsoons, trade winds, and tides (charting the tides in the English Channel for the first time); laid the foundations for

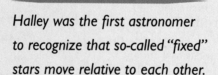

Halley was the first astronomer to recognize that so-called "fixed" stars move relative to each other.

the systematic study of the Earth's magnetic field; studied evaporation and how inland waters become salty; and even designed an underwater diving bell. He served as a British diplomat, advising the emperor of Austria and squiring the future czar of Russia around England (avidly discussing, we are told, both the importance of science and the quality of local brandy).

For a brief time Halley was a deputy comptroller of the British Mint (where he uncovered some local graft) and served as a Navy captain for scientific voyages (on one of which he had to deal with a mutiny). In 1703 he became a professor of geometry at Oxford and in 1720 was appointed Astronomer Royal of England. He continued observing the Earth and the sky and publishing his ideas for another 20 years, until death claimed him at age 85.

TABLE 12.2
Some Interesting Comets

Name	Period	Significance
Great Comet of 1577	Long	Tycho Brahe showed it was beyond the Moon.
Great Comet of 1843	Long	Brightest recorded comet; visible in daytime.
Daylight Comet of 1910	Long	Brightest comet of 20th century (so far).
West	Long	Nucleus broke into pieces in 1976.
Hyakutake	Long	Passed within 15 million km of Earth in 1996.
Hale–Bopp	Long	Very bright in April 1997.
Swift–Tuttle	133 years	Parent comet of Perseid meteor shower.
Halley	76 years	First comet found to be periodic; explored by spacecraft in 1986.
Biela	6.7 years	Broke up in 1846 and not seen again.
Giacobini–Zinner	6.5 years	First comet to be explored by spacecraft (1985).
Encke	3.3 years	Shortest known period.
Shoemaker–Levy 9	Changed	Broke into pieces in 1992; collided with Jupiter in 1994.

and remained visible for several weeks, even in urban areas (see the figure that opens this chapter).

The Comet's Nucleus

When we look at a comet, all we see is its temporary atmosphere of gas and dust, illuminated by sunlight. Since the escape velocity from such small bodies is very low, the atmosphere is rapidly escaping all the time; it must be replenished by new material, which has to come from somewhere. The source is the small, solid **nucleus** inside, usually hidden by the glow from the much larger atmosphere surrounding it. The nucleus is the real comet, the fragment of ancient material responsible for the atmosphere and the tail (Figure 12.13).

The modern theory of the physical and chemical nature of comets was first proposed by Harvard astronomer Fred Whipple in 1950 (Figure 12.14). Before Whipple's work, many astronomers thought that a comet's nucleus might be a loose aggregation of solids of meteoritic nature, the sort of orbiting "gravel bank" model we discussed for the asteroid Mathilde. Whipple proposed instead that the nucleus is a solid object a few kilometers across, composed in substantial part of water ice (but with other ices as well) mixed with silicate grains and dust. This proposal became known as the "dirty snowball" model.

The water vapor and other volatiles that escape from the nucleus when it is heated can be detected in the comet's head and tail, and we can use spectroscopy to analyze what atoms and molecules the nucleus ice consists of. We are somewhat less certain of the nonicy component,

FIGURE 12.14
Fred Whipple   Whipple, the father of modern comet studies, photographed at Harvard in 1957.   *(Smithsonian Astrophysical Observatory)*

however. No large fragments of solid matter from a comet have ever survived passage through the Earth's atmosphere to be studied as meteorites. Some very fine, microscopic grains of comet dust have been collected in the Earth's upper atmosphere, however, and have been studied in the laboratory (Figure 12.15). The spacecraft that encountered Comet Halley in March 1986 also carried dust detectors. From these various investigations it seems that much of the "dirt" in the dirty snowball consists of tiny bits of dark, primitive hydrocarbons and silicates, rather like the material thought to be present on the dark, primitive asteroids.

Since the nuclei of comets are small and dark, they are difficult to study. Just measuring their diameters has been

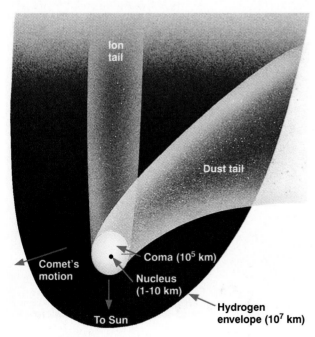

FIGURE 12.13
Parts of a Comet   This schematic illustration shows the main parts of a comet. Note that the different structures are not to scale.

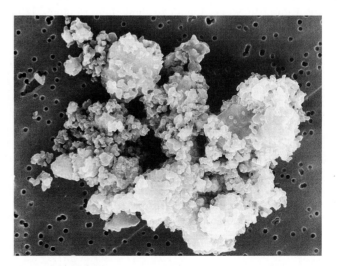

FIGURE 12.15
Captured Comet Dust   A particle that is believed to be a tiny fragment of cometary dust, collected in the upper atmosphere of the Earth.   *(Donald Brownlee, University of Washington)*

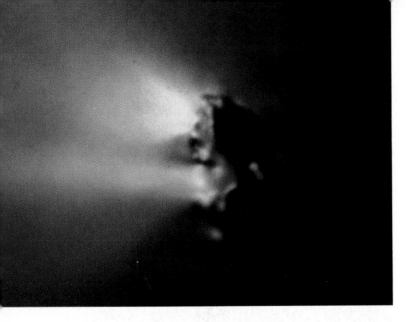

FIGURE 12.16
Comet Halley Nucleus This historic photograph of the black, irregularly shaped nucleus of Comet Halley was obtained by the Giotto spacecraft from a distance of about 1000 km. The bright areas are jets of material escaping from the surface. Details as small as 1 km can be made out. *(Max Planck Institut für Aeronomie, and Ball Aerospace Corporation, courtesy of Harold Reitsema)*

a problem. Our only direct measurements of a comet's nucleus were obtained in 1986, when three spacecraft swept past Comet Halley at close range. The Soviet VEGA 1 and VEGA 2 were the first to arrive, on March 6 and 9, 1986. Each plunged deeply into the inner atmosphere and dust cloud of the comet, passing within about 8000 km of the nucleus. Both VEGA craft were severely damaged by dust impacts, losing most of their solar cells and suffering the loss of several instruments at the time of closest approach.

However, the battering they received was not in vain. The trajectory data for the VEGA craft were provided to the European Space Agency to allow them to target their Giotto spacecraft for an even closer encounter, just 605 km from the comet's nucleus, on March 14, 1986. Giotto measured the dimensions of Comet Halley to be 6 km by 10 km and obtained a beautiful photograph of the dust-covered nucleus itself (Figure 12.16).

The Comet's Atmosphere

The spectacular activity that allows us to see comets is caused by the evaporation of cometary ices heated by sunlight. Beyond the asteroid belt, where comets spend most of their time, these ices are solidly frozen. But as a comet approaches the Sun, it begins to warm up. If water (H_2O) is the dominant ice, significant quantities vaporize as temperatures rise toward 200 K, which happens somewhat beyond the orbit of Mars. The evaporating H_2O in turn releases the dust that was mixed with the ice. Since the comet's nucleus is so small, its gravity cannot hold back either the gas or the dust, both of which flow away into space at speeds of about 1 km/s.

The comet continues to absorb energy as it approaches the Sun. A great deal of this energy goes into the evaporation of its ice, as well as into heating the surface. However, observations of many comets indicate that the evaporation is not uniform and that most of the gas is released in sudden spurts, perhaps confined to a few areas of the surface. Such jets were observed directly on the surface of Comet Halley by the spacecraft that photographed it in 1986; the jets turned out to resemble volcanic plumes or geysers (Figure 12.16). Most of the comet's surface is apparently inactive, with the ice buried under a layer of black silicates and carbon compounds.

The atmosphere of a comet is composed of the gas released from the nucleus, together with the dust and other solid material being carried along with it. Expanding into space at a speed of about 1 km/s, the atmosphere can reach an enormous size. The diameter of a comet's head (also called its *coma*) is often as large as Jupiter, and it can approach a diameter of a million kilometers (Figure 12.17).

The composition of the gas is primarily H_2O (about 80 percent in the case of Comet Halley), plus a few percent each of carbon dioxide (CO_2) and carbon monoxide (CO), along with small quantities of many additional gases, including hydrocarbons. Ultraviolet light from the Sun can break up the molecules of H_2O into oxygen and lighter hydrogen, which easily escapes. Modern telescopes have revealed huge hydrogen clouds, up to tens of millions of kilometers across, around comets. In fact, for a brief period

FIGURE 12.17
Head of Comet Halley Here we see the cloud of gas and dust that make up the head or coma of Comet Halley on January 20, 1986. On this scale, the nucleus (hidden inside the cloud) would be a dot too small to see. *(University of Arizona, courtesy of Uwe Fink)*

COMET HALLEY

FIGURE 12.18
Halley's Hydrogen Cloud A map of the distribution of hydrogen atoms around Comet Halley in February 1986 at about the time of its closest passage to the Sun. The map was made from data gathered by the ultraviolet spectrometer aboard the Pioneer Venus spacecraft in orbit around Venus. The white circle near the bottom left represents the size of the Sun to the same scale. *(I.A.F. Stewart, U. of Colorado and NASA)*

FIGURE 12.20
Comet Tails Comet Mrkos was photographed in 1957 with the wide-field telescope at Palomar Observatory. Note the smoother tail of dust curving to the right as individual dust particles spread out along the comet's orbit, and the straight ion tail pushed outward from the Sun by its wind of charged particles. *(Caltech/Palomar Observatory)*

in the spring of 1986, the hydrogen cloud around Comet Halley was the largest object in our solar system (Figure 12.18).

Many comets develop tails as they approach the Sun. A comet's tail is an extension of its atmosphere, consisting of the same gas and dust that make up its head. As early as the 16th century, observers realized that comet tails always point away from the Sun (Figure 12.19), not back along the comet's orbit. Newton proposed that comet tails are formed by a repulsive force of sunlight driving particles away from the head, an idea close to our modern view. In addition to sunlight, however, we now know that cometary gas is also repulsed by streams of ions (charged particles) emitted by the Sun (Figure 12.20).

12.4 ORIGIN AND EVOLUTION OF COMETS

The Oort Comet Cloud

Although comets are part of the solar system, observations show that they come initially from very great distances. By following their orbits, we can calculate that the *aphelia* (points farthest from the Sun) of new comets typically have values near 50,000 AU (more than a thousand times farther than Pluto). This clustering of aphelion distances was first noted by Dutch astronomer Jan Oort, who in 1950 proposed an idea for the origin of the comets that is still accepted today.

It is possible to calculate that a star's gravitational *sphere of influence*—the distance within which it can exert sufficient gravitation to hold onto orbiting objects—is about one-third of the distance to the nearest other stars. Stars in the vicinity of the Sun are spaced in such a way that the Sun's sphere of influence extends only a little beyond 50,000 AU, or about 1 LY. At such distances, objects in

FIGURE 12.19
Comet Tail Points Away from the Sun The orientation of a typical comet tail changes as the comet passes perihelion. Approaching the Sun, the tail is behind the incoming comet head, but on the way out, the tail precedes the head.

orbit about the Sun can be perturbed by the gravity of passing stars. Some of these perturbed objects can then take on orbits that bring them much closer to the Sun.

Oort suggested, therefore, that the new comets were examples of objects orbiting the Sun near the edge of its sphere of influence, whose orbits had been disturbed by nearby stars, eventually bringing them close to the Sun where we can see them. The reservoir of ancient icy objects from which the new comets are derived is now called the **Oort comet cloud.**

Astronomers estimate that there are about a trillion (10^{12}) comets in the Oort cloud. In addition, we estimate that about ten times this number of potential comets could be orbiting the Sun in the volume of space between the planets and the Oort cloud at 50,000 AU. These objects remain undiscovered because their orbits are too stable to permit any of them to be deflected inward close to the Sun. The total number of cometary objects could thus be on the order of 10 trillion (10^{13}), a very large number indeed.

What is the mass represented by 10^{13} comets? We can make an estimate if we assume something about comet sizes and masses. Let us suppose that the nucleus of Comet Halley is typical. Its observed volume is about 600 km^3. If the primary constituent is water ice with a density of about 1 g/cm^3, the total mass of Halley's nucleus must be about 6×10^{14} kg. This is about one ten billionth (10^{-10}) of the mass of the Earth.

If there are 10^{13} comets with this mass, their total mass would be equal to about 1000 Earths—greater than the mass of all the planets put together. Therefore, cometary material could be the most important constituent of the solar system after the Sun itself.

The Oort cloud spawns comets because they are deflected into the inner solar system by passing stars. In addition, astronomers have located a second source of comets just beyond the orbit of Pluto. This relatively nearby reservoir of cometary material, in the form of a flattened disk, is called the **Kuiper belt** after the Dutch-American astronomer who first suggested its existence. Astronomers have suspected for a number of years that some comets originate in this belt, but not until 1992 was the first member of this group of objects discovered within the belt itself. It turns out to be an exceedingly faint object (designated 1992QB1) with a diameter probably greater than 200 km (bigger than the kinds of comets we see come inward, but much smaller than planets). It has an orbit beyond Pluto and a period of revolution about the Sun of nearly 300 years (Figure 12.21).

More than 60 such objects have been found since then, some even larger than 1992QB1. (Of course, the largest objects reflect the most light and are thus the ones we are most likely to find.) One intriguing object, called 1996TL66, appears to be blacker than coal and larger than the state of Texas. About 40 percent of these *trans-Neptunian objects* share an orbital resonance with Neptune (they complete two orbits around the Sun for each three that Neptune makes). Since Pluto exhibits the same resonance, astronomers have nicknamed the smaller members

FIGURE 12.21
Kuiper Belt Object Two images of object 1992QB, taken in September 1992 with the 3.5-m New Technology Telescope at the European Southern Observatory. The object, circled on each image as it moved among the stars, was the first member of the Kuiper belt to be found. *(ESO)*

MAKING CONNECTIONS

Comet Hunting as a Hobby

When amateur astronomer David Levy, the codiscoverer of Comet Shoemaker–Levy 9, found his first comet, he had already spent 917 fruitless hours searching through the dark night sky. But the discovery of that first comet only whetted his appetite. Since then he has found 8 others on his own and 13 more working with others. Despite this impressive record, he ranks only fourth in the record books for number of comet discoveries. But David is still young and hopes to break that record someday.

All around the world, dedicated amateur observers spend countless nights scanning the sky for new comets. Astronomy is one of the very few fields of science where amateurs can still make a meaningful contribution, and the discovery of a comet is one of the most exciting ways

they can establish their place in astronomical history. Don Machholz, a California amateur (and comet hunter) who has been making a study of comet discoveries, reports that between 1975 and 1995, 38 percent of all comets discovered were found by amateurs. Those 20 years yielded 67 comets for amateurs, or almost four per year. That might sound pretty encouraging to new comet hunters, until they learn that the average number of hours the typical amateur spent searching for a comet before finding one was about 420. Clearly, this is not an activity for impatient personalities.

What do comet hunters do if they think they have found a new comet? First, they must check the object's location in an atlas of the sky, to make sure it really is a comet. Since the first sighting of a comet usually occurs when it is still far from the Sun, and before it sports a significant tail, it will only look like a small, fuzzy patch. And, through most amateur telescopes, so will nebulae (clouds of cosmic gas and dust) and galaxies (distant groupings of stars). Next they must check that they have not come across a comet that is already known, in which case they will only get a pat on the back instead of fame and glory. Then they must reobserve or rephotograph it some time later to see if its motion in the sky is appropriate for comets. Often comet hunters who think they have made a discovery get another comet hunter elsewhere in the country to confirm it. If everything checks out, the place they contact is the Central Bureau for Astronomical Telegrams at the Harvard–Smithsonian Center for Astrophysics in Cambridge, Massachusetts (whose Web site is cfa-www.harvard.edu/cfa/ps/cbat.html). If the discovery is confirmed, the Bureau will send the news out to astronomers and observatories around the world.

Amateur astronomer David Levy. *(Andrew Fraknoi)*

---

of the group *Plutinos.* Some astronomers speculate that Pluto is just the largest example of this group and may share a common origin with its smaller cousins.

Observations with the Hubble Space Telescope suggest that thousands of additional objects in the Kuiper belt could be identified with long-exposure photographs. These are probably icy objects, but they are too faint for us to determine their sizes or chemical properties. Clearly, we are a long way from completing any inventory of the smaller members of the solar system.

The Fate of Comets

Any comet we see today will have spent nearly its entire existence in the Oort cloud or Kuiper belt, at a temperature

near absolute zero. But once a comet enters the inner solar system, its previously uneventful life history begins to accelerate. It may, of course, survive its initial passage near the Sun and return to the cold reaches of space where it spent the previous 4.5 billion years. At the other extreme, it may impact the Sun or come so close that it is destroyed on its first perihelion passage. Sometimes, however, the new comet does not come that close to the Sun, but instead interacts with one or more of the planets.

A comet coming within the gravitational influence of a planet has three possible fates. It can (1) impact the planet, ending the story at once; (2) be speeded up and ejected, leaving the solar system forever; or (3) be perturbed into an orbit with a shorter period. In the last case, its fate is sealed. Each time it approaches the Sun, it loses part of its

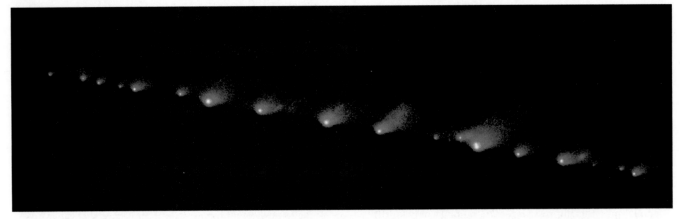

FIGURE 12.22
Jupiter Breaks Up a Comet   Comet Shoemaker–Levy 9 as photographed by the Hubble Space Telescope in May 1994, just two months before impact with Jupiter. Some 21 distinct objects can be seen stretching over a distance of 1.1 million km.   *(NASA/STScI)*

material and also has a significant chance of collision with a planet. Once the comet is in a short-period orbit, its lifetime starts being measured in thousands, not billions, of years.

A few comets end their lives catastrophically, by breaking apart (sometimes for no apparent reason). Especially spectacular was the fate of a faint comet called Shoemaker–Levy 9, which broke into about 20 pieces (Figure 12.22) when it passed close to Jupiter in July 1992. The fragments of Shoemaker–Levy were actually captured into a very elongated, two-year orbit around Jupiter, more than doubling the number of known jovian satellites. This was only a temporary enrichment of Jupiter's family, however,

because in July 1994 the comet fragments crashed into Jupiter, releasing energy equivalent to millions of megatons of TNT.

As each cometary fragment streaked into the jovian atmosphere at a speed of 60 km/s, it disintegrated and exploded, launching a hot fireball that carried the comet dust as well as atmospheric gases to high altitudes. These fireballs were clearly visible in profile, with the actual point of impact just beyond the jovian horizon as viewed from the Earth (Figure 12.23). As each explosive plume fell back into Jupiter, a region of the upper atmosphere larger than the Earth was heated to incandescence and glowed brilliantly for about 15 min, as seen with infrared-sensitive

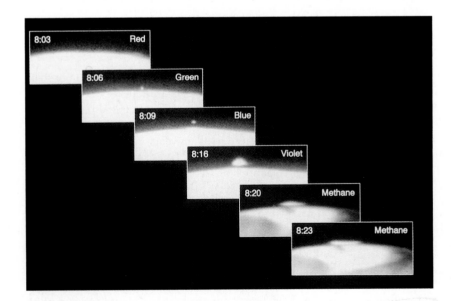

FIGURE 12.23
Comet Impact on Jupiter   This sequence of images shows the last fragment of Comet Shoemaker–Levy 9 to hit Jupiter. Although the initial impact is on the edge of Jupiter as seen from Earth, the giant planet's rotation soon brings the fireball caused by the impact into view. The time of each observation and the filter through which it was taken are shown in each box.   *(NASA/STScI)*

telescopes. Dark clouds of debris settled into the strato-sphere of Jupiter, producing long-lived "bruises" (each larger than the Earth) that could be easily seen through even small telescopes and looked spectacular in large ones (Figure 12.24). Millions of people all over the world peered at Jupiter through telescopes or followed the event via television and the internet. Seeing these impact explosions on Jupiter helps us to appreciate the disaster that would happen to our planet if we were hit by a comet or asteroid (see Figure 7.20 for an illustration).

For comets that do not meet so dramatic an end, measurements of the amount of gas and dust in their atmospheres permit us to estimate the total losses during one orbit. Typical loss rates are up to a million tons per day from an active comet near the Sun, adding up to some tens of millions of tons per orbit. At that rate, a typical comet will be gone after a few thousand orbits. This will probably be the fate of Comet Halley in the long run.

FIGURE 12.24

Impact Dust Cloud on Jupiter   One group of features resulting from the impact of Comet Shoemaker–Levy 9 with Jupiter, seen with the Hubble Space Telescope 105 minutes after the impact that produced the dark rings (the compact black dot came from another fragment). The inner edge of the diffuse outer ring is about the same size as the Earth. Later, the winds on Jupiter blended these features into a broad spot that remained visible for more than a month. *(NASA/STScI)*

SURFING THE WEB

⌨ Asteroid and Comet Impact Hazards Page [impact.arc.nasa.gov/]
Informative site about the dangers of asteroids and comets hitting the Earth and what we might do about it. Included are congressional and scientific position papers, a list of objects that pass near the Earth, useful background information, a bibliography, and annotated links to related Web sites around the world.

⌨ Comet Hale–Bopp Home Page [www.jpl.nasa.gov/comet/]
This page, maintained by Ron Baalke at JPL, is an excellent place to start your search for information and images about the bright comet that graced our skies in 1997. There are links to over 5000 images, plus animations, background material, and news of research discoveries.

⌨ Comet Observation Home Page [encke.jpl.nasa.gov]
Veteran comet observer Charles Morris offers a lot of useful information about comets and comet observations on this site.

⌨ Comet Shoemaker–Levy Home Page [www.jpl.nasa.gov/sl9/sl9.html]
This is *the* place to start if you want more information about the collision of the over 20 pieces of Comet Shoemaker–Levy 9 with Jupiter in July 1994. Ron Baalke of JPL has assembled a massive amount of information, including background material, discoveries, links to 1444 images from 64 observatories, animations, and links to others SL9 sites around the world. (If you followed up every link on this page, you could seriously flunk your astronomy course for lack of doing any other work.)

⌨ Kuiper Belt Home Page [www.ifa.hawaii.edu/~jewitt/kb.html]
David Jewitt of the University of Hawaii keeps an up-to-date list of all the objects that have recently been discovered in the Kuiper belt and some useful background material for instructors (including a bibliography).

⌨ Near-Earth Asteroid Rendezvous Mission [near.jhuapl.edu/]
The NEAR mission involves flying by the asteroid Mathilde and then matching courses with asteroid Eros. The official mission site features background information, images, status reports, and links.

⌨ Stardust Mission Home Page [stardust.jpl.nasa.gov/]
This NASA mission will encounter Comet Wild 2 in 2004, collecting comet dust and returning it to Earth in 2006. Read all about the ambitious plans, and what we hope to learn, at this site.

SUMMARY

The solar system includes many bodies that are much smaller than the planets and their larger satellites. The rocky ones are called **asteroids,** and the icy ones are called **comets**.

12.1   Ceres is the largest asteroid; about 15 are larger than 250 km, and 100,000 are larger than 1 km. Most are in the **asteroid belt,** between 2.2 and 3.3 AU from the Sun. The presence of asteroid families in the belt indicates that many asteroids are the remnants of asteroid collisions and fragmentation. The asteroids include both primitive and differentiated objects. Most asteroids are classed as C-type, meaning they are composed of carbonaceous materials. Dominating the inner belt are S-type (stony) asteroids, with a few M-type (metallic) ones. Vesta is rare in having a volcanic (basaltic) surface. We now have spacecraft images of several asteroids and the NEAR spacecraft has a rendezvous to orbit the Earth-approaching asteroid Eros.

12.2   The Trojan asteroids are dark, primitive objects gathered into two swarms controlled by Jupiter's gravity at 5.2 AU from the Sun. Of great interest are the Earth-approaching asteroids, called **Near-Earth Objects (NEOs),** estimated to number about 2000 (down to 1 km in diameter). These are on unstable orbits, and on time scales of 10^8 years, they will either impact one of the terrestrial planets or be ejected. Most of them probably come from the asteroid belt, but some may be dead comets. Radar images of Toutatis allowed

us a close view of one of these without a spacecraft encounter.

12.3   Halley first showed that some comets are on closed orbits and return periodically to the Sun. The heart of a comet is its **nucleus,** a few kilometers in diameter and composed of volatiles (primarily H_2O) and solids (including both silicates and carbonaceous materials). Whipple first suggested this "dirty snowball" model in 1950; it has been confirmed by spacecraft studies of Comet Halley. As the nucleus approaches the Sun, its volatiles evaporate (perhaps in localized jets or explosions) to form the comet's head or atmosphere, which escapes at about 1 km/s. The atmosphere streams away from the Sun to form a long **tail.**

12.4   Oort proposed in 1950 that comets are derived from what we now call the **Oort comet cloud,** which surrounds the Sun out to about 50,000 AU (near the limit of the Sun's gravitational sphere of influence) and contains between 10^{12} and 10^{13} comets. Some comets can be found in the **Kuiper belt,** a disk-shaped region beyond the orbit of Neptune. Comets are primitive bodies left over from the formation of the outer solar system. Once a comet is diverted into the inner solar system, it typically survives no more than about 1000 perihelion passages before losing all its volatiles. Some comets die spectacular deaths: Shoemaker–Levy 9, for example, broke into 20 pieces before colliding with Jupiter in 1994.

INTER-ACTIVITY

A   Your group is a Congressional committee charged with evaluating the funding for an effort to find all the NEOs (Near Earth Objects) that are larger than $1/2$ km across. Make a list of reasons why it would be useful to humanity to find such objects. What should we (could we) do if we found one that will hit the Earth in a few years?

B   Many cultures considered comets bad omens. Many legends associate comets with the death of kings, losses in war, ends of dynasties. Did any members of your group ever hear about such folktales? Discuss why comets may have gotten this bad reputation.

C   Because asteroids have a variety of compositions and a low gravity that makes the removal of materials quite easy, some people have suggested that mining asteroids

may be a way to get needed resources in the future. Make a list of materials in asteroids (and comets that come to the inner solar system) that may be valuable to a space-faring civilization. What are the pros and cons of undertaking mining operations on these small worlds?

D   As discussed in the last box of the chapter, amateur comet hunters typically spend over 400 hours of time scanning the skies with their telescopes to find a comet. That's a lot of time to spend (usually alone, usually far from city lights, usually in the cold, and always in the dark). Discuss with members of your group whether you can see yourself being this dedicated. Why do people undertake such quests? Do you envy their dedication?

REVIEW QUESTIONS

1.   Why are asteroids and comets important to our understanding of solar system history?

2.   Describe the main differences between C-type and S-type asteroids.

3.   Compare asteroids of the asteroid belt with the Earth-approaching asteroids. What are the main differences between the two groups?

4.   Describe the nucleus of a typical comet, and compare it with an asteroid of similar size.

5.   Describe the origin and eventual fate of the comets we see from the Earth.

6.   What evidence do we have for the existence of the Kuiper belt of comets?

7.   How did Comet Shoemaker–Levy 9 end its life in 1994?

8. Give at least two reasons why today's astronomers are so interested in the discovery of additional Earth-approaching asteroids.

9. Suppose you were designing a spacecraft that would match course with an asteroid and follow along its orbit. What sorts of instruments would you put on board, and what would you like to learn?

10. Suppose you were designing a spacecraft that would match course with a comet and move with it for a while. What sorts of instruments would you put on board, and what would you like to learn?

11. Suppose a comet were discovered approaching the Sun, one whose orbit would cause it to collide with the Earth 20 months later, after perihelion passage. (This is approximately the situation described in the science-fiction novel *Lucifer's Hammer* by Larry Niven and Jerry Pournelle.) What could we do? Would there be any way to protect ourselves from a catastrophe?

12. We believe that chains of comet fragments like Comet Shoemaker–Levy 9's have collided not only with the jovian planets, but occasionally with their satellites. What sort of features would you look for on the outer planet satellites to find evidence of such collisions?

PROBLEMS

13. What is the period of revolution about the Sun for an asteroid with a semimajor axis of 3 AU in the middle of the asteroid belt?

14. What is the period of revolution for a comet with aphelion at 5 AU and perihelion at the orbit of the Earth?

15. Suppose the Oort comet cloud contains 10^{12} comets with an average diameter of 10 km each. Calculate the mass of a comet 10 km in diameter, assuming it is composed mostly of water ice with a density of 1 g/cm^3. Next calculate the total mass of the comet cloud. Finally, compare this mass with those of the Earth and Jupiter.

16. The calculation in Problem 15 refers to the known Oort cloud, the source for the comets we see. If, as some astronomers suspect, there are ten times this many cometary objects in the solar system, how does the total mass of cometary matter compare with the total mass of the planets?

17. If the Oort comet cloud contains 10^{12} comets and 10 new comets are discovered coming close to the sun each year, what percentage of the comets have been used up since the beginning of the solar system?

SUGGESTIONS FOR FURTHER READING

On Asteroids

Burnham, R. "Here's Looking at Ida" in *Astronomy*, Apr. 1994, p. 38.

Cunningham, C. "The Captive Asteroids" in *Astronomy*, June 1992, p. 41. Describes the Trojans.

Durda, D. "All in the Family" in *Astronomy*, Feb. 1993, p. 36. Discusses asteroid families.

Kowal, C. *Asteroids: Their Nature and Utilization*, 2nd ed. 1996, John Wiley/Praxis. Introductory book by the astronomer who discovered Chiron.

McFadden, L. and Chapman, C. "Near-Earth Objects: Interplanetary Fugitives" in *Astronomy*, Aug. 1992, p. 30.

Morrison, D. "The Spaceguard Survey: Protecting the Earth from Cosmic Impacts" in *Mercury*, Sept./Oct. 1992, p. 103.

Ostro, S. "Radar Reveals a Double Asteroid" in *Astronomy*, Apr. 1990, p. 38.

Zimmerman, R. "Ice Cream Sundaes and Mashed Potatoes" in *Astronomy*, Feb. 1999, p. 54. On the NEAR mission.

On Comets

Aguirre, E. "The Great Comet of 1997" in *Sky & Telescope*, July 1997, p. 50. On Comet Hale–Bopp.

Benningfield, D. "Where Do Comets Come From?" in *Astronomy*, Sept. 1990, p. 28.

Bortle, J. "A Halley Chronicle" in *Astronomy*, Oct. 1985, p. 98. A chronology of each pass.

Brandt, J. and Chapman, R. *Rendezvous in Space*. 1992, W. H. Freeman. Introduction by two leading comet experts.

Gore, R. "Halley's Comet '86: Much More Than Met the Eye" in *National Geographic*, Dec. 1986, p. 758.

Levy, D. "How to Discover a Comet" in *Astronomy*, Dec. 1987, p. 74.

Levy, D. *The Quest for Comets*. 1994, Plenum Press. Personal story of comet discovery and comet science by an amateur astronomer who has found many comets.

Sagan, C. and Druyan, A. *Comet*. 1986, Random House. Very good presentation of historical material and human connections.

Sky & Telescope, March 1987, was a special issue about what we learned from Halley's Comet in 1986.

Spencer, J. and Mitton, J., eds. *The Great Comet Crash: The Collision of Comet Shoemaker–Levy 9 and Jupiter*. 1995, Cambridge U. Press. Good, nontechnical summary.

Stern, A. "Chiron: Interloper from the Kuiper Disk?" in *Astronomy*, Aug. 1994, p. 26.

Weissman, P. "Comets at the Solar System's Edge" in *Sky & Telescope*, Jan. 1993, p. 26.

Weissman, P. "The Oort Cloud" in *Scientific American*, Sept. 1998, p. 84.

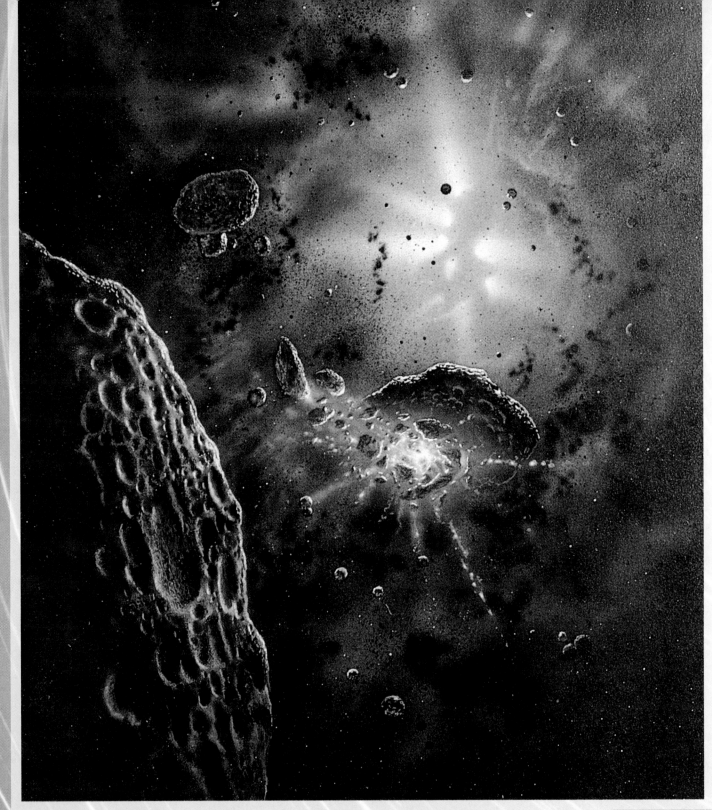

Collisions between planetesimals during the formative stages of the solar system. Many meteorites are fragments of these planetesimals, dating back to the origin of our planetary system.   *(Painting by Don Dixon)*

13 Cosmic Samples and the Origin of the Solar System

THINKING AHEAD

Imagine you are a scientist examining a sample of rock that had fallen from space a few days earlier and you find within it some of the chemical building blocks of life. How could you determine if those "organic" materials were extraterrestrial or merely the result of earthly contamination?

"I could more easily believe that two Yankee professors would lie than that stones would fall from the heavens."

Thomas Jefferson, responding to a report of a Connecticut meteorite fall in 1807.

We conclude our survey of the solar system with a discussion of its origin and evolution. Some of these ideas were introduced in Chapter 6; we now return to them, using the information we have learned about individual planets and smaller objects. But first, we look at another crucial way that astronomers learn about the ancient history of the solar system: by examining samples of primitive matter, the debris of the processes that formed the solar system some 4.5 billion years ago. Unlike the Apollo Moon rocks, these samples of cosmic material come to us free of charge—they literally fall from the sky. We call this material cosmic dust and meteorites.

FIGURE 13.1
Meteor Trails Two meteors can be seen against a starry background that includes the Milky Way. *(P. Parviainen, Polar Image)*

meteors per hour. Over the entire Earth, the total number of meteors bright enough to be visible must total about 25 million per day.

The typical bright meteor is produced by a particle with a mass of less than 1 g—no larger than a pea. The light you see comes from the much larger region of glowing gas surrounding this little grain of interplanetary material. Because of its high speed, the energy in a pea-sized meteor is as great as that of an artillery shell fired on Earth, but it is all dispersed high in the Earth's atmosphere.

If a particle the size of a golf ball strikes our atmosphere, it produces a much brighter trail called a fireball (Figure 13.2). A piece as large as a bowling ball has a fair chance of surviving its fiery entry if its approach speed is not too high. The total mass of meteoric material entering the Earth's atmosphere is estimated to be about 100 tons per day (which seems like a lot, if you imagine it all falling on your home, but is not very significant when spread out over the whole planet).

Meteor Showers

Many—perhaps most—of the meteors that strike the Earth can be associated with specific comets. Some of these periodic comets still return to our view; others have long ago fallen apart, leaving only a trail of dust behind them. The many dust particles from a given comet retain approximately the orbit of their parent, continuing to move together through space (spreading out over the orbit with time). When the Earth, in its travels around the Sun, crosses such a dust stream, we see a sudden burst of

13.1 METEORS

As we saw in the last chapter, the ices in comets evaporate when they get close to the Sun, spraying millions of tons of rock and dust into the inner solar system. The Earth is surrounded by this material, sweeping through thin clouds and streams of dust in its orbit about the Sun. As each of the larger dust or rock particles enters the Earth's atmosphere, it creates a brief fiery trail; this is often called a *shooting star*, but it is properly known as a **meteor.**

Observing Meteors

Meteors are the result of solid particles that enter the Earth's atmosphere from interplanetary space (Figure 13.1). Since the particles move at speeds of many kilometers per second, friction with the air vaporizes them at altitudes between 80 and 130 km. The resulting flashes of light fade out within a few seconds. These "shooting stars" got their name because at night their luminous vapors look like stars moving rapidly across the sky. To be visible, a meteor must be within about 200 km of the observer. On a typical dark, moonless night, an alert observer can see half a dozen

FIGURE 13.2
Fireball When a larger piece strikes the Earth's atmosphere, it can make a bright fireball. Note that during the long exposure needed to catch a fireball, the Earth's motion has made the stars into streaks. *(Photo © 1986 John Sanford)*

FIGURE 13.3

The Radiant of a Meteor Shower   The tracks of the meteors converge to a point in the distance, just as long, straight railroad tracks appear to do.

× Radiant

Meteor tracks

(a)

(b)

meteor activity that usually lasts several hours; these events are called **meteor showers.**

The dust particles that produce meteor showers are moving together in space before they encounter the Earth. Thus, as we look up at the atmosphere, their parallel paths seem to diverge from a place in the sky called the *radiant.* This is the direction in space from which the meteor stream is moving at us, just as long railroad tracks seem to diverge from a single spot on the horizon (Figure 13.3). Meteor showers are often designated by the constellation in which this radiant is located: For example, the Perseid meteor shower has its radiant in the constellation of Perseus. The characteristics of some of the more famous meteor showers are summarized in Table 13.1.

The meteoric dust is not always evenly distributed along the orbit of the comet, so during some years more

TABLE 13.1
Major Annual Meteor Showers

Shower Name	Date of Maximum	Associated Comet	Comet's Period (years)
Quadrantid	January 3	Unknown	—
Lyrid	April 21	Thatcher	415
Eta Aquarid	May 4	Halley	76
Delta Aquarid	July 30	Unknown	—
Perseid	August 11	Swift–Tuttle	105
Draconid	October 9	Giacobini–Zinner	7
Orionid	October 20	Halley	76
Taurid	October 31	Encke	3
Leonid	November 16	Tempel–Tuttle	33
Geminid	December 13	Phaethon*	1.4

* Phaethon is an Earth-approaching asteroid, not a comet.

meteors are seen when the Earth intersects the dust stream, and in other years fewer. A very clumpy distribution is associated with the Leonid meteors, which in 1833 and again in 1866 (after an interval of 33 years—the period of the comet) yielded the most spectacular showers (sometimes called *meteor storms*) ever recorded (Figure 13.4). During the impressive Leonid shower on November 17, 1966, up to 100 meteors were observed per second in some southwestern states.

Some scientists and engineers are concerned about the effect a meteor storm might have on space stations in orbit around the Earth, and in 1994 NASA postponed a Shuttle flight because of concern over a predicted meteor storm. The 1998 Leonids passed without any damage to the many spacecraft in orbit (although a number had been put in "safe mode," with sensitive instruments pointed away from the direction of the shower and instrument covers in place). Still, we continue to be concerned about meteor clumps, since our space activities are placing more and more targets for meteor impact around our planet.

The most dependable meteor shower at present is the Perseid shower, which appears each year for about three nights near August 11. In the absence of bright moonlight,

you can see meteors at a rate of about one per minute during a typical Perseid shower. Astronomers estimate that the total combined mass of the particles in the Perseid swarm is nearly a billion tons; the comet that gave rise to the particles, called Swift–Tuttle, must originally have had at least that much mass. However, if its initial mass were comparable to the mass measured for Comet Halley (see Chapter 12), then Swift–Tuttle would have contained several hundred billion tons, suggesting that only a very small fraction of the original cometary material survives in the meteor stream.

No shower meteor has ever survived its flight through the atmosphere and been recovered for laboratory analysis. However, there are other ways to investigate the nature of these particles and thereby to gain additional insight into the comets from which they are derived. Analysis of the flight paths of meteors shows that most of them are very light or porous, with densities typically less than 1.0 g/cm^3. If you placed a fist-sized lump of meteor material on a table in the Earth's gravity, it might well fall apart under its own weight.

Such light particles break up very easily in the atmosphere, accounting for the failure of even relatively large shower meteors to reach the ground. Comet dust is apparently fluffy, rather inconsequential stuff, as we can also infer from the tiny comet particles recovered in the Earth's atmosphere (see Figure 12.15). This fluff, by its very nature, cannot reach the Earth's surface intact. However, more substantial fragments, from asteroids, do make it into our laboratories, as we will see in the next section.

13.2 METEORITES: STONES FROM HEAVEN

Any fragment of interplanetary debris that survives its fiery plunge through the Earth's atmosphere is called a **meteorite.** Meteorites fall only very rarely in any one locality, but over the entire Earth hundreds fall each year. These rocks from the sky carry a remarkable record of the formation and early history of the solar system.

Extraterrestrial Origin of Meteorites

While occasional meteorites have been found throughout history, their extraterrestrial origin was not accepted by scientists until the beginning of the 19th century. Before that, these strange stones were either ignored or considered to have a supernatural origin.

The falls of the earliest recovered meteorites are lost in the fog of mythology. A number of religious texts speak of stones from heaven, which sometimes arrived at opportune moments to smite the enemies of the authors of those texts. At least one sacred meteorite has survived in the form of the Ka'aba, the holy black stone in Mecca that is revered by Islam as a relic from the time of the Patriarchs.

FIGURE 13.4
A Meteor Storm A woodcut of the great meteor shower or storm of 1833, shown here with considerable artistic license.

Showering with the Stars

Observing a meteor shower is one of the easiest and most enjoyable astronomy activities for beginners. The best thing about it is that you don't need a telescope or binoculars—in fact, they would positively get in your way! What you do need is a site far from city lights, with an unobstructed view of as much sky as possible. While the short bright lines in the sky made by individual meteors could, in theory, be traced back to a radiant point (as shown in Figure 13.3), the quick blips of light that represent the end of the meteor could happen anywhere above you.

> The trick of observing meteor showers is not to restrict your field of view, but to lie back and scan the sky alertly.

So the trick of observing meteor showers is not to restrict your field of view, but to lie back and scan the sky alertly. Try to select a good shower (see the list in Table 13.1) and a night when the Moon will not be bright at the time you are observing. The Moon, street lights, car headlamps, and bright flashlights will all get in the way of your seeing the faint meteor streaks.

You will also see more meteors after midnight, when you are on the hemisphere of the Earth that faces forward—in the direction of the Earth's revolution around the Sun. Before midnight, you are observing from the "back side" of the Earth, and the only meteors you see will be those that traveled fast enough to catch up with the Earth's orbital motion.

When you've gotten away from all the lights, give your eyes about fifteen minutes to get "dark adapted"—that is, for the pupil of your eye to open up as much as possible. (This adaptation is the same thing that happens in a dark movie theater. When you first enter, you can't see a thing; but eventually, as your pupils open wider, you can see pretty clearly by the faint light of the screen.)

Seasoned meteor observers find a hill or open field and make sure to bring warm clothing, a blanket, and a thermos of hot coffee or chocolate with them. (It's also nice to take along someone with whom you enjoy sitting in the dark.) Don't expect to see fireworks or a laser show—meteor showers are subtle phenomena, best approached with a patience that reflects the fact that some of the dust you are watching burn up may first have been gathered into its parent comet more than 4.5 billion years ago, as the solar system was just forming.

The modern scientific history of the meteorites begins in the late 18th century, when a few scientists suggested that some strange-looking stones were of such peculiar composition and structure that they were probably not of terrestrial origin. The general acceptance that indeed "stones fall from the sky" occurred after the French physicist Jean Baptiste Biot investigated a well-observed fall in 1803, in which many meteoritic stones were found, still warm, on the ground.

Meteorites sometimes fall in groups or showers. Such a fall may result when a group of pieces has been moving together in space before colliding with the Earth, but more likely the different stones are fragments of a single larger object that broke up during its violent passage through the atmosphere. It is important to remember that such a *shower of meteorites* has nothing to do with a *meteor shower*. No meteorites have ever been recovered in association with meteor showers. Whatever the ultimate source of the meteorites, they do not appear to come from the comets or their associated particle streams.

Meteorite Falls and Finds

Meteorites are found in two ways. First, sometimes bright meteors (fireballs) are observed to penetrate the atmosphere to very low altitudes. If we search the area beneath the point where the fireball burned out, we may find one or more remnants that reached the ground. Observed *meteorite falls*, in other words, may lead to the recovery of fallen meteorites. (A few meteorites have even hit buildings or, very rarely, people—see Making Connections: *Some Striking Meteorites.*)

Second, people sometimes discover unusual-looking rocks that turn out to be meteoritic; these are termed *meteorite finds.* Now that the public has become meteorite-conscious, many unusual fragments, not all of which turn out to be from space, are sent to experts each year. Some scientists divide these objects into two categories: "meteorites" and "meteorwrongs." Outside Antarctica (see the next paragraph), genuine meteorites turn up at an average rate of 25 or so per year. Most of these end up in natural history museums (Figure 13.5) or specialized meteoritical laboratories throughout the world.

Since the 1980s, a new source from the Antarctic has dramatically increased our knowledge of meteorites. More than ten thousand meteorites have been recovered from the Antarctic ice as a result of the peculiar motion of the ice in some parts of that continent (Figure 13.6). Meteorites that fall in regions where ice accumulates are buried and then carried slowly with the motion of the ice to other

MAKING CONNECTIONS

Some Striking Meteorites

Although meteorites fall regularly onto the Earth's surface, few of them have much of an impact (!) on human civilization. There is so much water and uninhabited land on our planet that rocks from space typically fall where no one even sees them come down. But given the number of meteorites that land each year, you may not be surprised that a few have struck buildings, cars, and even people.

John McAuliffe, deputy chief of the Wethersfield, Connecticut Fire Department, examines the hole made by a meteorite that crashed into the home of Robert and Wanda Donahue on November 8, 1982. *(© 1982, 1992 Dan Haar/The Hartford Courant)*

In November 1982, for example, Robert and Wanda Donahue of Wethersfield, Connecticut, were watching *M°A°S°H* on television when a 6-lb meteorite came thundering through their roof, making a hole in the living room ceiling. After bouncing back up into the attic, it finally came to rest under their dining room table.

Eighteen-year-old Michelle Knapp of Peekskill, New York, got quite a surprise one morning in October 1992. She had just purchased her very first car, her grandmother's 1980 Chevy Malibu. But she awoke to find its rear end mangled and a crater in the family driveway—thanks to a 3-lb meteorite. Michelle was not sure whether to be devastated by the loss of her car or thrilled by all the media attention.

In June 1994, José Martín and his wife were driving from Madrid, Spain, to a golfing vacation when a fist-sized meteorite crashed through the windshield, bounced off the dashboard, broke José's little finger, and then landed in the car's back seat. Before Martín, the most recent person known to have been struck by a meteorite was Annie Hodges of Sylacauga, Alabama. In November 1954, she was napping on a couch when a meteorite came through the roof, bounced off a large radio set, and hit her first on the arm and then on the leg.

In a recent compilation, amateur astronomer and meteorite historian Christopher Spratt found 61 recorded instances between 1790 and 1990 of meteorites striking human habitats or machinery (and we presume there have been other cases that have gone unreported or unrecorded). Although his compilation makes for an impressive list, two centuries are a lot of time and the Earth is a big place. The chance of you or your property being hit by a meteorite is so small that it is certainly not worth losing sleep over.

FIGURE 13.5
Meteorite Find This 1906 photo shows a 15-ton iron meteorite found in the Willamette Valley in Oregon. Although known to Native Americans in the area, it was "discovered" by an enterprising local farmer in 1902, who proceded to steal it and put it on display. It was eventually purchased for the American Museum of Natural History and is now on display in the Hayden Planetarium in New York City, the largest iron meteorite in the United States.
(J. O. Wheeler; courtesy Department of Library Services, American Museum of Natural History)

FIGURE 13.6
Antarctic Meteorite An iron meteorite lying on the Antarctic ice just before being added to our collection. *(NASA)*

FIGURE 13.7
A Variety of Meteorite Types At the top left is a piece of the Allende carbonaceous meteorite, with white inclusions that may date back to before the formation of the solar nebula. At the top right is a fragment of the iron meteorite responsible for the formation of Meteor Crater in Arizona (see Chapter 12). At bottom left is a piece of the Imilac stony-iron meteorite, a beautiful mixture of green olivine crystals and metallic iron. At bottom right is part of the Mern primitive stony meteorite. *(Chip Clark)*

areas where the ice is gradually worn away. After thousands of years, the rock again finds itself on the surface, along with other meteorites carried to these same locations. The ice thus concentrates the meteorites that have fallen both in a large area and over a long period of time.

Meteorite Classification

The meteorites in our collections have a wide range of compositions and histories, but traditionally they have been placed into three broad classes. First are the **irons,** composed of nearly pure metallic nickel–iron. Second are the **stones,** the term used for any silicate or rocky meteorite. Third are the much rarer **stony-irons,** made (as the name implies) of mixtures of stone and metallic iron (Figure 13.7).

Of these three types, the irons and stony-irons are the most obviously extraterrestrial in origin because of their metallic content. Pure iron almost never occurs naturally on Earth; it is generally found here as an *oxide* (chemically combined with oxygen) or other mineral ore. Therefore, if you ever come across a chunk of metallic iron, it is sure to be either man-made or a meteorite.

The stones are much more common than the irons but more difficult to recognize. Often laboratory analysis is required to demonstrate that a particular sample is really of extraterrestrial origin, especially if it has lain on the ground for some time and been subject to weathering. The most scientifically valuable stones are those collected immediately after they fall, or the Antarctic samples preserved in a nearly pristine state by the ice.

Table 13.2 summarizes the frequencies of occurrence of the different classes of meteorites among the fall, find, and Antarctic categories.

Ages and Compositions of Meteorites

It was not until the ages of meteorites were measured and techniques were developed for the detailed analysis of their compositions that scientists appreciated their true significance. The meteorites include the oldest and most primitive materials available for direct study in the laboratory. The ages of stony meteorites can be determined from the careful measurement of radioactive isotopes and their

TABLE 13.2
Frequency of Occurrence of Meteorite Classes

Class	Falls (%)	Finds (%)	Antarctic (%)
Primitive stones	88	51	85
Differentiated stones	8	1	12
Irons	3	42	2
Stony-irons	1	5	1

decay products, as described in Chapter 6. Almost all meteorites have radioactive ages between 4.48 and 4.56 billion years, as old as any ages we have measured in the solar system. The few exceptions are igneous rocks that have been ejected from cratering events on the Moon or Mars.

The average age for all of the old meteorites, calculated using the best data and most accurate values now available for radioactive half-lives, is 4.54 billion years, with an uncertainty of less than 0.1 billion years. This value (which we round off to 4.5 billion years in this book) is taken to represent the *age of the solar system*—the time since the first solids condensed and began to form into larger bodies.

The traditional classification of meteorites into irons, stones, and stony-irons is easy to use because it is obvious from inspection which category a meteorite falls into (although it may be much more difficult to distinguish a meteoritic stone from a terrestrial rock). Much more significant, however, is the distinction between *primitive* and *differentiated* meteorites. The differentiated meteorites are fragments of larger parent bodies that were molten before they broke up, allowing the denser materials (such as metals) to sink to their centers. Like many rocks on Earth, they have been subject to a degree of chemical reshuffling, with the different materials becoming sorted out according to density. Differentiated meteorites include the irons, which come from the metal cores of their parent bodies, and the stony-irons, which probably originate in regions between a metal core and a stony mantle.

For information on the *earliest* history of the solar system, however, we turn to the primitive meteorites—those made of materials that have not been subject to great heat or pressure since their formation. The fiery passage of the meteorite through the air takes place so rapidly that the interior (below a burned crust a few millimeters thick) never even becomes hot. Therefore, a fragment of primitive interplanetary debris is still primitive (in our sense of the word) after it lands on the Earth. What are the parent bodies of these primitive meteorites?

A comparison of their spectra indicates that these parent bodies are almost certainly asteroids. Recall from Chapter 12 that asteroids are believed to be fragments left over from the formation process of the solar system; thus it makes sense that they should be the parent bodies of the primitive meteorites.

The great majority of the meteorites that reach the Earth are primitive stones. Many of them are composed of light-colored gray silicates with some metallic grains mixed in, but there is also an important group of darker stones called **carbonaceous meteorites.** As their name suggests, these meteorites contain carbon, as well as various complex organic compounds—chemicals based on carbon, which on Earth are the simple building blocks of life. In addition, some of them also contain chemically bound water, and many are depleted in metallic iron. The carbonaceous meteorites probably originate among the dark, carbonaceous C-type asteroids concentrated in the outer part of the asteroid belt.

The Allende and Murchison Primitive Meteorites

Next to the tiny dust particles from comets, the carbonaceous meteorites are the most primitive materials available for laboratory study. Two large carbonaceous meteorites that fell within a few months of each other have proved particularly valuable in probing the birth of the solar system. The Allende meteorite fell in Mexico (see Figure 13.7) and the Murchison meteorite in Australia, both in 1969. Like other meteorites, Allende and Murchison are named for the towns near which they fell.

The Murchison meteorite is known for the variety of organic chemicals it has yielded. Most of the carbon compounds in carbonaceous meteorites are complex, tar-like substances that defy exact analysis. Murchison also contains 16 amino acids (the building blocks of proteins), 11 of which are rare on Earth. The most remarkable thing about them is that they include equal numbers with right-handed and left-handed molecular symmetry. Amino acids can have either kind of symmetry, but all life on Earth has evolved using only the left-handed versions to make proteins. The presence of both kinds of amino acids clearly demonstrates that the ones in the meteorite had a nonterrestrial origin!

The presence of these naturally occurring amino acids and other complex organic compounds in Murchison—formed without the benefit of the sheltering environment of planet Earth—shows that a great deal of interesting chemistry must have taken place when the solar system was forming. Perhaps some of the molecular building blocks of life on Earth were first delivered by primitive meteorites and comets. This is an interesting idea, because our planet was probably much too hot for any organic materials to survive its earliest history. But after the Earth's surface cooled, the asteroid and comet fragments that pelted it could have refreshed its supply of organic materials.

The Allende meteorite is a rich source of information on the formation of the solar system because it contains many individual grains with varied chemical histories. As much as 10 percent of the material in Allende appears to be older than the solar system. In examining these materials, we may be looking at some dust grains formed by previous generations of stars that were not destroyed in the processes that gave rise to our own system. In this way, meteorites may enable us to trace the genealogy of our solar system back to times long before its birth.

13.3 FORMATION OF THE SOLAR SYSTEM

The comets, asteroids, and meteorites are surviving remnants from the processes that formed the solar system. (The planets and the Sun, of course, also are the products of the formation process, although the material in them has

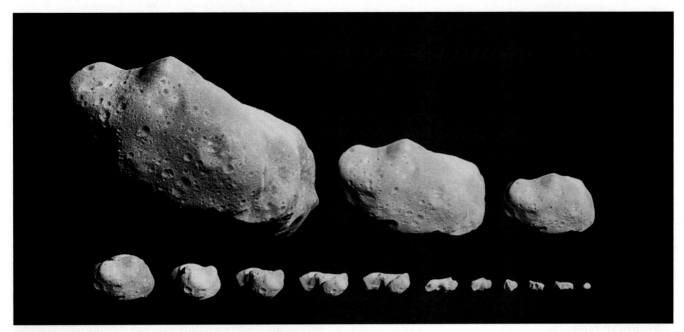

FIGURE 13.8

Approaching Ida In the 1990s the Galileo spacecraft sent back the first close-up images of asteroids (to add to our pictorial inventory of the solar system). This sequence of images was taken as the spacecraft approached the asteroid Ida in 1993 (from bottom right to top left). The asteroid rotates clockwise in 4 hours and 39 min when viewed from above, and the 14 images cover about one rotation. *(JPL/NASA)*

undergone a wide range of changes.) We are now ready to put together the information from the previous chapters in order to discuss what is known about the origin of the solar system.

Observational Constraints

There are certain basic properties of the planetary system that any viable theory of its formation must explain. These may be summarized under three categories: motion constraints, chemical constraints, and age constraints. We call them *constraints* because they place restrictions on our theories; unless a theory can explain the observed facts, it will not survive in the competitive marketplace of ideas that characterizes the endeavor of science.

There are many regularities to the motions in the solar system. We saw that the planets all revolve around the Sun in the same direction and approximately in the plane of the Sun's own rotation. In addition, most of the planets rotate in the same direction as they revolve and most of the satellites also move in counterclockwise orbits (when seen from the north). With the exception of the comets, the motions of the system members define a disk or frisbee shape. On the other hand, a full theory must be prepared to deal with the exceptions to these trends, such as the retrograde rotation of Venus.

In the realm of chemistry, we saw that Jupiter and Saturn have approximately the same composition—dominated by hydrogen and helium—as the Sun and the stars. Each of the other members of the planetary system is, to some degree, lacking in the light elements. A careful examination of the composition of solid solar-system objects shows a striking progression from the metal-rich inner planets, through those made predominantly of rocky materials, out to objects with ice-dominated compositions in the outer solar system. The comets in the Oort cloud and the Kuiper belt are also icy objects, whereas the asteroids (Figure 13.8) represent a transitional rocky composition with abundant dark, carbon-rich material.

As we saw in Chapter 6, this general chemical pattern can be interpreted as a temperature sequence, with the inner parts of the system strongly depleted in materials that could not condense (form a solid) at the high temperatures found near the Sun. Again, however, there are important exceptions to the general pattern. For example, it is difficult to explain the presence of water on Earth and Mars if these planets formed in a region where the temperature was too hot for ice to condense, unless the ice or water was brought in later from cooler regions.

As far as age is concerned, we discussed that radioactive dating demonstrates that some rocks on the surface of the Earth have been present for at least 3.8 billion years, and that certain lunar samples are 4.4 billion years old. In addition, the primitive meteorites all have radioactive ages near 4.5 billion years. The age of these unaltered building blocks is considered the age of the planetary system. The

FIGURE 13.9 ▶

Steps in Forming the Solar System   An artist has drawn a very rough sketch of the steps in the formation of the solar system from the solar nebula. As the nebula shrinks, its rotation causes it to flatten into a disk. Much of the material is concentrated in the hot center, which will ultimately become a star. Away from the center, solid particles can condense as the nebula cools, giving rise to planetesimals, the building blocks of the planets and satellites.

(1) The solar nebula contracts

(2) As the nebula shrinks, its motion causes it to flatten

(3) The nebula is a disk of matter with a concentration near the center

Formation of the protosun.   Solid particles condense as the nebula cools, giving rise to the planetesimals, which are the building blocks of the planets

similarity of the measured ages tells us that planets formed and their crusts cooled within a few hundred million years (at most) of the beginning of the solar system. Further, detailed examination of primitive meteorites indicates that they are made primarily from material that condensed or coagulated out of a hot gas; few identifiable fragments appear to have survived from before this hot-vapor stage 4.5 billion years ago.

The Solar Nebula

All of the foregoing constraints are consistent with the general idea introduced in Chapter 6—that the solar system formed 4.5 billion years ago out of a rotating cloud of hot vapor and dust called the solar nebula, with an initial composition similar to that of the Sun today. As the solar nebula collapsed under its own gravity, material fell toward the center, where things became more and more concentrated and hot. Increasing temperatures in the shrinking nebula vaporized most of the solid material that was originally present.

At the same time, the collapsing nebula began to rotate faster and faster through the conservation of angular momentum (see Chapter 2). One effect of this increased spin was to make it easier for material to fall in at the poles than at the equator (where the spinning tends to push things outward). Thus, more material fell inward along the axis of rotation, eventually giving the nebula a disk shape (Figure 13.9). The existence of this disk-shaped rotating nebula explains the primary motions in the solar system just discussed. Observations of very young stars support these ideas; a number of such stars are surrounded by flattened disks of the type we are describing here.

Picture the solar nebula at the end of the collapse phase, when it was at its hottest. With no more gravitational energy (from material falling in) to heat it, most of the nebula began to cool. The material in the center, however, where it was hottest and most crowded, formed a *star* that was able to maintain high temperatures in its immediate neighborhood by producing its own energy.

The temperature within the nebula therefore decreased with increasing distance from the Sun, much as the planets' temperatures vary with position today. As the nebula cooled, the gases interacted chemically to produce compounds; eventually these compounds condensed into liquid droplets or solid grains. This is similar to the process by which raindrops on Earth condense from moist air as it

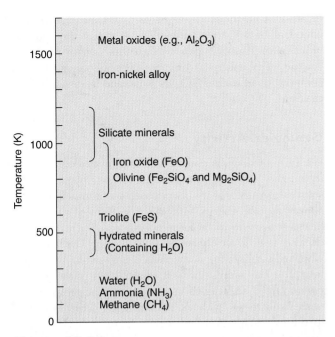

FIGURE 13.10

The Chemical Condensation Sequence in the Solar Nebula
The scale at left shows temperature; at right are the materials that would condense out at each temperature under the conditions expected to prevail in the nebula. *(Adapted from a diagram published by John Lewis, University of Arizona)*

rises over a mountain. Let's look in more detail at how material condensed at different places in the maturing nebula.

In the 1970s, geochemists proposed a detailed theory of what substances could condense at what temperatures in the solar nebula. We call the results of their calculations the *chemical condensation sequence* (Figure 13.10). The first materials to form grains were the metals and various rock-forming silicates. As the temperature dropped, these were joined throughout much of the solar nebula by sulfur compounds and by carbon- and water-rich silicates, such as those now found abundantly among the asteroids. However, in the inner parts of the nebula, the temperature never dropped low enough for such materials as ice or carbonaceous organic compounds to condense, so they are lacking on the innermost planets.

Far from the Sun, cooler temperatures allowed the oxygen to combine with hydrogen and condense in the form of water (H_2O) ice. Beyond the orbit of Saturn, carbon and nitrogen combined with hydrogen to make ices such as methane (CH_4) and ammonia (NH_3). This chemical condensation sequence explains the basic chemical composition differences among various regions of the solar system.

Formation of the Terrestrial Planets

The grains that condensed in the solar nebula rather quickly joined into larger and larger aggregates, until most

of the solid material was in the form of *planetesimals* a few kilometers to a few tens of kilometers in diameter (see Section 6.4). Some planetesimals still survive today as the comets and asteroids. Others have left their imprint on the cratered surfaces of many of the worlds we studied in earlier chapters. A substantial step up in size is required, however, to go from planetesimals to planets.

Some planetesimals were large enough to attract their neighbors gravitationally and thus to grow by the process called **accretion.** While the intermediate steps are not well understood, ultimately several dozen centers of accretion seem to have grown in the inner solar system. Each of these attracted surrounding planetesimals until it had acquired a mass similar to that of Mercury or Mars. At this stage we may think of these objects as *protoplanets*—"not quite ready for prime time" planets.

Each of these protoplanets continued to grow by the accretion of planetesimals. When planetesimals were colliding with other planetesimals, the energies involved were not necessarily large enough to cause either body to melt. Now, however, every incoming planetesimal was accelerated by the much larger gravity of the protoplanet. This means it struck with great energy—sufficient to melt both the projectile and a part of the impact area. Soon the entire protoplanet was heated to above the melting temperature of rocks. The result was planetary differentiation, with heavier metals sinking toward the core and lighter silicates rising toward the surface. As they were heated, the protoplanets lost some of their more volatile constituents (the lighter gases), leaving more of the heavier elements and compounds behind.

Formation of the Giant Planets

In the outer solar system, where the building blocks included ices as well as silicates, the protoplanets grew to be much larger, with masses ten times that of the Earth. These protoplanets of the outer solar system were so large that they were able to attract and hold the surrounding gas. As the hydrogen and helium rapidly collapsed onto their cores, the giant planets were heated by the energy of contraction. Such heating also begins the process of forming stars. But these giant planets were far too small to raise their central temperatures and pressures to the point where nuclear reactions could begin (and it is such reactions that give us our definition of a star). After glowing dull red for a few thousand years, the giant planets gradually cooled to their present state.

The collapse of nebular gas onto the cores of the giant planets explains how these objects came to have about the same hydrogen-rich composition as the Sun itself. The process was most efficient for Jupiter and Saturn; hence their compositions are most nearly "cosmic." Much less gas was captured by Uranus and Neptune, which is why these two planets have compositions dominated by the icy and rocky building blocks that made up their large cores, rather than by hydrogen and helium.

Further Evolution of the System

All of the processes we have just described, from the collapse of the solar nebula to the formation of protoplanets, took place within a few million years. However, the story of the formation of the solar system was not complete at this stage; there were many planetesimals and other debris that did not initially accumulate to form the planets. What was their fate?

Recall from Chapter 12 that we have evidence that the comets visible to us are merely the tip of the cosmic iceberg (if you'll pardon the pun). Most comets are believed to be in the Oort cloud, far from the region of the planets. Additional comets are in the Kuiper belt, which stretches beyond the orbit of Neptune. Our models show that these icy pieces probably formed near the present orbits of Uranus and Neptune, but were ejected from their initial orbits by the gravitational influence of the giant planets.

In the inner parts of the system, remnant planetesimals and perhaps several dozen protoplanets continued to whiz about. Over the vast span of time we are discussing, collisions among these objects were inevitable. Giant impacts at this stage probably stripped Mercury of part of its mantle and crust, reversed the rotation of Venus, and broke the Earth apart to create the Moon.

Smaller-scale impacts also added mass to the inner protoplanets. Because the gravity of the giant planets could "stir up" the orbits of the planetesimals, the material impacting on the inner protoplanets could have come from almost anywhere within the solar system. In contrast to the previous stage of accretion, therefore, this new material did not represent just a narrow range of compositions (from "local" planetesimals).

Much of the debris striking the inner planets was ice-rich material that had condensed in the outer part of the solar nebula. As this comet-like bombardment progressed, the Earth accumulated the water and various organic compounds that would later be critical to the formation of life. Mars and Venus probably also acquired water and organic materials from the same source.

Gradually, as the planets swept up or ejected the remaining debris, most of the leftover planetesimals disappeared. In the region between Mars and giant Jupiter, however, stable orbits are possible where small bodies can avoid impacting the planets or being ejected from the system. The planetesimals (and their fragments) that survive in this special location are what we now call the asteroids.

13.4 PLANETARY EVOLUTION

The era of giant impacts was probably confined to the first 100 million years of solar system history, ending by about 4.4 billion years ago. Shortly thereafter, the planets cooled and began to assume their present aspects. Up until about 4 billion years ago they continued to acquire volatiles, and their surfaces were heavily cratered from the remaining debris that hit them. However, as external influences declined, all of the terrestrial planets, as well as the satellites of the outer planets, began to follow their own evolutionary courses. The nature of this evolution depended on the composition of each object, its mass, and its distance from the Sun.

Geological Activity

We have seen a wide range in the level of geological activity on the terrestrial planets and icy satellites. Internal sources of such activity (as opposed to pummelling from above) require energy, either in the form of primordial heat left over from the formation of a planet or from the decay of radioactive elements in the interior. The larger the planet or satellite, the more likely it is to retain its internal heat and the more slowly it cools—this is the "baked potato effect" mentioned in Chapter 6. Therefore we are more likely to see evidence of continuing geological activity on the surface of larger worlds (Figure 13.11). Jupiter's satellite Io is an interesting exception to this rule; we saw that it has an unusual source of heat—the gravitational flexing of its interior by the tidal pull of Jupiter.

The Moon, the smallest of the terrestrial worlds, was internally active until about 3.3 billion years ago, when its major volcanism ceased. Since that time its mantle has cooled and become solid, and today even internal seismic activity has declined to almost zero. The Moon is a geologically dead world. Although we know much less about Mercury, it seems likely that this planet, too, ceased most volcanic activity about the same time the Moon did.

Mars represents an intermediate case, and it has been much more active than the Moon. The southern hemisphere crust had formed by 4 billion years ago, and the northern hemisphere volcanic plains seem to be contemporary with the lunar maria. However, the Tharsis bulge formed somewhat later, and activity in the large Tharsis volcanoes has apparently continued intermittently to the present.

Earth and Venus are the largest and most active terrestrial planets. Our planet experiences global plate tectonics driven by convection in its mantle. As a result, our surface is continually reworked, and most of the Earth's surface material is less than 200 million years old. Venus has generally similar levels of volcanic activity, but unlike the Earth, it has not experienced plate tectonics. Most of its surface appears to be roughly 500 million years old. We did see that the surface of our sister planet is being modified by a kind of "blob tectonics"—where hot material from below puckers and bursts through the surface, leading to coronae, pancake volcanoes, and other such features. A better understanding of the geological differences between Venus and Earth is a high priority for planetary geologists.

The geological evolution of the icy satellites has been distinct from that of the terrestrial planets. The energy sources were the same, but the materials are different. On

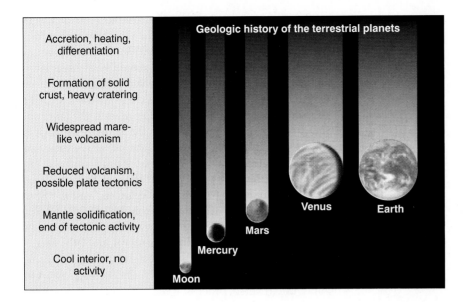

Geologic history of the terrestrial planets

Accretion, heating, differentiation

Formation of solid crust, heavy cratering

Widespread mare-like volcanism

Reduced volcanism, possible plate tectonics

Mantle solidification, end of tectonic activity

Cool interior, no activity

Moon

Mercury

Mars

Venus

Earth

FIGURE 13.11
Stages in the Geological History of a Terrestrial Planet In this graph, time increases downward along the left axis and the stages are labeled. Each planet is shown roughly in its present stage. The smaller the planet, the more quickly it passes through these stages.

these outer worlds, we see evidence of low-temperature volcanism, with the silicate lava of the inner planets being supplemented by sulfur and sulfur compounds on Io, and replaced by water and other ices on other outer-planet satellites.

Elevation Differences

The mountains on the terrestrial planets have very different origins. On the Moon and Mercury, the major mountains are ejecta thrown up by the large basin-forming impacts that took place billions of years ago. Most large mountains on Mars are volcanoes, produced by repeated eruptions of lava from the same vents. There are similar (but smaller) volcanoes on Earth and Venus. However, the highest mountains on Earth and Venus are the result of compression and uplift of the surface. On the Earth, this crustal compression results from collisions of one continental plate with another.

It is interesting to compare the maximum heights of the volcanoes on Earth, Venus, and Mars (Figure 13.12). On Venus and Earth, the maximum elevation differences between these mountains and their surroundings are about

10 km. Olympus Mons, in contrast, towers 26 km above its surroundings, and nearly 30 km above the lowest elevation areas on Mars (Figure 13.13).

One reason Olympus Mons is so much larger than its terrestrial counterparts is that the crustal plates on Earth never stop long enough to let a really large volcano grow. Instead, the moving plate creates a long row of volcanoes like the Hawaiian Islands. On Mars (and perhaps Venus) the crust remains stationary with respect to the underlying hot spot, and so the volcano can continue to grow for hundreds of millions of years.

A second difference relates to the force of gravity on the three planets. The surface gravity on Venus is nearly the same as that on Earth, but on Mars it is only about one third as great. In order for a mountain to survive, its internal strength must be great enough to support its weight against the force of gravity. Volcanic rocks have known strengths, and we can calculate that on Earth 10 km is about the limit. For instance, when new lava is added to the top of Mauna Loa in Hawaii, the mountain slumps downward under its own weight. The same height limit applies on Venus, where the force of gravity is the same as ours. On Mars, however, with its lesser surface gravity,

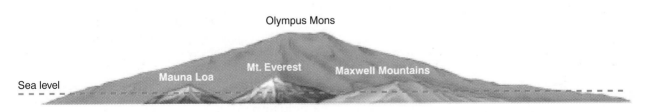

Olympus Mons

Sea level

Mauna Loa   Mt. Everest   Maxwell Mountains

FIGURE 13.12
The Highest Mountains on Mars, Venus, and Earth Mountains can rise taller on Mars because the surface gravity is less and there are no moving plates. The vertical scale is exaggerated by a factor of 3 to make comparison easier.

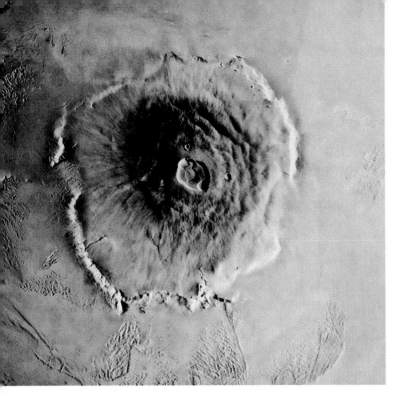

FIGURE 13.13

Olympus Mons   The largest martian volcano is seen here from above on this spectacular composite image created from many Viking orbiter photographs. The volcano is nearly 500 km wide at its base, and 26 km high.   *(NASA/USGS, courtesy of Alfred McEwen)*

much greater elevation differences can be supported, which helps explain why Olympus Mons is more than twice as high as the tallest mountains of Venus or Earth.

The same kind of calculation that determines the limiting height of a mountain can be used to ascertain the largest body that can have an irregular shape. Gravity, if it can, pulls all objects into the most "efficient" shape (where all the outside points are equally distant from the center). All of the planets and larger satellites are nearly spherical, due to the force of their own gravity having pulled them into a sphere. But the smaller the object, the greater is the departure from spherical shape that the strength of its rocks can support. For silicate bodies, the limiting diameter is about 400 km; larger objects will always be approximately spherical, while smaller ones can have almost any shape (as we see in photographs of asteroids, such as Figure 13.8).

Atmospheres

The atmospheres of the planets were formed by a combination of gas escaping from their interiors and the impacts of volatile-rich debris from the outer solar system. Each of the terrestrial planets must have originally had similar atmospheres, but Mercury was too small and too hot to retain its gas. The Moon probably never had an atmosphere, since the material composing it was depleted in volatiles.

The predominant volatile gas on the terrestrial planets is now carbon dioxide (CO_2), but initially there were probably also hydrogen-containing gases. In this more chemically reduced environment, there should have been large amounts of carbon monoxide (CO) and perhaps also traces of ammonia (NH_3) and methane (CH_4). Ultraviolet light from the Sun destroyed the reducing gases in the inner solar system, however, and most of the hydrogen escaped, leaving behind the oxidized atmospheres we see today on Earth, Venus, and Mars.

The fate of water was different for each of these three planets, depending on its size and distance from the Sun. Early in its history, Mars apparently had a thick atmosphere with abundant liquid water, but it could not retain those conditions. As we saw in Chapter 9, the CO_2 necessary for a substantial greenhouse effect was lost, the temperature dropped, and eventually the water froze. On Venus the reverse process took place, with a runaway greenhouse effect leading to the permanent loss of water. Only the Earth managed to maintain the delicate balance that permits liquid water to persist.

With the water gone, Venus and Mars each ended up with an atmosphere of about 96 percent carbon dioxide and a few percent nitrogen. On Earth, the presence first of water and then of life led to a very different kind of atmosphere. The CO_2 was removed to be deposited in marine sediment, while proliferation of life that could do photosynthesis eventually led to the release of enough oxygen to overcome natural chemical reactions that eliminate this gas from the atmosphere. As a result, we on Earth find ourselves with a great deficiency of CO_2 and with the only planetary atmosphere containing free oxygen.

In the outer solar system, Titan is the only satellite with a substantial atmosphere. Apparently this object contained sufficient volatiles—such as ammonia, methane, and nitrogen—to form an atmosphere. It was also large and cold enough to retain these gases. Thus, as we saw in Chapter 11, today Titan's atmosphere consists primarily of nitrogen. Compared with those on the inner planets, temperatures on Titan are too low for either carbon dioxide or water to be in vapor form. With these two common volatiles frozen solid, it is perhaps not too surprising that nitrogen has ended up as the primary atmospheric constituent.

We see that nature, starting with one set of chemical constituents, can fashion a wide range of final atmospheres appropriate to the conditions and history of each world. The atmosphere we have on Earth is the result of many eons of evolution and adaptation. And, as we saw, it can be changed by the actions of the life-forms that inhabit the planet.

Conclusion

We now come to the end of our study of the planetary system. Although we have learned a great deal about the other planets during the past few decades of spacecraft exploration, much remains unknown. Now that we have adopted the comparative approach to study these different worlds and the processes that form them and their environments, we see how limited our knowledge really is. More recently, we have had some surprises in store when we found planets around a number of nearby stars. Even such generalizations that giant planets form only in the outer parts of a planetary system may need re-examination in light of some of these new discoveries, as we will discuss in Chapter 15.

For now, we end by noting that we do not even understand our own planet well enough to assess the consequences of the changes humans are inadvertently inflicting on the atmosphere and oceans. The exploration of the solar system is one of the greatest human adventures, and it will continue, if only so that we can learn to appreciate and protect our own planet.

SURFING THE WEB

Meteor Showers Site
[medicine.wustl.edu/~kronkg/index.html]
Amateur astronomer Gary Kronk, author of a book on meteor showers, has set up a site with all the information anyone could want on major and minor showers, background and historical explanations, meteor legends, fireballs, observing hints, and links.

Sky On Line Meteor Page
[www.skypub.com/sights/meteors/meteors.shtml]
Another excellent site for serious or casual meteor observers, with lots of up-to-date information on upcoming showers, hints

for viewing, and links to organizations of meteor watchers around the world.

Mars Meteorites
[www.jpl.nasa.gov/snc/index.html]
Good resource with information and links about meteorites from Mars.

Nine Planets Meteors, Meteorites, and Impacts Page [seds.lpl.arizona.edu/nineplanets/nineplanets/meteorites.html]
A quick overview with a long list of links.

SUMMARY

13.1 When a fragment of interplanetary dust strikes the Earth's atmosphere, it burns up to create a **meteor.** Streams of dust particles traveling through space together produce **meteor showers,** in which we see meteors diverging from a spot in the sky called the *radiant* of the shower. Many meteor showers recur each year and are associated with particular comets.

13.2 **Meteorites** are the debris from space (mostly asteroid fragments) that survive to reach the surface of the Earth. Meteorites are called *finds* or *falls* according to how they are found; the most productive source today is the Antarctic ice cap. Meteorites are classified as **irons, stony-irons,** or **stones** according to their composition. Most stones are primitive objects, dated to the origin of the solar system 4.5 billion years ago. The most primitive are the **carbonaceous meteorites,** such as Murchison and Allende. These can contain a number of organic molecules.

13.3 A viable theory of solar system formation must take into account motion constraints, chemical constraints, and age constraints. Meteorites, comets, and asteroids are survivors of the *solar nebula* out of which the solar system formed. This nebula was the result of the collapse of an interstellar cloud of gas and dust, which contracted (conserving its angular momentum) to form the *protosun* surrounded by a thin, spinning disk of dust and hot vapor.

Condensation in the disk led to the formation of *planetesimals*, which became the building blocks of the planets. **Accretion** of infalling materials heated the planets, leading to their differentiation. The giant planets were also able to attract and hold gas from the solar nebula. After a few million years of violent impacts, most of the debris was swept up or ejected, leaving only the asteroids and cometary remnants surviving to the present. All of these violent events had terminated by about 4.4 billion years ago.

13.4 After their common beginning, each of the planets evolved on its own path. The different possible outcomes are illustrated by comparison of the terrestrial planets (Earth, Venus, Mars, Mercury, and the Moon). All are rocky, differentiated objects. The level of geological activity is proportional to mass: greatest for Earth and Venus, less for Mars, and absent for the Moon and Mercury. Mountains can result from impacts, volcanism, or uplift. Whatever their origin, higher mountains can be supported on smaller planets where the surface gravity is less. All the terrestrial planets may have acquired atmospheric volatiles from comet impacts. The Moon and Mercury lost their atmospheres; most volatiles on Mars are frozen due to its greater distance from the Sun; and Venus retained CO_2 but lost H_2O when it developed a massive greenhouse effect. Only Earth still has liquid H_2O on its surface and hence can support life.

INTER-ACTIVITY

A Ever since the true (cosmic) origin of meteorites was understood, people have tried to make money selling them to museums and planetaria. More recently, a growing number of private collectors have been interested in purchasing meteorite fragments, and a network of dealers (some more reputable than others) has sprung up to meet this need. What does your group think of all this? Who should own a meteorite? The person on whose land it falls, the person who finds it, or the local, state, or federal government where it falls? Why? Should there be any limit to what people charge for meteorites? Or should all meteorites be the common property of humanity? (If you can, try to do research on what the law is now in this area.)

B Your group has been formed to advise a very rich person who wants to buy some meteorites, but is afraid of being cheated and being sold some Earth rocks. What are some ways you would advise your client to make sure that the meteorites she buys are authentic?

C Your group is a committee set up to give advice to NASA about how to design satellites and telescopes in space so that the danger of meteor impacts is minimized. Remember that the heavier a satellite is, the harder (more expensive) it is to launch. What would you include in your recommendations?

D Discuss what you would do if you suddenly find that a small meteorite had crashed in or near your home. Whom would you call first, second, third? What would you do with the sample? (And would any damage to your home be covered by your insurance?)

E A friend of your group really wants to see a meteor shower. The group becomes a committee to assist her in fulfilling this desire. What time of year would be best? What equipment would you recommend she gets? What advice would you give her?

REVIEW QUESTIONS

1. A friend of yours sees a meteor shower and becomes concerned that the stars he knows (such as the ones in the Big Dipper) could self-destruct like shooting stars. How would you put his mind at rest?

2. In what ways are meteorites different from meteors? What is the probable origin of each?

3. How is Comet Halley connected to meteor showers?

4. What do we mean by primitive material? How can we tell if a meteorite is primitive?

5. Describe the solar nebula, and outline the sequence of events within the nebula that gave rise to the planetesimals.

6. Why do the giant planets and their satellites have compositions different from those of the terrestrial planets?

7. Explain the role of impacts in planetary evolution, including both giant impacts and more modest ones.

8. Why are some planets more geologically active than others?

9. Summarize the origins and evolutions of the atmospheres of Venus, Earth, and Mars.

THOUGHT QUESTIONS

10. What sorts of methods do scientists use to distinguish a meteorite from terrestrial material?

11. Explain why we do not believe meteorites come from comets, or meteors from asteroids.

12. Why do iron meteorites represent a much higher percentage of finds than of falls?

13. Why is it more useful to classify meteorites according to whether they are primitive or differentiated, rather than whether they are stones, irons, or stony-irons?

14. Which meteorites are the most useful for defining the age of the solar system? Why?

15. Suppose a new primitive meteorite is discovered (sometime after it falls in a field of soybeans) and analysis reveals that it contains a trace of amino acids, all of which show the same rotational symmetry (unlike the Murchison meteorite). What might you conclude from this finding?

16. How do we know when the solar system formed? Usually we say that the solar system is about 4.5 billion years old. To what does this age correspond?

17. We have seen how Mars can support greater elevation differences than can Earth or Venus. According to the same arguments, the Moon should have higher mountains than any of the other terrestrial planets, yet we know it does not. What is wrong with applying the same line of reasoning to the mountains on the Moon?

PROBLEMS

18. Consider the differentiated meteorites. We think the irons are from the cores, the stony-irons from the interfaces between mantles and cores, and the stones from the mantles of their differentiated parent bodies. If these parent bodies were like the Earth, what fraction of the meteorites would you expect to consist of irons, stony-irons, and stones? Is this consistent with the observed numbers of each?

19. Estimate the maximum height of the mountains on a hypothetical planet similar to the Earth but with twice the surface gravity of our planet.

20. The angular momentum of an object is proportional to the square of its size divided by its period of rotation (D^2/P). If angular momentum is conserved, then any change in size must be compensated for by a proportional change in period, in order to keep D^2/P a constant. Suppose that the solar nebula began with a diameter of 10,000 AU and a rotation period of 1 million years. What would be its rotation period when it had shrunk to the size of Pluto's orbit? Of Jupiter's orbit? Of the Earth's orbit?

SUGGESTIONS FOR FURTHER READING

On Meteors

Bagnall, P. "Watching Halley's Debris" in *Astronomy*, May 1992, p. 78. Describes the two meteor showers caused by Comet Halley.

Bone, N. *Meteors.* 1993, Sky Publishing. A good manual and information source.

Kronk, G. "Meteor Showers" in *Mercury*, Nov./Dec. 1988, p. 162.

Litmann, M. *The Heavens on Fire.* 1998, Cambridge U. Press. Best introduction to meteor showers in history and science.

On Meteorites

Chaikin, A. "A Stone's Throw from the Planets" in *Sky & Telescope,* Feb. 1983, p. 122. Discusses meteorites from the Moon and Mars.

Dodd, R. *Thunderstones and Shooting Stars: The Meaning of Meteorites.* 1986, Harvard U. Press. A clear guide to the science of meteorites.

Gallant, R. "Saga of the Lump: The Pallas Meteorite" in *Sky & Telescope,* Jan. 1999, p. 50.

Hutchinson, R. *The Search for Our Beginnings.* 1983, Oxford U. Press. Focuses on the role of meteorites in helping us understand the solar system.

McSween, H. *Meteorites and Their Parent Planets.* 1987, Cambridge U. Press. A good general introduction by a geologist.

Norton, O. *Rocks from Space,* 2nd ed. 1998, Mountain Press Publishing. A guide for amateurs, with information on science, folklore, and collecting.

Schaefer, B. "Meteor[ite]s that Changed the World" in *Sky & Telescope,* Dec. 1998, p. 68. On famous cases, including Tukuska, Chicxulub, and and the Black Stone of the Ka'aba.

Spratt, C. and Stephens, S. "Against All Odds: Meteorites That Have Struck [the Earth]" in *Mercury,* Mar./Apr. 1992, p. 50.

On the Evolution of the Solar System

See the readings recommended in Chapter 6, plus

Jayawardhana, R. "Spying on Planetary Nurseries" in *Astronomy,* Nov. 1998, p. 62. On dust disks around other stars that may be the birthplaces of other planetary systems.

Kastings, J. "How Climate Evolved on the Terrestrial Planets" in *Scientific American,* Feb. 1988, p. 90.

Rubin, A. "Microscopic Astronomy" in *Sky & Telescope,* July 1995, p. 24. Discusses the study of solar system history from analysis of samples.

Wood, J. "Forging the Planets: The Origin of Our Solar System" in *Sky & Telescope,* Jan. 1999, p. 36. Good overview.

Yulsman, T. "From Pebbles to Planets" in *Astronomy,* Feb. 1998, p. 56. Comparing the origin of our system to others.

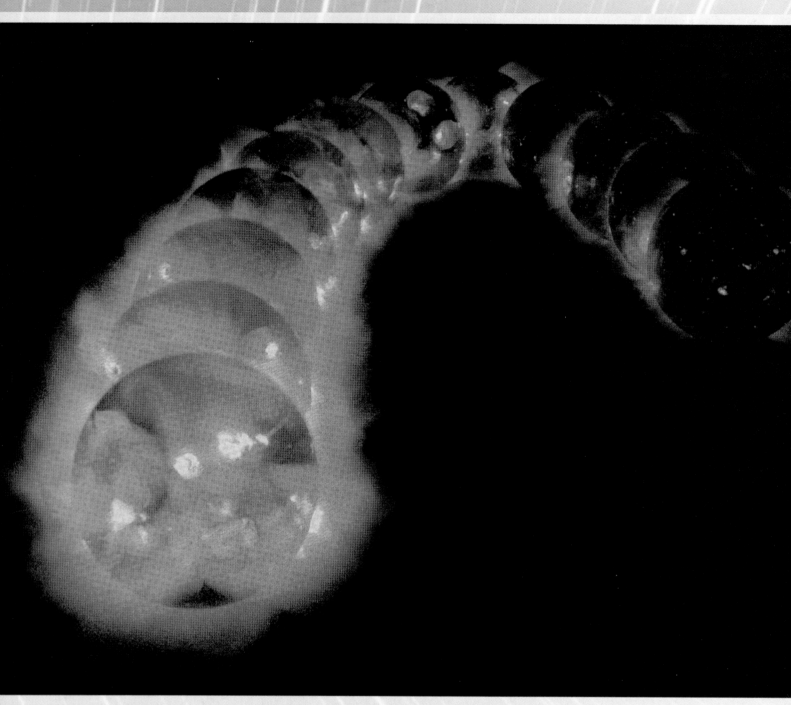

The Sun's Cycle in X-Rays   The level of solar activity varies with a period of approximately 11 years. These 12 x-ray images of the Sun's atmosphere were obtained by the Japanese Yohkoh satellite in 120-day increments between 1991 and 1995. This was the declining part of the solar cycle, and thus the first image (front left) shows the Sun near maximum activity, while the last image (right) shows the Sun near its activity minimum. The x-ray Sun appears completely different from the Sun we see with visible light. Only very hot gases can emit x rays. The Sun's thin outer atmosphere or corona, at a temperature of millions of degrees, is hot enough to emit x rays, while the much cooler surface of the Sun, at about 6000 degrees K, is not. As a result, an x-ray image reveals a bright glow for the corona (and for hot active regions at other depths) and a black disk for the surface of the Sun.   *(NASA/Yohkoh)*

14 The Sun: A Garden-Variety Star

> "The adventure of the Sun is the great natural drama by which we live, and not to have joy in it and awe of it, not to share in it, is to close a dull door on nature's sustaining and poetic spirit."
>
> Henry Beston, from "Midwinter" in *The Outermost House* (1928)

THINKING AHEAD

Can we trust the Sun? This may seem like a silly question—after all, the Sun has been rising in the east and setting in the west every day for billions of years. But the Sun has *not* put out exactly the same amount of heat during all that time—it has been both hotter and cooler than it is today. Since life on Earth ultimately depends on the Sun's heat and light, the way the Sun's energy output changes, and on what time scales, are questions of personal importance to those of us who share the Sun's cosmic neighborhood.

We now consider our star, the Sun. It is, by many measures, a rather ordinary star—not unusually hot or cold, old or young, large or small. Indeed, we are lucky that the Sun is typical. By studying it, we learn much that helps us understand the stars in general. Just as studies of the Earth help us understand observations of the more distant planets, so too the Sun serves astronomers as a guide to interpreting the messages contained in the light we receive from distant stars.

In this chapter we describe what the Sun looks like, how it changes with time, and how those changes affect the Earth. Some of the basic characteristics of the Sun are listed in Table 14.1, although many of the terms in that table may be unfamiliar to you until you've read further. Figure 14.1 shows what the Sun would look like if we could see all parts of it from the center to its outer atmosphere; the terms in the figure will become familiar to you as you read on.

TABLE 14.1
Solar Data

Datum	How Found	Value
Mean distance	Radar reflection from planets	I AU (149,597,892 km)
Maximum distance from Earth		1.521×10^8 km
Minimum distance from Earth		1.471×10^8 km
Mass	Orbit of Earth	333,400 Earth masses (1.99×10^{30} kg)
Mean angular diameter	Direct measure	31'59".3
Diameter of photosphere	Angular size and distance	109.3 × Earth diameter (1.39×10^6 km)
Mean density	Mass/volume	1.41 g/cm^3
Gravitational acceleration at photosphere (surface gravity)	GM/R^2	27.9 times Earth surface gravity = 273 m/s^2
Solar constant	Measure with instrument sensitive to radiation at all wavelengths	1370 W/m^2
Luminosity	Solar constant times area of spherical surface I AU in radius	3.8×10^{26} W
Spectral class	Spectrum	G2V
Effective temperature	Derived from luminosity and radius of Sun	5800 K
Rotation period at equator	Sunspots and Doppler shift in spectra taken at the edge of the Sun	24 d 16 h
Inclination of equator to ecliptic	Motions of sunspots	7°10'.5

 ## OUTER LAYERS OF THE SUN

The Sun, like all stars, is an enormous ball of extremely hot gas, shining under its own power. And we do mean enormous. The Sun has enough volume (takes up enough space) to hold about 1.3 million Earths. Most of the Sun's regions are shielded from our view by its outside layers, just as the Earth shows only its surface and atmosphere to our direct observation. The Sun has no solid surface, but it does have an extensive atmosphere—the only part of the Sun we can actually see. Astronomers divide the Sun's atmosphere into different regions—the photosphere, the chromosphere, and the corona—which we will examine in more detail shortly.

The Composition of the Sun

But first, we begin by asking what the Sun is made of. As explained in Chapter 4, we can use the *absorption line spectrum* to determine what elements are present. It turns out that the Sun contains the same elements as the Earth,

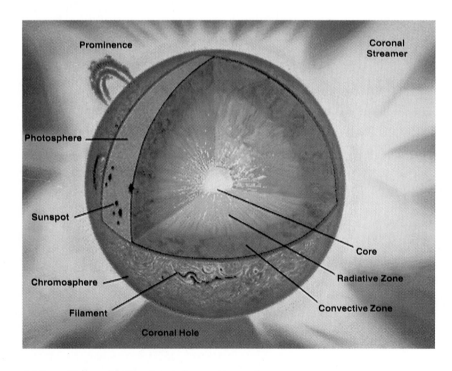

FIGURE 14.1
The Parts of the Sun This drawing shows the different parts of the Sun, from the hot core where the energy is generated, through regions where energy is transported outward, first by radiation and then by convection, and then out through the solar atmosphere. The parts of the atmosphere are labeled—the photosphere, chromosphere, and corona. The artist also shows some typical features in the atmosphere, such as coronal holes and prominences. *(NASA/SOHO)*

but *not* in the same proportions. About 73 percent of the Sun's mass is hydrogen, and another 25 percent is helium. All the other chemical elements (including those we know and love in our own bodies, such as carbon, oxygen, and nitrogen) make up only 2 percent of our star. The ten most abundant gases in the solar photosphere are listed in Table 14.2. Examine that table and notice how different the composition of the Sun's outer layer is from the Earth's crust, where we live. (In our planet's crust, the three most abundant elements are oxygen, silicon, and aluminum.) But the Sun is quite typical of the stars in the elements that dominate its makeup.

The fact that the Sun and the stars all have similar compositions and are made mostly of hydrogen and helium was first shown in a brilliant thesis in 1925 by Cecilia Payne-Gaposchkin, the first woman to get a PhD in astronomy in the United States (Figure 14.2). However, the idea that hydrogen and helium were the most abundant elements in stars was so unexpected and so shocking that she assumed that her analysis of the data must be wrong. At the time, she wrote, "The enormous abundance derived for these elements in the stellar atmosphere is almost certainly not real." It is often hard to accept new ideas that do not agree with what everyone "knows" to be right!

Before Payne's work, everyone assumed that the composition of the Sun and stars would be much like that of the Earth. It was three years after her thesis that other studies proved beyond doubt that the enormous abundance of hydrogen and helium in the Sun is indeed real. (And, as we will see, the composition of the stars is much more typical of the makeup of the universe than the odd concentration of heavier elements that characterizes our planet.)

Most of the elements found in the Sun are in the form of atoms, but more than 18 types of molecules, including water vapor and carbon monoxide, have been identified in the light from the Sun's cooler regions, such as the sunspots (see Section 14.2). The atoms and molecules contained in the Sun are all in the form of gases: The Sun is so hot that

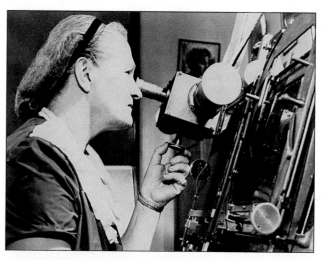

FIGURE 14.2
Cecilia Payne-Gaposchkin Her 1925 PhD thesis laid the foundations for understanding the composition of the Sun and the stars. Yet she was not given a formal appointment at Harvard, where she worked, until 1938 and was not appointed a professor until 1956. *(Harvard University Archives)*

no matter can survive as a liquid or a solid. Still, the Sun's gravity can compress the gases deep inside to a much higher density than the gases on Earth. The density is highest in the center of the Sun and decreases outward.

The Solar Photosphere

The Earth's air is generally transparent. But on a smoggy day it can become opaque, which prevents us from seeing through it past a certain point. Something similar happens in the Sun. Its outer atmosphere is transparent, allowing us to look a short distance through it. But when our line of sight reaches more opaque layers, we can no longer see through them. The **photosphere** (Figure 14.3) is where the Sun becomes opaque and marks the boundary past which we cannot see.

Beneath the photosphere, the gas—like all of the Sun's layers—absorbs and re-emits photons of energy that were produced in the solar interior. These photons are then quickly "grabbed" by other atoms and so remain trapped inside the Sun, where we cannot see them. Only photons absorbed and then re-emitted in the photosphere can escape from the Sun and reach the Earth.

Astronomers have found that the solar atmosphere changes from almost perfectly transparent to almost completely opaque in a distance of just over 400 km; it is this thin region that we call the photosphere, which comes from the Greek for "light sphere." When astronomers speak of the "diameter" of the Sun, they mean the size of the region surrounded by the photosphere.

The photosphere only looks sharp from a distance. If you were falling into the Sun, you would not feel any surface

TABLE 14.2
The Abundance of Elements in the Sun

Element	Percentage by Number of Atoms	Percentage by Mass
Hydrogen	92.0	73.4
Helium	7.8	25.0
Carbon	0.02	0.20
Nitrogen	0.008	0.09
Oxygen	0.06	0.8
Neon	0.01	0.16
Magnesium	0.003	0.06
Silicon	0.004	0.09
Sulfur	0.002	0.05
Iron	0.003	0.14

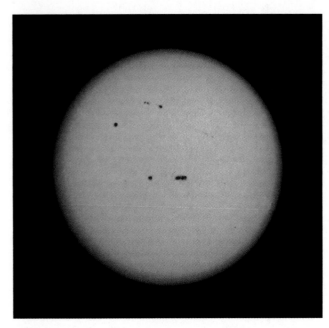

FIGURE 14.3
Sunspots This photograph shows several sunspots on the photosphere—the visible surface of the Sun. *(National Solar Observatory/National Optical Astronomy Observatories)*

but would just sense a gradual increase in the density of the gas surrounding you. It is much the same as falling through a cloud while skydiving. The cloud looks as if it has a sharp surface, but you do not feel a surface as you fall into it. (One big difference between these two scenarios, however, is temperature. The Sun is so hot that you would be vaporized long before you reached the photosphere. Skydiving in the Earth's atmosphere is much safer!)

From analysis of the light received from the Sun, astronomers can calculate how the temperature, density, and pressure of the gases vary through the photosphere. Things can change drastically in this thin layer. For example, as we descend over a distance of a little more than 300 km, the pressure and density increase by a factor of 10, while the temperature climbs from 4500 to 6000 K.

At a depth of 400 km, only 4 percent of the photons escape from the Sun. The other 96 percent are absorbed and re-emitted again, and through this process they gradually work their way outward, eventually reaching a region of low density with too few atoms to absorb the photons again. By the time they get to the top of the photosphere, 99.5 percent of the photons are escaping from the Sun, and these make up most of the sunlight we detect here on Earth.

We might note that the atmosphere of the Sun is not a very dense layer compared to the air in the room where you are reading this book. At a typical point in the photosphere, the pressure is less than 10 percent of the Earth's pressure at sea level, and the density is about one ten-thousandth of the Earth's atmospheric density at sea level.

The Chromosphere

The Sun's outer gases extend far beyond the photosphere (Figure 14.4). Because they are transparent to most visible radiation and emit only a small amount of light, these outer layers are difficult to observe. The region of the Sun's atmosphere that lies immediately above the photosphere is called the **chromosphere.** Until this century, the chromosphere was visible only when the photosphere was concealed by the Moon during a total solar eclipse (see Section 3.7). In the 17th century, several observers described what appeared to them as a narrow red "streak" or "fringe" around the edge of the Moon during a brief instant after the Sun's photosphere had been covered. The name *chromosphere,* from the Greek for "colored sphere," was given to this red streak.

Observations made during eclipses show that the chromospheric spectrum consists of bright emission lines, indicating that this layer is composed of hot, transparent gases emitting light at discrete wavelengths. The reddish color of the chromosphere arises from one of the strongest emission lines in the visible part of its spectrum—the bright red line caused by hydrogen, the element that, as we have already seen, dominates the composition of the Sun.

In 1868, observations of the chromospheric spectrum revealed a yellow emission line that did not correspond to any previously known element on Earth. Scientists quickly realized they had found a new element, and named it *helium* (after *helios,* the Greek word for "Sun"). It took until 1895 for helium to be discovered on our planet. Today, students are probably most familiar with it as the light gas used to inflate balloons, although it turns out to be the second most abundant element in the universe.

The chromosphere is about 2000 to 3000 km thick. The density of the chromospheric gases decreases as we go up, but the temperature increases from 4500 K at the photosphere to 10,000 K or so in the lower levels of the chromosphere. This temperature increase is a surprise, since throughout the rest of the Sun the temperature steadily decreases from the center outward.

The Transition Region

The increase in temperature does not stop with the chromosphere, however. There is a thin region in the solar atmosphere where the temperature changes from 10,000 K, typical of the chromosphere, to nearly a million degrees, characteristic of the corona. Appropriately, this part of the Sun is called the **transition region.** It is probably only a few tens of kilometers thick. Figure 14.5 summarizes how the temperature of the solar atmosphere changes from the photosphere to the corona.

This description makes the Sun sound rather like an onion, with smooth spherical shells, each one with a different temperature. For a long time, astronomers did indeed think of the Sun this way. However, we now know that

FIGURE 14.4

The Sun's Atmosphere   Composite image showing the three components of the solar atmosphere: the photosphere or surface of the Sun taken in ordinary light; the chromosphere, imaged in the light of the strong red spectral line of hydrogen (H-alpha), and the corona as seen with x rays.   *(David Alexander; NASA/Yohkoh)*

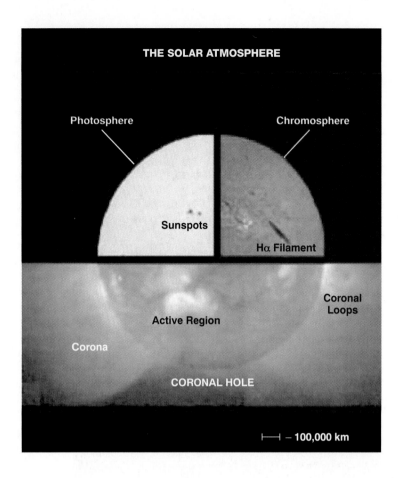

THE SOLAR ATMOSPHERE

Photosphere

Chromosphere

Sunspots

Hα Filament

Active Region

Coronal
Loops

Corona

CORONAL HOLE

⊢——⊣ – 100,000 km

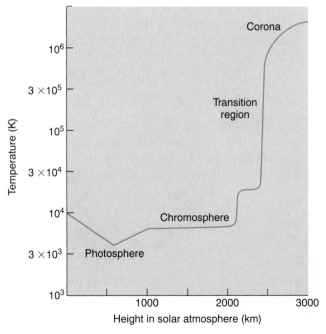

FIGURE 14.5

Temperatures in the Solar Atmosphere   On this graph, temperature is shown increasing upward and height above the photosphere is shown increasing to the right. Note the very rapid increase in temperature over a very short distance in the transition region between the chromosphere and the corona.

while this idea of layers—photosphere, chromosphere, transition region, corona—describes the big picture fairly well, the Sun's atmosphere is more complicated, with hot and cool regions intermixed.

For example, Figure 14.6 shows that the chromosphere contains many jet-like spikes of gas, called *spicules*, rising vertically through it. When viewed near the *limb* (or edge) of the Sun, so many are seen in projection that they give the effect of a forest. Individual spicules last only 10 min or so and consist of gas jets moving upward at about 30 km/s and rising to heights of 5000 to 20,000 km. The gas in the spicules has a temperature typical of the chromosphere (around 10,000 K) but these jet-like eruptions reach well into the much hotter corona. Similarly, clouds of carbon monoxide gas with temperatures colder than 4000 K have now been found in the much hotter chromosphere.

This complexity should not seem too surprising. We already know that temperatures in the Earth's atmosphere change as high and low pressure areas circulate and as the jet stream moves about. In the same way, the atmosphere of the Sun is constantly changing, with matter flowing up and down and changing temperature as it does so. Fortunately, the extremes on the Earth are small relative to the violent events on the Sun. Even hurricanes pale in comparison to "solar weather," as we shall see later in this chapter.

FIGURE 14.6
Spicules The spiky features are solar spicules, photographed in the light of the red Balmer line of hydrogen (called H-alpha). *(National Solar Observatory/National Optical Astronomy Observatories)*

The Corona

The outermost part of the Sun's atmosphere is called the **corona.** Like the chromosphere, the corona was first observed during total eclipses (Figure 14.7). Unlike the chromosphere, the corona has been known for many centuries: it was referred to by the Roman historian Plutarch and was discussed in some detail by Kepler. The corona extends millions of kilometers above the photosphere and emits half as much light as the full moon. The reason we don't see this light until an eclipse occurs is the overpowering brilliance of the photosphere. Just as bright city lights make it difficult to see faint starlight, so too the intense light from the photosphere hides the faint light from the corona. While the best time to see the corona is during a total solar eclipse, its brighter parts can now be photographed with special instruments even at other times, and it can be observed easily from orbiting spacecraft.

Studies of its spectrum show the corona to be very low in density. At the bottom of the corona there are only about 10^9 atoms per cubic centimeter, compared with about 10^{16} atoms per cubic centimeter in the upper photosphere and 10^{19} molecules per cubic centimeter at sea level in the Earth's atmosphere. The corona thins out very rapidly at greater heights, where it corresponds to a high vacuum by laboratory standards.

The corona is very hot. We know this because the spectral lines produced in it come from highly ionized atoms of iron, nickel, argon, calcium, and other elements. For example, astronomers observe lines of iron whose atoms have lost 16 electrons in the corona. Such a high degree of ionization requires a temperature of millions of degrees Kelvin. Indeed, because we don't routinely deal with such hot gases on Earth, for a while astronomers were unsure what caused the lines in the corona's spectrum. They thought the lines might come from a new element, which they named *coronium.* But in the 1930s this hypothetical element became unnecessary, because astronomers discovered that known elements, heated until many electrons are missing, could account for the coronal spectrum. Because of its high temperature, the corona is also very bright at x-ray wavelengths (Figure 14.8).

Why is the corona so hot? Observation and theory together have identified magnetic energy as the culprit. The Sun, like several of the planets, is a giant magnet, with a

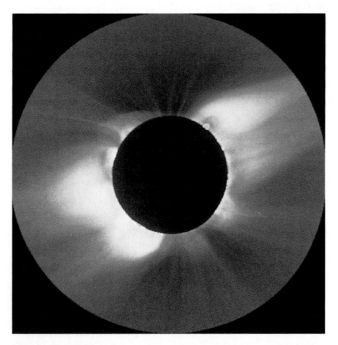

FIGURE 14.7
The Corona During an Eclipse This image of the Sun was taken during the solar eclipse on March 18, 1988. Since the light from the brilliant surface (photosphere) of the Sun is blocked by the Moon, it is possible to see the tenuous outer atmosphere of the Sun, which is called the corona. *(High Altitude Observatory/NCAR)*

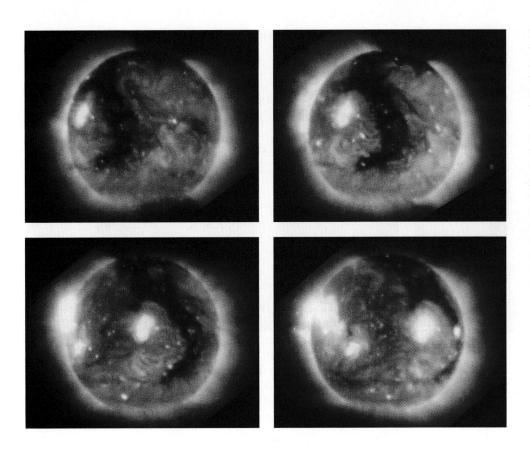

FIGURE 14.8
The Corona Seen in X Rays
X-ray images of the Sun taken from the Skylab satellite show hot coronal gas. The corona is patchy, with bright spots indicating regions where hot gas is concentrated. The long dark area with no x-ray emission is called a coronal hole. In these regions, hot gas streams away from the solar surface and out through the solar system. This stream of particles is called the solar wind. The four successive images of the Sun clearly show how the positions of solar features change as the Sun rotates. *(Courtesy Leon Golub, SAO)*

complicated magnetic field that can have a powerful effect on the thin, electrically charged gas in the outer regions of the Sun. The surface of the Sun is covered with magnetic loops, much like the loops of a carpet. If one loop in which the magnetic field has one polarity is jostled by turbulence into close proximity to a loop of the opposite polarity, the two loops can "reconnect," much as two bar magnets held side by side with their north poles in opposite directions, will snap together. The reconnection of the solar loops releases energy, which can accelerate and heat material that rises into the corona.

The Solar Wind

One of the most remarkable discoveries about the Sun's atmosphere is that it produces a stream of charged particles (mainly protons and electrons) that we call the **solar wind.** These particles flow outward from the Sun at a speed of about 400 km/s (about 900,000 mi/h)! The solar wind exists because the gases in the corona are so hot and moving so rapidly that they cannot be held back by solar gravity. (This wind was actually discovered by its effects on the charged tails of comets; in a sense, we can see the comet tails blow in the solar breeze the way wind socks flutter on Earth.)

By means of the solar wind, material actually leaves the Sun and travels outward into the solar system. Although the solar wind material is very, very rarified, the Sun has an enormous surface area. Astronomers estimate that the Sun is losing about 10 million tons of material each year. While this amount of lost mass seems large by Earth standards, it is completely insignificant for the Sun.

In visible light photographs the solar corona appears fairly uniform and smooth. X-ray pictures, however, show that the corona has loops, plumes, and both bright and dark regions (Figure 14.9). Sometimes large regions of the corona are relatively cool and quiet. These **coronal holes** are places of extremely low density and are usually (but not always) found in the polar regions of the Sun. They cause the empty spaces seen on some of the eclipse photographs of the solar corona (such as Figure 14.7).

Measurements show that hot coronal gas is present mainly where magnetic fields have trapped and concentrated it. The coronal holes lie between these concentrations of gas. The solar wind comes predominantly from these coronal holes, streaming through them into space unhindered by magnetic fields.

At the surface of the Earth, we are protected from the solar wind by our atmosphere and by the Earth's magnetic field. However, the magnetic field lines come into the Earth at the North and South Poles. Here charged particles from the solar wind can follow the field down to our upper atmosphere. As the Sun's particles strike atoms and molecules of air, they cause them to glow, producing beautiful curtains of light called the **aurora** (or the northern and southern lights). The most spectacular auroras occur at altitudes of 75 to 150 km. Strong activity on the Sun can

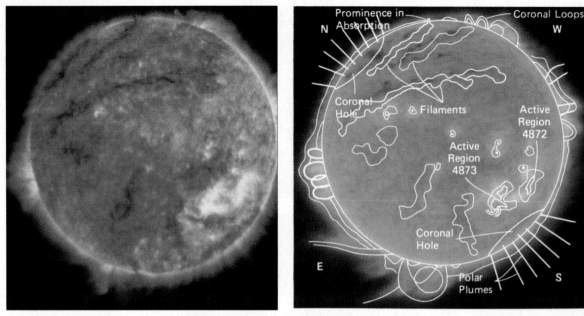

FIGURE 14.9

The X-Ray Sun   This x-ray/extreme ultraviolet image of the Sun was taken in 1987 from a sounding rocket 100 miles above White Sands Missile Range. The features shown in the photograph are identified in the drawing.   *(Art Walker/Stanford University and NASA)*

trigger "gusts" in the solar wind, which, in turn, can lead to truly dramatic auroral displays (Figure 14.10).

THE ACTIVE SUN INTRODUCED

Before the invention of the telescope, the Sun was thought to be an unchanging and perfect sphere. We now know, however, that the Sun is in a perpetual state of change—its surface is a seething, bubbling cauldron of hot gas. Areas that are darker and cooler than the rest of the surface come and go. Vast plumes of gas erupt into the chromosphere and corona. Occasionally, there are even giant explosions on the Sun that send enormous streamers of charged particles and energy hurtling toward the Earth. When they arrive, these can cause power outages and other serious effects on our planet.

FIGURE 14.10

The Aurora   Such a spectacular auroral display is the result of charged particles from the magnetosphere hitting the atmosphere near the Earth's magnetic poles.   *(John Warden/Alaska Stock Images)*

Art Walker: Doing Astronomy in Space

Since the Sun's corona is a rich source of high-energy photons, observations at x-ray wavelengths are a good way to learn more about our star. Unfortunately, x rays do not penetrate the Earth's atmosphere, so such observations must be done from rockets and satellites. Art Walker has been one of the pioneers in designing and flying instruments in space to learn more about the Sun.

Walker received a PhD in physics from the University of Illinois, supported in his graduate training by the Air Force reserve officer program. After receiving his degree, he began active service at the Air Force Weapons Laboratory, which was beginning a space research program in the physics of the Sun. (See Making Connections: *Solar Flares and Their Effects on Earth* to find out why the military is interested in the Sun.) Using a specially designed spectrometer aboard an Air Force satellite, he was able to take one of the first x-ray spectra of the Sun and identify several new emission lines in those spectra. Such measurements allowed him and his co-workers to probe the temperature, composition, and structure of the corona, and get a much better sense of what the Sun's atmosphere is like.

More recently, he has been working with a novel design for x-ray telescopes that makes them similar to some of the telescopes astronomers use on the ground. These instruments allow scientists to take very high-resolution x-ray images of the active Sun; they are also relatively easy to build and can be flown aboard sounding rockets whose paths are spectacular arcs traveling 100 or more miles above the Earth's surface.

In an interview, Walker emphasized one of the key challenges of doing astronomy from space: "In the laboratory, if you build an experiment and it doesn't quite work right, you can al-

---

Once the instrument is launched, it is out of your control.

---

ways tinker with it until you get the instrumentation to operate. In the case of space observations, you have to build the instrument anticipating all the things that might go wrong and eliminating each possibility of failure. Once

Art Walker *(Courtesy of Stanford University)*

the instrument is launched, it is out of your control, and you just have to hope that you have anticipated everything."

After working for the Aerospace Corporation, Walker became a Professor at Stanford University, where he has also served as Dean of the Graduate College. Perhaps his best-known student was Sally Ride, who received her PhD in astrophysics and went on to become America's first woman in space. Walker has chaired the Astronomy Advisory Committee for the National Science Foundation and also served on the Presidential Commission that investigated the accident that destroyed the Shuttle Challenger.

Photospheric Granulation

Observations show that the photosphere has a mottled appearance resembling grains of rice spilled on a dark tablecloth. This structure of the photosphere is now generally called **granulation** (Figure 14.11). Granules, which are typically 700 to 1000 km in diameter, appear as bright areas surrounded by narrow, darker regions. The lifetime of an individual granule is only 5 to 10 minutes.

FIGURE 14.11 ▶
Solar Granulation Near Sunspots   Each small, bright region is a rising column of hotter gas about 1000 km across. Cooler gas descends in the darker regions between the granules. The larger dark regions are sunspots.   *(National Solar Observatory/National Optical Astronomy Observatories)*

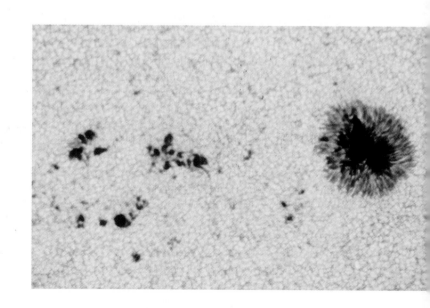

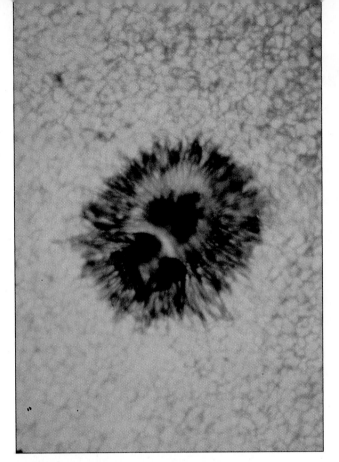

FIGURE 14.12

A Sunspot   This image of a sunspot, a cooler and thus darker region on the Sun, was taken with the McMath–Pierce Solar Telescope. The spot has a dark central region (called the umbra) surrounded by a less-dark region (the penumbra). Note that you can see the granulation on the Sun's surface.   *(W. Livingston, National Solar Observatories, National Optical Astronomy Observatories)*

The motions of the granules can be studied by examining the Doppler shifts (see Section 4.6) in the spectra of gases just above them. The bright granules are columns of hotter gases rising at speeds of 2 to 3 km/s from below the photosphere. As this rising gas reaches the photosphere, it spreads out and sinks down again into the darker regions between the granules. The centers of the granules are hotter than the intergranular regions by 50 to 100 K.

The granules, then, are the tops of *convection currents* of gas rising through the photosphere. Convection is the rising of a hot liquid or gas, transferring energy from the hotter layers underneath to the cooler layers above. As the gas cools, it falls back down, and the area looks darker from above. (You can see the same sort of hot currents rising in a pot of boiling water as you make more coffee in order to stay awake while reading your astronomy book.) Granulation is one piece of evidence that heat is flowing from the Sun's interior outward to the photosphere and beyond.

Sunspots

The most conspicuous of the photospheric features are the dark **sunspots** (Figure 14.12). Occasionally, these spots are large enough to be visible to the unaided eye, and we have records going back over a thousand years of observers who noticed them when haze or mist reduced the Sun's intensity. (We emphasize what your mother surely told you: Looking at the Sun for even a brief while can cause irreparable eye damage. This is the one area of astronomy where we don't encourage you to do your own observing without getting careful instructions or filters from your instructor!)

Today we know that sunspots are darker than the photosphere in which they are embedded because the gases in sunspots are as much as 1500 K cooler than the surrounding gases. Sunspots are nevertheless hotter than the surfaces of many stars. If they could be removed from the Sun, they would shine brightly. They appear dark only in contrast with the hotter, brighter surrounding photosphere.

Individual sunspots have lifetimes that range from a few hours to a few months. If a spot lasts and develops, it usually consists of two parts: an inner darker core, the *umbra*, and a surrounding less-dark region, the *penumbra*. Many spots become much larger than the Earth, and a few have reached diameters of 50,000 km. Frequently, spots occur in groups of 2 to 20 or more. The largest groups are very complex and may have over 100 spots. Like storms on the Earth, sunspots are not fixed in position, but they drift slowly compared with solar rotation, which carries them across the visible disk of the Sun.

By recording the apparent motions of the sunspots as the turning Sun carried them across its disk, Galileo in 1612 demonstrated that the Sun rotates on its axis (Figure 14.13) with a rotation period of approximately one month. Modern measurements show that the Sun's rotation period is about 25 days at the equator, 28 days at latitude 40°, and 36 days at latitude 80°, in a west-to-east direction (like the orbital motions of the planets). Note that the Sun, being composed of gas, does not have to rotate rigidly, the way a solid body does.

14.3   THE SUNSPOT CYCLE

Between 1826 and 1850, Heinrich Schwabe, a German pharmacist and amateur astronomer, kept daily records of the number of sunspots. What he was really looking for was a planet inside the orbit of Mercury, which he hoped to find by observing its dark silhouette as it passed between the Sun and the Earth. He failed to find the hoped for planet, but his diligence paid off with an even more important discovery—the **sunspot cycle.** He found that the number of sunspots varied systematically in cycles of about ten years.

What Schwabe observed was that, although individual spots are short-lived, the total number visible on the Sun at any one time is likely to be very much greater at certain times—the periods of sunspot maximum—than at other

FIGURE 14.13
Sunspots Rotate Across Sun's Surface This sequence of photographs of the Sun's surface tracks a large group of sunspots as the Sun turns. The series of exposures follows the rotation of the sunspots across the visible hemisphere of the Sun. The top sequence shows the Sun in ordinary light; the bottom sequence shows emission from the chromosphere. *(National Solar Observatory/National Optical Astronomy Observatories)*

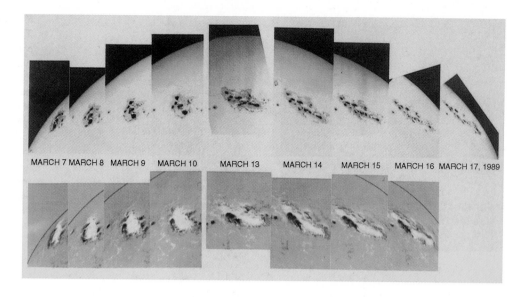

MARCH 7 MARCH 8 MARCH 9 MARCH 10 MARCH 13 MARCH 14 MARCH 15 MARCH 16 MARCH 17, 1989

times—the periods of sunspot minimum (Figure 14.14). We now know that sunspot maxima occur at an average interval of 11 years, but the intervals between successive maxima have ranged from as little as 8 years to as long as 16 years. During sunspot maxima, more than 100 spots can often be seen at once. During sunspot minima, sometimes no spots are visible. The Sun's activity is predicted to reach its next maximum in about 2001.

Magnetism and the Solar Cycle

It is the Sun's changing magnetic field that drives the sunspot cycle. The solar magnetic field is measured using a property of atoms called the *Zeeman effect*. Recall from Chapter 4 that an atom has many energy levels, and that spectral lines are formed when electrons shift from one level to another. If each energy level is precisely defined, then the difference between them is also quite precise. As an electron changes levels, the result is a sharp, narrow spectral line (either an absorption or emission line, depending on whether the electron's energy increases or decreases in the transition).

In the presence of a strong magnetic field, however, each energy level is separated into several levels very close to one another. The separation of the levels is proportional to the strength of the field. As a result, spectral lines formed in the presence of a field are not single lines, but a series of very closely spaced lines corresponding to the subdivision of the atomic energy levels. This splitting of lines in the presence of a magnetic field is termed the Zeeman effect.

Measurements of the Zeeman effect in the spectra of the light from sunspot regions (Figure 14.15) show them to have strong magnetic fields. Whenever sunspots are observed in pairs, or in groups containing two principal spots, one of the spots usually has the magnetic polarity of a

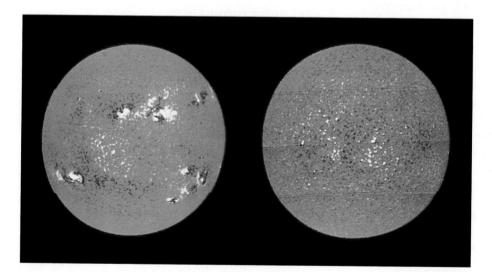

FIGURE 14.14
The Active and Quiet Sun Compared This comparison shows the number of sunspots and magnetic activity on the active (left) and quiet (right) Sun. The computer-generated images use yellow to indicate positive or north polarity, and dark blue for negative or south polarity. In the image of the active Sun, note that pairs of sunspots have opposite polarity. Note also that the polarity of the leading spot is different in the upper and lower hemispheres. At solar minimum there are no large sunspots, and the magnetic fields are weak. *(National Solar Observatory/National Optical Astronomy Observatories)*

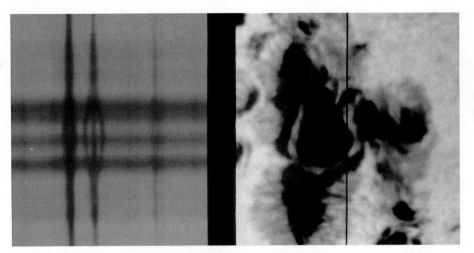

FIGURE 14.15
The Zeeman Effect These photographs show how magnetic fields in sunspots are measured by means of the Zeeman effect. The vertical black line in the picture at right indicates the position of the spectrograph slit through which light passed in order to obtain the spectrum in the picture at left. Note that the strongest spectral line in the left-hand picture is split into three components, which indicates that a strong magnetic field is present. *(National Optical Astronomy Observatories)*

north-seeking magnetic pole and the other has the opposite polarity. Moreover, during a given cycle the leading spots of pairs (or leading principal spots of groups) in the northern hemisphere all tend to have the same polarity, while those in the southern hemisphere all tend to have the opposite polarity.

During the next sunspot cycle, however, the polarity of the leading spots is reversed in each hemisphere. For example, if during one cycle the leading spots in the northern hemisphere all had the polarity of a north-seeking pole, the leading spots in the southern hemisphere would have the polarity of a south-seeking pole. During the next cycle, the leading spots in the northern hemisphere would have south-seeking polarity, while those of the southern hemisphere would have north-seeking polarity. Therefore, the sunspot cycle does not repeat itself in regard to magnetic polarity until two maxima have passed. The solar activity cycle, fundamentally a magnetic cycle, thus averages 22 years in length, not 11.

Why does the Sun's magnetic field change in strength and polarity in a nearly regular way? Calculations from detailed models of how the Sun works show that rotation and convection just below the solar surface can distort the magnetic fields. This causes them to grow and then decay, regenerating with opposite polarity approximately every 11 years. The calculations also show that as the fields grow stronger near solar maximum, they flow from the interior of the Sun toward its surface in the form of loops. When a large loop emerges from the solar surface, much like the loop formed by a snake as it slithers along, it creates region of sunspot activity.

The "ends" of the loop, where they penetrate the solar surface, have different polarities. This idea of magnetic loops offers a natural explanation of why the leading and trailing sunspots in an active region have opposite polarity. The leading sunspot coincides with one end of the loop and the trailing spot with the other end. Smaller loops form the magnetic carpet described earlier and appear to be responsible for heating the corona.

Magnetic fields also hold the key to explaining why sunspots are cooler and darker than the regions without strong magnetic fields. The forces produced by the magnetic field resist the motions of the bubbling columns of rising hot gases. Since these columns carry most of the heat from inside the Sun to the surface by means of convection, less heating occurs where there are strong magnetic fields. As a result, these regions are seen as darker, cooler sunspots.

 ACTIVITY ABOVE THE PHOTOSPHERE

To see those portions of the Sun lying above the photosphere, we must observe the *emission lines* from elements such as hydrogen and calcium. These lines are produced in the hotter, lower regions of gas in the chromosphere. Astronomers routinely photograph the Sun through monochromatic filters that pass light only at these special wavelengths. The hot corona, on the other hand, can be studied by observations of x rays.

Plages and Prominences

Pictures taken through filters that pass only the light of either calcium or hydrogen show bright "clouds" in the chromosphere around sunspots; these bright regions are known as **plages** (Figure 14.16). The plages actually contain all of the elements in the Sun, not just hydrogen and calcium. They are regions of higher temperature and density within the chromosphere. It just happens that the spectral lines of hydrogen and calcium produced by these clouds are strong and easy to observe at the higher temperatures.

Moving higher into the Sun's atmosphere, we come to the spectacular phenomena called **prominences** (Figure 14.17), which usually originate near sunspots. Eclipse observers often see prominences as red, flame-like protuberances rising above the eclipsed Sun and reaching high into the corona. Some, the *quiescent prominences*, are graceful loops that can remain nearly stable for many hours or even

days. They may extend to heights of tens of thousands of
kilometers above the solar surface. Others can move upward or have arches that surge slowly back and forth.

The relatively rare *eruptive prominences* appear to send
matter upward into the corona at speeds up to 700 km/s, and
the most active *surge prominences* may move as fast as
1300 km/s (almost 3 million mi/h). Some eruptive prominences have reached heights of over 1 million km above
the photosphere; the Earth would be completely lost inside one of these awesome displays (Figure 14.18).

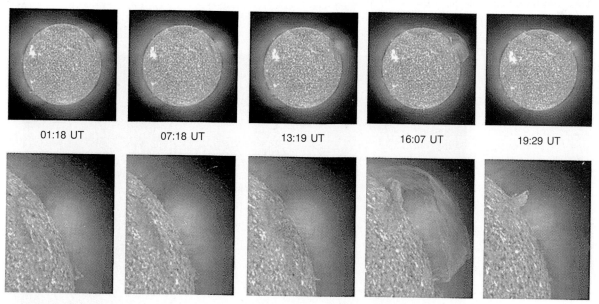

01:18 UT   07:18 UT   13:19 UT   16:07 UT   19:29 UT

FIGURE 14.18
An Eruptive Prominence A series of images from the SOHO satellite shows the
development of a huge eruptive prominence. The picture was taken in the light of a strong
line of ionized helium in the ultraviolet part of the spectrum (wavelength 30.4 nm). The
material in the erupting prominence is at temperatures of 60,000 to 80,000 K, and the
prominence eventually stretches more than 216,000 miles across, large enough to span 28
Earths. The abbreviation UT means universal time; you are seeing an eruption on the Sun
over the course of more than 18 hours. *(NASA/SOHO)*

MAKING CONNECTIONS

Solar Flares and Their Effects on Earth

When large numbers of particles are suddenly ejected from the Sun, the most obvious effect on the Earth is the appearance of a brilliant *aurora*. In March 1989 a gigantic flare, which occurred as the Sun approached maximum activity, was accompanied by a coronal mass ejection that produced an aurora visible as far south as Arizona (latitude 32°). Recall that auroras occur preferentially near the Earth's magnetic poles, because the charged particles from the Sun tend to flow down into the Earth's atmosphere along our magnetic field, which comes into the Earth near the poles. So seeing an aurora in the American southwest is really something!

The changes in the Earth's magnetic field caused by the sudden onslaught of charged particles from the Sun in turn generate changing electric currents near the surface of our planet. This effect is most noticeable in long power lines, but solar flares can even cause power station components to burn out. As a result of the solar explosion in March 1989, parts of Montreal and Quebec Province in Canada were without power for up to 9 hours. Other effects occurred as well. For example, because of the electrical interference, people found their automatic garage doors opening and closing for no apparent reason. Charged particles that reach the Earth from the Sun can also overload telephone circuits.

The short-wavelength radiation produced during solar flares heats the outer atmosphere of the Earth. In 1981 a very large solar flare occurred while the Space Shuttle Columbia was in orbit. The astronauts aboard found that the flare, which lasted for 3 hours, increased the temperature of the Earth's atmosphere at an altitude of 260 km from its normal value of 1200 K up to 2200 K. When the outer atmosphere is heated, it also expands, so that it reaches farther into space. As a consequence, friction between the atmosphere and spacecraft increases, dragging satellites to lower altitudes. At the time of the March 1989 flare, the system responsible for tracking some 19,000 objects orbiting the Earth temporarily lost track of 11,000 of them because their orbits were changed by the expansion of the Earth's atmosphere. During solar maximum, a number of satellites are brought to such a low altitude that they are destroyed by friction with the atmosphere.

The level of solar activity is a critical factor in calculating the lifetimes and orbits of satellites in near-Earth orbit. Flares could also be life-threatening to astronauts on a voyage to Mars. Obviously, it would be extremely valuable to have the ability to predict both the overall level of solar activity and the occurrence of individual flares. Solar astronomers are working very hard to learn how to make reliable predictions, but accurate forecasts of solar "weather" are proving to be an even more elusive goal than reliable forecasts of the weather on Earth.

A solar flare can result in the ejection of vast numbers of particles at speeds of thousands of kilometers per hour. *(National Solar Observatory/National Optical Astronomy Observatories)*

Flares

The most violent event on the surface of the Sun is a rapid eruption called a **solar flare**. A typical flare lasts for 5 to 10 min and releases a total amount of energy equivalent to that of perhaps a million hydrogen bombs. The largest flares (Figure 14.19 is a dramatic example) last for several hours and emit enough energy to power the entire United States at its current rate of electrical consumption for 100,000 years. Near sunspot maximum, small flares occur several times per day, and major ones may occur every few weeks.

Flares are often observed in the red light of hydrogen, but the visible emission is only a tiny fraction of the energy released when a solar flare explodes. At the moment of the explosion, the matter associated with the flare is heated to temperatures as high as 10 million K. At such high temperatures, a flood of x-ray and ultraviolet radiation is emitted.

Flares seem to occur when magnetic fields pointing in opposite directions release energy by interacting with and destroying each other—much as a stretched rubber band releases energy when it breaks. We have discussed this idea before, when we talked about heating the corona. What is different about flares is that its magnetic interactions cover a large volume in the solar corona and release a tremendous amount of electromagnetic radiation. In some cases, immense quantities of coronal material—mainly protons and electrons—may be also be ejected at high speeds (500–1000 km/s) into interplanetary space (Figure 14.19). Such *coronal mass ejections* can affect the Earth in several ways. (See Making Connections: *Solar Flares and Their Effects on Earth.*)

Active Regions

Sunspots, flares, and bright regions in the chromosphere and corona tend to occur together on the Sun. That is, they all tend to have similar longitudes and latitudes, but they are located at different heights in the atmosphere. Because they all occur together, they all vary with the sunspot cycle. For example, flares are more likely to occur near sunspot maximum, and the corona is much more conspicuous at that time (see image at the beginning of this chapter). A place on the Sun where these phenomena are seen is called an **active region** (Figure 14.20). Active regions are always associated with strong magnetic fields.

14.5 IS THE SUN A VARIABLE STAR?

The Sun rises faithfully every day at a time that can be precisely calculated. Each day it deposits energy on the Earth, warming it and sustaining life. But over the last decade, scientists have accumulated evidence that the Sun is not truly constant but varies over the centuries by a small amount—probably less than 1 percent. Still, that variation is enough to have profound effects on the Earth and its climate.

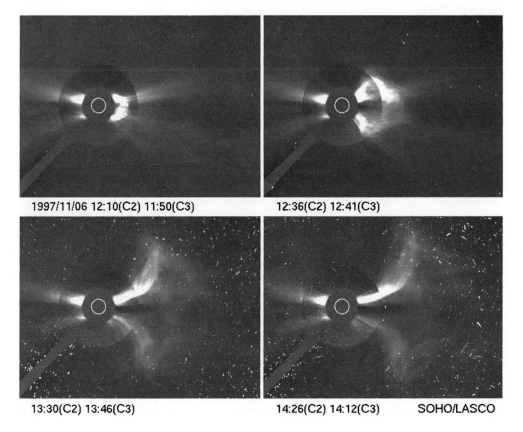

1997/11/06 12:10(C2) 11:50(C3)   12:36(C2) 12:41(C3)

13:30(C2) 13:46(C3)   14:26(C2) 14:12(C3)   SOHO/LASCO

FIGURE 14.19
A Flare on the Sun This sequence of composite images taken by SOHO spacecraft's coronagraphic instruments shows a large solar flare and coronal mass ejection. Many tons of solar particles are being blasted into space at speeds of millions of miles per hour. The white streaks and dots on the third and fourth images are produced by high-energy particles that reached the SOHO instruments, which were 155 million km away from the Sun at the time, in about 3 hours. The white circle indicates the size and position of the Sun. The blue disks (or coronagraphs) block the Sun, producing an artificial eclipse, so that it is possible to observe the faint corona. *(NASA/SOHO)*

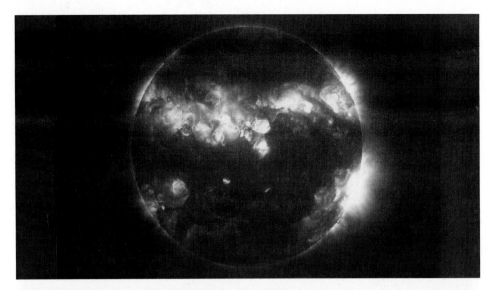

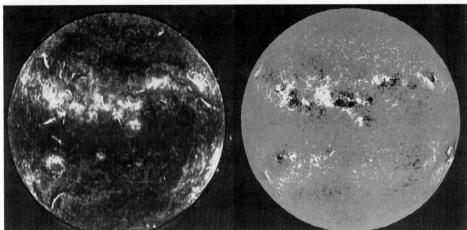

FIGURE 14.20
The Sun at Different
Wavelengths The three images of
the Sun seen here were taken at the
same time but at different
wavelengths. The top image shows x-
ray emission, which originates in the
hot corona. The image at lower left,
taken in the light of a strong line of
neutral helium, shows where
chromospheric emission is strong. In
the image at lower right, white and
black spots are regions of strong
magnetic field; white and black
correspond to opposite magnetic
polarities. You can see that coronal x-
ray emission, chromospheric emission,
and strong magnetic fields tend to
occur in the same locations, called
active regions. *(NASA/National Solar
Observatory/National Optical Astronomy
Observatories)*

Variations in the Number of Sunspots

We already know that the number of sunspots varies, with the time between sunspot maxima being about 11 years. However, the number of sunspots at maximum is not always the same. Considerable evidence shows that between the years 1645 and 1715, the number of sunspots, even at sunspot maximum, was much lower than it is now. This interval of extremely low activity was first noted by Gustav Spörer in 1887, and then by E. W. Maunder in 1890; it is now called the *Maunder Minimum.* The variation in the number of sunspots over the past four centuries is shown in Figure 14.21. Besides the Maunder Minimum in the 17th century, sunspot numbers were somewhat lower during the first part of the 19th century than they are now; this period is called the *Little Maunder Minimum.*

When the number of sunspots is high, the Sun is active in various other ways as well, and this activity affects the Earth directly. For example, there are more auroral displays when the sunspot number is high. Auroras are caused by the impact of energetic charged particles from the Sun on the Earth's magnetosphere, and the Sun is more likely to eject particles when it is active and the sunspot number is high. Historical accounts indicate that auroral activity was abnormally low throughout the several decades of the Maunder Minimum.

The best quantitative evidence of long-term (over several decades) variations in the level of solar activity comes from studies of the radioactive isotope carbon-14. The Earth is constantly bombarded by *cosmic rays*—high-energy charged particles that include protons and nuclei of heavier elements. The rate at which cosmic rays from sources outside the solar system reach the upper atmosphere depends on the level of solar activity. When the Sun is active, its charged particles, streaming out into the solar system, carry the Sun's strong magnetic field with them. This magnetic field shields the Earth from incoming cosmic rays. At times of low activity, when the Sun's magnetic field is weak, cosmic rays reach the Earth in larger numbers.

When the energetic cosmic-ray particles impact the upper atmosphere, they produce several different radioactive isotopes (see Section 4.4). One such isotope is carbon-14,

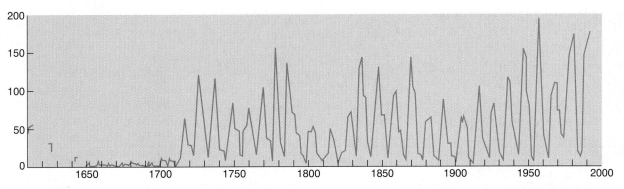

FIGURE 14.21

The Numbers of Sunspots as They Change with Time Note the absence of sunspots
from 1645 to 1715.

which is produced when nitrogen is struck by high-energy cosmic rays. The rate of carbon-14 production is higher when the activity of the Sun is lower and the solar magnetic field does not shield the Earth from bombardment by cosmic rays.

Some of the radioactive carbon then forms carbon dioxide molecules, which are ultimately incorporated into trees through photosynthesis. By measuring the amount of radioactive carbon in tree rings, we can estimate the historical levels of solar activity. Correlations with visual estimates of sunspot numbers over the past 300 years indicate that the carbon-14 estimates of solar activity are indeed valid. Because it takes an average of about 10 years for a carbon dioxide molecule to be absorbed from the atmosphere or ocean into plants, this technique cannot provide data on the 11-year solar cycle. It can be used, however, to look for long-term changes in the level of solar activity.

Estimates of the amount of carbon-14 in tree rings now extend continuously back about 8000 years. Variations in solar activity levels have occurred throughout this period, and at different times the Sun has been both more and less active than it is now. The measurements confirm that the amount of carbon-14 was unusually high, with solar activity correspondingly low, during both the Maunder Minimum and the Little Maunder Minimum. Activity was also low from 1410 to 1530 and from 1280 to 1340. Between about 1100 and 1250, the level of solar activity may have been even higher than it is now.

Solar Variability and the Earth's Climate: Historical Evidence

There is now good evidence that these changes in solar activity affect the Earth's climate. The Maunder Minimum was a time of exceptionally low temperatures in Europe—so low that this period is described as a Little Ice Age. The river Thames in London froze at least 11 times during the 17th century, ice appeared in the oceans off the coasts of southeast England, and low summer temperatures led to

short growing seasons and poor harvests (Figure 14.22). The global climate also appears to have been unusually cool from 1400 to 1510; this period was one of low solar activity as well.

These changes in climate can have profound impacts on human history. For example, Norse explorers from Norway first colonized Iceland and then reached Greenland by 986 A.D. From there, they were able to make repeated visits to the northeastern coasts of North American, including Newfoundland, between about 1000 and 1350 A.D. (The ships of the time did not allow the Norse explorers to make it all the way to North America directly but only from Greenland, which served as a station for further exploration.)

Most of Greenland is covered by ice, and the Greenland station was never self-sufficient; rather it depended on imports of food and other goods from Norway for its survival. When a little ice age began in the 13th century, voyaging became very difficult, and support of the Greenland colony was no longer possible. The last known contact with it was made by a ship from Iceland blown off course in 1410. When European ships again began to visit Greenland in 1577, the entire colony had disappeared.

Note how the estimated dates for these patterns of migration follow what we know about solar activity. Solar activity was unusually high between 1100 and 1250, which includes the time when the first European visits were made to North America. Activity was low from 1280 to 1340 and there was a little ice age, which was about the time that regular contact with North America and between Greenland and Europe stopped.

Solar Variability and the Earth's Climate: The Mechanism

The evidence then is that the Earth is on average cooler when solar activity is low. But why? Does the Sun actually put out less heat at times of low activity? Only in last few years have we obtained precise measurements over a

FIGURE 14.22
Europe's Little Ice Age   During the "Little Ice Age" in Europe, bodies of water froze every winter. This painting by Robert van den Hoecke from 1649 is entitled "Skating in the Town Moat of Brussels"   *(Kunsthistorisches Museum, Vienna; photo by Erich Lessing, Art Resources)*

whole solar cycle from space so that we can answer this question. It turns out, surprisingly, that the Sun is a tenth of 1 percent or so brighter during solar maximum. This is just the opposite of what we might have guessed since we might have expected the Sun to be fainter when it is covered with a large number of dark sunspots. The source of the extra radiation at sunspot maximum has not been identified but probably comes from bright regions, like plages, which are more abundant at solar maximum.

People who compute models of the Earth's atmosphere have been trying to understand how small changes in the Sun's luminosity might cause noticeable changes in the temperature of the Earth. For example, we observe that the ultraviolet radiation from the Sun is about 1 percent higher at solar maximum. Models of our atmosphere indicate that this extra radiation can create more *ozone* in the stratosphere. Ozone can absorb more sunlight and increase the temperature of the stratosphere. This in turn can affect the patterns of the winds aloft, modify the paths of storms, and change the temperature in lower layers of the Earth's atmosphere by differing amounts at different latitudes. However, according to the best estimates we have so far, the unusually cold temperatures at the time of the Maunder Minimum would have required approximately a 1 percent drop in solar luminosity, which is larger than the variations we see today.

We can get additional clues about the likelihood that the solar luminosity might at times vary by as much as 1 percent by looking at activity cycles in stars that are similar to the Sun. It turns out that the Sun's current variations (just 0.1 percent of the total energy emitted) are unusually small. The luminosity of most stars varies by 0.3 percent and some by as much as 1 percent. Therefore, it seems likely that the Sun too might sometimes vary by larger amounts than it does currently, and that the energy output at the time of the Maunder Minimum might indeed have been low enough to cause unusually cold temperatures on Earth. The observations of higher variability in the energy output of most other stars that are like the Sun also suggest that the stability of solar behavior—and the stability of the Earth's climate—over the past three centuries may be unusual.

All of these ideas are very new and must still be confirmed by further observations and modeling. And phenomena on Earth, such as volcanoes and greenhouse warming from the burning of fossil fuels, can also affect the global climate in ways we are only now coming to understand. Nevertheless, the evidence is growing that changes in the amount of energy emitted by the Sun do affect the climate of the Earth and the lives of the creatures who live there. That's one key reason for continuing to study the short- and long-term behavior of our star.

Observing the Sun

Looking directly at the Sun is very dangerous! Even a brief exposure can burn your retina and cause severe eye damage. Looking at the Sun directly through an unfiltered telescope is even worse, because the telescope concentrates the Sun's radiation and hence damages your eye even more quickly. But there are indirect ways of observing the Sun that are both safe and instructive.

Not long after Galileo began training his first telescope on the sky, his assistant and disciple Benedetto Castelli came up with a way of observing sunspots, whose nature was a big controversy in those days. Castelli projected the image of the Sun made by the telescope onto a sheet of paper, where the sunspots could be sketched safely. Using this technique, Galileo was able to show that sunspots were not small, dark planets blocking sunlight, as some had suggested, but dark areas on the Sun moving around as the Sun rotated. You can try this experiment for yourself.

> Castelli projected the image of the Sun made by the telescope onto a sheet of paper, where sunspots could be sketched safely.

Set up a small telescope on a steady mount so that it points at the Sun. Remember, you have to do this by trial and error; *never look through the telescope at the sun to see how well you're doing!* Instead, watch the telescope's shadow; it will be smallest when the telescope is pointing directly at the Sun. Then you'll be able to see an image of the Sun on a sheet of white cardboard or other light surface. Sketch the sunspots if there are any. Repeat the observation over the course of a week or two, and, orienting your sketches the same way, trace how the spots have moved over the face of the Sun.

Putting a cardboard collar around the telescope makes the image easier to see by casting a shadow on it and thus making the sphere of the Sun stand out. If you observe the Sun for a while, don't be surprised if its image moves across and off the paper over time. This is caused by the turning of the Earth, which makes the Sun appear to rise and set. Also, don't be upset if you don't see any sunspots, especially if you are observing during the solar activity minimum.

Looking through a telescope at the Sun is dangerous, but you can always view the Sun safely with a small telescope by projecting its image on a sheet of white cardboard.

SURFING THE WEB

🖥 General Information Sites about the Sun:

• The Nine Planets Site: [seds.lpl.arizona.edu/nineplanets/nineplanets/sol.html]
• Views of the Solar System Site: [www.hawastsoc.org/solar/eng/sun.htm]

🖥 The Stanford SOLAR Center
[solar-center.stanford.edu/]
An excellent place to begin your exploration of the Sun, with a treasure trove of solar information, including background, SOHO and other spacecraft results, images, links, activities, and even a section on solar folklore.

🖥 NASA Goddard Solar Flares Pages
[hesperia.gsfc.nasa.gov/sftheory/yohkoh.htm]
An introduction to solar flares, why we must study them using spacecraft, and what we learned from the Japanese Yohkoh (Sunbeam) mission.

🖥 Images of the Sun Today
[umbra.nascom.nasa.gov/images/latest.html]
Want to see what the Sun looks like *today*? This handy site collects images from all over the Web in many wavelengths and has handy links to the sites that store images of the Sun.

🖥 Sites for Space Missions About the Sun

• Ulysses (flies over the Sun's poles): [ulysses.jpl.nasa.gov/ULSHOME.html]
• SOHO (an observatory with many instruments): [sohowww.nascom.nasa.gov/]
• TRACE (extreme-UV observations): [www.lmsal.com/TRACE/welcome.html]
• Yohkoh Soft X-ray Telescope: [www.lmsal.com/SXT/homepage.html]

14.1 The Sun, our star, is surrounded by a number of layers that make up the solar atmosphere. In order of increasing distance from the center of the Sun, they are the **photosphere,** with a temperature that ranges from 4500 K to about 6800 K; the **chromosphere,** with a typical temperature of 10^4 K; the **transition region,** a zone that may be only a few kilometers thick, where the temperature increases rapidly from 10^4 K to 10^6 K; and the **corona,** with temperatures of a few million degrees Kelvin. Solar wind particles stream out into the solar system through **coronal holes.** Hydrogen and helium together make up 98 percent of the mass of the Sun, whose composition is much more characteristic of the universe at large than is a planet like the Earth.

14.2 The Sun's surface is mottled with upwelling currents seen as hot, bright **granules. Sunspots** are dark regions where the temperature is up to 1500 K cooler than in the surrounding photosphere. Their motion across the Sun's disk allows to calculate how fast the Sun turns on its axis. The Sun rotates more rapidly at its equator, where the rotation period is about 25 days, than near the poles, where the period is slightly greater than 36 days.

14.3 The number of visible sunspots varies according to a **sunspot cycle** that averages 11 years in length. Spots frequently occur in pairs. During a given 11-year cycle, all leading spots in the northern hemisphere have the same magnetic polarity, while all leading spots in the southern hemisphere have the opposite polarity. In the subsequent 11-year cycle, the polarity reverses. For this reason, the magnetic activity cycle of the Sun is often said to last for 22 years.

14.4 Sunspots, **solar flares, prominences,** and bright regions, including **plages,** tend to occur in **active regions—** that is, in places on the Sun with the same latitude and longitude but at different heights in the atmosphere. These active regions are connected with the Sun's powerful magnetic field.

14.5 Over long periods of time (100 years or more), there are changes in the level of solar activity and the number of sunspots seen at solar maximum. For example, the number of sunspots was unusually low from 1645 to 1715, a period now called the Maunder Minimum. There is strong historical evidence that the Earth is cooler when the number of sunspots is unusually low for several decades.

INTER-ACTIVITY

A Have your group make a list of all the ways the Sun affects your life on Earth. How long a list can you come up with? (Be sure you consider the everyday effects as well as the unusual effects due to solar activity.)

B Long before the nature of the Sun was fully understood, astronomer (and planet discoverer) William Herschel (1738–1822) proposed that the hot Sun may have a cool interior and may be inhabited. Have your group discuss this proposal and come up with modern arguments against it.

C In the text we discuss how the migration of Europeans to North American was apparently affected by climate change. If the Earth were to become significantly hotter,

either because of changes in the Sun or because of greenhouse warming, one effect would be an increase in the rate of melting of the polar ice caps. How would this affect modern civilization?

D Suppose we experience another Maunder Minimum on Earth, with a drop in the average temperatures. Have your group discuss how this would affect civilization and international politics. Make a list of the most serious effects you can think of.

E Watching sunspots move across the disk of the Sun is one way to show that our star rotates on its axis. Can your group come up with other ways to show the Sun's rotation?

REVIEW QUESTIONS

1. Describe the main differences between the composition of the Earth and that of the Sun.

2. Make a sketch of the Sun's atmosphere showing the locations of the photosphere, chromosphere, and corona. What is the approximate temperature of each of these regions?

3. Why do sunspots look dark?

4. What is the Zeeman effect, and what does it tell us about the Sun?

5. Describe three different types of solar activity.

6. How does activity on the Sun affect the Earth?

7. Which aspects of the Sun's activity cycle have a period of about 11 years? Which vary during intervals of about 22 years?

8. Summarize the evidence indicating that over several decades or more there have been variations in the level of solar activity.

9. Use the data in Table 14.1 to confirm that the density of the Sun is 1.4 g/cm^3. What kinds of materials have similar densities? One such material is ice. How do you know that the Sun is not made of ice?

10. If the rotation period of the Sun is determined by observing the apparent motions of sunspots, must any correction be made for the orbital motion of the Earth? If so, explain what the correction is and how it arises. If not, explain why the Earth's orbital revolution does not affect the observations.

11. Suppose an (extremely hypothetical) elongated sunspot forms that extends from a latitude of 30° to a latitude of 40° along a fixed line of longitude. How will the appearance of that sunspot change as the Sun rotates?

12. Suppose you live in northern Canada and an extremely strong flare is reported on the Sun. What precautions might you take? What could compensate you for your troubles?

13. Give some reasons for why is it difficult to determine whether or not small changes in the amount of energy radiated by the Sun have an effect on the Earth's climate?

PROBLEMS

14. Suppose you observe a major solar flare while astronauts are orbiting the Earth in the Shuttle. Use the data in Section 14.1 to calculate how long it will be before the charged particles ejected from the Sun during the flare reach the Shuttle.

15. Table 14.2 shows that 92 percent of the atoms in the Sun are hydrogen but only 73.4 percent of the mass of the Sun is made up of hydrogen. Explain this difference.

16. Nearly all the light from the Sun emerges from a layer that is only about 400 km thick. What fraction is this of the radius of the Sun? Suppose we could see light emerging directly from a layer that was 300,000 km thick. Would the Sun appear to have a sharp edge?

17. Suppose an eruptive prominence rises at a speed of 150 km/s. If it does not change speed, how far from the photosphere will it extend after 3 hours? How does this distance compare with the diameter of the Earth?

18. From the Doppler shifts of the spectral lines in the light coming from the east and west edges of the Sun, it is found that the radial velocities of the two edges differ by about 4 km/s. Find the approximate period of rotation of the Sun.

19. From the information in Figure 14.19, calculate the speed of the particles sent toward the spacecraft by this coronal mass ejection.

SUGGESTIONS FOR ADDITIONAL READING

Akasofu, S. "The Shape of the Solar Corona" in *Sky & Telescope*, Nov. 1994, p. 24.

Baliunas, S. and Soon, W. "The Sun–Climate Connection" in *Sky & Telescope*, Dec. 1996, p. 38.

Bartusiak, M. "The Sunspot Syndrome" in *Discover*, Nov. 1989, p. 44.

Eddy, J. "The Case of the Missing Sunspots" in *Scientific American*, May 1977, p. 80.

Emslie, A. "Explosions in the Solar Atmosphere" in *Astronomy*, Nov. 1987, p. 18. Discusses solar flares.

Frank, A. "Blowin' in the Solar Wind" in *Astronomy*, Oct. 1998, p. 60. On results from the SOHO spacecraft.

Friedman, H. *Sun and Earth*. 1986, W. H. Freeman. Contains good sections about the Sun's effects on the Earth.

Golub, L. "Heating the Sun's Million-Degree Corona" in *Astronomy*, May 1993, p. 27.

Hufbauer, K. *Exploring the Sun: Solar Science Since Galileo*. 1991, Johns Hopkins U. Press. A good history.

Jaroff, L. "Fury on the Sun" in *Time*, July 3, 1989, p. 46. Nice introduction to solar activity and the Sun's interior.

Kippenhahn, R. *Discovering the Secrets of the Sun*. 1994, John Wiley. Excellent modern introduction for the beginner.

Lang, K. "SOHO Reveals the Secrets of the Sun" in *Scientific American*, Mar. 1997.

Nichols, R. "Solar Max: 1980–1989" in *Sky & Telescope*, Dec. 1989, p. 601.

Schaefer, B. "Sunspots That Changed the World" in *Sky & Telescope*, Apr. 1997, p. 34. Historical events connnected with sunspots and solar activity.

Verschuur, G. "The Day the Sun Cut Loose" in *Astronomy*, Aug. 1989, p. 48. Examines events surrounding a huge flare.

Wentzel, D. *The Restless Sun*. 1989, Smithsonian Institution Press. Well-written summary by a noted astronomer and educator.

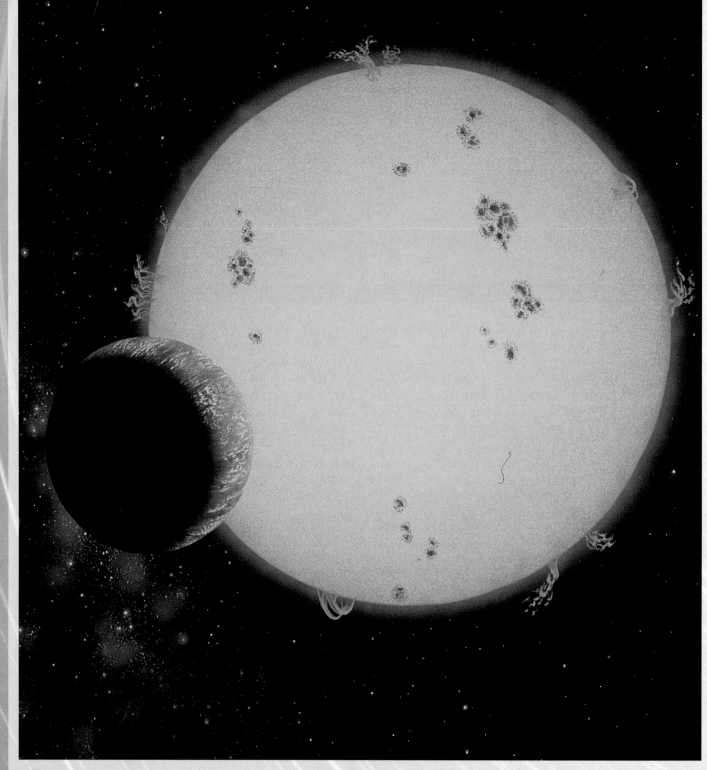

This artist's concept shows what the giant planet discovered orbiting the star 51 Pegasi might look like close up. As explained in the chapter, this planet was the first of over a dozen jovian planets found around other stars whose orbit turned out smaller than the orbit of Mercury in our own system. The planet around 51 Peg is at a distance of only about 7 million km from its star, taking a mere 4.2 days to complete its orbit. At that distance, the surface temperature of the planet would be so hot that (as astronomer Geoff Marcy has quipped) "A roast chicken might be done in a microsecond." The artist has shown prominences and sunspots on 51 Pegasi, evidence of an active atmosphere that might extend a significant way to the giant planet. She also shows the planet with bands like Jupiter, although our measurements can only allow us to estimate the mass of the planet, not its density, and thus we have no idea what sorts of materials the planet is made of. (Painting by Lynette Cook.)

15 The Search for Planets and Life Elsewhere

THINKING AHEAD

Suppose that one day we receive a message from a civilization around another star. In a series of pictures transmitted across space, they tell us a bit about their world and themselves. Do we reply? And if we do, will all humanity speak with a single voice, or will we send a bewildering variety of answers—from different countries, religions, and ethnic groups? Who speaks for Earth?

"We shall not cease from exploration, And the end of all our exploring, Will be to arrive where we started, And know the place for the first time."

T. S. Eliot, "Little Gidding" (from *The Four Quartets* in *The Collected Poems of T. S. Eliot,* 1934, 1936, Harcourt Brace & World)

Our voyages have taken us to dozens of planets and satellites, and we have traced their development over billions of years of time. Beyond our own solar system, the universe extends to distances that astound our imaginations. Our Sun is but one star among millions of billions, more numerous than the grains of sand on all the beaches of the world. As we have learned more about the universe, we have naturally wondered whether there might be other creatures out there who also take astronomy courses and think about worlds beyond their own.

It is time then to sum up where our voyage has taken us and to ask three remaining questions. First, we wonder if there are planets circling any other stars, or whether the solar system we have been studying is somehow unique. Second, we would like to know if any of these distant planets could bear some form of life. And, perhaps most profound of all, we ask if any such alien life might evolve, as we did, to contemplate the mystery of its own existence.

OUR PLACE IN THE UNIVERSE

Before we can think about life elsewhere, we need to consider how life was able to develop on Earth. Let's begin by examining the history of the atoms in your body before they became a part of you. In this section, we briefly summarize some information that was not covered in the text so far, but comes from several other branches of astronomy.

Where Were the Atoms in Your Body Billions of Years Ago?

The universe was born in a "big bang" about 13 to 15 billion years ago. After the initial hot, dense fireball of creation cooled sufficiently for atoms to exist, all matter consisted of hydrogen and helium (with a very small amount of lithium). The hydrogen atoms in the water and in the other hydrogen-rich molecules in your body formed at this early time; they are the oldest atoms that are part of you. But it is not possible to make an organism as complex and interesting as you with only the first three elements.

More complex atoms had to be "cooked" in the only places in the universe hot enough to do the job—the centers of stars. (Fusing together simpler nuclei to make more complex ones is what stars "do for a living" and how they produce the energy we see coming from them.) Thus, before something like you could evolve, several generations of stars had to go through their cycles of birth, life, and death.

Astronomers know from observations of distant parts of the universe that stars must have formed within the first billion years or so, because spectra of distant (and thus ancient) groups of stars already show the presence of some of the heavier elements. As the eons passed, generations of stars produced more and more of the heavier elements. The most massive stars not only produced the greatest variety of new nuclei, but then had the courtesy to explode (Figure 15.1), scattering the newly minted atoms into space.

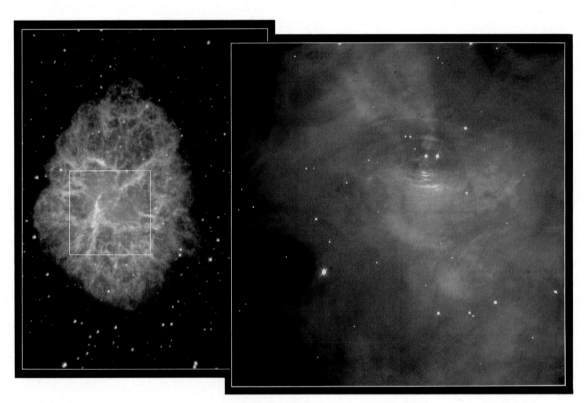

FIGURE 15.1
A Star That Exploded The Crab Nebula is the remnant of an exploding star, first seen in July 1054. Now almost 11 LY across, this object still glows with tremendous energy in many bands of the electromagnetic spectrum. It is powered inside by the compressed whirling remnant (corpse) of the original star, whose beams of energy still stir and excite the atoms thrown out by the explosion. The left-hand image, taken with a ground-based telescope, shows the full nebula. The right-hand image, taken with the Hubble Space Telescope, is a close-up of the Crab's central region. The stellar corpse can be seen as the left member of a pair of stars near the center of the frame. The Hubble was able to observe wisps of material streaming away from it at half the speed of light! Such exploding stars recycle new elements made during the star's life into the general supply of raw material from which new generations of stars and planets then form. *(J. Hester, P. Scowen, and NASA)*

FIGURE 15.2
A Comet Comet Hyakutake, captured in 1996 by amateur astrophotographer Robert Provin of California State University, Northridge. This 12-min exposure was taken on a clear night in the Mojave Desert with a camera and telephoto lens mounted on a telescope tripod. As the moving comet was held steady in the camera, the stars appeared to streak. *(R. Provin)*

years of Earth history, that life slowly evolved and became more complex. The course of evolution was punctuated by occasional planet-wide changes caused by collisions with those planetesimals (or their fragments) that had not been incorporated into the Sun or one of its accompanying worlds. Mammals may owe their domination of Earth's surface to just such a collision 65 million years ago.

Through many twisting turns, the course of evolution on Earth produced a creature with self-consciousness, able to ask questions about its own origins and place in the cosmos (Figure 15.3). Like most of the Earth, this creature is composed of atoms that were forged in earlier generations of stars—in this case assembled rather cleverly into brains, kidneys, fingers, and faces. We might say that through the thoughts of human beings, the matter in the universe can become aware of itself.

The atoms in your body are merely on loan to you from the lending library of atoms that make up our local corner of the universe. Atoms of many kinds circulate through your body and then leave it with the breath you inhale and exhale and the food you eat and excrete. Even the atoms that take up more permanent residence in your tissues will not be part of you much longer than you are alive. Ultimately, you will return your atoms to the vast reservoir of Earth, where they will be incorporated into other structures and even other living things in the millennia to come.

This picture of cosmic evolution, of our descent from the stars, has been obtained through the efforts of scientists in many fields over many decades. Some of its details are still tentative and incomplete, but we feel reasonably confident in its broad outlines. While we do not claim to

Over the years, thanks to these stellar explosions, the gas between the stars became increasingly enriched with heavier elements. In the cooler outer layers of old stars, atoms frequently combined into solid particles that we call interstellar dust. The next generations of stars and planets, containing atoms of carbon, nitrogen, silicon, iron, and the rest of the familiar elements, then formed from reservoirs of enriched gas and dust. One of the most remarkable discoveries of modern astronomy is that life on Earth is mostly composed of just those elements that stars find easiest to make.

About 5 billion years ago, a cloud of gas and dust in this cosmic neighborhood began to collapse under its own weight. Out of this cloud formed the Sun and its planets, together with all the smaller bodies, such as comets, that also orbit the Sun (Figure 15.2). The third planet from the Sun, as it cooled, developed an atmosphere that served to moderate temperature extremes and allow the formation of large quantities of liquid water on its surface. The chemicals available on the cooling Earth may have been further enriched by the addition of complex molecules frozen in the nuclei of comets that collided with our planet.

The chemical variety and moderate conditions on Earth eventually led to the formation of self-reproducing molecules and the beginnings of life. Over the billions of

FIGURE 15.3
A Young Human Human beings have the intellect to wonder about their planet and what lies beyond it. Through them, the universe becomes aware of itself. *(Photo by A. Fraknoi)*

know the full story of how the universe and we ourselves evolved over the course of cosmic history, it is remarkable how much we have been able to learn in the short time we have had instruments to probe the physical nature of the planets and stars.

The Copernican Principle

Our study of astronomy has shown us that we have always been wrong in the past whenever we have claimed that Earth is somehow unique. Copernicus and Galileo showed us that Earth is not the center of the solar system, but merely one of a number of bodies orbiting the Sun. Our study of the stars has demonstrated that the Sun itself is a rather undistinguished star, one among many billions of others. There seems nothing special about our position in the Milky Way Galaxy either, and nothing surprising about our Galaxy's position in the universe. (While this book has not concerned itself with the large-scale structure of the cosmos, you may recall from the Prologue that the universe is organized into vast islands or groups of stars, which we call galaxies.)

The recent discovery of planets around other stars (see section 15.3) confirms our idea that the formation of planets is probably a natural consequence of the formation of many kinds of stars and can happen around many other stars. While our current techniques of finding planets allow us to identify only Jupiter-mass bodies, there is no reason to believe that other planetary systems could not contain planets like the Earth as well.

Philosophers of science sometimes call this idea—that there is nothing special about our place in the universe—*the Copernican principle*. Although it may be tempting to consider ourselves the central and unique focus of all creation, no evidence for such a belief is found in any of the observations discussed in this book.

Most scientists, therefore, would be surprised if life were limited to our planet, and had started nowhere else. There are billions of stars in our Galaxy old enough for life to have developed on a planet around them, and there are billions of other galaxies as well. Astronomers and biologists have long conjectured that a series of events similar to those on the early Earth has probably led to living organisms on planets around other stars. And, where conditions are right, such life may well have evolved to become what we would call intelligent—that is, aware of and interested in its own cosmic history. (In this sense, we must conclude—with tongue firmly in cheek—that taking an astronomy class is the supreme example of intelligent behavior in the universe!)

Such arguments from the Copernican principle, however interesting they may be for philosophers, are nonetheless insufficient for scientists. Science demands data. We would like to find actual evidence for the existence of life, and even of intelligent life, elsewhere. Despite the sensationalistic claims of UFOs and alien abductions in the tabloid media, no such evidence has yet been found. But because many scientists feel that such a discovery would be a defining moment in the history of the human species, a number of searches for extraterrestrial life have already been carried out, and others will soon be underway.

15.2 ASTROBIOLOGY

In the last decade of the 20th century, many discoveries in both astronomy and biology have stimulated the development of a new scientific discipline called **astrobiology.** Astrobiology is the study of life in the universe: its origin, distribution, and ultimate fate. It brings together astronomers, biochemists, environmentalists, and biologists to work on the same problems from their own perspectives.

Among the problems they are trying to solve are determining the conditions under which life arose on Earth and understanding the extraordinary adaptability of life on our planet. (Life on Earth thrives at temperatures from below freezing to above the boiling point of water, and under conditions that vary from the cold dry Antarctic to dark locales deep under the ocean). Astrobiologists are also involved in planning the continuing search for evidence of life on Mars and on Jupiter's moon Europa, and in trying to understand where habitable zones might be found around other stars.

The Building Blocks of Life

While no unambiguous evidence has yet been found for life beyond the Earth, its chemical building blocks have been detected in a wide range of extraterrestrial environments. Meteorites, such as the one found near Murchison, Australia, have yielded a variety of amino acids (the molecular building blocks of proteins) whose chemical structures mark them as having an extraterrestrial origin (see Chapter 13). When we examine the gas and dust around comets such as Comet Halley, we also find a number of *organic molecules*—those that on Earth are associated with the chemistry of life.

One of the most interesting results of modern radio astronomy (see Chapter 5) has been the discovery of organic molecules in giant gas and dust clouds between stars. Over 100 different molecules have been identified in these reservoirs of cosmic raw material, including formaldehyde ("embalming fluid"), ethyl alcohol, acetic acid (vinegar), and others we know as important "stepping stones" in the development of life on Earth. Using radio telescopes and radio spectrometers, astronomers can measure the abundances of various chemicals in these clouds. We find organic molecules most readily in regions where the interstellar dust is most abundant—and it turns out these are precisely the regions where star formation (and probably planet formation) happens most easily (Figure 15.4).

Starting in the early 1950s, scientists have tried to duplicate in their laboratories the chemical pathways that led to life on our planet. In a series of experiments pioneered

FIGURE 15.4
A Cloud of Gas and Dust   This cloud of gas and dust in the constellation of Scorpius is the sort of region where complex molecules are found. It is also the sort of cloud where new stars form from the reservoir of gas and dust in the cloud. Radiation from a group of hot stars (off the picture to the bottom left) called the Scorpius OB Association is "eating into" the cloud, sweeping it into an elongated shape and causing the reddish glow seen at its tip.

(Photo by David Malin, © Anglo-Australian Observatory)

by Stanley Miller and Harold Urey at the University of Chicago, biochemists have simulated conditions on the early Earth and been able to produce many of the fundamental building blocks of life, including those that go into forming proteins and nucleic acids (Figure 15.5). While their accomplishment is still far from making life in the laboratory, it does show that the first steps on the long road to life may not be as difficult as once thought.

This then is the good news—that the building blocks for life seem to be relatively easy to make, both on Earth and in space. But it is a giant step from these building blocks to a functioning cell that can extract energy from its environment and reproduce itself. Even the simplest molecules of RNA and DNA—the genetic material that is central to the reproduction and chemical operation of a cell—contain millions of molecular units, each arranged in a precise sequence. The fact is that we do not understand how life originated. We hypothesize that life will arise whenever conditions are appropriate. But this hypothesis is another form of the Copernican Principle—that something that happened on Earth is likely to have happened elsewhere in the universe. Until we actually find extraterrestrial life, we cannot really know how well our hypothesis fits the actual workings of nature.

Life in the Solar System

In 1996, an enormous amount of public and media attention was focused on a suggestion by an interdisciplinary team of scientists that life may have formed and flourished on our neighbor planet Mars. They had done detailed laboratory analysis of an ancient martian rock sample recovered in Antarctica, a 4.5-billion-year-old meteorite called ALH 84001. Their work showed that this bit of martian crust (probably knocked into space by a powerful impact) had experienced wet conditions about 3.5 billion years ago. At that time, water flowed on Mars (see Chapter 9) and left traces of carbonate minerals and some organic compounds embedded in the potato-sized piece of rock that eventually fell to Earth. The team further claimed that tiny structures

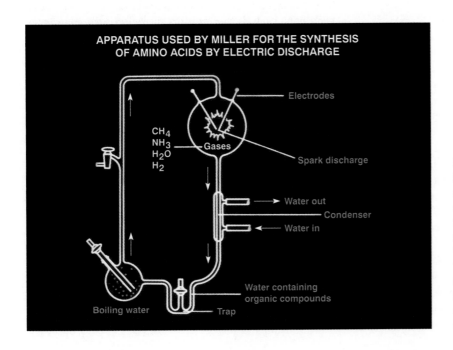

FIGURE 15.5
Simulating the Early Earth   The Miller–Urey experiment, performed in 1953, simulated conditions on the early Earth. An "atmosphere" consisting of methane, ammonia, water vapor, and hydrogen was subjected to electrical sparks to simulate lightning. Water at the bottom of the apparatus provided an "ocean" into which materials synthesized in the atmosphere could fall. When its contents were analyzed, the "ocean" was found to contain a variety of amino acids, the building blocks of proteins.

(F. Drake and the Astronomical Society of the Pacific)

FIGURE 15.6

"Black Smoker" on the Ocean Floor   Starting in 1977, marine geologists and biologists have been discovering that superheated and mineral-rich water from the interior of the Earth can emerge from vents on the ocean floor. As you can see here, when the hot vent water mixes with the much colder ocean water, the dissolved minerals can fall out of solution and make a kind of "black smoke." To the scientists' surprise, the regions around these black smokers were full of life; more than 350 new species have been discovered close to them. In the pitch darkness of the ocean floor, these forms of life derive their energy not from sunlight, but from the heat and chemistry of the vents.   (Peter Ryan/Scripps/Science Photo Library/Photo Researchers, Inc.)

seen in the rock at extremely high magnification might be the fossilized remains of ancient microbial life.

If there are indeed fossils in ancient martian rocks, this would be one of the most spectacular discoveries in the history of science. As we write in 1999, it appears to most scientists that this meteorite was contaminated with terrestrial chemicals during the millennia it lay trapped in the Antarctic ice, and that the "fossils" are too small to represent anything once living. But even if these specific features turn out to be of inorganic origin, the work on ALH 84001 has already verified that wet, apparently Earth-like conditions once existed on Mars (as earlier spacecraft observations had suggested).

Even before the controversy over ALH 84001, NASA had been planning a renewed search for life on Mars. As we discussed in Chapter 9, a series of spacecraft (called the Mars Surveyor program) will land on the martian surface, explore areas that appear to have once had liquid water,

and eventually return samples for analysis in terrestrial labs. The first of these Mars rocks should be available for study by the year 2008. Coming directly to Earth under controlled sterile conditions, they will not be subject to the contamination that happens to natural meteorites. While no one expects to find living microorganisms in the Mars rocks, they will be very carefully analyzed for fossil indications of ancient microbial life.

If there is evidence from these rocks or elsewhere that Mars once had life, then scientists will accelerate their search for possible survivors. They will search for life-forms that may have evolved to deal with the deteriorating climate of Mars, perhaps by finding some oasis or refuge that is warmer and wetter than most of the martian surface. The most likely source of liquid water on Mars today is deep below the surface, where extensive *aquifers* (layers with groundwater in them) may exist. Perhaps someday astronauts on Mars will drill deep wells down to this layer of liquid water and finally encounter living alien life.

Another location of great interest to astrobiologists is Europa. As we saw in Chapter 11, the Galileo spacecraft images of this Moon-sized world strongly indicate the presence of a global ocean of liquid water, beneath a thick ice crust. As the only location in the solar system beyond Earth where there could be large bodies of liquid water, Europa beckons us to further exploration. As we speculated in Chapter 11, perhaps future spacecraft will find microorganisms in the europan seas that resemble (in function if not in detailed molecular structure) the life that flourishes near hot springs that well up from vents in the deep oceans of Earth (Figure 15.6).

You may wonder why we are focusing on microbial life and not on searching for more advanced plants and animals. One reason is that even on Earth, microbial life dominates. Most of the species on our planet are microbial, and most of the biomass consists of life-forms too small for the human eye to see. Furthermore, it is Earth's microbes that have managed to adapt to our planet's most extreme environments, such as the interiors of rocks or the margins of boiling, acidic hot springs. Besides, multicellular creatures are relative newcomers to the roster of life on Earth. For its first three billion years of existence, all terrestrial life was microbial.

15.3   PLANETS BEYOND THE SOLAR SYSTEM

While our robot probes are still confined to the worlds of our own solar system, we certainly do not want to limit our search for life to our own neighborhood. Can we gather evidence about the possibility of life among the stars? We can begin by asking whether other stars are accompanied by one or more planets, as our Sun is.

Our theories of how star systems form predict that planets should accompany the birth of stars under a variety of conditions. However, until a few years ago, we lacked any proof of the existence of planets in any other star systems. This is because, compared to stars, planets out there are extremely dim and thus very hard to detect.

Finding Planets

Recently, a variety of new astronomical techniques have allowed us to search for planets, with some dramatic successes. These techniques still cannot provide a direct image of a planet, since, from our distant vantage point, the faint light reflected from a planet's surface is lost in the bright glare of its star. The kinds of searches possible with current technology look for the effects of planets *on their stars*. In particular, we are measuring subtle changes in the star's motion through space caused by the pull of surrounding planets.

To understand how this approach works, consider a single Jupiter-like planet in orbit about a star. Both the planet and the star in such systems actually revolve about their *common center of mass*. Remember from our discussion in Chapter 2 that gravity is a mutual attraction. The star and the planet each exert a force on the other, and we can find a stable point between them about which both objects move. The sizes of their orbits around this center of mass are inversely proportional to their masses.

Suppose the planet has a mass about one thousandth that of the star; then the size of the star's orbit is one thousandth the size of the planet's. In other words, while the star will be much easier to see than the planet, its orbital motion will be quite small. To get a sense of how difficult observing such motion might be, let's first see how hard Jupiter (the most massive planet in our solar system) would be to detect from the distance of a nearby star. Consider an alien astronomer trying to observe our own system from Alpha Centauri, the closest star system to our own (about 4.3 light-years away).

The diameter of Jupiter's apparent orbit viewed from this distance is 10 seconds of arc (arcsec), and that of the Sun's orbit is 0.010 arcsec. (A second of arc is 1/3600 of a degree.) Even if they had the best telescopes we presently use, these alien astronomers could not detect faint Jupiter directly. But if they could measure the apparent position of the Sun to sufficient precision, they would see it describe an orbit of diameter 0.010 arcsec with a period equal to that of Jupiter, which is 12 years. In other words, if they watched the Sun for 12 years, they would see it wiggle back and forth in the sky by this minuscule fraction of a degree. From the observed motion and the period of the "wiggle," they could deduce the mass of Jupiter and its distance using Kepler's laws.

Measuring positions in the sky this accurately is extremely difficult, but just beginning to be within our technical capabilities. So far, there have been no confirmed planet detections using this technique. New NASA missions are being planned, however, that will someday make such measurements from space.

The technique we have been describing relies on measuring the changes in the position of a star as its orbit moves it back and forth in the sky (across our line of sight). But as the star and planet orbit each other, part of their motion will be *in* our line of sight—i.e., toward us or away from us. Such motion (as discussed in Chapter 4) can be measured using the *Doppler effect* and the star's spectrum. As the star moves back and forth in orbit around the system's center of mass in response to the gravitational tug of an orbiting planet, the lines in its spectrum will shift back and forth (Figure 15.7).

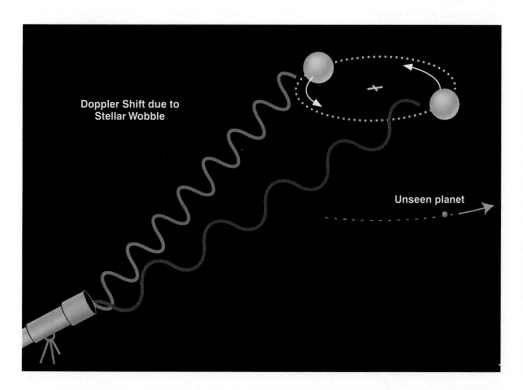

Doppler Shift due to Stellar Wobble

Unseen planet

FIGURE 15.7

Doppler Shift Due to Stellar Wiggle The motion of a star and a planetary companion around a common center of mass can be detected as a cyclical change in the Doppler shift in the star's spectrum. When the star is moving away from us, the lines in its spectrum show a tiny redshift; when it is moving toward us they show a tiny blueshift. The change in color (wavelength) has been exaggerated here for educational purposes. In reality, the Doppler shifts we measure are extremely small and require sophisticated equipment to be detected. *(Diagram courtesy of G. Marcy, San Francisco State University)*

FIGURE 15.8
European Planet Hunting Team Didier Queloz and Michel Mayor of the Geneva Observatory were the first to discover a giant planet around a Sun-like star. They are continuing their work using telescopes in Europe and in Chile and have found several other planets since 1995. *(M. Mayor & D. Queloz)*

Let's again consider the example of the Sun. Its *radial velocity* (motion toward or away from us) changes by about 13 m/s with a period of 12 years because of the gravitational pull of Jupiter. (The change becomes slightly more (15 m/s) if the effects of Saturn are also included.) This corresponds to about 30 mi/h, roughly the speed at which many of us drive around town. Detecting motion at this level in a star's spectrum presents an enormous technical challenge, but not an insurmountable one. Several groups of astronomers around the world have been actively engaged in searching for planets using specialized spectrographs designed for this purpose.

The first planet discovered using this technique was found in 1995 by Michel Mayor and Didier Queloz of the Geneva Observatory (Figure 15.8), using an advanced-design spectrograph at the 1.9-m telescope of the Haute-Provence Observatory in France. This planet orbits a star resembling our Sun called 51 Pegasi, about 40 LY away. (The star can be found in the sky near the great square of Pegasus, the flying horse of Greek mythology, one of the best-known star patterns.) To everyone's surprise, the planet takes a mere 4.2 days to orbit around the star. Remember that Mercury, the innermost planet in our solar system, takes 88 days to go once around.

This means the planet must be very close to 51 Pegasi, circling it about 7 million km away. At that distance, the energy of the star should heat the planet's surface to a temperature close to 1000 C (a bit hot for future tourism). From its motion, astronomers calculate that it has at least half the mass of Jupiter, making it clearly a jovian and not a terrestrial-type planet. The existence of a giant planet so close to its star was a big surprise.

With a 'year' lasting only four Earth days, the effects of the new planet become evident in the spectrum of its star after only a few nights of observation. Within a week of the initial discovery, Geoffrey Marcy and Paul Butler of San Francisco State University (Figure 15.9) confirmed the planet's existence using the 3-m telescope at the Lick Observatory in California. The two California astronomers and their students had been observing the spectra of about

120 nearby Sun-like stars for over eight years with a spectrograph that can measure radial velocity changes to within 3 m/s. Ironically, they had not included 51 Pegasi in their initial search program because it had been mistakenly listed in the U.S. version of a star catalogue as not quite resembling the Sun.

Since that initial discovery, the rate of progress has been breathtaking. As we write this paragraph, 20 giant planets have been discovered orbiting other stars (Table 15.1), most of them by the team led by Marcy and Butler, and there will surely be still more by the time you read this. Think about it: We now know more planets outside our own solar system than within it! About 6 percent of the solar-like stars studied to date have at least one giant planet orbiting relatively close to them. The minimum masses of these planets range from about 0.4 to 11 times the mass of Jupiter. (Astronomers only know the *minimum* mass of the planets, because to determine the exact mass using the Doppler shift and Kepler's laws, we must also know the angle at which the planet's orbit is oriented to our view—something we don't have an independent way of knowing under most circumstances. Still, if the minimum mass is as large as the ones listed in our table, we know for sure that we are dealing with planets significantly more massive than our Earth.)

Each of the stars had shown the presence of only one giant planet thus far. But in spring 1999, Marcy's team announced (and another U.S. team confirmed) the discovery of the first planetary system outside our own—around the star Upsilon Andromedae, about 44 LY away. An inner planet takes 4.6 days to orbit—very close, like the planet around 51 Pegasi. But two outer planets take 242 and 1267 days to go around, putting them (if they were in the solar system) between Venus and Earth and just inside the asteroid belt. Hints of other multiple-planet systems are al-

FIGURE 15.9
American Planet Hunting Team Paul Butler and Geoff Marcy were both at San Francisco State University when they confirmed the discovery of the planet around 51 Pegasi and went on to discover most of the planets that have been found around other stars so far. They have separated for now, with Butler going to the Anglo-Australian observatory to look for planets in the Southern Hemisphere and Marcy continuing the search in the Northern Hemisphere. *(San Francisco State University)*

TABLE 15.1
18 Stars with Planets Known as of February 1999
(The planets are shown in order of increasing orbital period.)

Name of Star[1]	Period of Planet's Orbit (days)	Semimajor Axis (AU)	Minimum Planet Mass (Jupiters)	Estimated Distance (LY)
HD 187123	3.1	0.04	0.5	160
Tau Bootis	3.3	0.04	3.6	50
HD 75289	3.5	0.05	0.4	95
51 Pegasi	4.2	0.05	0.5	50
Upsilon Andromedae	4.6	0.05	0.6	44
	242	0.83	2.1	44
	1267	2.5	4.6	44
HD 217107	7.11	0.07	1.3	64
55 Cancri	14.7	0.12	0.8	45
Gliese 86 (HD 13445)	15.8	0.11	4.9	36
HD 195019	18.3	0.14	3.5	120
Rho Coronae Borealis	39.6	0.23	1.1	55
HD 168443	58	0.28	5.0	124
Gliese 876	60.9	0.21	2.1	15
HD 114762	84	0.41	11.0	91
70 Virginis	117	0.47	6.8	60
HD 210277	437	1.15	1.4	70
16 Cygni B	803	1.7	1.7	70
47 Ursae Majoris	1093	2.1	2.4	45
14 Herculis (Gliese 614)	≈1600	≈2.5	3.3	55

[1]You can see from this table that astronomers do not have a single system for naming stars. Some brighter stars are denoted by a number or Greek letter plus the name of the constellation in which they are found, such as Tau Bootis or 55 Cancri. Others are simply called by their numbers in various catalogs of stars, such as Gliese 614 or HD 75289.

ready starting to appear in the data, suggesting that it may be easy to form planets and that our galaxy could be teeming with them.

Explaining the Planets We Found

The orbits of many of these newly discovered planets are, however, not at all what was expected. Thirteen of the 20 giant planets in our table are closer to their parent stars than is Mercury to the Sun. Some astronomers are calling these planets "hot Jupiters." The name sounds funny, but it has an important point to make: We think of Jupiters as cold worlds, far from their stars!

According to theories of planet formation, giant planets cannot form close to a star. Remember the steps in the formation of a giant planet (from Chapter 13): first the solid particles—the ice and dust—in the disk surrounding a newly forming star collide and stick together to form objects (planetesimals) ranging in size from small stones to mountains. These planetesimals then coagulate to form cores about the size of the terrestrial planets. The cores accrete gases from the surrounding disk to build giant planets like Jupiter and Saturn.

And now you see the problem. Many of the giant planets around other stars are found in regions so hot that there should not have been any ice or dust with which to build the planetary core (to say nothing of an abundant supply of

volatile gases). So are our theories—based on the only planet examples we had until recently, the ones in our own solar system—simply wrong? Or did the giant planets in other star systems form farther out and then migrate closer to the parent star? Right now, calculations seem to favor the second of these possibilities.

Picture a planet with the mass of Jupiter forming where "it should" (based on our models), at a distance of several AU from its parent star, from a vast disk of gas and dust. Interacting with the material in the enveloping disk can cause the planet to lose some of its orbital energy and slowly spiral inward toward the star. The trick is to come up with a way to get the wandering planet to stop that inward spiral before it simply falls into the star and disappears. One possibility is that many Jupiters form in such a disk, and one after another they spiral into the star, and the only ones that survive are those that form just when the disk material is finally all used up. Since the disk lifetime is typically several million years, and a planet will spiral into the star in a few hundred thousand years, it is possible for several generations of planets to simply vanish.

The discovery of the three giant planets orbiting Upsilon Andromedae presents additional challenges for theorists. While one giant planet may plausibly form in the right place and at the right time to survive the migration inward, it's harder to explain how three such planets will do so.

We need to make one other point before we leave the "hot Jupiters." The Doppler-shift technique is most likely to find those planets that are *massive* and *close to their stars.* After all, it is these planets that can pull their parent stars most noticeably around their centers of mass. And it is these planets that produce the shortest "wiggle period" in the motion of their stars. (As we saw, if a planet takes only five days to orbit its star, then the star's wiggle can be fully analyzed if we observe it for only five days. On the other hand, a planet that takes 30 years to go around its star—as Saturn does around the Sun—will take 30 years to make its star wiggle through one cycle.)

Thus, once we get over our astonishment that hot Jupiters exist at all, we should not be surprised that the first planets found using this technique are massive ones close to their stars. As the years go on, the teams using the Doppler technique will be searching "farther" out—for jovian planets that take longer to orbit and make smaller wiggles in the motion of their stars.

Habitable Planets Circling Other Stars

If giant planets can spiral in from several AU and be swallowed by their stars in the early days of a planetary system, then terrestrial planets that are formed even closer to the parent star are that much more likely to be destroyed. For this reason astronomers certainly wonder whether systems like our own and, in particular, planets like the Earth might be rare. Were Jupiter and Saturn formed just as the material in the disk surrounding the protosun was cleared away, so that they could remain in their original orbits and not migrate inward? Can habitable terrestrial planets therefore survive only in those systems where the timing of early events is just right? We can't answer such questions until we can detect planets with masses comparable to that of the Earth around other stars.

Right now, the techniques we have available do not allow us to make measurements precise enough to detect the effects of a low-mass planet like our own. (The Doppler shift the Earth would cause in a star's spectrum, for example, is too small for us to be able to detect.) Still, astronomers are already planning for new instruments and techniques for making the kinds of measurements it would take to show us Earths around other stars.

One approach is to use an indirect technique called *transit photometry.* In any large sample of stars with planets circling them, a few of the systems are aligned so that the planets pass in front of their star on each orbit, as seen from the Earth. If we could measure the brightness of the star with great precision, we could detect the tiny drop in light as the planet transits (moves across the stellar disk) and blocks a bit of its radiation. Observing from space (where very high precision can be obtained), we might someday use this approach to detect planets as small as the Earth, and thus determine the distributions of planetary sizes and orbits.

The best possible evidence for an Earth-like planet elsewhere would be an image. After all, "seeing is believing" is a very human prejudice. But imaging a distant planet is a formidable challenge indeed. Suppose, for example, that you were a great distance away and wished to detect reflected light from the Earth. The Earth intercepts and reflects less than one-billionth of the Sun's radiation, so its apparent brightness in visible light is less than a billionth that of the Sun. But the faintness of potential planets is not the biggest problem.

The real difficulty is that the faint light from a planet is swamped by the blaze of radiation from its parent star. If you are near-sighted, try looking at streetlights at night with your glasses off. (If your eyes are good, you can achieve the same effect by squinting.) You will see a halo of light surrounding every light. Bright stars seen through a telescope also appear to be surrounded by a halo of light. In this case the problem is not that the telescope is near-sighted, but rather that slight imperfections in its optics and atmospheric blurring prevent the star's light from coming to focus in a completely sharp point. Planets, if any, would lie within this halo, and their faint light could not be seen in the glare.

Overcoming this problem is of one of NASA's major goals for the next century. One technique is to build infrared interferometers (see Chapter 4 for a discussion of radio interferometers) in space. Again, we need to go into space to escape the blurring effects of the Earth's atmosphere. The infrared is the optimum wavelength range in which to observe because planets get brighter in the infrared while stars get fainter, thereby making it easier to detect a planet against the glare of its star. Interferometry is an efficient way to obtain *high resolution* (make out finer detail)—exactly what we need to observe the star and the nearby planet as two separate objects. Special techniques can be used to artificially suppress the light from the central star and make it easier to see the planet itself.

Once astronomers actually image an Earth-like planet, the next step would be to measure its spectrum and thus determine the composition of its atmosphere. The spectrum might even indicate whether life is present. In our own atmosphere, oxygen is produced by photosynthesis and methane by the decay of organic matter. If life were absent, neither element would be present in our atmosphere. So the discovery of methane and oxygen in the atmosphere of an Earth-like planet would be strong evidence that life as we know it is present. (Of course, there might be forms of life that we cannot yet imagine that would produce other elements, so the absence of oxygen and methane may not mean that life is absent.) It is not at all far-fetched to think that such experiments might be performed sometime in the next century.

For the first time in human history, we know that our planetary system is not the only one. By the time your children or grandchildren take their first college astronomy courses (we rely on you to recommend astronomy to them!), we may even know about other Earths out there. And, as we outlined in section 15.2, new evidence is encouraging us to think that the chemical steps that lead to life may be easier than we once thought, and may occur in a wide range of enviroments. Given these developments, we are naturally interested in asking whether life around some other star might have evolved to be "intelligent" and thus be interested in communicating with other life-forms who share a self-awareness and curiosity about the universe. And if there is such an intelligent species out there, how could we make contact with its members?

15.4 THE SEARCH FOR EXTRATERRESTRIAL INTELLIGENCE

Looking for extraterrestrial intelligence is similar to making contact with people who live in a remote part of the Earth. If students in the United States want to converse with students in Australia, for example, they have two choices. Either one group gets on an airplane and travels to meet the other, or they communicate via some message medium (today, probably by telephone, fax, e-mail, or short-wave radio). Given how expensive airline tickets are, most students would probably select the message route.

In the same way, if we want to get in touch with intelligent life around other stars, we can travel, or we can try to exchange messages. Because of the great distances involved, interstellar space travel is either very slow or very expensive. The fastest spacecraft the human species has built so far would take almost 80,000 years to get to the nearest star. While we could certainly design a faster craft, the more quickly we require it to travel, the greater the energy cost involved. To reach neighboring stars in less than a human life span, we would have to travel close to the speed of light. In that case, however, the expense would become truly astronomical.

Interstellar Travel

The late Bernard Oliver, vice president of the Hewlett-Packard Corporation and an engineer with an abiding interest in life elsewhere, made a revealing calculation about the costs of rapid space travel. Since we do not know what sort of technology we (or other civilizations) might someday develop, Oliver considered a trip to the nearest star in a spaceship with a "perfect engine"—one that would convert its fuel into energy with 100 percent efficiency. (No future technology can possibly do better. In reality, nature is unlikely to yield efficiency even close to the perfect value; just think how much of the energy released by the fuel in a car is wasted.) Even with a perfect engine, the energy cost of a single round-trip journey at 70 percent the speed of light turns out to be equivalent to about *500,000 years' worth of total U.S. electrical energy consumption!*

In case you are wondering why this figure is so high, you must remember that the voyagers could not depend on finding "gas stations" open at their destination. Therefore they would have to carry the fuel for the return legs of the journey with them, and getting that huge mass of fuel up to 70 percent the speed of light would be very expensive. The important thing about Oliver's calculation is that it does not depend on present-day technology (since it assumes a perfect engine), but only on the known laws of science. What it shows is that no matter who does the traveling, it is very expensive to go fast enough to get to the stars within the course of a single human life.

This is one reason astronomers are so skeptical about claims that unidentified flying objects (UFOs) are spaceships from extraterrestrial civilizations. Given the distance and expense involved, it seems unlikely that the dozens of UFOs (and, recently, even UFO abductions) reported each year could all be visitors from other stars so fascinated by Earth civilization that they are willing to expend fantastically large amounts of energy or time to reach us.

In fact, a sober evaluation of UFO reports often converts them to IFOs (identified flying objects), or NFOs (not-at-all flying objects). While some are hoaxes, others are natural phenomena such as ball lightning, fireballs, bright planets, or even flocks of birds with reflective bellies. Still others are human craft, such as private planes with some lights missing, or classified military airplanes. It is also interesting that not a single UFO has ever left behind any physical evidence that can be tested in a laboratory and shown to be of nonterrestrial origin.[1]

Some visionaries have suggested that we might someday overcome the long-time-and-large-energy demands of interstellar travel by hollowing out asteroids and sending them to the stars with large colonies of people on board (equipped with the provisions needed for a really long trip). While the original settlers would never see the stars, their far-future descendants might arrive and start a settlement on an Earth-like planet somewhere. This is, for now, only the stuff of science fiction.

Messages on Spacecraft

In the real world we do have four spacecraft—two Pioneers and two Voyagers—which, having finished their program of planetary exploration, are now leaving the solar system. At their coasting speeds, they will take hundreds of thousands or millions of years to get anywhere close to another star. On the other hand, they were the first products of human technology to leave our home system, and so we wanted to put messages on board to show where they came from.

Each Pioneer carries a plaque with a pictorial message engraved on a gold-anodized aluminum plate (Figure 15.10). The Voyagers, launched in 1977, have audio and

---

[1]If you are interested in pursuing the topic of what UFOs are and aren't, we recommend the following books: Klass, P. *UFO Abductions: A Dangerous Game* (1988, Prometheus Books); Peebles, C. *Watch the Skies: A Chronicle of the Flying Saucer Myth* (1994, Smithsonian Institution Press); and Shaeffer, R. *The UFO Verdict: Examining the Evidence* (1981, Prometheus Books). Also see the Web sites at the end of this chapter.

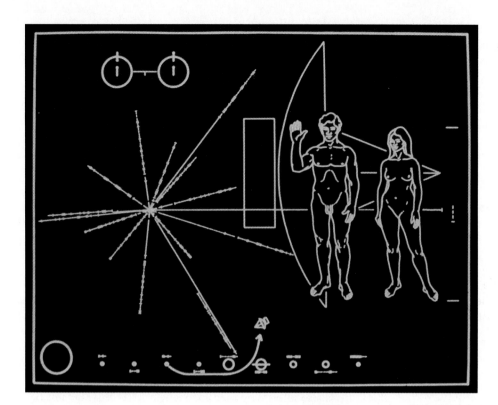

FIGURE 15.10

Interstellar Message This is the image engraved on the plaques aboard the Pioneer 10 and 11 spacecraft. The human figures are drawn in proportion to the spacecraft, which is shown behind them. The Sun and planets in the solar system can be seen at the bottom, with the trajectory that the spacecraft followed. The lines and markings in the left center show the positions and pulse periods for a number of stellar corpses (called pulsars) that give off regular pulses of radio waves, which might help locate the spacecraft's origins. *(NASA)*

video records attached (Figure 15.11), on which were included over 100 photographs and a selection of music from around the world. (Included among the excerpts from Bach, Beethoven, folk music, and tribal chants is one piece of rock and roll—"Johnny B. Goode," by Chuck Berry.) Given the vastness of space between stars in our section of the Galaxy, it is very unlikely that anyone will ever receive these messages. They are more like a message in a bottle thrown into the sea by a shipwrecked sailor, with no realistic expectation of its being found soon, but a slim hope that perhaps someday, somehow, someone will know of the sender's fate.

The Voyager Message

An excerpt from the Voyager record includes these words:

"We cast this message into the cosmos. It is likely to survive a billion years into our future, when our civilization is profoundly altered. . . . If [another] civilization intercepts Voyager and can understand these recorded contents, here is our message:

This is a present from a small, distant world, a token of our sounds, our science, our images, our music, our thoughts, and our feelings. We are attempting to survive our time so we may live into yours. We hope, someday, having solved the problems we face, to join a community of galactic civilizations. This record represents our hope and our determination, and our good will in a vast and awesome universe."—Jimmy Carter, President of the United States of America, June 16, 1977

Communicating with the Stars

If direct visits to stars are unlikely, we must turn to the other alternative for making contact—exchanging messages. Here the news is a lot better. We already know (and have learned to use) a messenger—electromagnetic radiation—that moves through space at the fastest speed in the universe. Traveling at the speed of light, radiation reaches the nearest star in only four years, and does so at a fraction of the cost of sending material objects. These advantages are so clear and obvious that we assume they will occur to any other species of intelligent beings who develop technology.

However, we have access to a wide spectrum of electromagnetic radiation, ranging from the longest-wavelength radio waves to the shortest-wavelength gamma rays. Which would be the best for interstellar communication? It would not be smart to select a wavelength that is easily absorbed by interstellar gas and dust, or one that is unlikely to penetrate the atmosphere of a planet like ours. Nor would we want to pick a wave that has lots of competition for attention in our neighborhood. For example, it would be difficult (and quite dumb) for us to put together a signal from our civilization in the visible-light region of the spectrum. How could it compete with our extremely strong local source of light, the Sun?

One final criterion makes the selection an easy one: We want the radiation to be inexpensive to produce in large quantities. When we consider all these requirements, radio waves win hands down. Being the lowest-frequency (and lowest-energy) band of the spectrum, they are not very expensive to produce (permitting us to use them extensively for communications on the Earth). They are not significantly absorbed by interstellar dust and gas. With some exceptions, they easily pass through the Earth's atmosphere, and through the atmospheres of the other planets we are acquainted with. And because the Sun does not put out a large quantity of radio waves, a radio message has a realistic chance of being "heard" above the local noise.

The Cosmic Haystack

For these reasons, many astronomers have decided that the radio band is probably the best place in the spectrum for communication among intelligent civilizations. Having made such a decision, however, we still have many questions and a daunting task ahead of us. Shall we *send* a message, or try to *receive* one? Obviously, if every civilization decides to receive only, then no one will be sending and everyone will be disappointed. On the other hand, it may be appropriate for us to *begin* by listening, since we are likely to be among the most primitive civilizations in the Galaxy who are interested in exchanging messages.

We do not make this statement to insult the human species (which, with certain exceptions, we are rather fond of). Instead, we base it on the fact that humans have had the ability to receive (or send) a radio message across interstellar distances for only a few decades. Compared with the ages of the stars and the Galaxy, this is a mere instant. If there are civilizations out there who are ahead of us in development by even a short time (in the cosmic sense),

they are likely to have a head start of many, many years. If there are civilizations behind us, chances are they are sufficiently far behind that they have not yet developed radio communications.

In other words, we, who have just started, may well be the "youngest" species in the Galaxy with this capability. Just as the youngest members of a community are often told to be quiet and listen to their elders for a while before they say something foolish, so we may want to begin our exercise in extraterrestrial communication by listening.

Even restricting our activities to listening, however, leaves us with an array of challenging questions. For example, you know from your own experience with radio transmissions that a typical signal comes in on only one channel (that is, it is carried by one small frequency band of radio waves). The owners of your favorite station are confident—because there aren't that many channels on the radio dial—that you will find the station despite this fact (although a few actually transmit over two different frequencies, one on the AM band and one on the FM band). Many of us, when we first arrive in a new city, scan up and down the radio band until we find the stations we like. But if we have only an AM radio in our car, and the stations playing our favorite music are all FM, then we are out of luck.

In the same way, it would be very expensive (and perhaps even ill mannered) for an extraterrestrial civilization to broadcast on a huge number of channels. Most likely, they would select one or a few channels for their particular

(a)

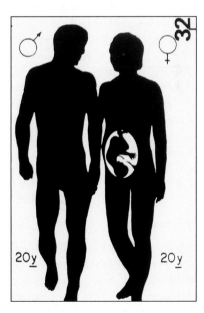

(b)

FIGURE 15.11

The Voyager Record   (a) Encoded onto a gold-coated copper disk, the Voyager record contains 118 photographs, 90 minutes of music from around the world, greetings in almost 60 languages, and other audio material. It is a summary of the sights and sounds of Earth. (b) One of the images encoded onto the record. Originally, the team devising the record wanted to send a photograph, from a medical book, of a nude man and a pregnant woman. However, NASA was concerned about offending some people on Earth, and so artist Jon Lomberg drew a silhouette version, which allowed the fetus inside the woman to be shown as well.   *(NASA)*

TABLE 15.2
The Cosmic Haystack Problem: Some Questions About an Extraterrestrial Message

- From what direction (which star) is the message coming?
- On what channels (or frequencies) is the message being broadcast?
- How wide in frequency is the channel?
- How strong is the signal (can our radio telescopes detect it)?
- Is the signal continuous, or does it shut off at times (as, for example, a lighthouse beam does when it turns away from us)?
- Does the signal drift (change) in frequency because of the changing relative motion of the source and the receiver?
- How is the message encoded in the signal (how do we decipher it)?
- Can we even recognize a message from a completely alien species? Might it take a form we don't at all expect?

message. But the radio band of the electromagnetic spectrum contains an astronomically large number of possible channels. How can we know in advance which one they have selected, and how they have coded their message into the radio signal?

Another problem that may make it hard for you to find your favorite type of radio station has relevance for intergalactic communication. If your radio has a poor antenna, it may not pick up the signal from a weak station some distance away. You may not learn about the existence of that station (and others like it) until you buy better equipment. The same will be true for interstellar transmissions. If an extraterrestrial civilization's signal is just too weak for our present-day radio telescopes, they may be broadcasting their little alien hearts out, but we will miss their signal completely.

Table 15.2 summarizes these and other factors that scientists must grapple with when trying to tune into radio messages from distant civilizations. Because their success depends on either guessing right about so many factors, or else searching through all the possibilities for each factor, some scientists have compared their quest to looking for a needle in a haystack. Thus they like to say that the list of factors in Table 15.2 defines the *cosmic haystack problem.*

Radio Searches

Although the cosmic haystack problem seems daunting, many other research problems in astronomy also require a large investment of time, equipment, and patient effort. And, as several astronomers have pointed out, if we don't search, we're sure not to find anything. Thus several groups of radio astronomers have undertaken searches for extraterrestrial messages during the last three decades.

The very first search for such radio signals was conducted by astronomer Frank Drake in 1960, using the 85-ft antenna at the National Radio Astronomy Observatory (Figure 15.12). Called Project Ozma, after the queen of the exotic "Land of Oz" in the children's stories of L. Frank

Baum, his experiment involved looking at about 7200 channels and two nearby stars over a period of 200 hours. Although he found nothing, Drake demonstrated the feasibility of such a search, and set the stage for the more sophisticated projects that followed. (It is interesting to note that what took 200 hours in 1960 could be done with today's automated systems in about a thousandth of a second.)

Since 1960, more than 60 radio searches have been carried out by scientists around the world, each exploring a minuscule region of the cosmic haystack. Although a number of interesting signals have been found, none has met the crucial test of being detected more than once, so it could be checked. Scientists are continuing several of the searches, always trying to improve their equipment and beat the odds against finding that elusive needle of a message.

In 1992, NASA began the most comprehensive search for radio messages ever undertaken, only to have Congress

FIGURE 15.12
Project Ozma A 25th-anniversary photo of some members of the Project Ozma team standing in front of the 85-ft radio telescope with which the 1960 search for extraterrestrial messages was performed. Frank Drake is in the back row, second from the right.
(National Radio Astronomy Observatory)

(a)

(b)

FIGURE 15.13
Project Phoenix   (a) The 64-m radio telescope at Parkes, Australia, was used in 1995 to search for radio signals from possible extraterrestrial civilizations around 200 stars. (b) Phoenix project scientists Jill Tarter and Peter Backus are shown at the telescope controls during their observing run.   *(Photos by Seth Shostak, SETI Institute)*

cut the funding for the project after less than a year. Using private donations, the nonprofit SETI (Search for Extra-Terrestrial Intelligence) Institute has undertaken to continue the search. They now call it *Project Phoenix,* since the program has risen from the ashes of its funding crisis.

Using modern electronics and computers, the Phoenix system can "listen in" on tens of millions of channels simultaneously (Figure 15.13). Its software searches for promising signals (such as continuous waves or pulses) and alerts the experimenters if an interesting signal persists on any one channel, or moves between channels because of Doppler shift. The intent of the project (which resumed in 1995) is to search 2 billion channels for each of about 1000 nearby stars. To make it onto their list, a star must be roughly similar to the Sun and at least 3 billion years old. At that age it could have had enough time for intelligent life to develop on any Earth-like planets around it. About half of the candidate stars have now been observed. Other SETI programs search only a selected range of channels, but sweep the sky in the hope of encountering a message. So far, no signal has been found; the search will continue using other large radio telescopes and (eventually) much improved receivers.

What If We Succeed?

No one can predict when or whether such searches will be successful. It may well be that civilizations technologically far in advance of our own use other forms of communication that we are not yet aware of. After all, 150 years ago we did not have an inkling of the possibilities of radio commu-

nication, while today it is difficult to imagine our civilization without it. On the other hand, we would never dream of giving a preschooler a book like the one you are reading. Young children learning to read are given very simple books until they have mastered the basics. We hope that advanced civilizations remember their own youth and send out messages that even youngsters like us can find and interpret.

What will happen if we do find a radio signal that is unambiguously the product of an extraterrestrial intelligence? The existence of the signal itself will be of tremendous philosophical importance, demonstrating that we are not alone in the cosmos. But unless we are able to interpret the message, there may not be much practical value in the discovery. If we can eventually work out a method of mutual communication, however, an interesting question arises: Who will speak for planet Earth?

Suppose we know that a star 35 LY away has a technological civilization around it, and that civilization has given us a kind of code by which we can make ourselves understood. (An easy way to begin interacting might be to send pictures.) Who decides what to send? Does the whole planet try to agree on one set of messages, or can any individual or group send a separate communication? There may be many countries, religious groups, cultural organizations, corporations, and individuals who can afford a radio antenna and are interested in getting their message out. Among ourselves, we rarely speak with one voice; should we try to do so in addressing the universe? Confronting such questions may be a good test of whether there is intelligent life on Earth.

Jill Tarter: Trying to Make Contact

The year 1997 was a remarkable one for Jill Cornell Tarter, one of the world's leading scientists in the SETI field. The SETI Institute announced that she would be the recipient of its first endowed chair (the equivalent of an endowed research professorship) named in honor of Bernard Oliver. The National Science Foundation approved a proposal by a group of scientists and educators she headed to develop an innovative hands-on high school curriculum based on the ideas of cosmic evolution (the topics of this chapter). At roughly the same time, she was being besieged with requests for media interviews as news reports identified her as the model for Ellie Arroway, the protagonist of *Contact*, Carl Sagan's best-selling novel about SETI. (The book was made into a high-budget science fiction

Jill Tarter   *(Seth Shostak/SETI Institute)*

Being the only woman in a group is not a new situation to Tarter, who often found herself the only woman in her advanced science or math classes.

film, starring Jodie Foster, who talked with Tarter before taking the role.)

As Tarter is quick to point out, "Carl Sagan wrote a book about a woman who does what I do, not about me." Still, as the only woman in such a senior position in the small field of SETI, she was the center of a great deal of public attention. (However, colleagues and reporters warned that this was nothing compared with the attention she would get if her search for radio signals from other civilizations recorded a success.)

Being the only woman in a group is not a new situation to Tarter, who often found herself the only woman in her advanced science or math classes. Her father encouraged both in her interest in science and her "tinkering," and as an undergraduate at Cornell University, she majored in engineering physics. That training now comes in very handy in putting together and maintaining the complex devices that automatically scan for signals from other civilizations.

Switching to astrophysics for her graduate studies, she did a PhD thesis, which, among other topics, considered the formation of failed stars—those whose mass was not sufficient for the nuclear reactions that power the stars to begin inside them. Tarter coined the term "brown dwarf" for these small, dim objects, and it has remained the name astronomers use ever since. Later, her postgraduate work involved thinking about "dark matter," so-far-unseen material that seems (from its gravitational pull) to be a very substantial component of galaxies and groups of galaxies. Now that she is working full time searching for extraterrestrial civilizations, Tarter likes to joke that her entire career seems to have involved searching for the most elusive things in the universe.

It was while she was still in graduate school that Stuart Bowyer, one of her professors at Berkeley, gave her a study on the possibilities of SETI (chaired, as it turns out, by Bernard Oliver). Bowyer asked her if she wanted to be involved in a small experiment to siphon off a bit of radiation from a radio telescope as astronomers used it year in and year out and see if there was any hint of an intelligently coded radio message buried in the radio noise. Her engineering and computer programming skills became essential to the project, and soon she was hooked on the search for life elsewhere. Years later, as Tarter directs Project Phoenix, she may one day (like Ellie Arroway in the film) be the first person on Earth to know that humanity is not alone in the cosmos.

Conclusion

Whether or not we ultimately turn out to be the only intelligent species in our part of the Galaxy, our exploration of the cosmos will surely continue. A humble acknowledgment of how much we have left to learn is one of the fundamental hallmarks of science. We should, however, feel exhilarated about how much we have already managed to discover and curious about what else we might find out.

Our progress report on the ideas of astronomy ends here, but we hope that your interest in the universe does not. We hope you will keep up with developments in astronomy through the media, or by going to an occasional public lecture by a local scientist. Who, after all, can even guess all the amazing things that future research projects will reveal about both the universe and our connection with it?

The Astrobiology Web
[www.reston.com/astro/index.html]
This private Web site, maintained by a space enthusiast, collects a vast amount of information about many aspects of astrobiology in a well-organized way.

NASA Astrobiology
[astrobiology.arc.nasa.gov/index.html]
NASA's Ames Research Center is taking a leading role in astrobiology and its Web site can point you to NASA's plans for missions, as well as useful background information.

Extra-solar Planets Encyclopedia
[www.obspm.fr/planets]
Jean Schneider of the Paris Observatory keeps a detailed catalog of confirmed and unconfirmed planets and brown dwarfs, with basic data, useful background information, and references for each.

Searching for Extra-solar Planets [cannon.sfsu.edu/~gmarcy/planetsearch/planetsearch.html]
A site about the search for planets being conducted by Geoff Marcy and Paul Butler (which has been responsible for the majority of the planet discoveries so far), with charts, summaries, papers, popular articles, links, and an excellent graphic summary.

SETI Institute Home Page [www.seti-inst.edu]
This is the site for the main scientific organization searching for signals from extraterrestrial civilizations and trying to understand the formation and prevalence of life beyond Earth. You can find information on *Project Phoenix*, other SETI searches, the science background, and the social implications of success.

The Planetary Society's Search
[seti.planetary.org/]
This membership and advocacy organization helps pay for several SETI projects. Here you can find out more about the searches they sponsor and the quest to find life in the universe in general.

SERENDIP Home Page [sag-www.ssi.berkeley.edu/serendip/]
The SERENDIP Project cleverly piggybacks a search for signals from advanced civilizations to regular radio-astronomy observations on different major telescopes. Their home page is designed for the layperson to read.

SUMMARY

15.1 The universe began with only the very simplest elements. More complex nuclei are produced via nuclear fusion at the hot centers of stars; some stars explode at the end of their lives and thus recycle the newly formed elements into space. Our solar system formed about 4.5 billion years ago from a cloud of gas and dust enriched by several generations of heavier-element production in stars. The *Copernican Principle*—which suggests that there is nothing special about our place in the universe—implies that if life could develop on Earth, it should be able to develop in other places as well.

15.2 A new discipline called **astrobiology** combines the tools of astronomy and biology to understand how life might start elsewhere and to develop the means of finding it. Organic molecules, the building blocks of life, have been detected in meteorites, comets, and vast clouds of gas and dust among the stars. A number of complex organic molecules have been built up out of simpler chemical building blocks in laboratory experiments on Earth, indicating that the steps leading to life may not be as difficult as once thought. In 1996, scientists thought they had found fossil evidence of life in a rock that fell to Earth from Mars. However, after further analysis, most scientists now believe that the structures in the meteorite were not evidence of life after all. Because Mars was once warmer and wetter, however, such evidence may yet come our way as we explore Mars further. Europa is also a possible place to search for life in our solar system.

15.3 Because planets are dim and lost in the glare of their parent stars, they are very hard to image. Recently, however, we have discovered 20 planets around 18 stars (as of May 1999) by looking for the tiny Doppler shift the orbiting planet causes in the motion of the parent star around the system's center of mass. This technique most easily finds massive planets close to their stars, and 13 of the 18 are giant planets closer to their stars than Mercury is to our Sun. Explaining the existence of such planets is a challenge for our theories. Someday, new instruments and new techniques will help us determine whether Earth-like planets also exist out there.

15.4 Because they are so far away, traveling to the stars is either very slow or extremely expensive (in terms of energy required.) Despite the many UFO reports and the tremendous media publicity, there is no evidence that any of these are related to extraterrestrial spacecraft. The spacecraft we have sent beyond the planets (two Pioneers and two Voyagers) contain a plaque or recording with information about humanity, but these are not expected to arrive anywhere near a star for millions of years. Scientists have determined that the best way to communicate with any intelligent civilizations out there is by using electromagnetic waves, and radio waves seem best suited to the task. Since 1960, over 60 programs have searched for radio messages from civilizations around other stars, but they have only begun to comb the many different possible stars, frequencies, signal types, and other factors that make up what we call the *cosmic haystack problem*. If we do find a signal someday, deciding whether to answer may be one of the greatest challenges humanity will face.

INTER-ACTIVITY

A If one of the rocks from Mars does turn out to have unambiguous signs of ancient life that formed on Mars, what does your group think would be the implications of such a discovery for science and for our view of life elsewhere? Would such a discovery have any long-term effects on your own thinking?

B Suppose we receive a radio message from a civilization around another star (which shows clear evidence of intelligence, even if we can't quite decipher all the details of what they send). What does your group think the implications of this discovery would be? How would your own thinking or philosophy be affected by such a discovery?

C Your group is a subcommittee of scientists examining the issue of whether any of the "hot Jupiters" (giant planets closer to their stars than Mercury is to the Sun) could have life on or near them. Can you come up with places on or near such planets where life could develop or with forms of life that might survive there?

D A radio message has been received from a civilization around a star 40 light years away, which contains (in pictures) quite a bit of information about the beings that sent the message. The President of the United States has appointed your group to a high-level commission to advise whether humanity should answer the message (which was not particularly directed at us, but comes from a beacon that, like a lighthouse, sweeps out a circle in space.) How would you advise the President? Does your group agree or do you have a majority or minority view to present?

E If there is no evidence that so-called UFO's are extraterrestrial visitors, why does your group think that TV shows, newspapers, and movies spend so much time and effort publicizing the point of view that UFO's *are* craft from other worlds? Who stands to gain by exaggerating stories of unknown lights in the sky or simply fabricating stories that alien visitors are already here? Do you think scientists should simply ignore all the media publicity or should they try to respond?

F Suppose someday we do send colonies toward the stars in large hollowed-out asteroid cities? Make a list of some of the problems (astronomical, psychological, etc.) that the colonists might encounter on their long voyage.

G Suppose your group had been the team planning the sights and sounds of Earth for the Voyager spacecraft audio/video record. What would you have put on the record to represent our planet to another civilization?

H Let's suppose Earth civilization has decided to broadcast a radio (or TV) message announcing our existence to other possible civilizations among the stars. Your group is part of a large task-force of scientists, communications specialists, and people from the humanities charged with deciding the form and content of our message. What would you recommend? For example, would it be useful to start with one of Shakespeare's plays coded as a very powerful FM broadcast? Why or why not?

REVIEW QUESTIONS

1. What is the Copernican principle? Make a list of scientific discoveries that confirm it.

2. Where in the solar system (and beyond) have scientists found evidence of organic molecules?

3. Give a short history of the atoms that are now in your little finger, going back to the beginning of the universe.

4. Why is it so hard to see planets around other stars and so easy to see them around our own?

5. What technique did astronomers use to discover giant planets around other stars? Explain how this technique works and what it can tell us about the planets we discover.

6. How might we find planets like the Earth around other stars in the future?

7. Why is taking an image of a planet around another star so difficult? How might we be able to do so in the future?

8. Why is traveling between the stars difficult?

9. What are the advantages to using radio waves for communication between civilizations that live around different stars? List as many as you can.

10. What is the "cosmic haystack problem"? List as many of its components as you can think of.

11. In what ways is Project Phoenix an improvement over the first search ever conducted for radio signals from extraterrestrial civilizations?

12. Would a human being have been possible during the first generation of stars that formed right after the Big Bang? Why or why not?

13. How would you demonstrate that a rock found in Antarctica is from Mars and not from Earth?

14. Why were giant planets close to their stars the first ones to be discovered? Why has the same technique not been used yet to discover giant planets at the distance of Saturn?

15. Suppose you wanted to take an image of a planet around another star. Would it be easier to observe in visible light or in infrared? Why? Which would be easier to find: a planet located at 1 AU or at 5 AU from its star?

16. What kind of evidence do you think would convince astronomers that an extraterrestrial spacecraft has landed on the Earth? (*Hint:* See question 13)

17. Suppose astronomers find that another star (closely resembling the Sun) has a planet with the Earth's mass at roughly the Earth's distance. What experiments might you suggest to help us discover whether this planet has life on it?

18. An astronomer has just found a new planet around a Sun-like star. If its orbit has a period of 2.4 years, use Kepler's third law to determine the size of its semimajor axis in astronomical units. (See Chapter 2 to remind yourself about the law.)

19. Suppose astronomers discover a radio message from a civilization whose planet orbits a star 35 light years away. Their message encourages us to send them a radio answer. Suppose the governing bodies on *each* planet take two Earth years to decide that they will answer and to frame a message. How long after we get their first message can we begin listening for their reply to our reply? (What would be a way to cut down this interval between messages, once we are in regular communication?)

20. Think of our Milky Way Galaxy as a flat disk of diameter 100,000 LY. Suppose we are one of a thousand civilizations interested in communicating via radio waves that are randomly distributed around the Galaxy. How far away would the nearest such neighboring civilization be?

On the Search for Planets Elsewhere

Boss, A. *Looking for Earths: The Race to Find New Solar Systems.* 1998, J. Wiley.

Croswell, K. *Planet Quest: The Epic Discovery of Alien Solar Systems.* 1997, Free Press.

Goldsmith, D. *Worlds Unnumbered: The Search for Extra-solar Planets.* 1997, University Science Books.

Halpern, P. *The Quest for Alien Planets: Exploring Worlds Outside the Solar System.* 1997, Plenum. All four of the above books are responsible summaries of the new discoveries about planets elsewhere.

Marcy, G., and Butler, R. "The Diversity of Planetary Systems" in *Sky & Telescope,* Mar. 1998, p. 30. A progress report on planet discoveries by the leading team.

McInnis, D. "Wanted: Life-Bearing Planets" in *Astronomy,* Apr. 1998, p. 38. On future instruments that could find planets more like the Earth.

Stephens, S. "Planet Hunters" in *Astronomy,* July 1998, p. 58. Profile of Geoff Marcy and Paul Butler, who found the largest number of extra-solar planets.

On the Search for Life Elsewhere

Croswell, K. "Interstellar Trekking" in *Astronomy,* June 1998, p. 46. Science perspective on how to travel to the stars.

deDuve, C. *Vital Dust: The Origin and Evolution of Life on Earth.* 1995, Basic Books. A Nobel laureate recounts our modern understanding of how life emerged on our planet.

Dick, S. *The Biological Universe: The 20th Century Extraterrestrial Life Debate.* 1996, Cambridge U. Press. Fine historical survey of our thinking about life elsewhere.

Lemonick, M. *Other Worlds: The Search for Life in the Universe.* 1998, Simon & Schuster. *Time* magazine senior science correspondent summarizes the new ideas and discoveries.

LePage, A. & MacRobert, A. "SETI Searches Today" in *Sky & Telescope,* Dec. 1998, p. 44. On the scientific searches for signals from other civilizations.

Pendleton, Y. and Farmer, J. "Life: A Cosmic Imperative?" in *Sky & Telescope,* July 1997, p. 42. On the rise of life and environments that make it possible or impossible.

Shilling, G. "The Chance of Finding Aliens: Re-evaluating the Drake Equation" in *Sky & Telescope,* Dec. 1998, p. 36.

Shostak, S. "When ET Calls Us" in *Astronomy,* Sept. 1997, p. 36. What do we do when we receive a message?

Treiman, A. "Microbes in a Martian Meteorite: The Evidence Has Grown More Cloudy" in *Sky & Telescope,* Apr. 1999, p. 52.

APPENDIX 1

Astronomy on the World Wide Web

In just a few short years, the World Wide Web has become a planet-wide library, full of information, images, opinion, and argument. Astronomers, always eager to use technological tools like computers, jumped on the Web bandwagon early, and there are thousands of sites dedicated to astronomical ideas.

This great wealth of astronomy pages also presents a problem: Unlike a textbook, which is written by (and checked by) experts in their fields, Web sites can be posted by anyone. Thus there are many Web sites with misinformation, exaggeration, misunderstanding, and downright nonsense about the universe. Some of these are offered by well-meaning but inexperienced enthusiasts, while others are by charlatans or hoaxers out to make a quick buck or promote a hidden cause.

Throughout *Voyages* we have recommended reliable Web sites for each chapter and topic as we went along. In this appendix, we list some more general sites that are especially good at guiding astronomy beginners to helpful information, the latest images, or reliable organizations. Most of the sites we list include links to other sites that their Webmasters have found useful.

Note that Web addresses and sites can change faster than any mere mortal can follow. For the latest information on Web sites we recommend, check the *Voyages* Web site, at: **www.saunderscollege.com/astro/fraknoi**

1. Magazines that Cover Astronomy

Astronomy **Magazine** [www.astronomy.com] has the largest circulation of any magazine devoted to the universe and is designed especially for astronomy hobbyists and armchair astronomers. Their site feature many sections on the hobby of astronomy and on things you can do to become involved.

Sky & Telescope **Magazine** [www.skypub.com] is an older and somewhat higher-level magazine for astronomy hobbyists. More of their articles are written by astronomers, and they have better news and resources sections. Their Web site is especially rich in observational guides, software and equipment reviews, and listings of astronomy clubs around the country.

Scientific American **Magazine** [www.sciam.com/] offers one astronomy article about every second issue. These articles, a number of which are reproduced on their Web site, are at a slightly higher level than *Voyages,* but—often being written by the astronomers who have done the work being described—are authoritative and up-to-date.

Mercury **Magazine** [www.aspsky.org/subpages/merc.html] is a popular-level magazine published by the Astronomical

Society of the Pacific that features articles and columns for astronomers, educators, and astronomy enthusiasts. They often deal with more philosophical issues and the influence of astronomy on other areas of human thought.

2. Organizations that Deal with Astronomy for the Public

The Astronomical Society of the Pacific [www.aspsky.org] is the largest and oldest national astronomy organization in the U.S. The membership of the Society, founded in 1889, includes astronomers, educators, and astronomy enthusiasts. Their site features a non-profit catalog of interesting astronomy items (posters, videos, software, etc.) and also a large quantity of information related to astronomy education.

The Planetary Society [www.planetary.org], founded by the late Carl Sagan, works to encourage planetary exploration and the search for life elsewhere. Its Web site has many pages of information in these two areas, and also features a catalog of useful educational materials.

The Astronomical League [www.astroleague.org/] is the umbrella organization of astronomy clubs in the U.S. Their sites have good lists of astronomy clubs around the country, activities and hints for astronomy hobbyists, and information on regional and national programs. (Another good list of astronomy clubs can be found through *Sky & Telescope* magazine at: www.skypub.com/resources/directory/directory.html.)

3. Sources for the Latest Astronomy Images

Anglo-Australian Observatory Image Collection [www.aao.gov.au/images.html]
A marvelous library of images (focusing on nebulae and galaxies) taken using large Australian telescopes. Many are by David Malin, who has been acknowledged as one of the finest astronomical photographers of our time. Includes captions and ordering information.

Astronomy Picture of the Day [antwrp.gsfc.nasa.gov/apod/astropix.html]
On this popular site, astronomers Robert Nemiroff and Jerry Bonell feature one relatively new celestial image each day with a brief non-technical caption. Over the years, some of the best astronomical images have been featured here and an index is available on site.

European Southern Observatory [www.eso.org/outreach/gallery/]

This growing album contains images from the large telescopes in the southern hemisphere run by a consortium of European countries. With the advent of the Very Large Telescope, there will be an increasing number of important new images and results on this site.

History of Physics Visual Archives [www.aip.org/history/esva]

A database of images from the history of physics and astronomy, which can be searched and ordered via the Internet.

Hubble Space Telescope [oposite.stsci.edu/pubinfo/pictures.html]

All the magnificent Hubble images, all with captions, and many with detailed background information, can be found at this site. You can see the latest images, the ones the staff considers the Hubble's "greatest hits," or search for objects of interest to you.

NASA Image Exchange [nix.nasa.gov]

An online gallery and database of the images in the collections of NASA centers around the country. This site includes only a small number of celestial images, but many terrestrial ones.

National Optical Astronomy Observatories Image Gallery [www.noao.edu/image_gallery]

NOAO, whose Director is *Voyages* co-author Sidney Wolff, includes a number of major telescopes in the U.S. and the Southern Hemisphere and is a national facility, where any astronomer can apply for time. Some of the best images from NOAO instruments are collected at this site.

NGC Images on the Net [www.aspsky.org/html/resources/ngc.html]

The New Galactic Catalog (NGC) is a list of nebulae, star clusters, and galaxies, assembled by J.L.E. Dryer in 1888, and its numbering system is still used by astronomers to denote such objects. This site, assembled by the staff of the Astronomical Society of the Pacific, is an index to the locations of images of deep space objects around the Web, organized by NGC number.

Planetary Photojournal [photojournal.jpl.nasa.gov]

This site, run by the Jet Propulsion Laboratories and the U.S. Geological Survey Branch of Astro-geology in Flagstaff, is one of the most useful resources on the Web. (We may not be entirely objective, however, since two of the authors of *Voyages* served as advisors for the project to set up the site!) It currently features over 2,000 of the best images from planetary exploration, with detailed captions and excellent indexing. You can dial up images by world, feature name, date, or catalog number, and download images in a number of popular formats. The one problem with the site is "NASA chauvinism"; only NASA mission images are currently included.

Planetary Picture List [seds.lpl.arizona.edu/nineplanets/nineplanets/picturelist.html]

Bill Arnett, the Webmaster of the Nine Planets site (see under Chapter 6) also keeps this impressive catalog of solar system images on the Web, organized by object. The lists tell you what format each image can be found in, and how much space it takes up.

NASA Photo Gallery [www.nasa.gov/gallery/photo]

We list this last because it is not a site with images, but rather a catalog of *other* NASA sites that do feature images.

The list includes a wide range of sites with earth and space photography, plus maps and some simulations.

4. Debunking Pseudo-science (like astrology, UFO's, etc.)

Astronomical Pseudo-science: A Skeptic's Resource List [www.aspsky.org/html/astro/pseudobib.html]

An annotated list of written and Web resources for dealing with astrology, UFO's, the Face on Mars, ancient astronauts, and many similar fringe topics.

Debunking Pseudo-science (in General): [www.csicop.org/]

For students or instructors who want assistance in dealing with topics that mix up science with new age philosophy, paranormal beliefs, conspiracy theories, and utter silliness, the premier organization is the awkwardly named, but very effective Committee for the Scientific Investigation of Claims of the Paranormal (CSICOP). Composed of scientists, educators, magicians, legal experts, doctors, and others who are tired of the uncritical acceptance of pseudo-science by the media and the public, CSICOP seeks to provide the skeptical, rational side of many controversial topics, such as UFO's, astrology, psychic power, the Bermuda Triangle, the Face on Mars, crop circles, etc. Their Web site is a treasure trove of information, anecdotes, experiments, and links.

5. Miscellaneous Sites We Have Found Useful

Astronomy Plus: Interdisciplinary Approaches to Astronomy [www.saunderscollege.com/astro/fraknoi/teaching/interdisc.html]

A bibliography of readings organized by subject; it includes the relationship between astronomy and art, music, poetry, psychology, the environment, law, stamp collecting, historical events, etc.

Science Fiction Stories with Good Astronomy and Physics [www.saunderscollege.com/astro/fraknoi/stories/stories.html]

A listing of novels and stories organized by topic in astronomy, with a brief descriptions.

Astronomical Observing Tips [www.skypub.com/tips/tips.html]

This series of excellent articles from *Sky & Telescope* magazine introduces beginners to observational astronomy. It includes good advice on getting started with just your eyes, binoculars, small telescopes, star wheels, photography, and much more.

History of Astronomy Site [www.astro.uni-bonn.de/~pbrosche/astoria.html]

Professor Wolfgang Dick maintains this comprehensive Web site on behalf of the German Working Group for the History of Astronomy, and it contains a wonderfully complete series of links to sites about all aspects of astronomical history. Among its best features is the ability to select the name of a historical astronomer and find links that provide instant biographical information in most cases.

Astronomical Software Listing [www.skypub.com/resources/software/commercial.html]

Astronomy educator John Mosley has organized a wealth of information on commercially published astronomy software, with brief annotations, ordering information, and references to full reviews in *Sky & Telescope* magazine.

AstroWeb: Astronomy/Astrophysics on the Internet
[www.cv.nrao.edu/fits/www/astronomy.html]
This site, maintained by a consortium of astronomy institutions (including the Space Telescope Science Institute and the National Radio Astronomy Observatory) is a collection of almost 3,000 astronomy-related links, organized by subject. Some links are to highly technical sites, some to pages that are just right for beginners. You can check if an astronomer has a personal Web page, look up what astronomy departments or observatories have put on the Web, and delve further into many specialized areas of astronomy.

Space Calendar [www.jpl.nasa.gov/calendar] is an amazing listing of space- and astronomy-related events, including upcoming launches, significant anniversaries, and things to see in the sky. Most entries in the calendar have a Web link where you can find more information.

APPENDIX 2

Sources of Astronomical Information

I. Popular-level Astronomy Magazines

Astronomy **Magazine** (Kalmbach Publishing, P. O. Box 1612, Waukesha, WI 53187; http://www.astronomy.com). The astronomy publication with the largest circulation in the world; it is very colorful and basic.

Griffith Observer (Griffith Observatory, 2800 E. Observatory Rd., Los Angeles, CA 90027; http://www.griffithobs. org). A small magazine specializing in historical topics, edited with humor and verve.

Mercury **Magazine** (Astronomical Society of the Pacific, 390 Ashton Ave., San Francisco, CA 94112; http://www.aspsky. org). The magazine of the largest general membership astronomy organization in the U.S., with many interesting features, particularly about the influence of astronomy on other fields.

Planetary Report (The Planetary Society, 65 N. Catalina Ave., Pasadena, CA 91106; http:// www.planetary.org). The magazine of a large organization that promotes the exploration of the solar system and the search for life elsewhere.

Stardate Magazine (McDonald Observatory, RLM 15.308, University of Texas, Austin, TX 78712; http://stardate. utexas.edu). Accompanies the popular Public Radio program on astronomy.

Sky & Telescope (P.O. Box 9111, Belmont, MA 02178; http://www.skypub.com) Found in most libraries, this is the popular astronomy magazine "of record" with many excellent articles on astronomy and amateur astronomy.

The following magazines (found in most libraries) contain frequent reports on astronomical developments: *Discover, National Geographic, Science News,* and *Scientific American.*

2. Organizations for Astronomy Enthusiasts

American Association of Variable Star Observers, 25 Birch St., Cambridge, MA 02138 (617-354-0484); http://www.aavso.org. An organization of amateur astronomers devoted to making serious observations of stars whose brightness changes.

Astronomical League, c/o Berton Stevens, 2112 Kingfisher Lane E., Rolling Meadows, IL 60008; http://www. astroleague.org. Umbrella organization of amateur astronomy clubs in the U.S., with many activities. If you write to this all-volunteer organization, be sure to enclose a stamped self-addressed envelope.

Astronomical Society of the Pacific, 390 Ashton Ave., San Francisco, CA 94112 (415-337-1100); http://www.aspsky. org. This international organization brings together scientists, teachers, and people with a general interest in astronomy; its name is merely a reminder of its origins on the West Coast of the U.S. in 1889. Write for their catalog of interesting astronomy materials or information about their national meetings and astronomy expos.

Committee for the Scientific Investigation of Claims of the Paranormal (CSICOP), P. O. Box 703, Buffalo, NY 14226 (716-636-1425); http://www.csicop.org. An organization of scientists, educators, magicians, and other skeptics that seeks to inform the public about the rational perspective on such pseudosciences as astrology, UFOs, psychic power, etc. Publishes *The Skeptical Inquirer* magazine, full of great debunking articles, and holds meetings and workshops around the world.

International Dark Sky Association, c/o David Crawford, 3545 N. Stewart, Tucson, AZ 85716; http://www.darksky. org/ida/index.html. Small non-profit organization devoted to fighting light pollution and educating politicians, lighting engineers, and the public about the importance of not spilling light where it will interfere with astronomical observations.

The Planetary Society, 65 N. Catalina Ave., Pasadena, CA 91106 (818-793-1675); http://www.planetary.org. Large national membership organization founded by Carl Sagan and others; lobbies for more planetary exploration and SETI; publishes a colorful magazine and has a catalog of reasonably priced slides, videos, and gift items.

The Royal Astronomical Society of Canada, 136 Dupont St., Toronto, Ontario M5R 1V2 (416-924-7973); http://www. rasc.ca. The main organization of amateur astronomers in Canada, with local centers in each province. Write for information on contacting the center near you.

3. Selected Institutions with Services and Materials for the Public

Jet Propulsion Laboratory, Public Information Office, 4800 Oak Grove Dr., Pasadena, CA 91109; http://www.jpl. nasa.gov. This NASA center will usually respond to public inquiries about planetary missions with information pamphlets and lithographs.

NASA Headquarters, Public Information Branch, Washington, DC 20546; http://www.nasa.gov. You can obtain information about NASA missions and projects. Leave plenty of time to get a response.

Space Telescope Science Institute, Public Information Office, 3700 San Martin Dr., Baltimore, MD 21218; http://oposite.stsci.edu. Will often respond to intelligently worded requests for information or pictures from the Hubble Space Telescope.

APPENDIX 3

Glossary

Note: This glossary covers all of astronomy and thus includes many terms that are not in this book.

absolute brightness (magnitude) *See* luminosity.

absolute zero A temperature of $-273°C$ (or 0 K), where all molecular motion stops.

absorption spectrum Dark lines superimposed on a continuous spectrum.

accelerate To change velocity; to speed up, slow down, or change direction.

accretion Gradual accumulation of mass, as by a planet forming by the building up of colliding particles in the solar nebula or gas falling into a black hole.

accretion disk A disk of matter spiraling in toward a massive object; the disk shape is the result of conservation of angular momentum.

active galactic nucleus A galaxy is said to have an active nucleus if unusually violent events are taking place in its center, emitting large quantities of electromagnetic radiation. Seyfert galaxies and quasars are examples of galaxies with active nuclei.

active region Areas on the Sun where magnetic fields are concentrated; sunspots, prominences, and flares all tend to occur in active regions.

albedo The fraction of incident sunlight that a planet or minor planet reflects.

alpha particle The nucleus of a helium atom, consisting of two protons and two neutrons.

amplitude The range in variability, as in the light from a variable star.

angular diameter Angle subtended by the diameter of an object.

angular momentum A measure of the momentum associated with motion about an axis or fixed point.

antapex (solar) Direction away from which the Sun is moving with respect to the local standard of rest.

Antarctic Circle Parallel of latitude 66°30′S; at this latitude the noon altitude of the Sun is 0° on the date of the summer solstice.

antimatter Matter consisting of antiparticles: antiprotons (protons with negative rather than positive charge), positrons (positively charged electrons), and antineutrons.

aperture The diameter of an opening, or of the primary lens or mirror of a telescope.

apex (solar) The direction toward which the Sun is moving with respect to the local standard of rest.

aphelion Point in its orbit where a planet (or other body) is farthest from the Sun.

apogee Point in its orbit where an Earth satellite is farthest from the Earth.

apparent brightness A measure of the observed light received from a star or other object at the Earth; i.e., how bright an object appears in the sky, as contrasted with its luminosity.

Arctic Circle Parallel of latitude 66°30′N; at this latitude the noon altitude of the Sun is 0° on the date of the winter solstice.

array (interferometer) A group of several telescopes that is used to make observations at high angular resolution.

association A loose group of young stars whose spectral types, motions, or positions in the sky indicate that they have probably had a common origin.

asterism An especially noticeable star pattern in the sky, such as the Big Dipper.

asteroid An object orbiting the Sun that is smaller than a major planet, but that shows no evidence of an atmosphere or of other types of activity associated with comets. Also called a minor planet.

asteroid belt The region of the solar system between the orbits of Mars and Jupiter in which most asteroids are located. The main belt, where the orbits are generally the most stable, extends from 2.2 to 3.3 AU from the Sun.

astrology The pseudoscience that deals with the supposed influences on human destiny of the configurations and locations in the sky of the Sun, Moon, and planets; a primitive belief system that had its origin in ancient Babylonia.

astronomical unit (AU) Originally meant to be the semimajor axis of the orbit of the Earth; now defined as the semimajor axis of the orbit of a hypothetical body with the mass and period that Gauss assumed for the Earth. The semimajor axis of the orbit of the Earth is 1.000000230 AU.

atom The smallest particle of an element that retains the properties that characterize that element.

atomic mass unit *Chemical:* one-sixteenth of the mean mass of an oxygen atom. *Physical:* one-twelfth of the mass of an atom of the most common isotope of carbon. The atomic mass unit is approximately the mass of a hydrogen atom, 1.67×10^{-27} kg.

atomic number The number of protons in each atom of a particular element.

atomic weight The mean mass of an atom of a particular element in atomic mass units.

aurora Light radiated by atoms and ions in the ionosphere, mostly in the magnetic polar regions.

autumnal equinox The intersection of the ecliptic and celestial equator where the Sun crosses the equator from north to south. A time when every place on Earth has 12 hours of daylight and 12 hours of night.

axis An imaginary line about which a body rotates.

azimuth The angle along the celestial horizon, measured eastward from the north point to the intersection of the horizon with the vertical circle passing through an object.

Balmer lines Emission or absorption lines in the spectrum of hydrogen that arise from transitions between the second (or first excited) and higher energy states of the hydrogen atom.

bands (in spectra) Emission or absorption lines, usually in the spectra of chemical compounds, so numerous and closely spaced that they coalesce into broad emission or absorption bands.

bar A force of 100,000 newtons acting on a surface area of 1 square meter is equal to 1 bar. The average pressure of the Earth's atmosphere at sea level is equal to 1.013 bars.

barred spiral galaxy Spiral galaxy in which the spiral arms begin from the ends of a "bar" running through the nucleus rather than from the nucleus itself.

barycenter The center of mass of two mutually revolving bodies.

basalt Igneous rock, composed primarily of silicon, oxygen, iron, aluminum, and magnesium produced by the cooling of lava. Basalts make up most of Earth's oceanic crust and are also found on other planets that have experienced extensive volcanic activity.

big bang theory A theory of cosmology in which the expansion of the universe is presumed to have begun with a primeval explosion.

binary star Two stars revolving about each other.

binding energy The energy required to separate completely the constituent parts of an atomic nucleus.

blackbody A hypothetical perfect radiator, which absorbs and re-emits all radiation incident upon it.

black dwarf A presumed final state of evolution for a low-mass star, in which all of its energy sources are exhausted and it no longer emits significant radiation.

black hole A collapsed massive star (or other collapsed body) whose velocity of escape is equal to or greater than the speed of light; thus no radiation can escape from it.

Bohr atom A particular model of an atom, invented by Niels Bohr, in which the electrons are described as revolving about the nucleus in circular orbits.

brown dwarf An object intermediate in size between a planet and a star. The approximate mass range is from about 1/100 of the mass of the Sun up to the lower mass limit for self-sustaining nuclear reactions, which is 0.08 solar mass.

carbonaceous meteorite A primitive meteorite made primarily of silicates but often including chemically bound water, free carbon, and complex organic compounds. Also called carbonaceous chondrites.

carbon-nitrogen-oxygen (CNO) cycle A series of nuclear reactions in the interiors of stars involving carbon as a catalyst, by which hydrogen is transformed to helium.

cassegrain focus An optical arrangement in a reflecting telescope in which light is reflected by a second mirror to a point behind the primary mirror.

CBR *See* cosmic background radiation.

CCD *See* charge-coupled device.

cD galaxy A supergiant elliptical galaxy frequently found at the center of a cluster of galaxies.

celestial equator A great circle on the celestial sphere 90° from the celestial poles; where the celestial sphere intersects the plane of the Earth's equator.

celestial meridian An imaginary line on the celestial sphere passing through the north and south points on the horizon and through the zenith.

celestial poles Points about which the celestial sphere appears to rotate; intersections of the celestial sphere with the Earth's polar axis.

celestial sphere Apparent sphere of the sky; a sphere of large radius centered on the observer. Directions of objects in the sky can be denoted by their position on the celestial sphere.

center of gravity Center of mass.

center of mass The average position of the various mass elements of a body or system, weighted according to their distances from that center of mass; that point in an isolated system that moves with constant velocity, according to Newton's first law of motion.

cepheid variable A star that belongs to a class of yellow supergiant pulsating stars. These stars vary periodically in brightness, and the relationship between their periods and luminosities is useful in deriving distances to them.

Chandrasekhar limit The upper limit to the mass of a white dwarf (equals 1.4 times the mass of the Sun).

charge-coupled device (CCD) An array of electronic detectors of electromagnetic radiation, used at the focus of a telescope (or camera lens). A CCD acts like a photographic plate of very high sensitivity.

chemical condensation sequence The calculated chemical compounds and minerals that would form at different temperatures in a cooling gas of cosmic composition; used to infer the composition of grains that formed in the solar nebula at different distances from the protosun.

chromosphere That part of the solar atmosphere that lies immediately above the photospheric layers.

circular satellite velocity The critical speed that a revolving body must have in order to follow a circular orbit.

circumpolar zone Those portions of the celestial sphere near the celestial poles that are either always above or always below the horizon.

closed universe A model of the universe in which the curvature of space is such that straight lines eventually curve back upon themselves; in this model, the universe expands from a big bang, stops, and then contracts to a big crunch.

cluster of galaxies A system of galaxies containing several to thousands of member galaxies.

color index Difference between the magnitudes of a star or other object measured in light of two different spectral regions, for example, blue minus visual $(B - V)$ magnitudes.

coma (of comet) The diffuse gaseous component of the head of a comet; i.e, the cloud of evaporated gas around a comet nucleus.

comet A small body of icy and dusty matter that revolves about the Sun. When a comet comes near the Sun, some of its material vaporizes, forming a large head of tenuous gas, and often a tail.

compound A substance composed of two or more chemical elements.

conduction The transfer of energy by the direct passing of energy or electrons from atom to atom.

conservation of angular momentum The law that the total amount of angular momentum in a system remains the same (in the absence of any force not directed toward or away from the point or axis about which the angular momentum is referred).

constellation One of 88 sectors into which astronomers divide the celestial sphere; many constellations are named after a prominent group of stars within them that represents a person, animal, or legendary creature from ancient mythology.

continental drift A gradual movement of the continents over the surface of the Earth due to plate tectonics.

continuous spectrum A spectrum of light composed of radiation of a continuous range of wavelengths or colors rather than only certain discrete wavelengths.

convection The transfer of energy by moving currents in a fluid.

core (of a planet) The central part of a planet, consisting of higher density material.

corona (of Galaxy) A region lying above and below the plane of the Galaxy out to much greater distances than the material that gives off electromagnetic radiation.

corona (of Sun) Outer atmosphere of the Sun.

coronal hole A region in the Sun's outer atmosphere where visible coronal radiation is absent.

cosmic background radiation (CBR) The microwave radiation coming from all directions that is believed to be the redshifted glow of the Big Bang.

cosmic rays Atomic nuclei (mostly protons) that are observed to strike the Earth's atmosphere with exceedingly high energies.

cosmological constant A term in the equations of general relativity that represents a repulsive force in the universe. The cosmological constant may or may not be zero.

cosmological principle The assumption that, on the large scale, the universe at any given time is the same everywhere — isotropic and homogeneous.

cosmology The study of the organization and evolution of the universe.

crater A circular depression (from the Greek word for cup), generally of impact origin.

crescent moon One of the phases of the Moon when it appears less than half full.

critical density In cosmology, the density that provides enough gravity to bring the expansion of the universe just to a stop after infinite time.

crust The outer layer of a terrestrial planet.

dark matter Nonluminous mass, whose presence can be inferred only because of its gravitational influence on luminous matter. Dark matter may constitute as much as 99 percent of all the mass in the universe. The composition of the dark matter is not known.

dark nebula A cloud of interstellar dust that obscures the light of more distant stars and appears as an opaque curtain.

declination Angular distance north or south of the celestial equator.

degenerate gas A gas in which the allowable states for the electrons have been filled; it behaves according to different laws from those that apply to "perfect" gases and resists further compression.

density The ratio of the mass of an object to its volume.

deuterium A "heavy" form of hydrogen, in which the nucleus of each atom consists of one proton and one neutron.

differential galactic rotation The rotation of the Galaxy, not as a solid wheel, but so that parts adjacent to one another do not always stay close together.

differentiation (geological) Gravitational separation or segregation of materials of different density into layers in the interior of a planet or satellite.

disk (of Galaxy) The central plane or "wheel" of our Galaxy, where most of the luminous mass is concentrated.

dispersion Separation, from white light, of different wavelengths being refracted by different amounts.

Doppler effect Apparent change in wavelength of the radiation from a source due to its relative motion away from or towards the observer.

Earth-approaching asteroid An asteroid with an orbit that crosses the Earth's orbit or that will at some time cross the Earth's orbit as it evolves under the influence of the planets' gravity. *See also* near-Earth object.

eccentricity (of ellipse) Ratio of the distance between the foci to the major axis.

eclipse The cutting off of all or part of the light of one body by another; in planetary science, the passing of one body into the shadow of another.

eclipsing binary star A binary star in which the plane of revolution of the two stars is nearly edge-on to our line of sight, so that the light of one star is periodically diminished by the other passing in front of it.

ecliptic The apparent annual path of the Sun on the celestial sphere.

effective temperature *See* temperature (effective).

ejecta Material excavated from an impact crater, such as the blanket of material surrounding lunar craters and crater rays.

electromagnetic force One of the four fundamental forces or interactions of nature; the force that acts between charges and binds atoms and molecules together.

electromagnetic radiation Radiation consisting of waves propagated through the building up and breaking down of electric and magnetic fields; these include radio, infrared, light, ultraviolet, x rays, and gamma rays.

electromagnetic spectrum The whole array or family of electromagnetic waves, from radio to gamma rays.

electron A negatively charged subatomic particle that normally moves about the nucleus of an atom.

element A substance that cannot be decomposed, by chemical means, into simpler substances.

elementary particle One of the basic particles of matter. The most familiar of the elementary particles are the proton, neutron, and electron.

ellipse A curve for which the sum of the distances from any point on the ellipse to two points inside (called the foci) is always the same.

ellipticity The ratio (in an ellipse) of the major axis minus the minor axis to the major axis.

emission line A discrete bright line in the spectrum.

emission nebula A gaseous nebula that derives its visible light from the fluorescence of ultraviolet light from a star in or near the nebula.

emission spectrum A spectrum consisting of emission lines.

energy level A particular level, or amount, of energy possessed by an atom or ion above the energy it possesses in its least energetic state; also used to refer to the states of energy an electron can have in an atom.

epicycle A circular orbit of a body in the Ptolemaic system, the center of which revolves about another circle (the deferent).

equator A great circle on the Earth, 90° from (or equidistant from) each pole.

equinox One of the intersections of the ecliptic and celestial equator; one of the two times during the year when the length of the day and night are the same.

equivalence principle Principle that a gravitational force and a suitable acceleration are indistinguishable within a sufficiently local environment.

escape velocity The velocity a body must achieve to break away from the gravity of another body and never return to it.

eucrite meteorite One of a class of basaltic meteorites believed to have originated on the asteroid Vesta.

event horizon The surface through which a collapsing star passes when its velocity of escape is equal to the speed of light; that is, when the star becomes a black hole.

excitation The process of imparting to an atom or an ion an amount of energy greater than it has in its normal or least-energy state.

exclusion principle *See* Pauli exclusion principle.

extinction Reduction of the light from a celestial body produced by the Earth's atmosphere, or by interstellar absorption.

extragalactic Beyond our own Milky Way Galaxy.

eyepiece A magnifying lens used to view the image produced by the objective of a telescope.

fall (of meteorites) Meteorites seen in the sky and recovered on the ground.

fault In geology, a crack or break in the crust of a planet along which slippage or movement can take place, accompanied by seismic activity.

field A mathematical description of the effect of forces, such as gravity, that act on distant objects. For example, a given mass produces a gravitational field in the space surrounding it, which produces a gravitational force on objects within that space.

find (of meteorites) A meteorite that has been recovered but was not seen to fall.

fireball A spectacular meteor, seen for more than an instant in the sky.

fission The breakup of a heavy atomic nucleus into two or more lighter ones.

flare A sudden and temporary outburst of light from an extended region of the Sun's surface.

fluorescence The absorption of light of one wavelength and re-emission of it at another wavelength; especially the conversion of ultraviolet into visible light.

flux The rate at which energy or matter crosses a unit area of a surface.

focal length The distance from a lens or mirror to the point where light converged by it comes to a focus.

focus (of ellipse) One of two fixed points inside an ellipse from which the sum of the distances to any point on the ellipse is a constant.

focus (of telescope) Point where the rays of light converged by a mirror or lens meet.

forbidden lines Spectral lines that are not usually observed under laboratory conditions because they result from atomic transitions that are highly improbable.

force That which can change the momentum of a body; numerically, the rate at which the body's momentum changes.

Fraunhofer line An absorption line in the spectrum of the Sun or of a star.

Fraunhofer spectrum The array of absorption lines in the spectrum of the Sun or a star.

frequency Number of vibrations per unit time; number of waves that cross a given point per unit time (in radiation).

fusion The building up of heavier atomic nuclei from lighter ones.

galactic cannibalism The process by which a larger galaxy strips material from a smaller one.

galactic cluster An "open" cluster of stars located in the spiral arms or disk of the Galaxy.

galaxy A large assemblage of stars; a typical galaxy contains millions to hundreds of billions of stars.

Galaxy The galaxy to which the Sun and our neighboring stars belong; the Milky Way is light from remote stars in the disk of the Galaxy.

gamma rays Photons (of electromagnetic radiation) of energy higher than those of x rays; the most energetic form of electromagnetic radiation.

general relativity theory Einstein's theory relating acceleration, gravity, and the structure (geometry) of space and time.

geocentric Centered on the Earth.

giant (star) A star of large luminosity and radius.

giant molecular cloud Large, cold interstellar clouds, with diameters of dozens of light years and typical masses of 10^5 solar masses; found in the spiral arms of galaxies, these clouds are where massive stars form.

globular cluster One of about 120 large spherical star clusters that form a system of clusters centered on the center of the Galaxy.

grand unified theories (GUTs) Physical theories that attempt to describe the four interactions (forces) of nature as different manifestations of a single force.

granite The type of igneous silicate rock that makes up most of the continental crust of the Earth.

granulation The rice-grain-like structure of the solar photosphere; granulation is produced by upwelling currents of gas that are slightly hotter, and therefore brighter, than the surrounding regions, which are flowing downward into the Sun.

gravity The mutual attraction of material bodies or particles.

gravitational energy Energy that can be released by the gravitational collapse, or partial collapse, of a system; i.e., by particles that fall in toward the center of gravity.

gravitational lens A configuration of celestial objects, one of which provides one or more images of the other by gravitationally deflecting its light.

gravitational redshift The redshift of electromagnetic radiation caused by a gravitational field. The slowing of clocks in a gravitational field.

great circle Circle on the surface of a sphere that is the curve of intersection of the sphere with a plane passing through its center.

greenhouse effect The blanketing (absorption) of infrared radiation near the surface of a planet by, for example, carbon dioxide in its atmosphere.

ground state The lowest energy state of an atom.

H or H$_0$ *See* Hubble constant.

H I region Region of neutral hydrogen in interstellar space.

H II region Region of ionized hydrogen in interstellar space.

half-life The time required for half of the radioactive atoms in a sample to disintegrate.

halo (of galaxy) The outermost extent of our Galaxy or another, containing a sparse distribution of stars and globular clusters in a more or less spherical distribution.

heavy elements In astronomy, usually those elements of greater atomic number than helium.

helio- Prefix referring to the Sun.

heliocentric Centered on the Sun.

helium flash The nearly explosive ignition of helium in the triple-alpha process in the dense core of a red giant star.

Herbig-Haro (HH) object Luminous knots of gas in an area of star formation, which are set to glow by jets of material from a protostar.

hertz A unit of frequency: one cycle per second. Named for Heinrich Hertz, who first produced radio radiation.

Hertzsprung–Russell (H–R) diagram A plot of luminosity against surface temperature (or spectral type) for a group of stars.

highlands (lunar) The older, heavily cratered crust of the Moon, covering 83 percent of its surface and composed in large part of anorthositic breccias.

homogeneous Having a consistent and even distribution of matter that is the same everywhere.

horizon (astronomical) A great circle on the celestial sphere 90° from the zenith; more popularly, the circle around us where the dome of the sky meets the Earth.

horoscope A chart used by astrologers, showing the positions along the zodiac and in the sky of the Sun, Moon, and planets at some given instant and as seen from a particular place on Earth — usually corresponding to the time and place of a person's birth.

Hubble constant Constant of proportionality between the velocities of remote galaxies and their distances. The Hubble constant is thought to lie in the range of 15 to 30 km/s per million LY.

Hubble law (or the law of the redshifts) The radial velocities of remote galaxies are proportional to their distances from us.

hydrostatic equilibrium A balance between the weights of various layers, as in a star or the Earth's atmosphere, and the pressures that support them.

hypothesis A tentative theory or supposition, advanced to explain certain facts or phenomena, which is subject to further tests and verification.

igneous rock Any rock produced by cooling from a molten state.

inclination (of an orbit) The angle between the orbital plane of a revolving body and some fundamental plane — usually the plane of the celestial equator or of the ecliptic.

inertia The property of matter that requires a force to act on it to change its state of motion; the tendency of objects to continue doing what they are doing in the absence of outside forces.

inertial system A system of coordinates that is not itself accelerated, but that either is at rest or is moving with constant velocity.

inflationary universe A theory of cosmology in which the universe is assumed to have undergone a phase of very rapid expansion during the first 10^{-30} s. After this period of rapid expansion, the Big Bang and inflationary models are identical.

infrared cirrus Patches of interstellar dust, which emit infrared radiation and look like cirrus clouds on the images of the sky.

infrared radiation Electromagnetic radiation of wavelength longer than the longest (red) wavelengths that can be perceived by the eye, but shorter than radio wavelengths.

interference A phenomenon of waves that mix together such that their crests and troughs can alternately reinforce and cancel one another.

international date line An arbitrary line on the surface of the Earth near longitude 180° across which the date changes by one day.

interstellar dust Tiny solid grains in interstellar space, thought to consist of a core of rock-like material (silicates) or graphite surrounded by a mantle of ices. Water, methane, and ammonia are probably the most abundant ices.

interstellar extinction The attenuation or absorption of light by dust in the interstellar medium.

interstellar medium or interstellar matter Interstellar gas and dust.

inverse-square law (for light) The amount of energy (light) flowing through a given area in a given time (flux) decreases in proportion to the square of the distance from the source of energy or light.

ion An atom that has become electrically charged by the addition or loss of one or more electrons.

ionization The process by which an atom gains or loses electrons.

ionosphere The upper region of the Earth's atmosphere in which many of the atoms are ionized.

ion tail (of comet) *See* plasma.

irregular galaxy A galaxy without rotational symmetry; neither a spiral nor an elliptical galaxy.

irregular satellite A planetary satellite with an orbit that is retrograde, or of high inclination or eccentricity.

isotope Any of two or more forms of the same element, whose atoms all have the same number of protons but different numbers of neutrons.

isotropic The same in all directions.

joule The metric unit of energy; the work done by a force of 1 newton (N) acting through a distance of 1 m.

jovian planet or giant planet Any of the planets Jupiter, Saturn, Uranus, and Neptune in our solar system, or planets of roughly that mass and composition in other planetary systems.

kinetic energy Energy associated with motion.

Kuiper belt A reservoir of cometary material just beyond the orbit of Pluto.

laser An acronym for *l*ight *a*mplification by *s*timulated *e*mission of *r*adiation; a device for amplifying a light signal at a particular wavelength into a coherent beam.

latitude A north-south coordinate on the surface of the Earth; the angular distance north or south of the equator measured along a meridian passing through a place.

law of areas Kepler's second law: the radius vector from the Sun to any planet sweeps out equal areas in the planet's orbital plane in equal intervals of time.

law of the redshifts *See* Hubble law.

leap year A calendar year with 366 days, inserted approximately every 4 years to make the average length of the calendar year as nearly equal as possible to the tropical year.

light or visible light Electromagnetic radiation that is visible to the eye.

light curve A graph that displays the time variation of the light from a variable or eclipsing binary star.

light year The distance light travels in a vacuum in one year; 1 LY = 9.46×10^{12} km, or about 6×10^{12} mi.

line broadening The phenomenon by which spectral lines are not precisely sharp but have finite widths.

line profile A plot of the intensity of light versus wavelength across a spectral line.

Local Group The cluster of galaxies to which our Galaxy belongs.

local standard of rest A coordinate system that shares the average motion of the Sun and its neighboring stars about the galactic center.

Local Supercluster The supercluster of galaxies to which the Local Group belongs.

longitude An east-west coordinate on the Earth's surface; the angular distance, measured east or west along the equator, from the Greenwich meridian to the meridian passing through a place.

luminosity The rate at which a star or other object emits electromagnetic energy into space.

luminosity class A classification of a star according to its luminosity within a given spectral class. Our Sun, a G2V star, has luminosity class V.

luminosity function The relative numbers of stars (or other objects) of various luminosities.

lunar eclipse An eclipse of the Moon.

Lyman lines A series of absorption or emission lines in the spectrum of hydrogen that arise from transitions to and from the lowest energy states of the hydrogen atoms.

Magellanic Clouds Two neighboring galaxies visible to the naked eye from southern latitudes.

magma Mobile, high-temperature molten state of rock, usually of silicate mineral composition and with dissolved gases and other volatiles.

magnetic field The region of space near a magnetized body within which magnetic forces can be detected.

magnetic pole One of two points on a magnet (or the Earth) at which the greatest density of lines of force emerge. A compass needle aligns itself along the local lines of force on the Earth and points more or less toward the magnetic poles of the Earth.

magnetosphere The region around a planet in which its intrinsic magnetic field dominates the interplanetary field carried by the solar wind; hence, the region within which charged particles can be trapped by the planetary magnetic field.

magnitude A measure of the amount of light flux received from a star or other luminous object.

main sequence A sequence of stars on the Hertzsprung–Russell diagram, containing the majority of stars, that runs diagonally from the upper left to the lower right.

major axis (of ellipse) The maximum diameter of an ellipse.

mantle (of Earth) The greatest part of the Earth's interior, lying between the crust and the core.

mare (pl. maria) Latin for "sea"; name applied to the dark, relatively smooth features that cover 17 percent of the Moon.

mass A measure of the total amount of material in a body; defined either by the inertial properties of the body or by its gravitational influence on other bodies.

mass extinction The sudden disappearance in the fossil record of a large number of species of life, to be replaced by new species in subsequent layers. Mass extinctions are indications of catastrophic changes in the environment, such as might be produced by a large impact on the Earth.

mass-light ratio The ratio of the total mass of a galaxy to its total luminosity, usually expressed in units of solar mass and solar luminosity. The mass-light ratio gives a rough indication of the types of stars contained within a galaxy and whether or not substantial quantities of dark matter are present.

mass-luminosity relation An empirical relation between the masses and luminosities of many (principally main-sequence) stars.

Maunder Minimum The interval from 1645 to 1715 when solar activity was very low.

mean solar day Average length of the apparent solar day.

merger (of galaxies) When galaxies (of roughly comparable size) collide and form one combined structure.

meridian (celestial) The great circle on the celestial sphere that passes through an observer's zenith and the north (or south) celestial pole.

meridian (terrestrial) The great circle on the surface of the Earth that passes through a particular place and the North and South Poles of the Earth.

Messier catalog A catalog of nonstellar objects compiled by Charles Messier in 1787 (includes nebulae, star clusters, and galaxies).

metals In general, any element or compound whose electron structure makes it a good conductor of electricity. In astronomy, all elements beyond hydrogen and helium.

metamorphic rock Any rock produced by the physical and chemical alteration (without melting) of another rock that has been subjected to high temperature and pressure.

metastable level An energy level in an atom from which there is a low probability of an atomic transition accompanied by the radiation of a photon.

meteor The luminous phenomenon observed when a small piece of solid matter enters the Earth's atmosphere and burns up; popularly called a "shooting star."

meteorite A portion of a meteoroid that survives passage through the atmosphere and strikes the ground.

meteoroid A particle or chunk of typically rocky or metallic material in space before any encounter with the Earth.

meteor shower Many meteors appearing to radiate from a common point in the sky caused by the collision of the Earth with a swarm of solid particles, typically from a comet.

micron Old term for micrometer (10^{-6} meter).

microwave Shortwave radio wavelengths.

Milky Way The band of light encircling the sky, which is due to the many stars and diffuse nebulae lying near the plane of the Galaxy.

minerals The solid compounds (often primarily silicon and oxygen) that form rocks.

minor planet *See* asteroid.

model atmosphere or photosphere The result of a theoretical calculation of the run of temperature, pressure, density, and so on, through the outer layers of the Sun or a star.

molecule A combination of two or more atoms bound together; the smallest particle of a chemical compound or substance that exhibits the chemical properties of that substance.

momentum A measure of the inertia or state of motion of a body; the momentum of a body is the product of its mass and velocity. In the absence of a force, momentum is conserved.

near-Earth object (NEO) A comet or asteroid whose path intersects the orbit of the Earth.

nebula Cloud of interstellar gas or dust.

neutrino A fundamental particle that has little or no rest mass and no charge but that does have spin and energy. Neutrinos rarely interact with ordinary matter.

neutron A subatomic particle with no charge and with mass approximately equal to that of the proton.

neutron star A star of extremely high density composed almost entirely of neutrons.

nonthermal radiation *See* synchrotron radiation.

nova A star that experiences a sudden outburst of radiant energy, temporarily increasing its luminosity by hundreds to thousands of times.

nuclear Referring to the nucleus of the atom.

nuclear bulge Central part of our or another galaxy.

nuclear transformation The change of one atomic nucleus into another, as in nuclear fusion.

nucleosynthesis The building up of heavy elements from lighter ones by nuclear fusion.

nucleus (of atom) The heavy part of an atom, composed mostly of protons and neutrons, and about which the electrons revolve.

nucleus (of comet) The solid chunk of ice and dust in the head of a comet.

nucleus (of galaxy) Central concentration of matter at the center of a galaxy.

occultation The passage of an object of large angular size in front of a smaller object, such as the Moon in front of a distant star or the rings of Saturn in front of the Voyager spacecraft.

Oort comet cloud The large spherical region around the Sun from which most "new" comets come; a reservoir of objects with aphelia at about 50,000 AU, or extending about a third of the way to the nearest other stars.

opacity Absorbing power; capacity to impede the passage of light.

open cluster A comparatively loose or "open" cluster of stars, containing from a few dozen to a few thousand members, located in the spiral arms or disk of the Galaxy; sometimes referred to as a galactic cluster.

open universe A model of the universe in which gravity is not strong enough to bring the universe to a halt; it expands forever. In this model the geometry of spacetime is such that if you go in a straight line, you not only can never return to where you started, but even more space opens up than you would expect from Euclidean geometry.

optical In astronomy: relating to the visible-light band of the electromagnetic spectrum. Optical observations are those made with visible light.

optical double star Two stars at different distances that are seen nearly lined up in projection so that they appear close together, but that are not really gravitationally associated.

orbit The path of a body that is in revolution about another body or point.

oscillation A periodic motion; in the case of the Sun, a periodic or quasi-periodic expansion and contraction of the whole Sun or some portion of it.

ozone A heavy molecule of oxygen that contains three atoms rather than the more normal two. Designated O_3.

parabola A conic section of eccentricity 1.0; the curve of the intersection between a circular cone and a plane parallel to a straight line in the surface of the cone.

parallax An apparent displacement of a nearby star that results from the motion of the Earth around the Sun; numerically, the angle subtended by 1 AU at the distance of a particular star.

parsec A unit of distance in astronomy, equal to 3.26 light years. At a distance of 1 parsec, a star has a parallax of one arcsecond.

Pauli exclusion principle Quantum mechanical principle by which no two particles of the same kind can have the same position and momentum.

peculiar velocity The velocity of a star with respect to the local standard of rest; that is, its space motion, corrected for the motion of the Sun with respect to our neighboring stars.

penumbra The outer, not completely dark part of a shadow; the region from which the source of light is not completely hidden.

perfect radiator or blackbody A body that absorbs and subsequently re-emits all radiation incident upon it.

periastron The place in the orbit of a star in a binary-star system where it is closest to its companion star.

perigee The place in the orbit of an Earth satellite where it is closest to the center of the Earth.

perihelion The place in the orbit of an object revolving about the Sun where it is closest to the Sun's center.

period-luminosity relation An empirical relation between the periods and luminosities of certain variable stars.

perturbation The disturbing effect, when small, on the motion of a body as predicted by a simple theory, produced by a third body or other external agent.

photochemistry Chemical changes caused by electromagnetic radiation.

photometry The measurement of light intensities.

photon A discrete unit of electromagnetic energy.

photosphere The region of the solar (or a stellar) atmosphere from which continuous radiation escapes into space.

pixel An individual picture element in a detector; for example, a particular silicon diode in a CCD.

plage A bright region of the solar surface observed in the monochromatic light of some spectral line.

Planck's constant The constant of proportionality relating the energy of a photon to its frequency.

planet Any of the nine largest bodies revolving about the Sun, or any similar bodies that may orbit other stars. Unlike stars, planets do not (for the most part) give off their own light, but only reflect the light of their parent star.

planetarium An optical device for projecting on a screen or domed ceiling the stars and planets and their apparent motions in the sky.

planetary nebula A shell of gas ejected from, and enlarging about, a certain kind of extremely hot star that is nearing the end of its life.

planetesimals The hypothetical objects, from tens to hundreds of kilometers in diameter, that formed in the solar nebula as an intermediate step between tiny grains and the larger planetary objects we see today. The comets and some asteroids may be leftover planetesimals.

plasma A hot ionized gas.

plate tectonics The motion of segments or plates of the outer layer of the Earth over the underlying mantle.

polar axis The axis of rotation of the Earth; also, an axis in the mounting of a telescope that is parallel to the Earth's axis.

Population I and II Two classes of stars (and systems of stars), classified according to their spectral characteristics, chemical compositions, radial velocities, ages, and locations in the Galaxy.

positron An electron with a positive rather than negative charge; an antielectron.

potential energy Stored energy that can be converted into other forms; especially gravitational energy.

precession (of Earth) A slow, conical motion of the Earth's axis of rotation, caused principally by the gravitational pull of the Moon and Sun on the Earth's equatorial bulge.

precession of the equinoxes Slow westward motion of the equinoxes along the ecliptic that results from precession.

pressure Force per unit area; expressed in units of atmospheres or pascals.

prime focus The point in a telescope where the objective focuses the light.

prime meridian The terrestrial meridian passing through the site of the old Royal Greenwich Observatory; longitude 0°.

primitive In planetary science and meteoritics, an object or rock that is little changed, chemically, since its formation, and hence representative of the conditions in the solar nebula at the time of formation of the solar system. Also used to refer to the chemical composition of an atmosphere that has not undergone extensive chemical evolution.

primitive meteorite A meteorite that has not been greatly altered chemically since its condensation from the solar nebula; called in meteoritics a chondrite (either ordinary chondrite or carbonaceous chondrite).

primitive rock Any rock that has not experienced great heat or pressure and therefore remains representative of the original condensates from the solar nebula — never found on any object large enough to have undergone melting and differentiation.

principle of equivalence Principle that a gravitational force and a suitable acceleration are indistinguishable within a sufficiently local environment.

prism A wedge-shaped piece of glass that is used to disperse white light into a spectrum.

prominence A phenomenon in the solar corona that commonly appears like a flame above the limb of the Sun.

proper motion The angular change per year in the direction of a star as seen from the Sun.

proton A heavy subatomic particle that carries a positive charge; one of the two principal constituents of the atomic nucleus.

proton–proton cycle A series of thermonuclear reactions by which nuclei of hydrogen are built up into nuclei of helium.

protoplanet or -star or -galaxy The original material from which a planet (or a star or galaxy) condensed.

pulsar A variable radio source of small angular size that emits very rapid radio pulses in very regular periods that range from fractions of a second to several seconds.

pulsating variable A variable star that pulsates in size and luminosity.

quantum efficiency The ratio of the number of photons incident on a detector to the number actually detected.

quantum mechanics The branch of physics that deals with the structure of atoms and their interactions with one another and with radiation.

quasar An object of very high redshift that looks like a star and is extragalactic and highly luminous; an active galactic nucleus.

radar The technique of transmitting radio waves to an object and then detecting the radiation that the object reflects back to the transmitter; used to measure the distance to, and motion of, a target object.

radial velocity The component of relative velocity that lies in the line of sight; motion toward or away from the observer.

radial velocity curve A plot of the variation of radial velocity with time for a binary or variable star.

radiant (of meteor shower) The point in the sky from which the meteors belonging to a shower seem to radiate.

radiation A mode of energy transport whereby energy is transmitted through a vacuum; also the transmitted energy itself.

radiation pressure The transfer of momentum carried by electromagnetic radiation to a body that the radiation impinges upon.

radioactive dating The technique of determining the ages of rocks or other specimens by the amount of radioactive decay of certain radioactive elements contained therein; something you do when you are desperate on Saturday night.

radioactivity (radioactive decay) The process by which certain kinds of atomic nuclei naturally decompose, with the spontaneous emission of subatomic particles and gamma rays.

radio galaxy A galaxy that emits greater amounts of radio radiation than average.

radio telescope A telescope designed to make observations in radio wavelengths.

reddening (interstellar) The reddening of starlight passing through interstellar dust, caused because dust scatters blue light more effectively than red.

red giant A large, cool star of high luminosity; a star occupying the upper right portion of the Hertzsprung–Russell diagram.

redshift A shift to longer wavelengths of light, typically a Doppler shift caused by the motion of the source away from the observer.

reducing In chemistry, referring to conditions in which hydrogen dominates over oxygen, so that most other elements form compounds with hydrogen. In very reducing conditions free hydrogen (H_2) is present and free oxygen (O_2) cannot exist.

reflecting telescope A telescope in which the principal optical component (objective) is a concave mirror.

reflection nebula A relatively dense dust cloud in interstellar space that is illuminated by reflected starlight.

refracting telescope A telescope in which the principal optical component (objective) is a lens or system of lenses.

refraction The bending of light rays passing from one transparent medium (or a vacuum) to another.

relativistic particle (or electron) A particle (electron) moving at nearly the speed of light.

relativity A theory formulated by Einstein that describes the relations between measurements of physical phenomena by two different observers who are in relative motion at constant velocity (the special theory of relativity) or that describes how a gravitational field can be replaced by a curvature of spacetime (the general theory of relativity).

resolution The degree to which fine details in an image are separated, or the smallest detail that can be discerned in an image.

resonance An orbital condition in which one object is subject to periodic gravitational perturbations by another, most commonly arising when two objects orbiting a third have periods of revolution that are simple multiples or fractions of each other.

retrograde (rotation or revolution) Backward with respect to the common direction of motion in the solar system; counterclockwise as viewed from the north, and going from east to west rather than from west to east.

retrograde motion An apparent westward motion of a planet on the celestial sphere or with respect to the stars.

revolution The motion of one body around another.

rift zone In geology, a place where the crust is being torn apart by internal forces, generally associated with the injection of new material from the mantle and with the slow separation of tectonic plates.

right ascension A coordinate for measuring the east-west positions of celestial bodies; the angle measured eastward along the celestial equator from the vernal equinox to the hour circle passing through a body.

rotation Turning of a body about an axis running through it.

RR Lyrae variable One of a class of giant pulsating stars with periods less than one day.

runaway greenhouse effect A process whereby the heating of a planet leads to an increase in its atmospheric greenhouse effect and thus to further heating, thereby quickly altering the composition of its atmosphere and the temperature of its surface.

satellite A body that revolves about a planet.

Schwarzschild radius *See* event horizon.

scientific method The procedure scientists follow to understand the natural world: (1) the observation of phenomena or the results of experiments; (2) the formulation of hypotheses that describe these phenomena and that are consistent with the body of knowledge available; (3) the testing of these hypotheses by noting whether or not they adequately predict and describe new phenomena or the results of new experiments; (4) the modification or rejection of hypotheses that are not confirmed by observations or experiment.

sedimentary rock Any rock formed by the deposition and cementing of fine grains of material.

seeing The unsteadiness of the Earth's atmosphere, which blurs telescopic images. Good seeing means the atmosphere is steady.

seismic waves Vibrations traveling through the Earth's interior that result from earthquakes.

seismology (solar) The study of small changes in the radial velocity of the Sun as a whole or of small regions on the surface of the Sun. Analyses of these velocity changes can be used to infer the internal structure of the Sun.

seismology (terrestrial) The study of earthquakes, the conditions that produce them, and the internal structure of the Earth as deduced from analyses of seismic waves.

semimajor axis Half the major axis of a conic section, such as an ellipse.

SETI The search for extraterrestrial intelligence, usually applied to searches for radio signals from other civilizations.

Seyfert galaxy A galaxy belonging to the class of those with active galactic nuclei; one whose nucleus shows bright emission lines; one of a class of galaxies first described by C. Seyfert.

shepherd satellite Informal term for a satellite that is thought to maintain the structure of a planetary ring through its close gravitational influence.

sidereal period The period of revolution of one body about another measured with respect to the stars.

sidereal time Time on Earth measured with respect to the stars, rather than the Sun; the local hour angle of the vernal equinox.

sidereal year Period of the Earth's revolution about the Sun with respect to the stars.

sign (of zodiac) Astrological term for any of 12 equal sections along the ecliptic, each of length 30°. Because of precession, these signs today are no longer lined up with the constellations from which they received their names.

singularity A theoretical point of zero volume and infinite density to which any object that becomes a black hole must collapse, according to the general theory of relativity.

SNC meteorite One of a class of basaltic meteorites now believed by many planetary scientists to be impact-ejected fragments from Mars.

solar activity Phenomena of the solar atmosphere: sunspots, plages, and related phenomena.

solar antapex Direction away from which the Sun is moving with respect to the local standard of rest.

solar apex The direction toward which the Sun is moving with respect to the local standard of rest.

solar eclipse An eclipse of the Sun by the Moon, caused by the passage of the Moon in front of the Sun. Solar eclipses can occur only at the time of new moon.

solar motion Motion of the Sun, or the velocity of the Sun, with respect to the local standard of rest.

solar nebula The cloud of gas and dust from which the solar system formed.

solar seismology The study of pulsations or oscillations of the Sun in order to determine the characteristics of the solar interior.

solar system The system of the Sun and the planets, their satellites, the minor planets, comets, meteoroids, and other objects revolving around the Sun.

solar time A time based on the Sun; usually the hour angle of the Sun plus 12 h.

solar wind A flow of hot charged particles leaving the Sun.

solstice Either of two points on the celestial sphere where the Sun reaches its maximum distances north and south of the celestial equator; time of the year when the daylight is the longest or the shortest.

spacetime A system of one time and three spatial coordinates, with respect to which the time and place of an event can be specified.

space velocity or space motion The velocity of a star with respect to the Sun.

spectral class (or type) The classification of stars according to their temperatures using the characteristics of their spectra; the types are O B A F G K M.

spectral line Radiation at a particular wavelength of light produced by the emission or absorption of energy by an atom.

spectral sequence The sequence of spectral classes of stars arranged in order of decreasing temperatures of stars of those classes.

spectrometer An instrument for obtaining a spectrum; in astronomy, usually attached to a telescope to record the spectrum of a star, galaxy, or other astronomical object.

spectroscopic binary star   A binary star in which the components are not resolved optically, but whose binary nature is indicated by periodic variations in radial velocity, indicating orbital motion.

spectroscopic parallax   A parallax (or distance) of a star that is derived by comparing the apparent magnitude of the star with its absolute magnitude as deduced from its spectral characteristics.

spectroscopy   The study of spectra.

spectrum   The array of colors or wavelengths obtained when light (or other radiation) from a source is dispersed, as in passing it through a prism or grating.

speed   The rate at which an object moves without regard to its direction of motion; the numerical or absolute value of velocity.

spicule   A jet of rising material in the solar chromosphere.

spiral arms   Arms (or long denser regions) of interstellar material and young stars that wind out in a plane from the central nucleus of a spiral galaxy.

spiral density wave   A mechanism for the generation of spiral structure in galaxies; a density wave interacts with interstellar matter and triggers the formation of stars. Spiral density waves are also seen in the rings of Saturn.

spiral galaxy   A flattened, rotating galaxy with pinwheel-like arms of interstellar material and young stars winding out from its nucleus.

spring tide   The highest tidal range of the month, produced when the Moon is near either the full or the new phase.

standard bulb   An astronomical object of known luminosity; such an object can be used to determine distances.

star   A sphere of gas shining under its own power.

star cluster   An assemblage of stars held together by their mutual gravity.

Stefan-Boltzmann law   A formula from which the rate at which a blackbody radiates energy can be computed; the total rate of energy emission from a unit area of a blackbody is proportional to the fourth power of its absolute temperature.

stellar evolution   The changes that take place in the characteristics of stars as they age.

stellar model   The result of a theoretical calculation of the physical conditions in the different layers of a star's interior.

stellar parallax   *See* parallax.

stellar wind   The outflow of gas, sometimes at speeds as high as hundreds of kilometers per second, from a star.

stony-iron meteorite   A type of meteorite that is a blend of nickel-iron and silicate materials.

stony meteorite   A meteorite composed mostly of stony material.

stratosphere   The layer of the Earth's atmosphere above the troposphere (where most weather takes place) and below the ionosphere.

strong nuclear force or strong interaction   The force that binds together the parts of the atomic nucleus.

subduction zone   In terrestrial geology, a region where one crustal plate is forced under another, generally associated with earthquakes, volcanic activity, and the formation of deep ocean trenches.

summer solstice   The point on the celestial sphere where the Sun reaches its greatest distance north of the celestial equator; the day with the longest amount of daylight.

Sun   The star about which the Earth and other planets revolve.

sunspot   A temporary cool region in the solar photosphere that appears dark by contrast against the surrounding hotter photosphere.

sunspot cycle   The semiregular 11-year period with which the frequency of sunspots fluctuates.

supercluster   A large region of space (more than 100 million LY across) where groups and clusters of galaxies are more concentrated; a cluster of clusters of galaxies.

supergiant   A star of very high luminosity and relatively low temperature.

supernova   An explosion that marks the final stage of evolution of a star. A Type I supernova occurs when a white dwarf accretes enough matter to exceed the Chandrasekhar limit, collapses, and explodes. A Type II supernova marks the final collapse of a massive star.

surface gravity   The weight of a unit mass at the surface of a body.

synchrotron radiation   The radiation emitted by charged particles being accelerated in magnetic fields and moving at speeds near that of light.

tail (of a comet)   *See* dust tail of a comet *and* plasma.

tangential (transverse) velocity   The component of a star's space velocity that lies in the plane of the sky.

tectonic   Geological features that result from stresses and pressures in the crust of a planet. Tectonic forces can lead to earthquakes and motion of the crust.

temperature   A measure of how fast the particles in a body are moving or vibrating in place; a measure of the average heat energy in a body.

temperature (Celsius; formerly centigrade)   Temperature measured on scale where water freezes at 0° and boils at 100°.

temperature (color)   The temperature of a star as estimated from the intensity of the stellar radiation at two or more colors or wavelengths.

temperature (effective)   The temperature of a blackbody that would radiate the same total amount of energy that a particular object, such as a star, does.

temperature (excitation)   The temperature of a star as estimated from the relative strengths of lines in its spectrum that originate from atoms in different stages of excitation.

temperature (Fahrenheit)   Temperature measured on a scale where water freezes at 32° and boils at 212°.

temperature (ionization)   The temperature of a star as estimated from the relative strengths of lines in its spectrum that originate from atoms in different stages of ionization.

temperature (Kelvin)   Absolute temperature measured in Celsius degrees, with the zero point at absolute zero.

temperature (radiation)   The temperature of a blackbody that radiates the same amount of energy in a given spectral region as does a particular body.

terrestrial planet   Any of the planets Mercury, Venus, Earth, or Mars; sometimes the Moon or Pluto are included in the list.

theory   A set of hypotheses and laws that have been well demonstrated to apply to a wide range of phenomena associated with a particular subject.

thermal energy   Energy associated with the motions of the molecules or atoms in a substance.

thermal equilibrium A balance between the input and out-flow of heat in a system.

thermal radiation The radiation emitted by any body or gas that is not at absolute zero.

thermonuclear energy Energy associated with thermonuclear reactions or that can be released through thermonuclear reactions.

thermonuclear reaction A nuclear reaction or transformation that results from encounters between particles that are given high velocities (by heating them).

tidal force A differential gravitational force that tends to deform a body.

tidal stability limit The distance—approximately 2.5 planetary radii from the center—within which differential gravitational forces (or tides) are stronger than the mutual gravitational attraction between two adjacent orbiting objects. Within this limit, fragments are not likely to accrete or assemble themselves into a larger object. Also called the Roche limit.

tide Deformation of a body by the differential gravitational force exerted on it by another body; in the Earth, the deformation of the ocean surface by the differential gravitational forces exerted by the Moon and Sun.

transition region The region in the Sun's atmosphere where the temperature rises very rapidly from the relatively low temperatures that characterize the chromosphere to the high temperatures of the corona.

triple-alpha process A series of two nuclear reactions by which three helium nuclei are built up into one carbon nucleus.

tropical year Period of revolution of the Earth about the Sun with respect to the vernal equinox.

Tropic of Cancer The parallel (circle) of latitude 23.5° N.

Tropic of Capricorn The parallel (circle) of latitude 23.5° S.

troposphere Lowest level of the Earth's atmosphere, where most weather takes place.

turbulence Random motions of gas masses, as in the atmosphere of a star.

21-cm line A line in the spectrum of neutral hydrogen at the radio wavelength of 21 cm.

ultraviolet radiation Electromagnetic radiation of wavelengths shorter than the shortest visible wavelengths; radiation of wavelengths in the approximate range 10 to 400 nm.

umbra The central, completely dark part of a shadow.

uncertainty principle Heisenberg uncertainty principle. It is fundamentally impossible to make simultaneous measurements of a particle's position and velocity with infinite accuracy.

universe The totality of all matter, radiation and space; everything accessible to our observations.

variable star A star that varies in luminosity.

velocity The speed and the direction a body is moving; e.g., 44 km/s toward the north galactic pole.

velocity of escape The speed with which an object must move in order to enter a parabolic orbit about another body (such as the Earth), and hence move permanently away from the vicinity of that body.

vernal equinox The point on the celestial sphere where the Sun crosses the celestial equator passing from south to north; a time in the course of the year when the day and night are roughly equal.

very-long-baseline interferometry (VLBI) A technique of radio astronomy whereby signals from telescopes thousands of kilometers apart combined to obtain very high resolution by letting waves from different sites interfere with each other.

visual binary star A binary star in which the two components are telescopically resolved.

void A region between clusters and superclusters of galaxies that appears relatively empty of galaxies.

volatile materials Materials that are gaseous at fairly low temperatures. This is a relative term, usually applied to the gases in planetary atmospheres and to common ices (H_2O, CO_2, and so on), but it is also sometimes used for elements such as cadmium, zinc, lead, and rubidium that form gases at temperatures up to 1000 K. (These are called volatile elements, as opposed to refractory elements.)

volume A measure of the total space occupied by a body.

watt A unit of power (energy per unit time).

wavelength The spacing of the crests or troughs in a wave.

weak nuclear force or weak interaction The nuclear force involved in radioactive decay. The weak force is characterized by the slow rate of certain nuclear reactions—such as the decay of the neutron, which occurs with a half-life of 11 min.

weight A measure of the force due to gravitational attraction.

white dwarf A star that has exhausted most or all of its nuclear fuel and has collapsed to a very small size; such a star is near its final stage of life.

Wien's law Formula that relates the temperature of a blackbody to the wavelength at which it emits the greatest intensity of radiation.

winter solstice Point on the celestial sphere where the Sun reaches its greatest distance south of the celestial equator; the time of the year with the shortest amount of daylight.

Wolf-Rayet star One of a class of very hot stars that eject shells of gas at very high velocity.

x rays Photons of wavelengths intermediate between those of ultraviolet radiation and gamma rays.

x-ray stars Stars (other than the Sun) that emit observable amounts of radiation at x-ray frequencies.

year The period of revolution of the Earth around the Sun.

Zeeman effect A splitting or broadening of spectral lines due to magnetic fields.

zenith The point on the celestial sphere opposite to the direction of gravity; or the direction opposite to that indicated by a plumb bob; the point directly above the observer.

zero-age main sequence Main sequence on the H–R diagram for a system of stars that have completed their contraction from interstellar matter and are now deriving all their energy from nuclear reactions, but whose chemical composition has not yet been altered by nuclear reactions.

zodiac A belt around the sky 18° wide centered on the ecliptic.

zone of avoidance A region near the Milky Way where obscuration by interstellar dust is so heavy that few or no exterior galaxies can be seen.

APPENDIX 4

Powers-of-Ten Notation

In astronomy (and other sciences), it is often necessary to deal with very large or very small numbers. In fact, when numbers become truly large in everyday life, such as the national debt in the U.S., we call them astronomical. Among the ideas astronomers must routinely deal with is that the Earth is 150,000,000,000 m from the Sun, and the mass of the hydrogen atom is 0.00000000000000000000000000167 kg. No one in his or her right mind would want to continue writing so many zeros!

Instead, scientists have agreed on a kind of shorthand notation, which is not only easier to write, but (as we shall see) makes multiplication and division of large and small numbers much less difficult. If you have never used this powers-of-ten notation or scientific notation, it may take a bit of time to get used to it, but you will soon find it much easier than keeping all those zeros.

Writing Large Numbers

The convention in this notation is that we generally have only one number to the left of the decimal point. If a number is not in this format, it must be changed. The number 6 is already in the right format, because for integers, we understand there to be a decimal point to the right of them. So 6 is really 6., and there is indeed only one number to the left of the decimal point. But the number 165 (which is 165.) has three numbers to the left of the decimal point, and is thus ripe for conversion.

To change 165 to proper form, we must make it 1.65 and then keep track of the change we have made. (Think of the number as a weekly salary, and suddenly it makes a lot of difference whether we have $165 or $1.65.) We keep track of the number of places we moved the decimal point by expressing it as a power of ten. So 165 becomes 1.65×10^2 or 1.65 multiplied by ten to the second power. The small raised 2 is called an exponent, and it tells us how many times we moved the decimal point to the left.

Note that 10^2 also designates 10 squared or 10×10, which equals 100. And 1.65×100 is just 165, the number we started with. Another way to look at scientific notation is that we separate out the messy numbers out front, and leave the smooth units of ten for the exponent to denote. So a number like 1,372,568 becomes 1.372568 times a million (10^6) or 1.372568 times 10 multiplied by itself 6 times. We had to move the decimal point six places to the left (from its place after the 8) to get the number into the form where there is only one digit to the left of the decimal point.

The reason we call this powers-of-ten notation is that our counting system is based on increases of ten; each place in our numbering system is ten times greater than the place to the right of it. As you have probably learned, this got started because human beings have ten fingers and we started counting with them. It is interesting to speculate that if we ever meet intelligent life-forms with only eight fingers, their counting system would probably be a powers-of-eight notation!

So, in the example we started with, the number of meters from the Earth to the Sun is 1.5×10^{11}. Elsewhere in the book, we mention that a string a light year long would fit around the Earth's equator 236 million or 236,000,000 times. In scientific notation, this would become 2.36×10^8. Now if you like expressing things in millions, as the annual reports of successful companies do, you might like to write this number as 236×10^6. However, the usual convention is to have only one number to the left of the decimal point.

Writing Small Numbers

Now take a number like 0.00347, which is also not in the standard (agreed-to) form for scientific notation. To put it into that format, we must make the first part of it 3.47 by moving the decimal point three places to the right. Note that this motion to the right is the opposite of the motion to the left that we discussed above. To keep track, we call this change negative and put a minus sign in the exponent. Thus 0.00347 becomes 3.47×10^{-3}.

In the example we gave at the beginning, the mass of the hydrogen atom would then be written as 1.67×10^{-27} kg. In this system, one is written as 10^0, a tenth as 10^{-1}, a hundredth as 10^{-2}, and so forth. Note that any number, no matter how large or how small, can be expressed in scientific notation.

Multiplication and Division

The powers-of-ten notation is not only compact and convenient, it also simplifies arithmetic. To multiply two numbers expressed as powers of ten, you need only multiply the numbers out front and then add the exponents. If there are no numbers out front, as in $100 \times 100,000$, then you just add the exponents (in our notation, $10^2 \times 10^5 = 10^7$). When there are numbers out of front, you do have to multiply them, but they are much easier to deal with than numbers with many zeros in them.

Some examples:

$$3 \times 10^5 \times 2 \times 10^9 = 6 \times 10^{14}$$
$$0.04 \times 6,000,000 = 4 \times 10^{-2} \times 6 \times 10^6$$
$$= 24 \times 10^4 = 2.4 \times 10^5$$

Note in the second example that when we added the exponents, we treated negative exponents as we do in regular arithmetic (-2 plus 6 equals 4). Also, notice that the first result we got had a 24 in it, which was not in the acceptable form, having two places to the left of the decimal point, and we therefore changed it to 2.4.

To divide, you divide the numbers out front and subtract the exponents. Here are several examples:

$$1,000,000 \div 1000 = 10^6 \div 10^3 = 10^{6-3} = 10^3$$
$$9 \times 10^{12} \div 2 \times 10^3 = 4.5 \times 10^9$$
$$2.8 \times 10^2 \div 6.2 \times 10^5 = 0.452 \times 10^{-3} = 4.52 \times 10^{-4}$$

If this is the first time that you have met scientific notation, we urge you to practice using it (you might start by solving the exercises below). Like any new language, the notation looks complicated at first, but gets easier as you practice it.

Exercises

1. On April 8, 1996, the Galileo spacecraft was 775 million kilometers from Earth. Convert this number to scientific notation. How many astronomical units is this? (An astronomical unit is the distance from the Earth to the Sun; see above, but remember to keep your units consistent!)
2. During the first six years of its operation, the Hubble Space Telescope circled the Earth 37,000 times, for a total of 1,280,000,000 km. Use scientific notation to find the number of kilometers in one orbit.
3. In a college cafeteria, a soybean-vegetable burger is offered as an alternative to regular hamburgers. If 489,875 burgers were eaten during the course of a school year, and 997 of them were veggie-burgers, what fraction of the burgers does this represent?
4. In a June 1990 Gallup poll, 27 percent of adult Americans thought that alien beings have actually landed on Earth. The number of adults in the U.S. in 1990 was (according to the census) about 186,000,000. Use scientific notation to determine how many adults believe aliens have visited the Earth.
5. In 1995, 1.7 million degrees were awarded by colleges and universities in the U.S. Among these were 41,000 PhD degrees. What fraction of the degrees were PhD's? Express this number as a percent. (Now find a job for all those PhD's!)
6. A star 60 lightyears away has been found to have a large planet orbiting it. Your uncle wants to know the distance to this planet in old-fashioned miles. If light travels 186,000 miles per second, and there are 60 seconds in a minute, 60 minutes in an hour, 24 hours in a day, and 365 days in a year, how many miles away is that star?

APPENDIX 5

Units Used in Science

In the American system of measurement (originally developed in England), the fundamental units of length, weight, and time are the foot, pound, and second, respectively. There are also larger and smaller units, which include the ton (2240 lb), the mile (5280 ft), the rod (16½ ft), the yard (3 ft), the inch (1/12 ft), the ounce (1/16 lb), and so on. Such units, whose origins in decisions by British royalty have been forgotten by most people, are quite inconvenient for conversion and arithmetic computation.

In science, therefore, it is more usual to use the metric system, which has been adopted in virtually all countries except the United States. Its great advantage is that every unit increases by a factor of ten, instead of the strange factors in the American system. The fundamental units of the metric system are

length: 1 meter (m)

mass: 1 kilogram (kg)

time: 1 second (s)

A meter was originally intended to be 1 ten-millionth of the distance from the equator to the North Pole along the surface of the Earth. It is about 1.1 yd. A kilogram is the mass that on Earth results in a weight of about 2.2 lb. The second is the same in metric and American units.

The most commonly used quantities of length and mass of the metric system are the following:

Length

1 kilometer (km) = 1000 meters = 0.6214 mile

1 meter (m) = 0.001 km = 1.094 yards = 39.37 inches

1 centimeter (cm) = 0.01 meter = 0.3937 inch

1 millimeter (mm) = 0.001 meter = 0.1 cm

1 micrometer (μm) = 0.000001 meter = 0.0001 cm

1 nanometer (nm) = 10^{-9} meter = 10^{-7} cm

To convert from the American system, here are a few helpful factors:

1 mile = 1.6093 km

1 inch = 2.5400 cm

Mass

Although we don't make the distinction very carefully in everyday life on Earth, strictly speaking the kilogram is a unit of mass (measuring how many atoms a body has) and the pound is a unit of weight (measuring how strongly the Earth's gravity pulls on a body).

1 metric ton = 10^6 grams = 1000 kg (and it produces a weight of 2.2046×10^3 lbs on Earth)

1 kg = 1000 grams (and it produces a weight of 2.2046 lbs on Earth)

1 gram (g) = 0.0353 oz (and the equivalent weight is 0.0022046 lb)

1 milligram (mg) = 0.001 g

And a weight of 1 lb is equivalent on Earth to a mass of 0.4536 kg, while a weight of 1 oz is produced by a mass of 28.3495 g.

Temperature

Three temperature scales are in general use:

1. Fahrenheit (F); water freezes at 32°F and boils at 212°F.
2. Celsius or centigrade° (C); water freezes at 0°C and boils at 100°C.
3. Kelvin or absolute (K); water freezes at 273 K and boils at 373 K.

All molecular motion ceases at −459°F = −273°C = 0 K, a temperature called *absolute zero*. Kelvin temperature is measured from this lowest possible temperature. It is the temperature scale most often used in astronomy. Kelvins have the same value as centigrade or Celsius degrees, since the difference between the freezing and boiling points of water is 100 degrees in each.

On the Fahrenheit scale, the difference between the freezing and boiling points of water is 180 degrees. Thus, to convert Celsius degrees or Kelvins to Fahrenheit, it is necessary to multiply by 180/100 = 9/5. To convert from Fahrenheit to Celsius degrees or Kelvins, it is necessary to multiply by 100/180 = 5/9.

The full conversion formulas are:

$$K = °C + 273$$
$$°C = 0.555 \times (°F - 32)$$
$$°F = (1.8 \times °C) + 32$$

°Celsius is now the name used for centigrade temperature; it has a more modern standardization but differs from the old centigrade scale by less than 0.1°.

APPENDIX 6

Some Useful Constants for Astronomy

Physical Constants

speed of light (c) = 2.9979×10^8 m/s

gravitational constant (G) = 6.672×10^{-11} N m^2/kg^2

Planck's constant (h) = 6.626×10^{-34} joule·s

mass of a hydrogen atom (m_H) = 1.673×10^{-27} kg

mass of an electron (m_e) = 9.109×10^{-31} kg

Rydberg constant (R) = 1.0974×10^7 per m

Stefan-Boltzmann constant (σ) =
    5.670×10^{-8} joule/(s·m^2·deg^4)

constant in Wien's law ($\lambda_{max}T$) = 2.898×10^{-3} m·deg

electron volt (energy) (eV) = 1.602×10^{-19} joules

energy equivalent of 1 ton TNT = 4.3×10^9 joules

Astronomical Constants

astronomical unit (AU) = 1.496×10^{11} m

light year (LY) = 9.461×10^{15} m

parsec (pc) = 3.086×10^{16} m = 3.262 LY

sidereal year (yr) = 3.158×10^7 s

mass of Earth (M_E) = 5.977×10^{24} kg

equatorial radius of Earth (R_E) = 6.378×10^6 m

obliquity of ecliptic (ϵ) = 23° 27′

surface gravity of Earth (g) = 9.807 m/s^2

escape velocity of Earth (v_E) = 1.119×10^4 m/s

mass of Sun (M_{Sun}) = 1.989×10^{30} kg

equatorial radius of Sun (R_{Sun}) = 6.960×10^8 m

luminosity of Sun (L_{Sun}) = 3.83×10^{26} watts

solar constant (flux of energy received at Earth)
    S = 1.37×10^3 watts/m^2

Hubble constant (H_0) =
    approximately 20 km/sec per million LY

APPENDIX 7

Physical Data for the Planets

Planet	Diameter (km)	Diameter (Earth = 1)	Mass (Earth = 1)	Mean Density (g/cm³)	Rotation Period (days)	Inclination of Equator to Orbit (°)	Surface Gravity (Earth = 1)	Velocity of Escape (km/s)
Mercury	4,878	0.38	0.055	5.43	58.6	0.0	0.38	4.3
Venus	12,104	0.95	0.82	5.24	−243.0	177.4	0.91	10.4
Earth	12,756	1.00	1.00	5.52	0.997	23.4	1.00	11.2
Mars	6,794	0.53	0.107	3.9	1.026	25.2	0.38	5.0
Jupiter	142,800	11.2	317.8	1.3	0.41	3.1	2.53	60
Saturn	120,540	9.41	94.3	0.7	0.43	26.7	1.07	36
Uranus	51,200	4.01	14.6	1.2	−0.72	97.9	0.92	21
Neptune	49,500	3.88	17.2	1.6	0.67	29	1.18	24
Pluto	2,200	0.17	0.0025	2.0	−6.387	118	0.09	1

Orbital Data for the Planets

Planet	Semimajor Axis AU	Semimajor Axis 10⁶ km	Sidereal Period Tropical Years	Sidereal Period Days	Mean Orbital Speed (km/s)	Orbital Eccentricity	Inclination of Orbit to Ecliptic (°)
Mercury	0.3871	57.9	0.24085	87.97	47.9	0.206	7.004
Venus	0.7233	108.2	0.61521	224.70	35.0	0.007	3.394
Earth	1.0000	149.6	1.000039	365.26	29.8	0.017	0.0
Mars	1.5237	227.9	1.88089	686.98	24.1	0.093	1.850
(Ceres)	2.7671	414	4.603		17.9	0.077	10.6
Jupiter	5.2028	778	11.86		13.1	0.048	1.308
Saturn	9.538	1427	29.46		9.6	0.056	2.488
Uranus	19.191	2871	84.07		6.8	0.046	0.774
Neptune	30.061	4497	164.82		5.4	0.010	1.774
Pluto	39.529	5913	248.6		4.7	0.248	17.15

Adapted from *The Astronomical Almanac* (U.S. Naval Observatory)

APPENDIX 8

Satellites of the Planets

Planet	Satellite Name	Discovery	Semimajor Axis (km × 1000)	Period (days)	Diameter (km)	Mass (10²⁰ kg)	Density (g/cm³)
Earth	Moon	—	384	27.32	3476	735	3.3
Mars	Phobos	Hall (1877)	9.4	0.32	23	1×10^{-4}	2.0
	Deimos	Hall (1877)	23.5	1.26	13	2×10^{-5}	1.7
Jupiter	Metis	Voyager (1979)	128	0.29	20	—	—
	Adrastea	Voyager (1979)	129	0.30	40	—	—
	Amalthea	Barnard (1892)	181	0.50	200	—	—
	Thebe	Voyager (1979)	222	0.67	90	—	—
	Io	Galileo (1610)	422	1.77	3630	894	3.6
	Europa	Galileo (1610)	671	3.55	3138	480	3.0
	Ganymede	Galileo (1610)	1,070	7.16	5262	1482	1.9
	Callisto	Galileo (1610)	1,883	16.69	4800	1077	1.9
	Leda	Kowal (1974)	11,090	239	15	—	—
	Himalia	Perrine (1904)	11,480	251	180	—	—
	Lysithea	Nicholson (1938)	11,720	259	40	—	—
	Elara	Perrine (1905)	11,740	260	80	—	—
	Ananke	Nicholson (1951)	21,200	631 (R)	30	—	—
	Carme	Nicholson (1938)	22,600	692 (R)	40	—	—
	Pasiphae	Melotte (1908)	23,500	735 (R)	40	—	—
	Sinope	Nicholson (1914)	23,700	758 (R)	40	—	—
Saturn	Unnamed	Voyager (1985)	118.2	0.48	15?	3×10^{-5}	—
	Pan	Voyager (1985)	133.6	0.58	20	3×10^{-5}	—
	Atlas	Voyager (1980)	137.7	0.60	40	—	—
	Prometheus	Voyager (1980)	139.4	0.61	80	—	—
	Pandora	Voyager (1980)	141.7	0.63	100	—	—
	Janus	Dollfus (1966)	151.4	0.69	190	—	—
	Epimetheus	Fountain, Larson (1980)	151.4	0.69	120	—	—
	Mimas	Herschel (1789)	186	0.94	394	0.4	1.2
	Enceladus	Herschel (1789)	238	1.37	502	0.8	1.2
	Tethys	Cassini (1684)	295	1.89	1048	7.5	1.3
	Telesto	Reitsema et al. (1980)	295	1.89	25	—	—
	Calypso	Pascu et al. (1980)	295	1.89	25	—	—
	Dione	Cassini (1684)	377	2.74	1120	11	1.4
	Helene	Lecacheux, Laques (1980)	377	2.74	30	—	—
	Rhea	Cassini (1672)	527	4.52	1530	25	1.3
	Titan	Huygens (1655)	1,222	15.95	5150	1346	1.9
	Hyperion	Bond, Lassell (1848)	1,481	21.3	270	—	—
	Iapetus	Cassini (1671)	3,561	79.3	1435	19	1.2
	Phoebe	Pickering (1898)	12,950	550 (R)	220	—	—
Uranus	Cordelia	Voyager (1986)	49.8	0.34	40?	—	—
	Ophelia	Voyager (1986)	53.8	0.38	50?	—	—
	Bianca	Voyager (1986)	59.2	0.44	50?	—	—
	Cressida	Voyager (1986)	61.8	0.46	60?	—	—
	Desdemona	Voyager (1986)	62.7	0.48	60?	—	—
	Juliet	Voyager (1986)	64.4	0.50	80?	—	—
	Portia	Voyager (1986)	66.1	0.51	80?	—	—
	Rosalind	Voyager (1986)	69.9	0.56	60?	—	—

(Table continues)

Planet	Satellite Name	Discovery	Semimajor Axis (km × 1000)	Period (days)	Diameter (km)	Mass (10^{20} kg)	Density (g/cm^3)
	Belinda	Voyager (1986)	75.3	0.63	60?	—	—
	Unnamed	Voyager (1986)	73.5	0.63	40?	—	—
	Puck	Voyager (1985)	86.0	0.76	170	—	—
	Miranda	Kuiper (1948)	130	1.41	485	0.8	1.3
	Ariel	Lassell (1851)	191	2.52	1160	13	1.6
	Umbriel	Lassell (1851)	266	4.14	1190	13	1.4
	Titania	Herschel (1787)	436	8.71	1610	35	1.6
	Oberon	Herschel (1787)	583	13.5	1550	29	1.5
	Caliban	Gladman, Nicholson,	7,000	562	60?	—	—
	Sycorax	et. al. (1997)	12,000	1,261	120?	—	—
Neptune	Naiad	Voyager (1989)	48	0.30	50	—	—
	Thalassa	Voyager (1989)	50	0.31	90	—	—
	Despina	Voyager (1989)	53	0.33	150	—	—
	Galatea	Voyager (1989)	62	0.40	150	—	—
	Larissa	Voyager (1989)	74	0.55	200	—	—
	Proteus	Voyager (1989)	118	1.12	400	—	—
	Triton	Lassell (1846)	355	5.88 (R)	2720	220	2.1
	Nereid	Kuiper (1949)	5,511	360	340	—	—
Pluto	Charon	Christy (1978)	19.7	6.39	1200	—	—

APPENDIX 9

Upcoming (Total) Eclipses

1. Total Eclipses of the Sun

Date	Duration of Totality (min)	Where Visible
2001 June 21	4.9	Southern Africa
2002 Dec. 4	2.1	South Africa, Australia
2003 Nov. 23	2.0	Antarctica
2005 April 8	0.7	South Pacific Ocean
2006 March 29	4.1	Africa, Asia Minor, U.S.S.R.
2008 Aug. 1	2.4	Arctic Ocean, Siberia, China
2009 July 22	6.6	India, China, South Pacific
2010 July 11	5.3	South Pacific Ocean
2012 Nov. 13	4.0	Northern Australia, South Pacific
2013 Nov. 3	1.7	Atlantic Ocean, Central Africa
2015 March 20	4.1	North Atlantic, Arctic Ocean
2016 March 9	4.5	Indonesia, Pacific Ocean
2017 Aug. 21	2.7	Pacific Ocean, U.S.A., Atlantic Ocean

Date	Duration of Totality (min)	Where Visible
2019 July 2	4.5	South Pacific, South America
2020 Dec. 14	2.2	South Pacific, South America, South Atlantic Ocean
2021 Dec. 4	1.9	Antarctica
2023 April 20	1.3	Indian Ocean, Indonesia
2024 April 8	4.5	South Pacific, Mexico, East U.S.A.
2026 Aug. 12	2.3	Arctic, Greenland, North Atlantic, Spain
2027 Aug. 2	6.4	North Africa, Arabia, Indian Ocean
2028 July 22	5.1	Indian Ocean, Australia, New Zealand
2030 Nov. 25	3.7	South Africa, Indian Ocean, Australia

2. Some Total Lunar Eclipses

2000 Jan. 21	2003 Nov. 9	2007 March 3
2000 July 16	2004 May 4	2007 Aug. 28
2001 Jan. 9	2004 Oct. 28	2008 Feb. 21
2003 May 16		

APPENDIX 10

The Chemical Elements

Element	Symbol	Atomic Number	Atomic Weight* (Chemical Scale)	Number of Atoms per 10^{12} Hydrogen Atoms
Hydrogen	H	1	1.0080	1×10^{12}
Helium	He	2	4.003	8×10^{10}
Lithium	Li	3	6.940	2×10^{3}
Beryllium	Be	4	9.013	3×10^{1}
Boron	B	5	10.82	9×10^{2}
Carbon	C	6	12.011	4.5×10^{8}
Nitrogen	N	7	14.008	9.2×10^{7}
Oxygen	O	8	16.00	7.4×10^{8}
Fluorine	F	9	19.00	3.1×10^{4}
Neon	Ne	10	20.183	1.3×10^{8}
Sodium	Na	11	22.991	2.1×10^{6}
Magnesium	Mg	12	24.32	4.0×10^{7}
Aluminum	Al	13	26.98	3.1×10^{6}
Silicon	Si	14	28.09	3.7×10^{7}
Phosphorus	P	15	30.975	3.8×10^{5}
Sulfur	S	16	32.066	1.9×10^{7}
Chlorine	Cl	17	35.457	1.9×10^{5}
Argon	Ar(A)	18	39.944	3.8×10^{6}
Potassium	K	19	39.100	1.4×10^{5}
Calcium	Ca	20	40.08	2.2×10^{6}
Scandium	Sc	21	44.96	1.3×10^{3}
Titanium	Ti	22	47.90	8.9×10^{4}
Vanadium	V	23	50.95	1.0×10^{4}
Chromium	Cr	24	52.01	5.1×10^{5}
Manganese	Mn	25	54.94	3.5×10^{5}
Iron	Fe	26	55.85	3.2×10^{7}
Cobalt	Co	27	58.94	8.3×10^{4}
Nickel	Ni	28	58.71	1.9×10^{6}
Copper	Cu	29	63.54	1.9×10^{4}
Zinc	Zn	30	65.38	4.7×10^{4}
Gallium	Ga	31	69.72	1.4×10^{3}
Germanium	Ge	32	72.60	4.4×10^{3}
Arsenic	As	33	74.91	2.5×10^{2}
Selenium	Se	34	78.96	2.3×10^{3}
Bromine	Br	35	79.916	4.4×10^{2}
Krypton	Kr	36	83.80	1.7×10^{3}
Rubidium	Rb	37	85.48	2.6×10^{2}
Strontium	Sr	38	87.63	8.8×10^{2}
Yttrium	Y	39	88.92	2.5×10^{2}
Zirconium	Zr	40	91.22	4.0×10^{2}
Niobium (Columbium)	Nb(Cb)	41	92.91	2.6×10^{1}
Molybdenum	Mo	42	95.95	9.3×10^{1}
Technetium	Tc(Ma)	43	(99)	—
Ruthenium	Ru	44	101.1	68
Rhodium	Rh	45	102.91	13
Palladium	Pd	46	106.4	51
Silver	Ag	47	107.880	20
Cadmium	Cd	48	112.41	63
Indium	In	49	114.82	7
Tin	Sn	50	118.70	1.4×10^{2}
Antimony	Sb	51	121.76	13
Tellurium	Te	52	127.61	1.8×10^{2}

* Where mean atomic weights have not been well determined, the atomic mass numbers of the most stable isotopes are given in parentheses.

(Table continues)

Element	Symbol	Atomic Number	Atomic Weight* (Chemical Scale)	Number of Atoms per 10^{12} Hydrogen Atoms
Iodine	I(J)	53	126.91	33
Xenon	Xe(X)	54	131.30	1.6×10^2
Cesium	Cs	55	132.91	14
Barium	Ba	56	137.36	1.6×10^2
Lanthanum	La	57	138.92	17
Cerium	Ce	58	140.13	43
Praseodymium	Pr	59	140.92	6
Neodymium	Nd	60	144.27	31
Promethium	Pm	61	(147)	—
Samarium	Sm(Sa)	62	150.35	10
Europium	Eu	63	152.00	4
Gadolinium	Gd	64	157.26	13
Terbium	Tb	65	158.93	2
Dysprosium	Dy(Ds)	66	162.51	15
Holmium	Ho	67	164.94	3
Erbium	Er	68	167.27	9
Thulium	Tm(Tu)	69	168.94	2
Ytterbium	Yb	70	173.04	8
Lutecium	Lu(Cp)	71	174.99	2
Hafnium	Hf	72	178.50	6
Tantalum	Ta	73	180.95	1
Tungsten	W	74	183.86	5
Rhenium	Re	75	186.22	2
Osmium	Os	76	190.2	27
Iridium	Ir	77	192.2	24
Platinum	Pt	78	195.09	56
Gold	Au	79	197.00	6
Mercury	Hg	80	200.61	19
Thallium	Tl	81	204.39	8
Lead	Pb	82	207.21	1.2×10^2
Bismuth	Bi	83	209.00	5
Polonium	Po	84	(209)	—
Astatine	At	85	(210)	—
Radon	Rn	86	(222)	—
Francium	Fr(Fa)	87	(223)	—
Radium	Ra	88	226.05	—
Actinium	Ac	89	(227)	—
Thorium	Th	90	232.12	1
Protactinium	Pa	91	(231)	—
Uranium	U(Ur)	92	238.07	1
Neptunium	Np	93	(237)	—
Plutonium	Pu	94	(244)	—
Americium	Am	95	(243)	—
Curium	Cm	96	(248)	—
Berkelium	Bk	97	(247)	—
Californium	Cf	98	(251)	—
Einsteinium	E	99	(254)	—
Fermium	Fm	100	(253)	—
Mendeleevium	Mv	101	(256)	—
Nobelium	No	102	(253)	—
Lawrencium	Lr	103	(262)	—
Rutherfordium	Rf	104	(261)	—
Dubnium	Db	105	(262)	—
Seaborgium	Sg	106	(263)	—
Bohrium	Bh	107	(262)	—
Hassium	Hs	108	(264)	—
Meitnerium	Mt	109	(266)	—
Ununnilium	Uun	110	(269)	—
Unununium	Uuu	111	(272)	—
Ununbium	Uub	112	(277)	—
Ununquadium	Uuq	114	(285)	—
Ununhexium	Uuh	116	(289)	—
Ununoctium	Uuo	118	(293)	—

* Where mean atomic weights have not been well determined, the atomic mass numbers of the most stable isotopes are given in parentheses.

APPENDIX 11

The Constellations

Constellation (Latin name)	Genitive Case Ending	English Name or Description	Abbreviation	Approximate Position α h	δ °
Andromeda	Andromedae	Princess of Ethiopia	And	1	+40
Antila	Antilae	Air pump	Ant	10	−35
Apus	Apodis	Bird of Paradise	Aps	16	−75
Aquarius	Aquarii	Water bearer	Aqr	23	−15
Aquila	Aquilae	Eagle	Aql	20	+5
Ara	Arae	Altar	Ara	17	−55
Aries	Arietis	Ram	Ari	3	+20
Auriga	Aurigae	Charioteer	Aur	6	+40
Boötes	Boötis	Herdsman	Boo	15	+30
Caelum	Caeli	Graving tool	Cae	5	−40
Camelopardus	Camelopardis	Giraffe	Cam	6	+70
Cancer	Cancri	Crab	Cnc	9	+20
Canes Venatici	Canum Venaticorum	Hunting dogs	CVn	13	+40
Canis Major	Canis Majoris	Big dog	CMa	7	−20
Canis Minor	Canis Minoris	Little dog	CMi	8	+5
Capricornus	Capricorni	Sea goat	Cap	21	−20
Carina*	Carinae	Keel of Argonauts' ship	Car	9	−60
Cassiopeia	Cassiopeiae	Queen of Ethiopia	Cas	1	+60
Centaurus	Centauri	Centaur	Cen	13	−50
Cepheus	Cephei	King of Ethiopia	Cep	22	+70
Cetus	Ceti	Sea monster (whale)	Cet	2	−10
Chamaeleon	Chamaeleontis	Chameleon	Cha	11	−80
Circinus	Circini	Compasses	Cir	15	−60
Columba	Columbae	Dove	Col	6	−35
Coma Berenices	Comae Berenices	Berenice's hair	Com	13	+20
Corona Australis	Coronae Australis	Southern crown	CrA	19	−40
Corona Borealis	Coronae Borealis	Northern crown	CrB	16	+30
Corvus	Corvi	Crow	Crv	12	−20
Crater	Crateris	Cup	Crt	11	−15
Crux	Crucis	Cross (southern)	Cru	12	−60
Cygnus	Cygni	Swan	Cyg	21	+40
Delphinus	Delphini	Porpoise	Del	21	+10
Dorado	Doradus	Swordfish	Dor	5	−65
Draco	Draconis	Dragon	Dra	17	+65
Equuleus	Equulei	Little horse	Equ	21	+10
Eridanus	Eridani	River	Eri	3	−20
Fornax	Fornacis	Furnace	For	3	−30
Gemini	Geminorum	Twins	Gem	7	+20
Grus	Gruis	Crane	Gru	22	−45
Hercules	Herculis	Hercules, son of Zeus	Her	17	+30
Horologium	Horologii	Clock	Hor	3	−60
Hydra	Hydrae	Sea serpent	Hya	10	−20
Hydrus	Hydri	Water snake	Hyi	2	−75

(Table continues)

Constellation (Latin name)	Genitive Case Ending	English Name or Description	Abbre-viation	α h	δ °
Indus	Indi	Indian	Ind	21	−55
Lacerta	Lacertae	Lizard	Lac	22	+45
Leo	Leonis	Lion	Leo	11	+15
Leo Minor	Leonis Minoris	Little lion	LMi	10	+35
Lepus	Leporis	Hare	Lep	6	−20
Libra	Librae	Balance	Lib	15	−15
Lupus	Lupi	Wolf	Lup	15	−45
Lynx	Lyncis	Lynx	Lyn	8	+45
Lyra	Lyrae	Lyre or harp	Lyr	19	+40
Mensa	Mensae	Table Mountain	Men	5	−80
Microscopium	Microscopii	Microscope	Mic	21	−35
Monoceros	Monocerotis	Unicorn	Mon	7	−5
Musca	Muscae	Fly	Mus	12	−70
Norma	Normae	Carpenter's level	Nor	16	−50
Octans	Octantis	Octant	Oct	22	−85
Ophiuchus	Ophiuchi	Holder of serpent	Oph	17	0
Orion	Orionis	Orion, the hunter	Ori	5	+5
Pavo	Pavonis	Peacock	Pav	20	−65
Pegasus	Pegasi	Pegasus, the winged horse	Peg	22	+20
Perseus	Persei	Perseus, hero who saved Andromeda	Per	3	+45
Phoenix	Phoenicis	Phoenix	Phe	1	−50
Pictor	Pictoris	Easel	Pic	6	−55
Pisces	Piscium	Fishes	Psc	1	+15
Piscis Austrinus	Piscis Austrini	Southern fish	PsA	22	−30
Puppis*	Puppis	Stern of the Argonauts' ship	Pup	8	−40
Pyxis* (= Malus)	Pyxidus	Compass of the Argonauts' ship	Pyx	9	−30
Reticulum	Reticuli	Net	Ret	4	−60
Sagitta	Sagittae	Arrow	Sge	20	+10
Sagittarius	Sagittarii	Archer	Sgr	19	−25
Scorpius	Scorpii	Scorpion	Sco	17	−40
Sculptor	Sculptoris	Sculptor's tools	Scl	0	−30
Scutum	Scuti	Shield	Sct	19	−10
Serpens	Serpentis	Serpent	Ser	17	0
Sextans	Sextantis	Sextant	Sex	10	0
Taurus	Tauri	Bull	Tau	4	+15
Telescopium	Telescopii	Telescope	Tel	19	−50
Triangulum	Trianguli	Triangle	Tri	2	+30
Triangulum Australe	Trianguli Australis	Southern triangle	TrA	16	−65
Tucana	Tucanae	Toucan	Tuc	0	−65
Ursa Major	Ursae Majoris	Big bear	UMa	11	+50
Ursa Minor	Ursae Minoris	Little bear	UMi	15	+70
Vela*	Velorum	Sail of the Argonauts' ship	Vel	9	−50
Virgo	Virginis	Virgin	Vir	13	0
Volans	Volantis	Flying fish	Vol	8	−70
Vulpecula	Vulpeculae	Fox	Vul	20	+25

* The four constellations Carina, Puppis, Pyxis, and Vela originally formed the single constellation, Argo Navis.

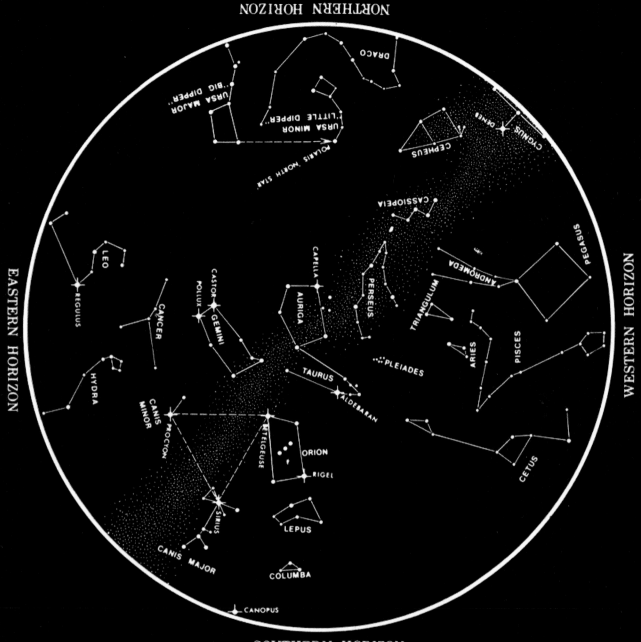

NORTHERN HORIZON

DRACO

BIG DIPPER
URSA MAJOR

LITTLE DIPPER
URSA MINOR

POLARIS NORTH STAR

CYGNUS
DENEB
CEPHEUS

CASSIOPEIA

PEGASUS

ANDROMEDA

CAPELLA

PERSEUS

TRIANGULUM

LEO

AURIGA

CASTOR
POLLUX GEMINI

ARIES

PISCES

REGULUS

CANCER

PLEIADES

HYDRA

TAURUS
ALDEBARAN

CANIS
MINOR

CETUS

PROCYON

BETELGEUSE

ORION

RIGEL

SIRIUS

LEPUS

CANIS MAJOR

COLUMBA

CANOPUS

EASTERN HORIZON

WESTERN HORIZON

SOUTHERN HORIZON

THE NIGHT SKY IN JANUARY

Latitude of chart is 34°N, but it is
practical throughout the continental
United States.

To use: Hold chart vertically and turn
it so the direction you are facing
shows at the bottom.

Chart time (Local Standard):

10 p.m. First of month

9 p.m. Middle of month

8 p.m. Last of month

Star Chart from GRIFFITH OBSERVER, Griffith Observatory, Los Angeles

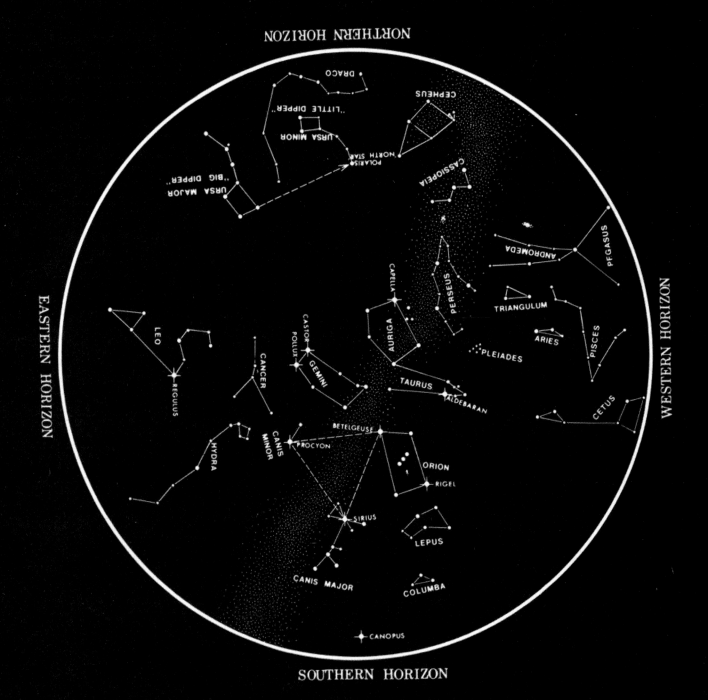

THE NIGHT SKY IN FEBRUARY

Latitude of chart is 34°N, but it is practical throughout the continental United States.

To use: Hold chart vertically and turn it so the direction you are facing shows at the bottom.

Chart time (Local Standard):

10 p.m. First of month

9 p.m. Middle of month

8 p.m. Last of month

THE NIGHT SKY IN MARCH

Latitude of chart is 34°N, but it is practical throughout the continental United States.

To use: Hold chart vertically and turn it so the direction you are facing shows at the bottom.

Chart time (Local Standard):

10 p.m. First of month

9 p.m. Middle of month

8 p.m. Last of month

Star Chart from *GRIFFITH OBSERVER*, Griffith Observatory, Los Angeles

NORTHERN HORIZON

CEPHEUS
CASSIOPEIA
DRACO
PERSEUS
VEGA
POLARIS "NORTH STAR"
CAPELLA
HERCULES
URSA MINOR "LITTLE DIPPER"
AURIGA
TAURUS
CORONA BOREALIS
BOOTES
URSA MAJOR "BIG DIPPER"
ALDEBARAN

EASTERN HORIZON
WESTERN HORIZON

SERPENS
CASTOR
GEMINI
ARCTURUS
POLLUX
BETELGEUSE
RIGEL
CANCER
ORION
LEO
PROCYON
CANIS MINOR
LIBRA
REGULUS
VIRGO
SIRIUS
SPICA
CORVUS
CANIS MAJOR
HYDRA

SOUTHERN HORIZON

THE NIGHT SKY IN APRIL

Latitude of chart is 34°N, but it is
practical throughout the continental
United States.

To use: Hold chart vertically and turn
it so the direction you are facing
shows at the bottom.

Chart time (Local Standard):

10 p.m. First of month

9 p.m. Middle of month

8 p.m. Last of month

NORTHERN HORIZON

CASSIOPEIA

CEPHEUS

DENEB

CYGNUS

POLARIS 'NORTH STAR'

CAPELLA

AURIGA

DRACO

LYRA

VEGA

"LITTLE DIPPER"

URSA MINOR

GEMINI

CASTOR

POLLUX

HERCULES

CORONA BOREALIS

BOOTES

"BIG DIPPER"

URSA MAJOR

CANCER

CANIS MINOR

PROCYON

SERPENS

OPHIUCHUS

ARCTURUS

LEO

REGULUS

VIRGO

SCORPIUS

ANTARES

LIBRA

SPICA

CORVUS

HYDRA

EASTERN HORIZON

WESTERN HORIZON

SOUTHERN HORIZON

THE NIGHT SKY IN MAY

Latitude of chart is 34°N, but it is practical throughout the continental United States.

To use: Hold chart vertically and turn it so the direction you are facing shows at the bottom.

Chart time (Local Standard):

10 p.m. First of month

9 p.m. Middle of month

8 p.m. Last of month

Star Chart from *GRIFFITH OBSERVER*, Griffith Observatory, Los Angeles

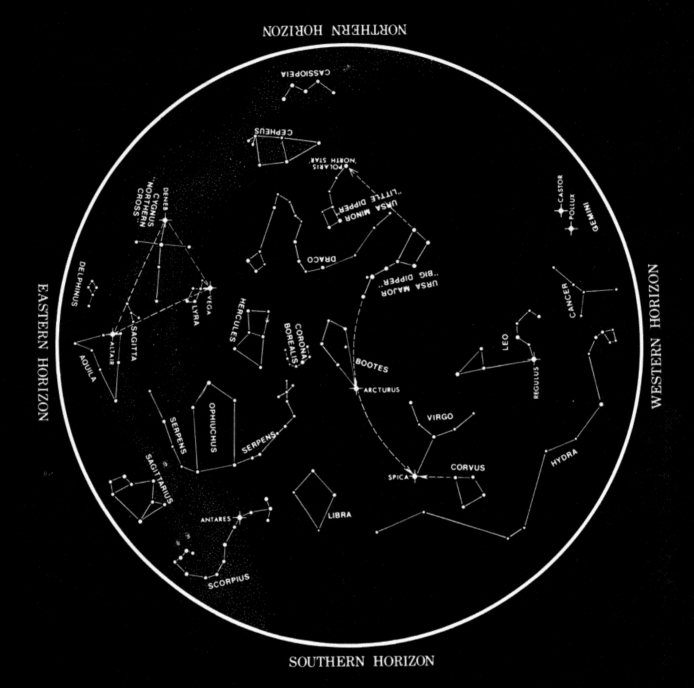

NORTHERN HORIZON

CASSIOPEIA

CEPHEUS

POLARIS "NORTH STAR"

CYGNUS "NORTHERN CROSS"

DENEB

"LITTLE DIPPER"

URSA MINOR

DRACO

CASTOR

POLLUX

GEMINI

DELPHINUS

VEGA

LYRA

HERCULES

CORONA BOREALIS

"BIG DIPPER"

URSA MAJOR

CANCER

EASTERN HORIZON

SAGITTA

ALTAIR

AQUILA

BOOTES

ARCTURUS

LEO

REGULUS

WESTERN HORIZON

SERPENS

OPHIUCHUS

SERPENS

VIRGO

SAGITTARIUS

SPICA

CORVUS

HYDRA

ANTARES

LIBRA

SCORPIUS

SOUTHERN HORIZON

THE NIGHT SKY IN JUNE

Latitude of chart is 34°N, but it is
practical throughout the continental
United States.

To use: Hold chart vertically and turn
it so the direction you are facing
shows at the bottom.

Chart time (Local Standard):

10 p.m. First of month

9 p.m. Middle of month

8 p.m. Last of month

Star Chart from *GRIFFITH OBSERVER*, Griffith Observatory, Los Angeles

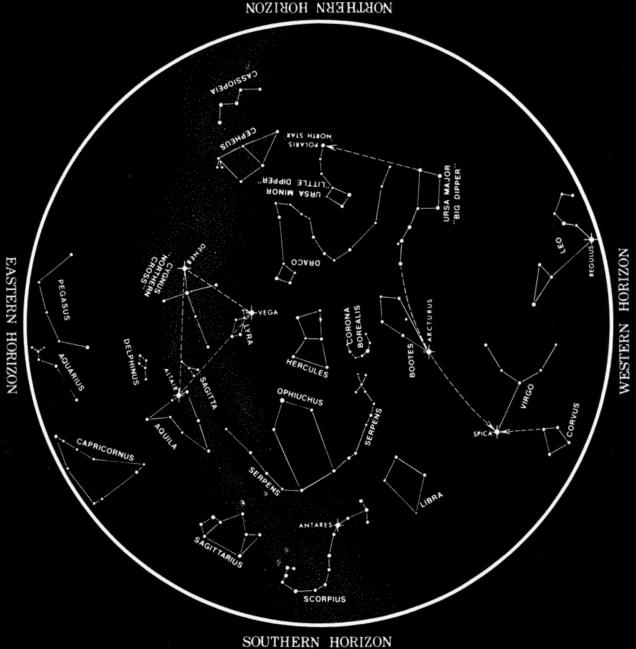

NORTHERN HORIZON

CASSIOPEIA

CEPHEUS

POLARIS NORTH STAR

URSA MINOR "LITTLE DIPPER"

URSA MAJOR "BIG DIPPER"

DRACO

DENEB

CYGNUS "NORTHERN CROSS"

VEGA

LYRA

CORONA BOREALIS

ARCTURUS

LEO

REGULUS

PEGASUS

DELPHINUS

HERCULES

BOOTES

EASTERN HORIZON

AQUARIUS

ALTAIR

SAGITTA

OPHIUCHUS

SERPENS

VIRGO

CORVUS

WESTERN HORIZON

AQUILA

CAPRICORNUS

SERPENS

SPICA

LIBRA

ANTARES

SAGITTARIUS

SCORPIUS

SOUTHERN HORIZON

THE NIGHT SKY IN JULY

Latitude of chart is 34°N, but it is
practical throughout the continental
United States.

To use: Hold chart vertically and turn
it so the direction you are facing
shows at the bottom.

Chart time (Local Standard):

10 p.m. First of month

9 p.m. Middle of month

8 p.m. Last of month

Star Chart from GRIFFITH OBSERVER, Griffith Observatory, Los Angeles

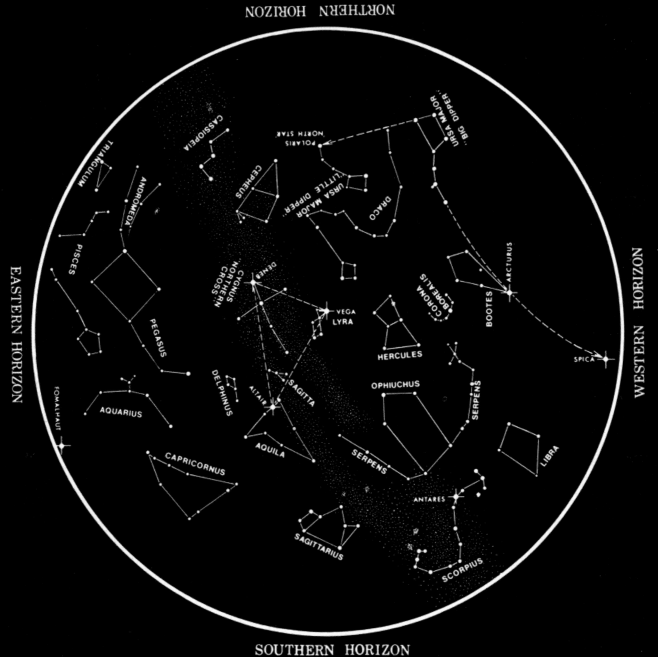

NORTHERN HORIZON

EASTERN HORIZON

WESTERN HORIZON

SOUTHERN HORIZON

THE NIGHT SKY IN AUGUST

Latitude of chart is 34°N, but it is practical throughout the continental United States.

To use: Hold chart vertically and turn it so the direction you are facing shows at the bottom.

Chart time (Local Standard):

10 p.m. First of month

9 p.m. Middle of month

8 p.m. Last of month

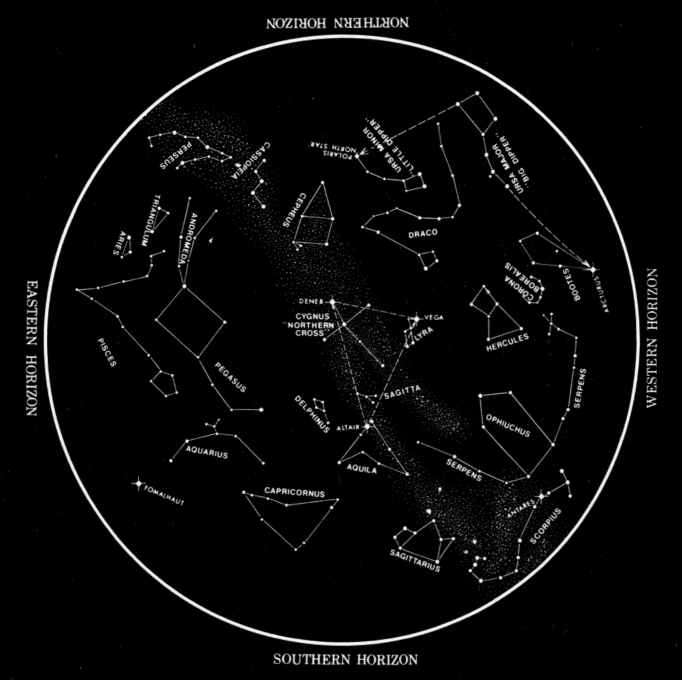

NORTHERN HORIZON

EASTERN HORIZON

WESTERN HORIZON

SOUTHERN HORIZON

THE NIGHT SKY IN SEPTEMBER

Latitude of chart is 34°N, but it is practical throughout the continental United States.

To use: Hold chart vertically and turn it so the direction you are facing shows at the bottom.

Chart time (Local Standard):

10 p.m. First of month

9 p.m. Middle of month

8 p.m. Last of month

Star Chart from *GRIFFITH OBSERVER*, Griffith Observatory, Los Angeles

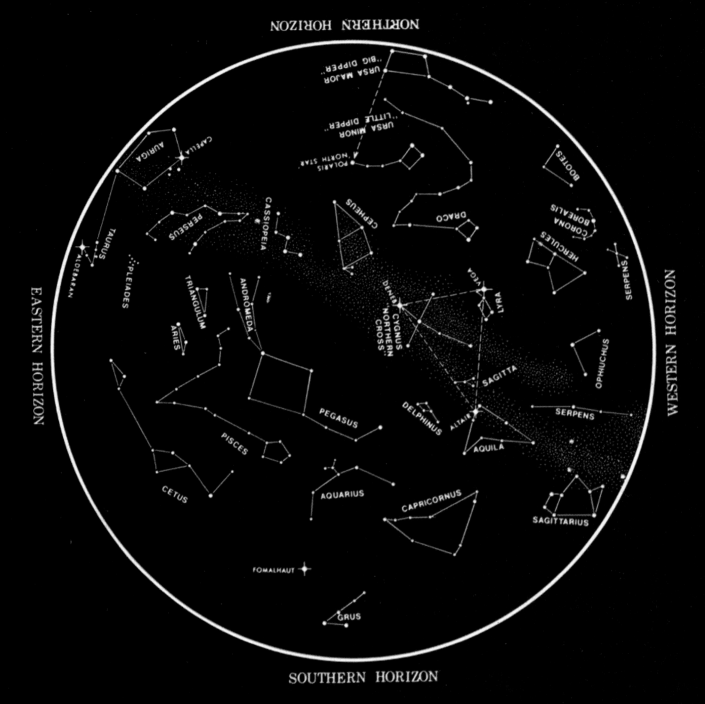

NORTHERN HORIZON

"BIG DIPPER"
URSA MAJOR

"LITTLE DIPPER"
URSA MINOR

POLARIS
"NORTH STAR"

BOÖTES

AURIGA
CAPELLA

DRACO

CORONA
BOREALIS

PERSEUS

CASSIOPEIA

CEPHEUS

HERCULES

SERPENS

TAURUS

ALDEBARAN

PLEIADES

TRIANGULUM

ANDROMEDA

VEGA
LYRA

DENEB
CYGNUS
"NORTHERN
CROSS"

OPHIUCHUS

ARIES

SAGITTA

PEGASUS

DELPHINUS

ALTAIR

SERPENS

PISCES

AQUILA

CETUS

AQUARIUS

CAPRICORNUS

SAGITTARIUS

FOMALHAUT

GRUS

SOUTHERN HORIZON

EASTERN HORIZON

WESTERN HORIZON

THE NIGHT SKY IN OCTOBER

Latitude of chart is 34°N, but it is practical throughout the continental United States.

To use: Hold chart vertically and turn it so the direction you are facing shows at the bottom.

Chart time (Local Standard):

10 p.m. First of month

9 p.m. Middle of month

8 p.m. Last of month

Star Chart from *GRIFFITH OBSERVER*, Griffith Observatory, Los Angeles

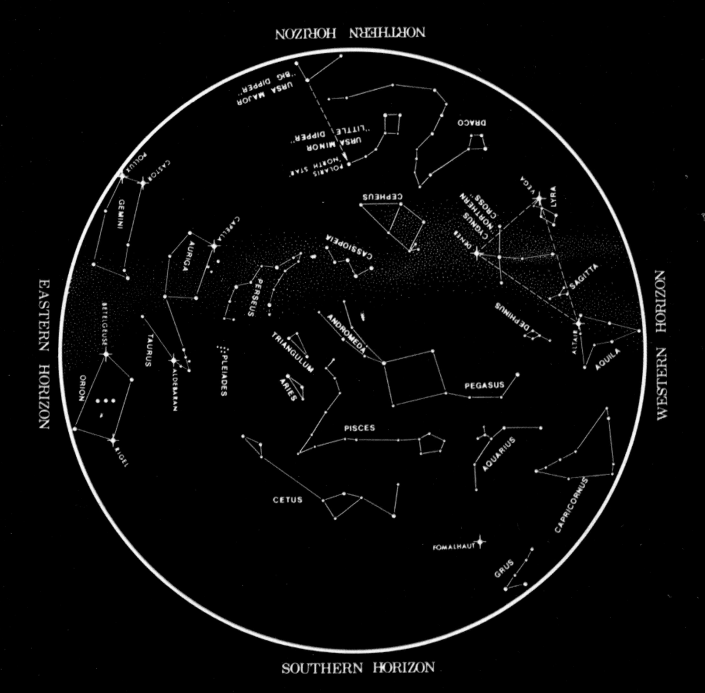

THE NIGHT SKY IN NOVEMBER

Latitude of chart is 34°N, but it is practical throughout the continental United States.

To use: Hold chart vertically and turn it so the direction you are facing shows at the bottom.

Chart time (Local Standard):

10 p.m. First of month

9 p.m. Middle of month

8 p.m. Last of month

Star Chart from *GRIFFITH OBSERVER*, Griffith Observatory, Los Angeles

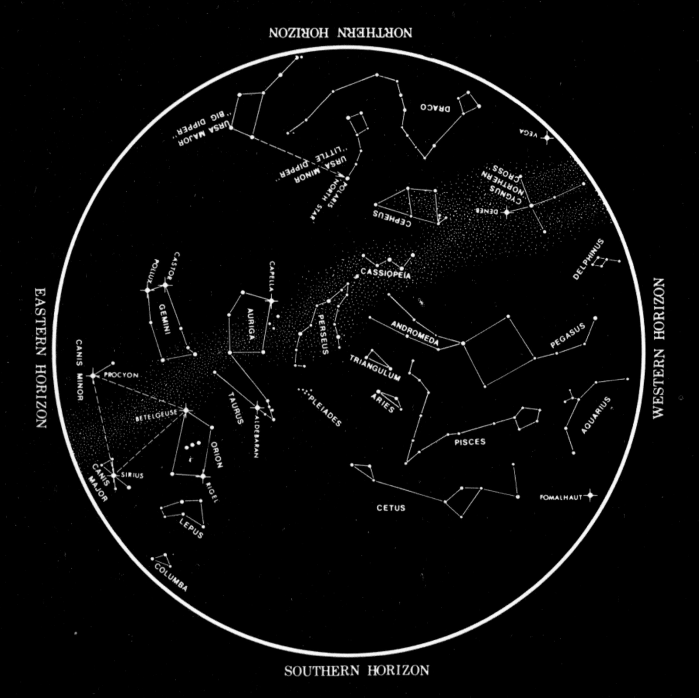

THE NIGHT SKY IN DECEMBER

Latitude of chart is 34°N, but it is practical throughout the continental United States.

To use: Hold chart vertically and turn it so the direction you are facing shows at the bottom.

Chart time (Local Standard):

10 p.m. First of month

9 p.m. Middle of month

8 p.m. Last of month

Star Chart from *GRIFFITH OBSERVER*, Griffith Observatory, Los Angeles

Index